Theory and Design of Linear Active Networks

THEORY AND DESIGN OF LINEAR ACTIVE NETWORKS

Sundaram Natarajan

Department of Electrical Engineering
Tennessee Technological University

MACMILLAN PUBLISHING COMPANY
NEW YORK

COLLIER MACMILLAN CANADA, INC.
TORONTO

COLLIER MACMILLAN PUBLISHERS
LONDON

Macmillan Publishing Company
866 Third Avenue, New York, NY 10022

Collier Macmillan Canada, Inc.

Printed in the United States of America

Printing number: 1 2 3 4 5 6 7 8 9 10 Year: 7 8 9 0 1 2 3 4 5

Library of Congress Cataloging-in-Publication Data

Natarajan, Sundaram.
Theory and design of linear active networks.

Bibliography: p.
Includes index.
1. Electric filters, Active. 2. Electric circuits, Linear. I. Title.
TK7872.F5N18 1987 621.3815′324 87-7703
ISBN 0-02-949730-2

To

the loving memory

of

my father

RM. Sundaram Chettiar

Contents

Preface

This book is based on course notes developed by the author for senior undergraduate and first-year graduate students. It is also intended as a reference book for practicing engineers. As a textbook, it can be used either in *filter design* courses or in courses on electronic circuits where the emphasis is on *operational amplifiers and their applications.* If it is used in a filter design course, selected topics can be taught in one semester or, with supplementary material on passive or digital filter designs available from other sources, it can be used for a two-semester sequence. If it is used in an electronic circuits course, as developed by the author at the University of Kentucky during the last four years, selected topics may be taught during one semester. However, by supplementing this material with nonlinear applications of operational amplifiers, it can be taught during two semesters.

The initial emphasis of this book is on applications of operational amplifiers. However, one major application of the operational amplifier is its use in active *RC* and switched capacitor filters, and these are the topics that are primarily dealt with in this book. Chapter 1 presents a detailed analysis of the operational amplifier. The practical problems that arise in operational amplifier circuits are also discussed, so that practicing engineers and students will be aware of the limitations of this component.

Since the most important application of the operational amplifier is in filter design and since filter design in turn requires certain approximation procedures, Chap. 2 deals with the *major approximation procedures.* Sufficient coverage is also given to *frequency transformations. Program listings* in BASIC are provided for obtaining the appropriate network function using any one of the major approximation procedures. These programs can be used on an IBM PC or any one of its compatibles. This means that charts and tables are not required for solving the approximation problems.

Active filter designers, and in general all electronic circuit designers, are concerned with the sensitivity of their designs and the effects of both tolerance and future variations in the component values of both active and passive elements. Therefore Chap. 3 deals with sensitivity, sensitivity measures, and the variations of some important quantities. Special emphasis is given to filter design. A figure of merit for active filters, called the *ripple factor*, is also derived in this chapter. This ripple factor serves two purposes in later chapters. It is used to compare various filter networks and to develop a filter design of a specific configuration with minimum sensitivity. Finally, this chapter deals with a simple predistortion technique that can be used to design filters that take into account the nonideal effects of both the op amps and the passive components used in the network. Practicing engineers will be particularly interested in this discussion.

Chapters 4, 5, and 6, discuss practical active *RC* filter configurations. These configurations involve both single- and multiple-amplifier filters. The nonideal properties of the op amps used can cause drastic changes in the performance of the designed networks. The majority of the books available include the analysis of these effects on selected circuits and leave to the designer the problems involved in designing the circuits, including the nonideal properties of the op amp, particularly its finite gain-bandwidth product. In this book, particularly in these three chapters in which the various networks are introduced, the discussion includes the effects of the nonideal properties of the op amps in the analysis. In addition, *methods for designing all the circuits with minimum active sensitivity that consider all the above effects* are also presented. These chapters also deal with the so-called *actively compensated* active filters, some of which have zero active pole sensitivities and some of which have much reduced sensitivities compared to their conventional counterparts. Last but not least, these chapters include appendixes that provide computer programs in BASIC for the IBM PC and its compatibles. These programs can be used to design all the important networks and take into consideration the nonideal properties of the op amps used in them. The above additional features are considered important for all filter designers, both electrical students and practicing engineers.

The design of *higher-order* filters is dealt with in Chap. 7. It considers cascade realizations, multiple-loop configurations, and *LC* simulations. With multiple-loop configurations and *LC* simulations, low-sensitivity higher-order filters can be designed. This chapter includes a discussion of the *dynamic range* problem as well. Chapter 8 is concerned with the recently emerging *switched capacitor filters*. This type of filter has rapidly become very popular because it is compatible with MOS large-scale integration.

A large number of references were used in writing this book. Many of them have been cited, but some have not. A bibliography has also been included. Other references can always be located by consulting the references cited. I wish to thank the authors whose results have been used and whose names may or may not appear in the text.

I am also grateful to B. Leon of the University of Kentucky, Lexington, who has always been a source of inspiration and encouragement and to P. K. Rajan of Tennessee Technological University, Cookville, and S. Yuvarajan of North Dakota State University, Fargo, who read the manuscript and suggested modifications. I would also like to acknowledge the contributions made by some of my former students during the development of the course material and by Vikie Brann who typed part of the manuscript.

Last but not the least, I express my gratitude to my wife Saroja and my sons Sundaram and Senthilvelu for their patience and understanding throughout the preparation of this book.

Sundaram Natarajan
Cookeville, TN
April 3, 1987

1

Operational Amplifiers

THE ADVENT OF INTEGRATED CIRCUIT ELEMENTS has completely changed the way in which electronic circuits and signal processing circuits are realized. A particular process can be completely integrated on a single chip using present-day technology. In the design of analog systems, one electronic component responsible for this dramatic and continuing change is the high-gain operational amplifier (op amp). The availability of high-performance and inexpensive op amps has influenced the design and construction of systems ranging from simple electronic circuits to sophisticated electronic equipment. In many systems the use of this component has lowered the cost and improved the performance characteristics of the circuits considerably, and it has made possible the design and implementation of circuits and systems previously considered too complex.

Before we consider the use of the op amp as a component of a signal processing circuit, in order to understand this device better we first discuss its circuitry. We shall also consider the problems associated with this component and seek solutions to some of them. In addition, this chapter will describe the limitations of this device, a knowledge of which is important to the circuit designer.

1.1 *An Overview*

The op amp has many terminals (as many as eight in some cases). Since it is an active device, two of these terminals are used for connecting the power

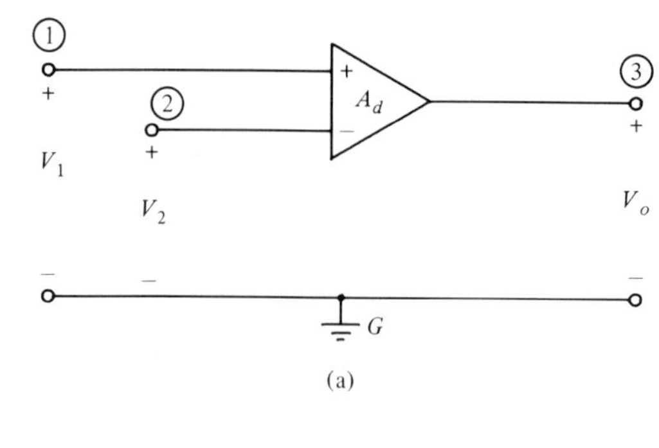

(a)

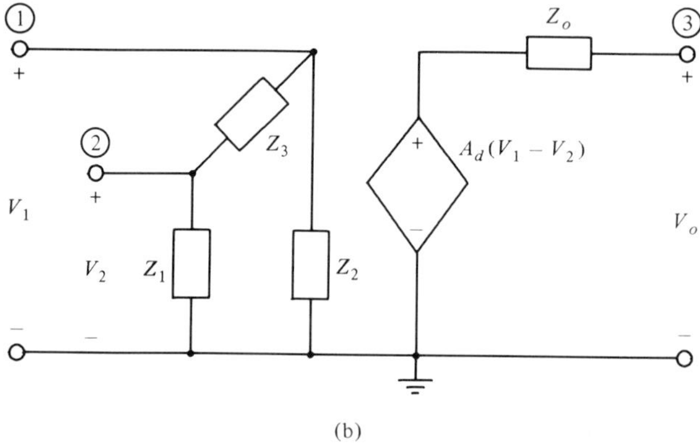

(b)

Figure 1.1 (*a*) Circuit symbol for an op amp. (*b*) Approximate equivalent circuit for an op amp.

supply to establish proper biasing for the transistors inside. There are a few other terminals in some op amps that are used to adjust the biasing so that, when the signal input to the op amp is zero, the output is also zero and no dc component appears in the output. When the op amp is a component of a circuit, and as far as small-signal analysis is considered, the above terminals do not play any role at all. In regard to its operation in a circuit, an op amp is considered a three-terminal device. Two of the terminals are input terminals, and the third is the output terminal. The circuit symbol for an op amp is shown in Fig. 1.1*a*, where terminals 1 and 2 are the noninverting and inverting input terminals, respectively, and terminal 3 is the single-ended output terminal. Basically, an op amp acts as a differential amplifier with a gain of A_d. The approximate equivalent circuit of this component is shown in Fig. 1.1*b*. In a practical op amp, the equivalent circuit parameters Z_1, Z_2, and Z_3 are the input impedances, however, their values are on the order of a few megaohms and in most circuit applications their influence can be ignored. Further, for low-frequency applications, the differential gain A_d is

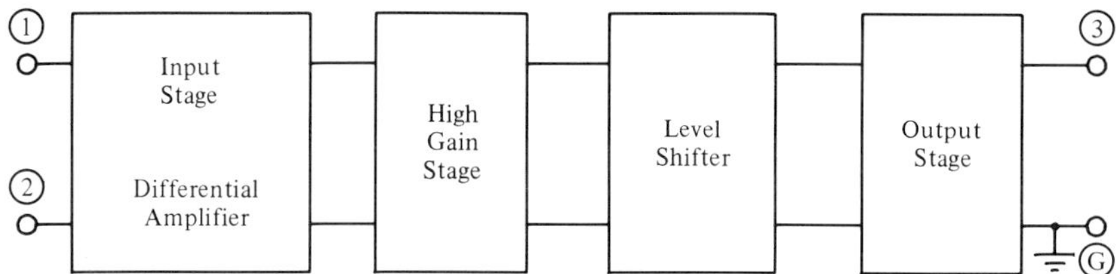

Figure 1.2 Basic stages of an op amp.

on the order of 10^5 or more for most practical op amps, and thus we consider it as though it tends to ∞. The output impedance Z_o is only a few ohms. Thus, when the op amp is considered ideal, the equivalent circuit parameters satisfy the conditions

$$Z_1 = Z_2 = Z_3 = \infty \qquad Z_o = 0$$

and

$$\frac{1}{A_d} = 0$$

Also, note that when $V_1 = V_2$, then $V_o = 0$.

A practical op amp is one in which the above expressions are approximately true. It is a complex device consisting of many electronic circuits connected together and can be considered a cascade connection of four different basic circuits as shown in Fig. 1.2. In an analysis of the entire op amp circuit, each basic circuit can be considered individually and represented by its equivalent.

Every electronic circuit should be analyzed twice. First, we analyze the biasing of the circuit. As we are mostly interested in amplifiers, we must make sure that proper dc biasing is obtained so that all the transistors are operating in the active region. Second, as far as the signal is concerned, we must use small-signal equivalent circuits for the transistors so that analysis of the circuits can be carried out to determine the salient parameters. In integrated circuit (IC) technology, for reasons that will become clear later, proper biasing is obtained by using what are known as current sources. We shall discuss biasing techniques in a later section after we consider the equivalent circuits of the transistors.

1.2 *Review of Bipolar Junction Transistors and Field Effect Transistors*

In this section we review some of the basic ideas related to the equivalent circuits of bipolar junction transistors (BJTs) and field effect transistors (FETs). For details, the reader is referred to references [1, 2]. The symbolic

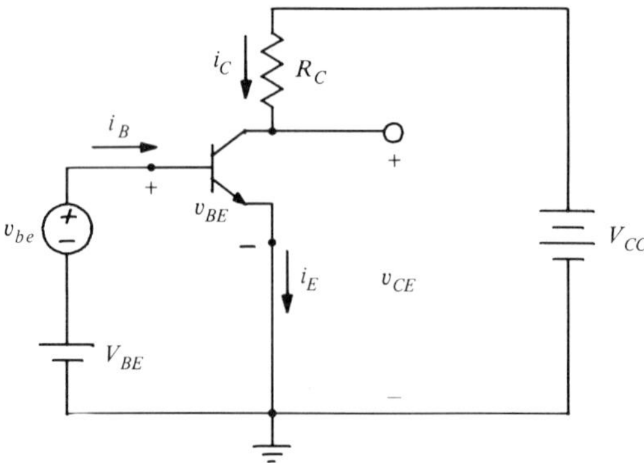

Figure 1.3 Biasing arrangement of an *npn* transistor.

notation for a BJT and the biasing voltages are shown in Fig. 1.3. The operation of a transistor amplifier is greatly influenced by the quiescent (or bias) current. The parameters of the small-signal equivalent circuit of a transistor depend on the biasing conditions. To operate this transistor as an amplifier, we need to bias it such that the base-emitter junction is forward-biased and the base-collector junction is reverse-biased. This will ensure that the transistor is biased in the active region. In the case of an *npn* transistor, the two batteries V_{BE} and V_{CC} as shown in Fig. 1.3 provide the biasing; R_c is called the load resistor; and v_{be} is the signal that is applied externally. Voltages and currents denoted by uppercase letters are dc quantities (for now—later they become phasors), which are just bias voltages and currents. Quantities denoted by lowercase letters with uppercase subscripts are instantaneous values. And quantities denoted by lowercase letters with lowercase subscripts are incremental signals, which we usually refer to as small signals. For example, voltages that appear across the base-emitter junction are V_{BE}, the biasing dc voltage, v_{BE}, the instantaneous voltage, and v_{be}, the small-signal voltage. In this circuit, both V_{BE} and v_{be} are externally applied. Instantaneous quantities are just the algebraic sum of the bias quantities and the small signals. For example, we can write equations such as

$$\begin{aligned} i_C &= I_C + i_c \\ v_{BE} &= V_{BE} + v_{be} \\ v_{CE} &= V_{CE} + v_{ce} \end{aligned}$$

which are obvious from the circuit in Fig. 1.3. With this understanding of the notation, we state that an *npn* transistor is in the active mode when $v_{BE} > 0$ and $v_{CB} \geq 0$. If $|v_{be}| \ll |V_{BE}|$ and $|v_{cb}| \ll |V_{CB}|$, then the transistor is in the active mode if $V_{BE} > 0$ and $V_{CB} \geq 0$. Likewise, a *pnp* transistor is in the active mode if $V_{BE} < 0$ and $V_{CB} \leq 0$. One main purpose of dc analysis of an amplifier circuit is to verify the conditions we have just mentioned in order to determine whether all the transistors are in the active mode.

Before we find the small-signal equivalent circuits of a transistor, let us review some of the fundamental concepts relating to a transistor. Referring to Fig. 1.3 and using Kirchhoff's current and voltage laws (KCL and KVL, respectively), we can immediately write

$$i_C + i_B = i_E \tag{1.1a}$$

and

$$v_{CB} + v_{BE} = v_{CE} \tag{1.1b}$$

The above relationships hold true for both direct current and small signals:

$$I_C + I_B = I_E \tag{1.2a}$$

$$V_{CB} + V_{BE} = V_{CE} \tag{1.2b}$$

and

$$i_c + i_b = i_e \tag{1.3a}$$

$$v_{cb} + v_{be} = v_{ce} \tag{1.3b}$$

Without considering the physical aspects of the working of a transistor, it is sufficient for us to know that the collector current i_C is related to the base-emitter voltage v_{BE} by the following exponential expression [2]:

$$i_C = I_{CO}\left(1 + \frac{v_{ce}}{V_A}\right)\varepsilon^{v_{BE}/V_T} \tag{1.4}$$

where I_{CO}, V_A, and V_T are the reverse saturation current, Early voltage, and thermal voltage, respectively. The thermal voltage V_T is defined as

$$V_T = \frac{KT}{q} \tag{1.5}$$

where K is Boltzmann's constant, q is the electronic charge, and T is the absolute temperature (in degrees Kelvin). Substituting values for K and q, we have

$$V_T = \frac{T}{11{,}600}$$

At room temperature the value of $V_T \simeq 25$ mV, which is the typical value used in the rest of this chapter. The reverse bias current I_{CO} is proportional to the area of the emitter-base junction. Typically, I_{CO} is in the range 10^{-12} to 10^{-15} A and is a function of the temperature. The ideal value of the Early voltage V_A is ∞, however, it ranges from 100 to 150 V. Since in most cases $|v_{ce}| \ll V_A$, we can approximate (1.4) as

$$i_C \simeq I_{CO}\varepsilon^{v_{BE}/V_T} \tag{1.6}$$

In most cases we can use (1.6) instead of (1.4).

Again, based on physical considerations, the base current i_B is also proportional to ε^{V_{BE}/V_T}. Thus we can also write

$$i_C = \beta i_B \tag{1.7}$$

where β is a constant for a given transistor. In fact, β is dependent on the bias current I_C and is also temperature-dependent. However, it is relatively constant with respect to I_C and therefore we assume that it is a constant. We should also mention that it is frequency-dependent, and since we are interested only in low-frequency applications, it is also assumed to be frequency-independent. Usually the value of β is specified by the manufacturer. Using (1.1) and (1.7), we have

$$i_E = \frac{\beta + 1}{\beta} i_C$$

or

$$i_C = \alpha i_E \tag{1.8}$$

where $\alpha = \beta/(\beta + 1) < 1$. Usually the value of β ranges from a few tens to a few hundreds. When $\beta \gg 1$, then $\alpha \simeq 1$. The value of α can be approximated to be unity if $\beta \gg 1$. The quiescent operating condition of a transistor can be established by suppressing the signal input. Under such conditions direct voltages and currents are the only quantities that are possible. For example, in the circuit in Fig. 1.3, if $v_{be} = 0$, then $i_b = i_c = v_{ce} = v_{cb} = 0$. Then the relationships between the bias currents and voltages can be established. For example, (1.2) gives some of them. The following relationships can also be established using (1.4), (1.7), and (1.8).

$$I_C = I_{CO} \varepsilon^{V_{BE}/V_T} \tag{1.9a}$$

$$I_C = \beta I_B \tag{1.9b}$$

$$I_C = \alpha I_E \tag{1.9c}$$

Example 1.1: Given that $I_{CO} = 10^{-14}$ A (a typical value), find I_C when $V_{BE} = 0.5$, 0.6, and 0.7 V. Also, find I_B when $\beta = 100$.

The results are given in the accompanying table.

V_{BE} (V)	I_C (mA)	I_B (μA)
0.5	0.0048	0.048
0.6	0.265	2.65
0.7	14.46	144.6

Note that the value of I_C is very small until about $V_{BE} = 0.5$ V. This voltage is called the *cut-in voltage*. ■

Example 1.2: For an *npn* transistor it is given that $V_{BE} = 0.7$ and $I_C = 1$ mA. Find V_{BE} when $I_C = 0.1$ mA and $I_C = 10$ mA.

$$I_{C1} = I_{CO}\varepsilon^{V_{BE1}/V_T}$$

$$I_{C2} = I_{CO}\varepsilon^{V_{BE2}/V_T}$$

Therefore

$$\frac{I_{C1}}{I_{C2}} = \varepsilon^{(V_{BE1} - V_{BE2})/V_T}$$

or

$$V_{BE2} = V_{BE1} + V_T \ln \frac{I_{C2}}{I_{C1}}$$

When $I_{C1} = 1$ mA and $V_{BE1} = 0.7$ V, we have, when $I_{C2} = 0.1$ mA,

$$V_{BE2} = 0.7 + 25 \times 10^{-3} \ln 0.1 = 0.642 \text{ V}$$

If $I_{C2} = 10$ mA,

$$V_{BE2} = 0.7 + 25 \times 10^{-3} \ln 10 = 0.758 \text{ V}$$

Note that, over a wide range of bias current I_C, the value of V_{BE} is close to 0.7 V. ■

In most cases (except in critical cases) we shall assume that the forward bias voltage of the base-emitter junction is 0.7 V. Though this is only an approximation, it is quite valid and does not substantially affect the computation of other bias currents and voltages and at the same time saves a lot of calculations.

Example 1.3: In the circuit in Fig. 1.4, assume that $\beta = 100$. Calculate I_C, I_E, I_B, and V_{CB}.

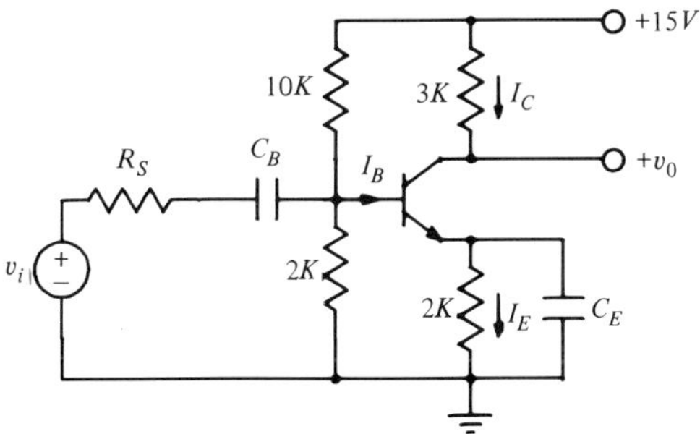

Figure 1.4 Circuit for Example 1.3.

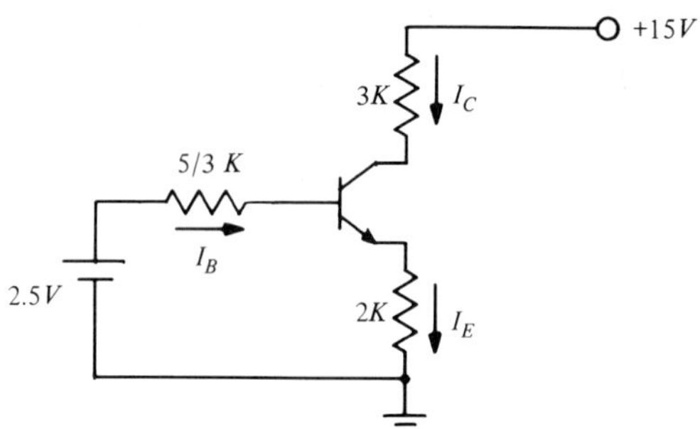

Figure 1.5 Equivalent circuit for Fig. 1.4 for dc analysis.

Note that the capacitances act as an open circuit for direct currents and voltages. Therefore, using Thevenin's equivalence for the base side, we have an equivalent circuit as shown in Fig. 1.5. Assume that all the currents are in milliamperes and all the resistances are in kiloohms. Applying KVL on the input side, we have

$$\tfrac{5}{3}I_B + 0.7 + 2I_E = 2.5$$

where I_B and I_E are in milliamperes.

Since

$$I_E = I_C + I_B = (\beta + 1)I_B$$

we have

$$I_B = \frac{1.8}{\frac{5}{3} + 202} = 8.838\ \mu\text{A}$$

and

$$I_C = 0.8838\ \text{mA}$$

Also,

$$I_E = 0.8926\ \text{mA}$$

Now

$$V_B = V_E + 0.7 = 2I_E + 0.7 = 2.485\ \text{V}$$

and

$$V_C = 15 - 3I_C = 12.349\ \text{V}$$

Since $V_C \geq V_B$, in an *npn* transistor, the transistor works in the active region. ■

We now return to the circuit in Fig. 1.3 and note that the emitter is grounded. This works as a common-emitter amplifier if $V_C \geq V_B$, and it can be used to find the small-signal equivalent circuit for the transistor. Using Taylor series approximation of the instantaneous value of i_C, we have, for small signals,

$$i_C(v_{be}, v_{ce}) \simeq i_C(0,0) + \left.\frac{\partial i_C}{\partial v_{be}}\right|_{v_{be}=v_{ce}=0} v_{be} + \left.\frac{\partial i_C}{\partial v_{ce}}\right|_{v_{be}=v_{ce}=0} v_{ce}$$

$$= I_C + \frac{I_C}{V_T} v_{be} + \frac{I_C}{V_A} v_{ce}$$

Since $i_C = I_C + i_c$, we have the small-signal collector current

$$i_c = \frac{I_C}{V_T} v_{be} + \frac{I_C}{V_A} v_{ce}$$

$$= g_m v_{be} + \frac{1}{r_{ce}} v_{ce} \tag{1.10}$$

where

$$g_m = \frac{I_C}{V_T} \quad \text{and} \quad r_{ce} = \frac{V_A}{I_C} \tag{1.11}$$

As per our definition of i_b, we have $i_b = i_c/\beta$. Therefore

$$i_b = \frac{g_m}{\beta} v_{be} + \frac{1}{\beta r_{ce}} v_{ce}$$

Rearranging the above equation, we have

$$v_{be} = \frac{\beta}{g_m} i_b - \frac{1}{g_m r_{ce}} v_{ce}$$

$$= r_{be} i_b - h_{re} v_{ce} \tag{1.12}$$

where

$$r_{be} = \frac{\beta}{g_m} = \frac{\beta V_T}{I_C} \quad \text{and} \quad h_{re} = \frac{1}{g_m r_{ce}} = \frac{V_T}{V_A} \tag{1.13}$$

Example 1.4: For the transistor in the circuit in Fig. 1.4 in Example 1.3, find the parameters g_m, r_{ce}, r_{be}, and h_{re}.

Assume a typical value of $V_A = 120$ V. In Example 1.3, we found that $I_C = 0.8838$ mA. Therefore, using (1.11) and (1.13), we have

$$g_m = \frac{0.8838}{25} = 35.35 \times 10^{-3} \text{ S}$$

$$r_{ce} = \frac{120}{0.8838} = 135.78 \text{ k}\Omega$$

$$r_{be} = \frac{100}{35.35 \times 10^{-3}} = 2.829 \text{ k}\Omega$$

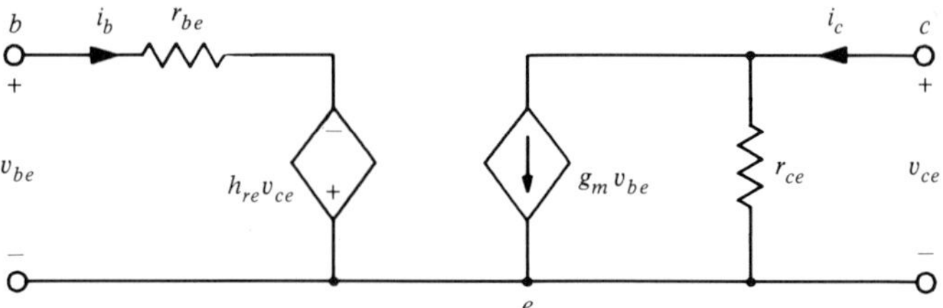

Figure 1.6 Approximate small-signal equivalent circuit for a transistor.

and

$$h_{re} = \frac{25 \times 10^{-3}}{120} = 208.3 \times 10^{-6}$$ ■

The equations describing the small-signal current i_c and the small-signal input voltage v_{be}, (1.10) and (1.12), can be represented using controlled sources as shown in Fig. 1.6. We call this an approximate equivalent circuit for the following reasons.

1. The silicon material between the base terminal and the base-emitter junction provides a resistance r_{bb} (usually on the order of a few tens of ohms).
2. For high-frequency applications, the junction capacitances must be included. Thus, between base and collector terminals and base and emitter terminals there are capacitances on the order of 1 to 3 pF.

Thus a more exact small-signal equivalent circuit is shown in Fig. 1.7. However, since we are interested in only low-frequency applications, we can ignore the effects of C_{bc} and C_{be}. Further, since h_{re} does not depend on the biasing current I_c and is usually much smaller than unity (see Example 1.4), we can make a further approximation that $h_{re} \simeq 0$. Further, the value of r_{bb} is usually on the order of a few tens of ohms and the value of r_{be} is a few

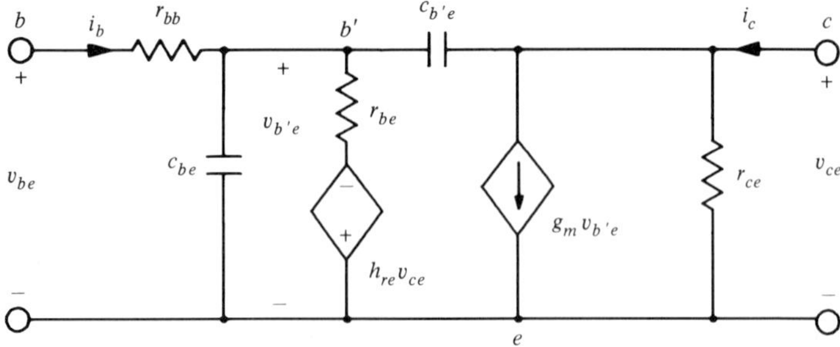

Figure 1.7 A more accurate small-signal equivalent circuit for a transistor.

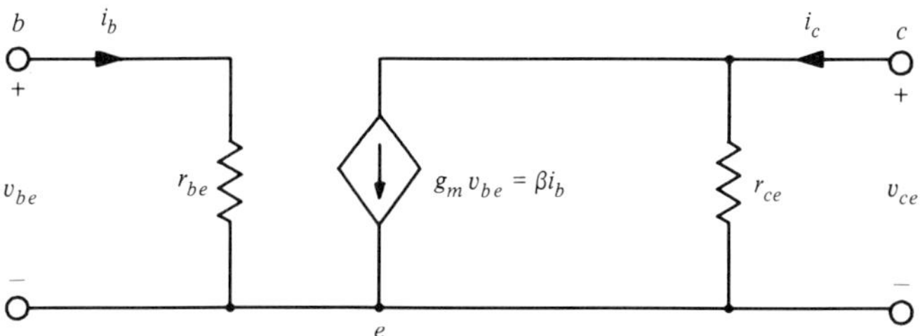

Figure 1.8 Approximate low-frequency small-signal model for a transistor.

kiloohms or at least many hundreds of ohms. Therefore we can also ignore the value of r_{bb} in comparison to r_{be}. Thus an approximate equivalent circuit, which we will use in most cases, is shown in Fig. 1.8. In this circuit again, depending on the load circuit connected to the collector, we may be able to ignore the effect of r_{ce}, which is on the order of 100 kΩ.

Before leaving the topic of equivalent circuits of BJTs, let us derive another useful form of an equivalent circuit. Since $i_e = i_c/\alpha$, from (1.10) we have

$$\begin{aligned} i_e &= \frac{g_m}{\alpha} v_{be} + \frac{1}{\alpha r_{ce}} v_{ce} \\ &= \frac{1}{r_e} v_{be} + \frac{1}{\alpha r_{ce}} v_{ce} \end{aligned} \tag{1.14}$$

where

$$r_e = \frac{\alpha}{g_m} = \frac{\alpha V_T}{I_C} = \frac{V_T}{I_E} \tag{1.15}$$

Since the value of r_e is a few tens of ohms and αr_{ce} is many tens of kiloohms, we can approximate (1.14) as

$$i_e \simeq \frac{v_{be}}{r_e} \tag{1.16}$$

Using (1.16) and the fact that $i_c = \alpha i_e$, we obtain another approximate equivalent circuit as shown in Fig. 1.9. Note that the parameters of small-signal equivalent circuits depend on the quiescent current I_C, and that one set of parameters evaluated for a particular bias current I_C cannot be used for all bias currents. This is one more reason why we have to analyze dc bias conditions in the circuit.

Example 1.5: Consider the circuit in Fig. 1.4 in Example 1.3. Assume that the source resistor $R_s = 10$ kΩ and C_B, C_E tend to ∞ for the signal. Find the voltage gain v_o/v_i and the input and output resistances of the amplifier. Use the results of Example 1.4 for the equivalent circuit.

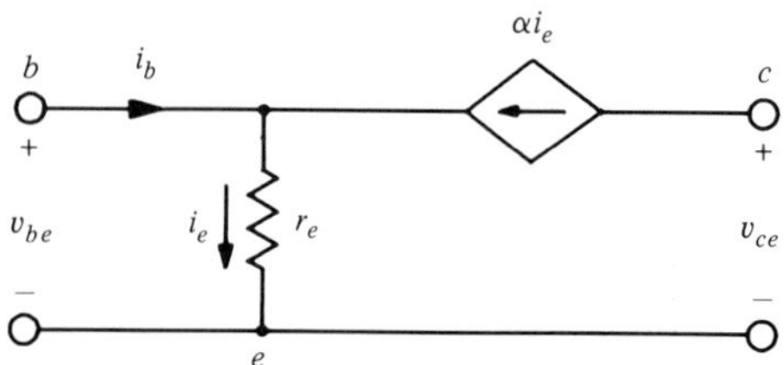

Figure 1.9 Approximate equivalent circuit for low-frequency applications.

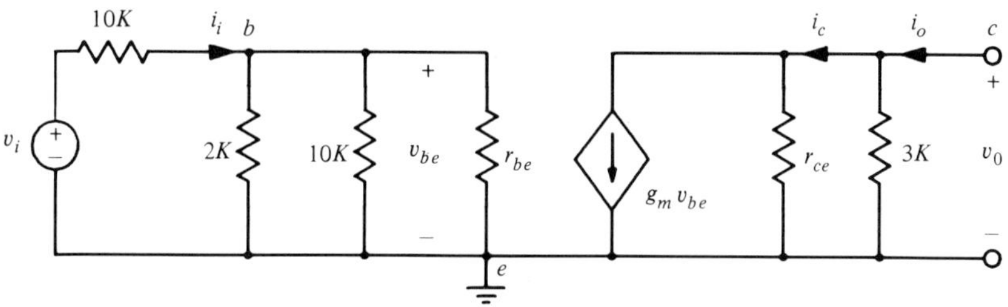

Figure 1.10 Small-signal equivalent circuit for the amplifier circuit shown in Fig. 1.4.

First we note that, for signal frequencies, C_B and C_E tend to ∞. This means that these two branches act as short circuits for the signal. Also, the node potential to which the power supply of 15 V is connected does not change. Therefore, as far as the signal is concerned, it is a ground terminal. Using these facts and the small-signal equivalent circuit in Fig. 1.8, we have the equivalent circuit for the entire circuit as shown in Fig. 1.10.

We derived the values of g_m, r_{be}, and r_{ce} in Example 1.4, and we will use them here. Let R_i be equal to 2 kΩ||10 kΩ||r_{be}, and thus we have

$$R_i = 1.0487 \text{ k}\Omega$$

Therefore

$$v_{be} = \frac{v_i \times 1.0487}{11.0487} = 94.92 \times 10^{-3} v_i$$

Let R_o be equal to r_{ce}||3 kΩ. Then,

$$R_o = 2.935 \text{ k}\Omega$$

and

$$\begin{aligned} v_o &= -g_m v_{be} R_o \\ &= (-45.17)(94.92 \times 10^{-3})(2.935 v_i) \\ &= -9.849 v_i \end{aligned}$$

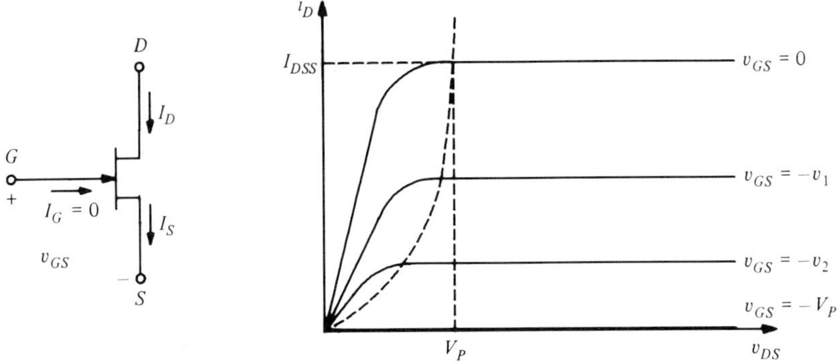

Figure 1.11 An *n*-channel FET and its output characteristics.

Therefore

$$\frac{v_o}{v_i} = -9.849$$

The negative sign indicates phase reversal of the output with respect to the input voltage v_i. The input resistance of the amplifier is v_{be}/i_i. This is just the value of R_i that we calculated earlier. The output resistance of the amplifier is v_o/i_o when $v_i \equiv 0$. With $v_i = 0$, we have $v_{be} = 0$, and therefore the current source becomes open-circuited. Then v_o/i_o is just $R_o = 2.935$ kΩ, as calculated earlier. ∎

The equivalent circuit for an FET in its active region will be derived next. An *n*-channel junction FET and its typical output characteristics are shown in Fig. 1.11. It is a voltage-controlled device, and the value of i_G is always on the order of picoamperes; therefore we assume that $i_G \equiv 0$. With small values of v_{DS}, when the reverse bias of v_{GS} is increased (negatively increasing), there is a particular potential difference v_{GS} at which the drain current stops. Such a condition is called the *pinch-off condition*, and the corresponding voltage v_{GS} is called the *pinch-off voltage*. That is,

$$V_P = v_{GS}\big|_{i_D \equiv 0}, \qquad \text{when } v_{DS} \text{ is small}$$

For an *n*-channel FET, the pinch-off voltage is negative. When $v_{DG} > |V_P|$, pinch-off occurs, and this is the active region for an FET. Thus an FET works in its active region if the drain voltage is higher than the gate voltage by $|V_P|$. In the active region, the output characteristics are almost horizontal lines whose heights are determined by v_{GS}. The transfer characteristic i_D versus the v_{GS} characteristic beyond pinch-off is plotted in Fig. 1.12. This curve can be approximated by square law as

$$i_D = I_{DSS}\left(1 - \frac{v_{GS}}{V_P}\right)^2 \tag{1.17}$$

Both I_{DSS} and V_P are constants for a particular device.

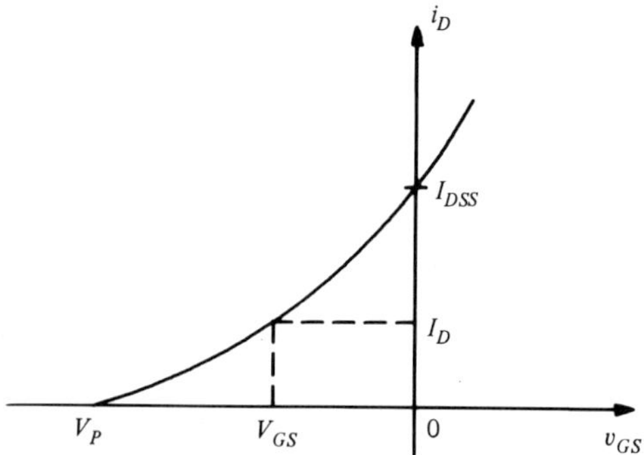

Figure 1.12 Transfer characteristic of an FET.

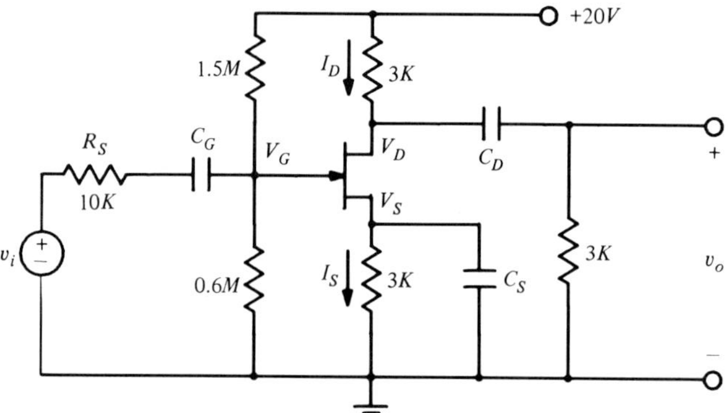

Figure 1.13 The FET amplifier circuit for Example 1.6.

Example 1.6: Consider the circuit shown in Fig. 1.13. Find the bias voltages and currents and thus determine whether the device works in the active region. $I_{DSS} = 12$ mA, and $V_p = -4$ V.

Since $I_G = 0$ and C_G acts as an open circuit for direct current, we have

$$V_G = \frac{0.6}{1.5 + 0.6}(20) = 5.714 \text{ V}$$

Also, note that $I_S = I_D$ and that C_S acts as an open circuit for direct current:

$$V_S = 3I_D$$

where I_D is in milliamperes.

Therefore

$$V_{GS} = 5.714 - 3I_D$$

Now, I_D can also be obtained as

$$I_D = I_{DSS}\left(1 - \frac{V_{GS}}{V_p}\right)^2$$

$$= I_{DSS}\left(1 - \frac{5.714 - 3I_D}{-4}\right)^2$$

$$= 12(2.429 - 0.75I_D)^2$$

The above equation is a quadratic equation. Therefore there are two solutions for I_D:

$$I_D = 2.616 \text{ mA} \qquad \text{and} \qquad 4.009 \text{ mA}$$

If $I_D = 4.009$ mA, V_{DS} will be negative. Therefore the appropriate solution for our problem is

$$I_D = 2.616 \text{ mA}$$

Then, of course,

$$V_S = 7.847 \text{ V} \qquad \text{and} \qquad V_D = 12.15 \text{ V}$$

Note that $V_D > V_G + |V_p|$. Therefore the FET is in the active region. Also, $V_{GS} = -2.133$ V. ■

In order to derive the small-signal equivalent circuit, consider a simple common-source amplifier as shown in Fig. 1.14. The signal v_{gs} is an externally applied signal. The FET remains in the active region when $v_D \geq |V_p| + v_{GS}$. Assume that this condition is satisfied. Now, the instanta-

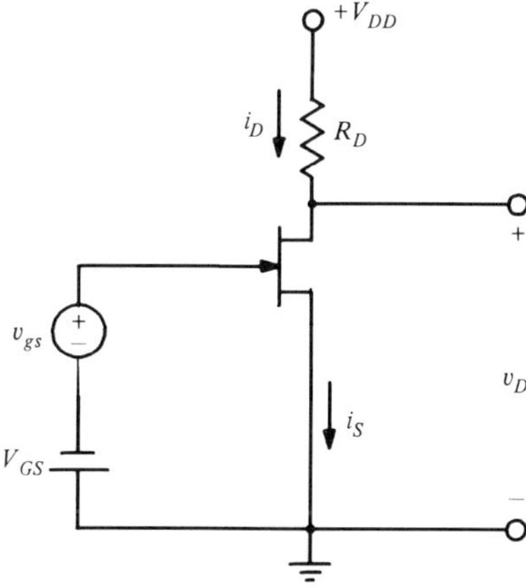

Figure 1.14 Common-source amplifier with biasing arrangements.

neous value of i_D is

$$i_D = I_{DSS}\left(1 - \frac{v_{GS}}{V_p}\right)^2 = I_{DSS}\left(1 - \frac{V_{GS} + v_{gs}}{V_p}\right)^2$$

Expanding i_D near the operating condition if $i_D|_{v_{gs}=0} = I_D$, we have (using a Taylor series)

$$\begin{aligned} i_D &\simeq I_{DSS}\left(1 - \frac{V_{GS}}{V_p}\right)^2 + \frac{2I_{DSS}}{-V_p}\left(1 - \frac{V_{GS}}{V_p}\right)v_{gs} \\ &= I_D + g_m v_{gs} \end{aligned}$$

where

$$g_m = \frac{2I_{DSS}}{-V_p}\left(1 - \frac{V_{GS}}{V_p}\right) \tag{1.18}$$

Separating the small signal i_d, we have

$$i_d = g_m v_{gs} \tag{1.19}$$

The above equation, along with $i_g \equiv 0$ (since $i_G \equiv 0$), can be represented as shown in Fig. 1.15a. Note that the parameter of the equivalent circuit is dependent on the bias voltage V_{GS} and in turn on I_D. That is,

$$g_m = \frac{2I_{DSS}}{-V_p}\sqrt{\frac{I_D}{I_{DSS}}} \tag{1.20}$$

A more accurate model of an FET should include an output resistor across the dependent current source in Fig. 1.15a. Further, if the equivalent circuit is used for high-frequency applications, it should also include junction capacitances. Thus a more accurate model is the one shown in Fig. 1.16a. The finite r_o is due to the finite slope of the output characteristics shown in Fig. 1.11, which were assumed to be horizontally flat. As in the case of BJTs, the value of r_o is inversely proportional to I_D in this case. The

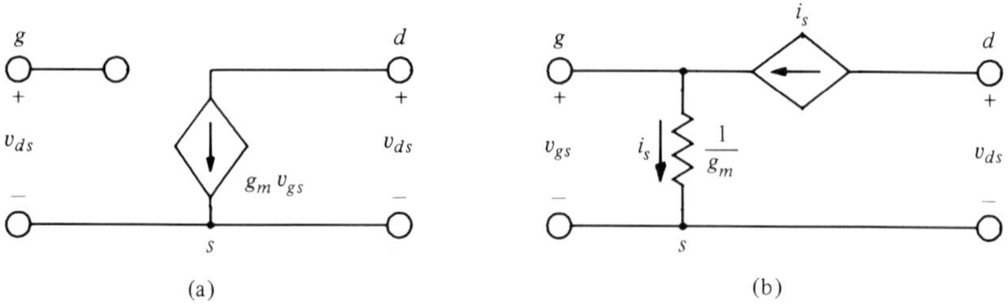

Figure 1.15 Approximate small-signal equivalent circuits for an FET.

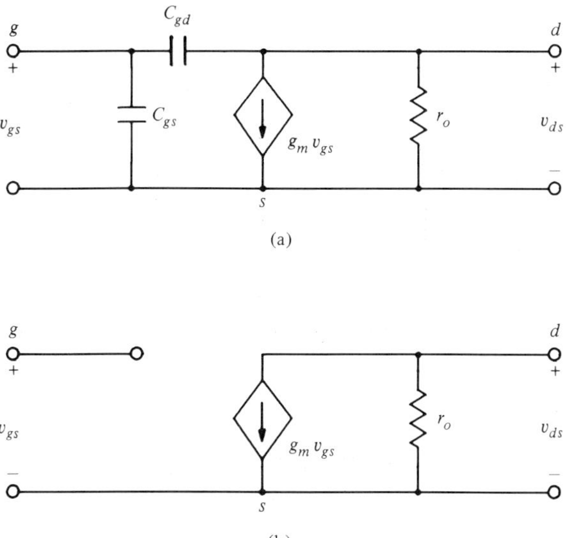

Figure 1.16 More accurate equivalent circuits for an FET. (*a*) For high frequency. (*b*) For low frequency.

typical value of r_o is many tens of kiloohms to a few hundred kiloohms. The values of C_{gd} and C_{gs} are on the order of 1 to 3 pF, and their influence can be ignored if we consider only low-frequency applications, as the case here. Therefore we will use the equivalent circuit in Fig. 1.16*b* in most of our applications. Sometimes r_o can be considered infinite, depending on the loading circuit connected to the drain of the FET.

Example 1.7: In the circuit in Fig. 1.13, assume that C_G, C_S, and C_D tend to ∞ for the signal frequencies and that their corresponding signal branches can be considered short circuits. Use the results of Example 1.6 to find v_o/v_i, the input and output resistances. Assume that r_o for the small-signal equivalent shown in Fig. 1.16*b* is 100 kΩ when $I_D =$ 1 mA.

Since r_o is inversely proportional to I_D, for the given circuit, and since $I_D = 2.616$ mA, the value of r_o is

$$r_o = \frac{100}{2.616} = 38.23 \text{ k}\Omega$$

The small-signal equivalent circuit for the entire circuit is as shown in Fig. 1.17. Referring to this figure, we find that

$$R_i = 1.5 \text{ M}\Omega \| 0.6 \text{ M}\Omega = 0.429 \text{ M}\Omega$$

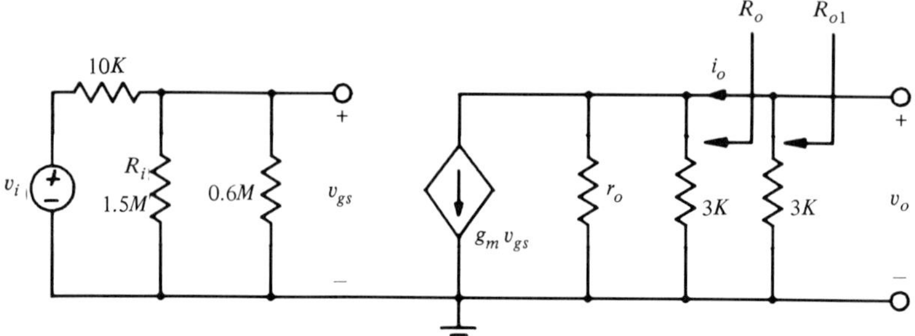

Figure 1.17 Small-signal equivalent of the entire circuit in Fig. 1.13.

Therefore

$$v_{gs} = \frac{(v_i)(0.429)}{0.429 + 0.01} = \frac{0.429}{0.439} v_i = 0.977 v_i$$

Using (1.20), we obtain the value of g_m, and it is 2.801 mA/V. Also, the value of $R_{o1} = 3\ \text{k}\Omega \| 3\ \text{k}\Omega \| 38.23\ \text{k}\Omega = 1.443\ \text{k}\Omega$. Thus

$$v_o = -g_m R_{o1} v_{gs} = -3.951 v_i$$

Therefore we have

$$\frac{v_o}{v_i} = -3.951$$

The negative sign indicates phase reversal of v_o with respect to the voltage v_i. The input resistance is R_i and is equal to 0.429 MΩ. The output resistance of the amplifier is v_o/i_o when $v_i = 0$. This resistance is $R_o = 2.782$ kΩ.

1.3 *Basic Differential Amplifiers*

The input stage of an operational amplifier is a differential amplifier. We first consider a differential amplifier using BJTs as shown in Fig. 1.18. Usually two power supplies are used so that the output can swing both positive and negative with respect to ground. Here R_c is the load resistance connected to each collector, and we assume that Q_1 and Q_2 are identical transistors. Before we find the output voltages v_{o1} and v_{o2} in terms of the two input voltages v_1 and v_2, we have to establish the bias currents because, as noted before, the equivalent circuit for small signals depends on the bias currents. The bias currents are found by making $v_1 = v_2 = 0$ in this circuit. This means that both B_1 and B_2 are at ground potential. Then, since the emitters are tied together, V_E must be approximately equal to

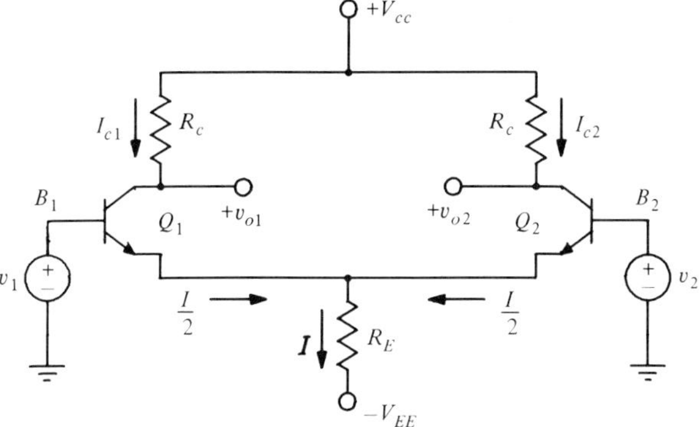

Figure 1.18 Simple differential amplifier using BJTs.

-0.7 V. Then, the direct current through R_E must be

$$I \simeq \frac{V_{EE} - 0.7}{R_E} \tag{1.21}$$

Because of the symmetry of the circuit, we have the emitter currents

$$I_{E1} = I_{E2} = \frac{I}{2} = \frac{V_{EE} - 0.7}{2R_E} \tag{1.22}$$

Furthermore, the collector currents I_{C1} and I_{C2} can be found to be

$$I_{C1} = I_{C2} = \alpha I_{E1} = \frac{\alpha(V_{EE} - 0.7)}{2R_E} \tag{1.23}$$

Given the β value of the transistor, we can find both collector currents at the bias point. The transistors Q_1 and Q_2 will work in the active region if $V_{CB} \geq 0$, that is, $V_C \geq 0$. Since $V_C = V_{CC} - I_{C1}R_C$, the transistors Q_1 and Q_2 will be in the active region if the following condition is satisfied:

$$R_E \geq \frac{\alpha R_C(V_{EE} - 0.7)}{2V_{CC}}$$

Once the value of I_C has been determined, then the parameters in the small-signal equivalent can be determined quite easily. The small-signal equivalent circuit for the entire circuit in Fig. 1.18 can be found by replacing each transistor with its corresponding equivalent circuit. Since both transistors are identical, the parameters are also the same. Thus the entire small-signal equivalent circuit is as shown in Fig. 1.19. From Fig. 1.19, we find that

$$v_{be1} = v_1 - v_e \qquad \text{and} \qquad v_{be2} = v_2 - v_e$$

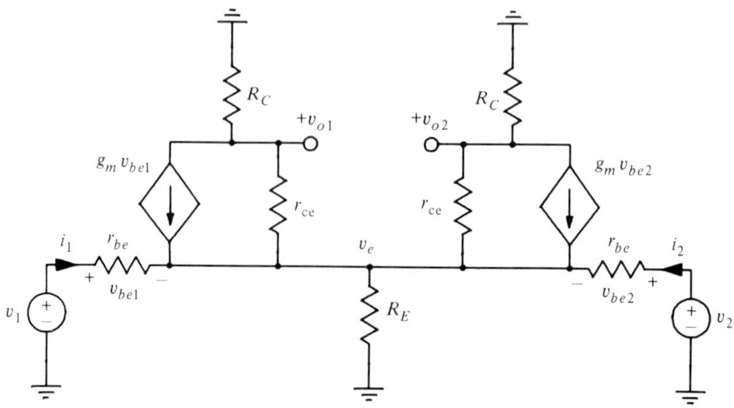

Figure 1.19 Small-signal equivalent of the circuit in Fig. 1.18.

and therefore

$$g_m v_{be1} = g_m(v_1 - v_e) \qquad \text{and} \qquad g_m v_{be2} = g_m(v_2 - v_e) \tag{1.24}$$

Using KCL at the nodes where v_e, v_{o1}, and v_{o2} have been marked, we can write three independent equations in terms of v_e, v_{o1}, and v_{o2}. Using (1.24) in these three equations, we obtain

$$v_e\left(2g_m + \frac{2}{r_{be}} + \frac{1}{R_E} + \frac{2}{r_{ce}}\right) - \frac{v_{o1}}{r_{ce}} - \frac{v_{o2}}{r_{ce}} = \left(g_m + \frac{1}{r_{be}}\right)(v_1 + v_2) \tag{1.25a}$$

$$-v_e\left(g_m + \frac{1}{r_{ce}}\right) + v_{o1}\left(\frac{1}{R_c} + \frac{1}{r_{ce}}\right) = -g_m v_1 \tag{1.25b}$$

$$-v_e\left(g_m + \frac{1}{r_{ce}}\right) + v_{o2}\left(\frac{1}{R_c} + \frac{1}{r_{ce}}\right) = -g_m v_2 \tag{1.25c}$$

Since $g_m r_{ce} = V_A/V_T \simeq 5000$, we can ignore the term $1/r_{ce}$ compared to g_m in all the above equations. Then, solving for v_e, we obtain

$$v_e = \frac{v_1 + v_2}{2} \frac{g_m R'_c r_{be} + R_c}{[R_c(1 + r_{be}/2R_E) + g_m R'_C r_{be}]} \tag{1.26}$$

where

$$\frac{1}{R'_C} = \frac{1}{R_c} + \frac{1}{r_{ce}} \tag{1.27}$$

From (1.25b) and (1.25c), we can easily find v_{o1} and v_{o2}:

$$v_{o1} = -g_m R'_C v_1 + g_m R'_C v_e \tag{1.28}$$

and

$$v_{o2} = -g_m R'_C v_2 + g_m R'_C v_e \tag{1.29}$$

Substituting (1.26) into (1.28) and (1.29), we can obtain the expressions for v_{o1} and v_{o2}. But first let us define two voltages. The *differential mode signal* v_d is defined as

$$v_d = v_1 - v_2 \tag{1.30}$$

And the *common-mode signal* v_{cm} is defined as

$$v_{cm} = \frac{v_1 + v_2}{2} \tag{1.31}$$

In terms of v_d and v_{cm}, we can solve for v_1 and v_2:

$$v_1 = \frac{v_d}{2} + v_{cm} \tag{1.32}$$

and

$$v_2 = -\frac{v_d}{2} + v_{cm} \tag{1.33}$$

Substituting (1.32) and (1.33) for v_1 and v_2, respectively, in (1.26), (1.28), and (1.29) and solving for v_{o1} and v_{o2} in terms of v_d and v_{cm}, we obtain

$$v_{o1} = -\frac{g_m R'_c}{2} v_d + \frac{g_m r_{be} R_c R'_c/(2R_E)}{g_m r_{be} R'_c + R_c[1 + r_{be}/(2R_E)]} v_{cm} \tag{1.34}$$

and

$$v_{o2} = \frac{g_m R'_c}{2} v_d + \frac{g_m r_{be} R_c R'_c/(2R_E)}{g_m r_{be} R'_c + R_c[1 + r_{be}/(2R_E)]} v_{cm} \tag{1.35}$$

In an ideal differential amplifier, either one of the outputs, for example v_{o2}, should be of the form

$$v_{o2} = A_d v_d \tag{1.36}$$

However, from (1.35) we see that v_{o2} is of the form

$$v_{o2} = A_d v_d + A_{cm} v_{cm} \tag{1.37}$$

where A_d and A_{cm} are known as the differential-mode and common-mode gains, respectively. Comparing (1.35) and (1.37), we have

$$A_d = \frac{g_m R'_c}{2} \tag{1.38}$$

and

$$A_{cm} = \frac{g_m r_{be} R_c R'_c/(2R_E)}{g_m r_{be} R'_c + R_c[1 + r_{be}/(2R_E)]} \tag{1.39}$$

When $r_{ce} \gg R_c$, which is usual in many cases, it is possible to simplify (1.38) and (1.39). If $r_{ce} \gg R_c$, then $R'_c \simeq R_c$, and therefore

$$A_d \simeq \frac{g_m R_c}{2} = \frac{I_c R_c}{2V_T} = \frac{\alpha R_c}{4R_E} \frac{V_{EE} - 0.7}{V_T} \tag{1.40}$$

and

$$A_{cm} \simeq \frac{\beta R_c}{2R_E(\beta + 1)} = \frac{\alpha R_c}{2R_E} \tag{1.41}$$

since $g_m r_{be} = \beta$ and $\beta + 1 \gg r_{be}/2R_E$. In an ideal differential amplifier A_{cm} should be zero. However, in practice it is not zero, as is obvious from (1.41). In this regard, a quantity called the *common-mode rejection ratio* (CMRR) is defined as

$$\begin{aligned} \rho_{cm} &= \frac{A_d}{A_{cm}} \\ &= 20 \log \left| \frac{A_d}{A_{cm}} \right| \quad \text{dB} \end{aligned} \tag{1.42}$$

By substituting approximate gains in (1.42), we have

$$\rho_{cm} \simeq 20 \log \left| \frac{V_{EE} - 0.7}{2V_T} \right| \tag{1.43}$$

One more quantity of interest is the input resistance faced by each of the sources v_1 and v_2. They are, referring to the equivalent circuit in Fig. 1.19,

$$R_{i1} = \left. \frac{v_1}{i_1} \right|_{v_2 = 0} \qquad \text{and} \qquad R_{i2} = \left. \frac{v_2}{i_2} \right|_{v_1 = 0}$$

Note that $i_1 = (v_1 - v_e)/r_{be}$. Therefore, after substituting $v_2 = 0$ into (1.26) and calculating i_1, we have

$$i_1 = \frac{v_1}{r_{be}} \left[1 - \frac{1}{2} \frac{g_m r_{be} R'_c + R_c}{R_c[1 + r_{be}/(2R_E)] + g_m R'_c r_{be}} \right]$$

Since $g_m r_{be} = \beta$ and $g_m r_{be} R'_C \gg R_c[(1 + r_{be})/(2R_E)]$, we have

$$i_1 \simeq \frac{v_1}{2r_{be}}$$

Therefore

$$R_{i1} = \left. \frac{v_1}{i_1} \right|_{v_2 = 0} \simeq 2r_{be}$$

Because of the symmetry of the circuit, we conclude that

$$R_{i2} = \left.\frac{v_2}{i_2}\right|_{v_1=0} \simeq 2r_{be}$$

Example 1.8: In the circuit in Fig. 1.18, assume that $V_{CC} = V_{EE} = 15$ V. Given that $R_c = 10$ kΩ, $R_E = 8.2$ kΩ, and $\beta_1 = \beta_2 = 100$, find A_d, A_{cm}, CMRR, and R_{i1}.

We shall calculate both A_d and A_{cm} using (1.38) and (1.39), as well as the approximate expressions (1.40) and (1.41). This will give us an idea about the approximations involved in obtaining (1.40) and (1.41). In order to use (1.38) and (1.39), we must calculate the parameters of the equivalent circuit. For this we need to find the value of I_C first. Since $\beta = 100$, $\alpha = 0.9901$. Therefore we have

$$I_{C1} = I_{C2} = \frac{15 - 0.7}{(2)(8.2)}(0.9901) = 0.863 \text{ mA}$$

Using this value of I_C, we can easily verify that both Q_1 and Q_2 are indeed in the active region. Assuming that $V_A = 120$ V and $V_T = 25$ mV (at room temperature), we find the values of g_m, r_{ce}, and r_{be} using (1.11) and (1.13):

$$g_m = \frac{I_C}{V_T} = \frac{0.863}{25} = 34.53 \text{ mA/V}$$

$$r_{ce} = \frac{V_A}{I_C} = 139 \text{ k}\Omega$$

$$r_{be} = \frac{\beta}{g_m} = 2.896 \text{ k}\Omega$$

Since $R_c = 10$ kΩ and $r_{ce} = 139$ kΩ, we use (1.27) and find that $R'_c = 9.329$ kΩ. Therefore

$$A_d = \frac{g_m R'_c}{2} = 161.1 = 44.14 \text{ dB}$$

and

$$A_{cm} = 0.6022 = -4.406 \text{ dB}$$

Using the approximate formulas (1.40) and (1.41), we obtain the approximate values of A_d and A_{cm}:

$$A_d \simeq 172.7 = 44.74 \text{ dB}$$

and

$$A_{cm} \simeq 0.6037 = -4.383 \text{ dB}$$

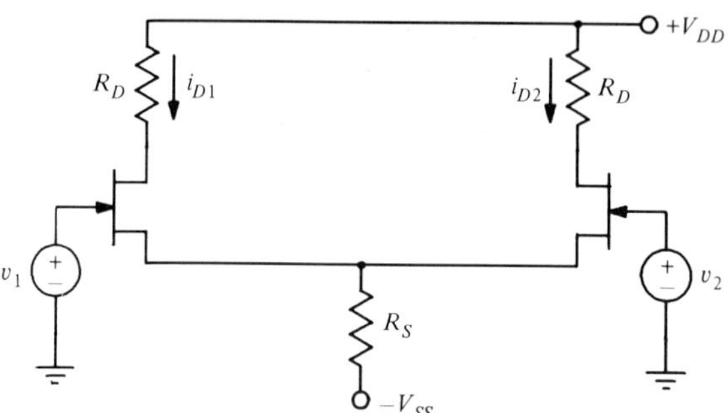

Figure 1.20 Differential amplifier using FETs.

We can calculate the exact value of ρ_{cm} and directly from the above values of A_d and A_{cm}:

$$\rho_{cm} = 48.55 \text{ dB}$$

Using the approximate expression (1.43), we have $\rho_{cm} = 49.13$. We can easily see from the above calculations that the approximate values are very close to the actual values and that, in practice, the approximate expressions are quite valid in most cases. Finally, the input resistances are

$$R_{i1} = R_{i2} = 2r_{be} = 5.792 \text{ k}\Omega$$ ■

The input resistances of an op amp are just the input resistances of the input stage, which is the differential amplifier. Such input resistances may be very low, as in the previous example, if we use BJTs. Some op amps use an input stage that employs FETs so that a high-input impedance in excess of 10^{10} Ω can be realized. Therefore we shall next consider a differential amplifier using FETs as shown in Fig. 1.20.

The dc biasing problem can be solved as in Example 1.6. Given the bias current I_D, we are able to determine the small-signal equivalent circuit. If the transistors are identical, then $I_{D1} = I_{D2} = I_D$, $g_{m1} = g_{m2} = g_m$, and $r_{o1} = r_{o2} = r_o$. When we use the small-signal equivalent circuit for the FET shown in Fig. 1.16, the small-signal equivalent circuit for the entire circuit in Fig. 1.20 is as shown in Fig. 1.21. Note that

$$v_{gs1} = v_1 - v_s \qquad \text{and} \qquad v_{gs2} = v_2 - v_s$$

Therefore

$$g_m v_{gs1} = g_m(v_1 - v_s) \qquad \text{and} \qquad g_m v_{gs2} = g_m(v_2 - v_s)$$

Using the above equations and KCL at the nodes where v_s, v_{o1}, and v_{o2}

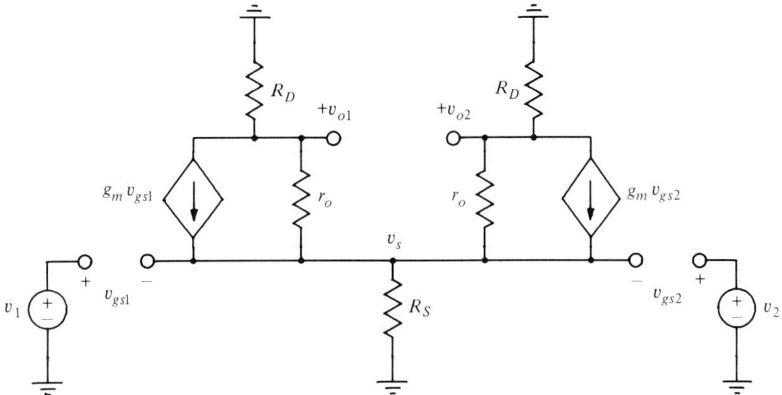

Figure 1.21 Small-signal equivalent circuit for the differential amplifier shown in Fig. 1.20.

have been marked, we obtain

$$v_s\left(2g_m + \frac{2}{r_o} + \frac{1}{R_s}\right) - \frac{1}{r_o}v_{o1} - \frac{1}{r_o}v_{o2} = 2g_m(v_1 + v_2)$$

$$-\left(g_m + \frac{1}{r_o}\right)v_s + \left(\frac{1}{R_D} + \frac{1}{r_o}\right)v_{o1} = -g_m v_1$$

$$-\left(g_m + \frac{1}{r_o}\right)v_s + \left(\frac{1}{R_D} + \frac{1}{r_o}\right)v_{o2} = -g_m v_2$$

Using the same definitions of differential-mode and common-mode voltages, v_d and v_{cm}, as in the previous case and solving the above three equations, we obtain the output voltages v_{o1} and v_{o2}:

$$v_{o1} = -\frac{g_m R'_D}{2}v_d + \frac{g_m R_D R'_D}{R_D + 2R'_D R_S g'_m}v_{cm}$$

and

$$v_{o2} = \frac{g_m R'_D}{2}v_d + \frac{g_m R_D R'_D}{R_D + 2R'_D R_S g'_m}v_{cm}$$

where $R'_D = R_D \| r_o$ and $g'_m = g_m + 1/r_o$.

The differential-mode and common-mode gains are

$$A_d = \frac{g_m R'_D}{2} \tag{1.44}$$

and

$$A_{cm} = \frac{g_m R_D R'_D}{R_D + 2R'_D R_S g'_m} \tag{1.45}$$

The CMRR is, then,

$$\rho_{cm} = 20 \log\left|\frac{A_d}{A_{cm}}\right|$$

$$= 20 \log\left|\frac{1}{2} + \frac{R'_D}{R_D} g'_m R_S\right| \tag{1.46}$$

Note that the input resistances of the differential amplifiers for both input signals v_1 and v_2 are infinite. However, in practice these resistances are in excess of 10^{10} Ω.

Example 1.9: Consider the network in Fig. 1.20 in which $V_{DD} = V_{SS} =$ 15 V, $R_D = 6.8$ kΩ, $I_{DSS} = 2$ mA, $V_P = -4$ V, and $R_S = 6.2$ kΩ. It is also given that r_o of each FET is 50 kΩ when $I_D = 1$ mA. Find the differential-mode and common-mode gains. Also, find the CMRR.

In order to find the parameters of the small-signal equivalent circuit, we must first establish the bias conditions. For this, we have to ground the input voltages v_1 and v_2. This means that

$$V_{G1} = V_{G2} \equiv 0$$

Let I_D be the drain current flowing through each FET. Then, the current flowing through the resistance R_S must be $2I_D$. Therefore

$$V_{S1} = V_{S2} = 2I_D R_S - V_{SS}$$
$$= 12.4 I_D - 15$$

where I_D is in milliamperes.

Since $V_{G1} = V_{G2} = 0$, we have

$$V_{G1S1} = V_{G2S2} = 15 - 12.4 I_D$$

Now, using (1.17), we have

$$I_D = I_{DSS}\left(1 - \frac{V_{G1S1}}{V_p}\right)^2$$

$$= 2\left(1 + \frac{15 - 12.4 I_D}{4}\right)^2$$

The appropriate solution of the above equation is

$$I_D = 1.275 \text{ mA}$$

Therefore

$$V_D = 15 - 6.8 \times 1.275 = 6.332 \text{ V}$$

Note that $V_D > V_G + |V_p| = 4$ V. Therefore both Q_1 and Q_2 are in the active region. To find the value of g_m, we need

$$V_{GS} = 15 - 12.4I_D = -0.8066$$

Therefore, using (1.20), we have

$$g_m = 0.7984 \text{ mA/V}$$

Since r_o is inversely proportional to I_D, we have

$$r_o = \frac{50}{1.275} = 39.22 \text{ k}\Omega$$

Then, $R'_D = R_D \| r_o = 5.795$ kΩ. Therefore

$$A_d = \frac{g_m R'_D}{2} = 2.313 = 7.285 \text{ dB}$$

and

$$A_{cm} = 0.4903 = -6.191 \text{ dB}$$

Therefore

$$\rho_{cm} = 13.48 \text{ dB}$$ ■

A single-stage amplifier will not satisfy the gain requirement of an op amp. Therefore we need another high-gain stage of amplification. Further, the common-mode rejection ratio is very poor. Usually we require a common-mode rejection ratio of more than 100 dB. If we use the circuit in Fig. 1.18 or 1.20 to increase the CMRR, we must increase the value of R_E or R_S, respectively. This can be seen from (1.41) and (1.45), respectively. In fact, to attain the theoretical limit of $\rho_{cm} = \infty$, we need to have a value of $R_E = R_S = \infty$, which is impossible. We cannot go on increasing the value of R_E or R_S without limit and without penalty. For example, consider the case of the differential amplifier in Fig. 1.18 using BJTs. Equation (1.43) clearly indicates that, in order to achieve a common-mode rejection ratio of 100 dB, we need a supply voltage of at least $V_{EE} \simeq 250$ V, which is impractical in the case of integrated circuits. To eliminate this and other associated undesirable effects, we can replace the biasing resistors R_E and R_S in these two circuits with a circuit that provides a large resistance R for small signals and an appropriate required direct current with a small dc voltage across it. Such circuits are called *current sources* (and *sinks*). A simple circuit for a current sink is shown in Fig. 1.22.

If V_{CC} is chosen appropriately and $V_{CB} \geq 0$, then the transistor will be in the active region. It is also possible to connect a load to the collector of the transistor. The bias current I_C can be set to have any required value by adjusting the reference current I_{ref}, which in turn is controlled by both R_F

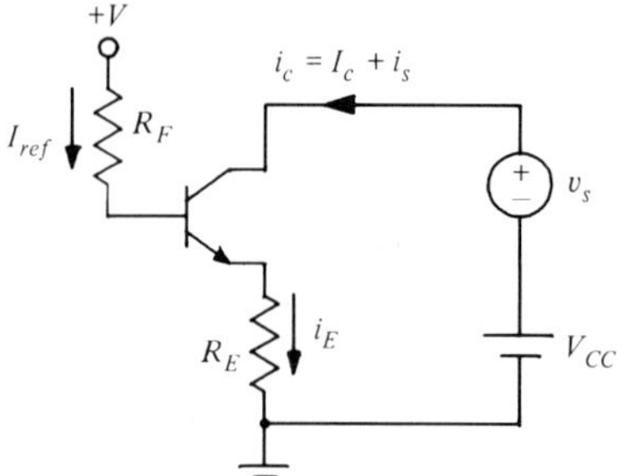

Figure 1.22 Simple current sink with a signal voltage v_s.

and R_E for a given V. The collector bias current is

$$I_C \simeq I_E = \frac{V - V_{BE}}{R_E + R_F/(\beta + 1)} \tag{1.47}$$

This current will be almost constant for a given voltage V. For a small-signal voltage v_s, the output resistance is the resistance of the small-signal equivalent circuit, looking into the collector terminal of the circuit. To calculate this resistance, the small-signal equivalent circuit is shown in Fig. 1.23. The output resistance of this circuit is $R_o = v_s/i_s$. Using simple circuit analysis and the fact that $\beta = g_m r_{be}$, we can show that

$$R_o = \frac{r_{ce}[(\beta + 1)R_E + R_F + r_{be}] + R_E(R_F + r_{be})}{R_E + R_F + r_{be}}$$

$$\simeq \frac{r_{ce}[(\beta + 1)R_E + R_F] + R_E R_F}{R_E + R_F} \tag{1.48}$$

since $r_{be} \ll R_F$. Since r_{ce} is on the order of 100 kΩ or more, the value of R_o will be at least a few hundred kiloohms. Consider the following example.

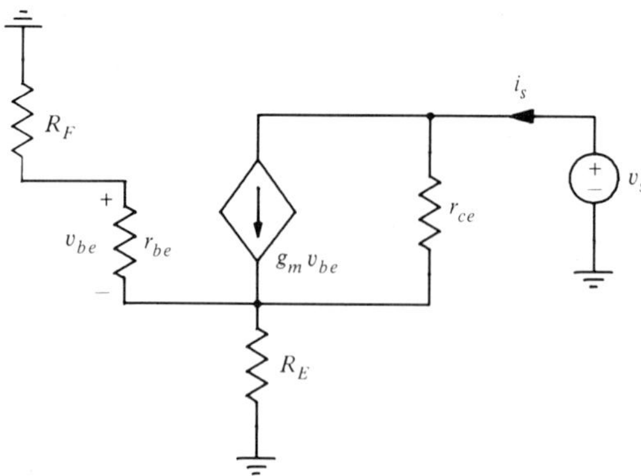

Figure 1.23 Small-signal equivalent of the circuit shown in Fig. 1.22.

Example 1.10: Given that, in the circuit in Fig. 1.22, $R_F = 390\ \text{k}\Omega$, $R_E = 10\ \text{k}\Omega$, and $V = V_{CC} = 15$ V, find the collector bias current I_C and the output resistance for small signals. Assume that $\beta = 100$ and $V_A = 120$ V.

Using (1.47), we have

$$I_E = \frac{15 - 0.7}{10 + 390/101} = 1.032 \text{ mA}$$

Therefore

$$I_C = \alpha I_E = \frac{\beta}{\beta + 1} I_E = 1.021 \text{ mA}$$

Next, we obtain

$$r_{ce} = \frac{V_A}{I_C} = \frac{120}{1.021}$$
$$= 117.5\ \text{k}\Omega$$

Now, using (1.48),

$$R_o \simeq \frac{117.5[(101)(10) + 390] + 3900}{400}$$
$$= 420.9\ \text{k}\Omega$$

■

Example 1.11: Consider a differential amplifier circuit using a current sink for biasing as shown in Fig. 1.24. The resistances R_C and R_E are both $10\ \text{k}\Omega$, and $R_B = 390\ \text{k}\Omega$. Find the differential-mode and common-mode gains. Also, find the CMRR and the input resistances for the signal

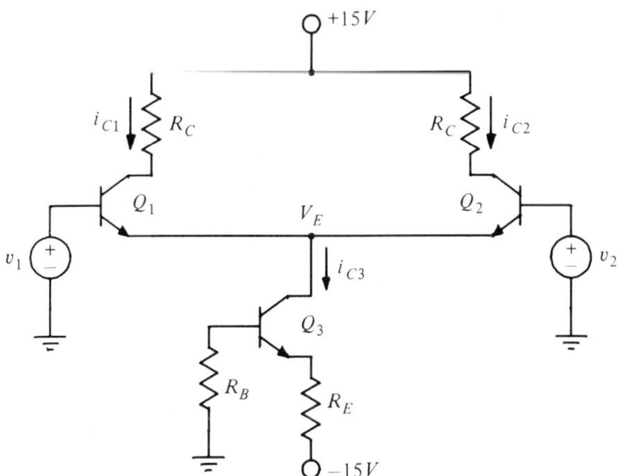

Figure 1.24 Improved differential amplifier.

sources. Assume that all the transistors are identical and that $\beta = 100$ and $V_A = 120$ V.

The biasing problem for the transistor Q_3 was solved in Example 1.10, and therefore

$$I_{C3} = 1.021 \text{ mA}$$

Since $I_{E1} = I_{E2} = I_{C3}/2$, we have

$$I_{C1} = I_{C2} = \alpha I_{E1} = \frac{\alpha}{2} I_{C3}$$
$$= 0.5057 \text{ mA}$$

We have to check whether all the transistors are in the active region. First, note that

$$V_{B1} = V_{B2} = 0 \text{ V}$$

and

$$V_E \simeq -0.7 \text{ V}$$
$$V_{B3} = -\frac{I_{C3}}{\beta} R_B = \frac{-1.021}{100}(390)$$
$$= -3.984 \text{ V}$$

Therefore $V_{C3} = -0.7 \text{ V} > V_{B3}$, and thus Q_3 is in the active region.

$$V_{C1} = V_{C2} = 15 - I_{C1} R_C$$
$$= 15 - 0.5057(10)$$
$$= 0.943 \text{ V} > V_{B1} = V_{B2}$$

This ensures that Q_1 and Q_2 are also in the active region.

The small-signal equivalent parameters of Q_1 and Q_2 are

$$g_{m2} = g_{m1} = \frac{I_C}{V_T} = \frac{0.5057}{25} \times 10^3 = 20.23 \text{ mA/V}$$

$$r_{ce2} = r_{ce1} = \frac{V_A}{I_C} = \frac{120}{0.5057} = 237.3 \text{ k}\Omega$$

$$r_{be1} = r_{be2} = \frac{\beta}{g_m} = 4.944 \text{ k}\Omega$$

The small-signal equivalent circuit in Fig. 1.24 is the same as the one shown in Fig. 1.19, where R_E must be replaced by the output resistance R_o, looking into the collector of Q_3, which was calculated to be 420.9 kΩ. This implies that (1.38) and (1.39), which we developed earlier for the equivalent circuit in Fig. 1.19, can still be used with $R_E = 420.9$ kΩ. Note, however, that we cannot use (1.40), (1.41), and (1.43) because these equations are for a particular biasing arrangement of the circuit in Fig. 1.18. To find the value of A_d, we need $R'_C = R_C \| r_{ce1}$. Therefore we first obtain

$$R'_C = 10\ \text{k}\Omega \| 237.3\ \text{k}\Omega$$
$$= 9.596\ \text{k}\Omega$$

Then we use (1.38) to find

$$A_d = \frac{(20.23)(9.596)}{2}$$
$$= 97.04 = 39.74\ \text{dB}$$

Since $g_m r_{be} = \beta$, we find from (1.39) that

$$A_{cm} = \frac{\beta R_c R'_c}{2\beta R'_E R_c + R_c(2R_E + r_{be})}$$

Substituting the various values, we have

$$A_{cm} = 0.01175 = -38.6\ \text{dB}$$

Therefore

$$\rho_{cm} = 39.74 - (-38.6) = 78.34\ \text{dB}$$ ■

This value of ρ_{cm} is much higher than that obtained in Example 1.8. It can be improved further by reducing the value of I_{C3}. The input resistance is $2r_{be}$ and is approximately 10 kΩ, which is not quite sufficient (a problem that we will address later). Note that by reducing the bias current it is possible to increase the input resistance. However, this technique tends to reduce the g_m value of the transistors, and this in turn tends to reduce the differential-mode gain. However, by connecting a large load resistance R_C, this can be compensated for. We cannot simply increase the load resistance R_C without limit as it will in turn need a large supply voltage. However, one solution we are already aware of is to connect a current source in place of R_C, which will provide a large resistance for small signals with a moderately small dc voltage drop across it. Such a current source, when used as a load, is called an *active load*.

Another method of increasing the input impedance without decreasing the bias current substantially is to use a compound transistor. This technique works both in integrated circuits and in circuits with discrete components. The circuit of a compound transistor, known as a *Darlington pair*, is shown in Fig. 1.25.

Adding one or two more transistors to the circuit is not really expensive, especially, in IC technology (which will be discussed later). The emitter of Q_1 feeds the base of Q_2. To study its properties for small-signal operation

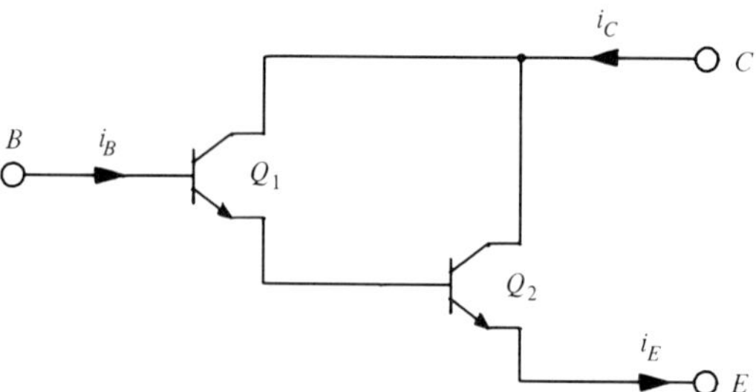

Figure 1.25 Compound transistor.

assume that both Q_1 and Q_2 are in the active region. Then, the small-signal equivalent circuit for the entire circuit will be as shown in Fig. 1.26. Usually, the resistance r_{ce1} is very high, hence we shall assume, for simplicity, that $r_{ce1} \to \infty$. Then one can show that

$$v_{be} = [r_{be1} + (\beta_1 + 1)r_{be2}]i_b \tag{1.49}$$

and

$$i_c = [\beta_2(\beta_1 + 1) + \beta_1]i_b + \frac{v_c}{r_{ce2}} \tag{1.50}$$

The above two equations can be represented in terms of the equivalent circuit shown in Fig. 1.27, where

$$r_{be} = r_{be1} + (\beta_1 + 1)r_{be2} \tag{1.51}$$

and

$$\beta = \beta_2(\beta_1 + 1) + \beta_1 \tag{1.52}$$

The circuit shown in Fig. 1.27 can represent the equivalent circuit of a transistor whose r_{be} and β values are, respectively, given by (1.51) and (1.52). Thus, essentially, the circuit in Fig. 1.25 acts as a transistor.

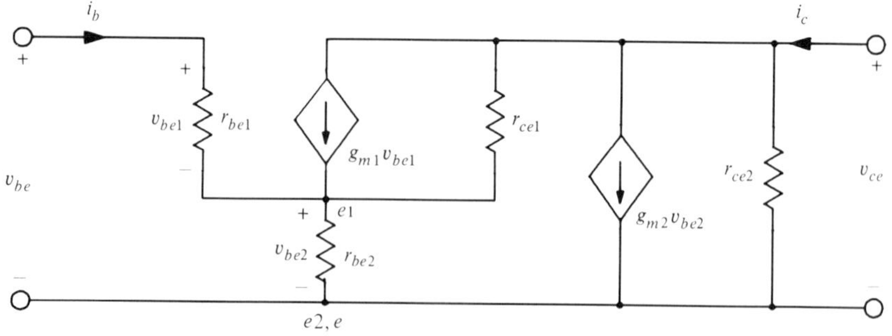

Figure 1.26 Equivalent circuit for a compound transistor.

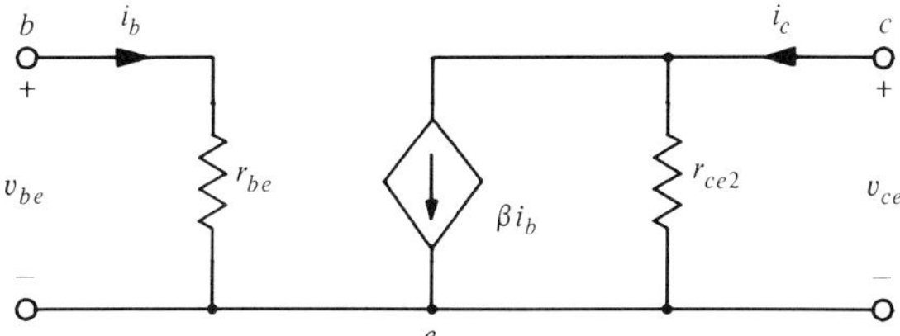

Figure 1.27 Approximate equivalent of the circuit shown in Fig. 1.26.

However, for this circuit, the r_{be} and β values will be substantially higher than the comparable values of β and r_{be} for single transistors. To understand this, assume that both transistors in Fig. 1.25 have β values of 100. Then the β value of the compound transistor will be about 10,200. This is really very high, and one can easily approximate the α value of compound transistors to be 1. To see that the effective value of r_{be} is really high, assume that we replace transistors Q_1 and Q_2 in the differential amplifier in Fig. 1.24 with compound transistors, where each transistor has the same parameters as those of the transistors in Fig. 1.24. A portion of this arrangement is shown in Fig. 1.28. Assume that Q_1, Q_2, Q_3, and Q_4 are identical transistors, each having a β value of 100. We found the value of I_C in Example 1.11 to be 1.021 mA. Therefore, in the circuit in Fig. 1.28, we have

$$I_{C3} = I_{C4} = \frac{\alpha I_C}{2} = 0.5057 \text{ mA}$$

and thus the value of r_{be3} is the same as that calculated earlier:

$$r_{be3} = 4.944 \text{ k}\Omega$$

Since $I_{E1} = I_{B3}$, we have

$$I_{E1} = \frac{I_{C3}}{\beta} = 5.057\,\mu\text{A}$$

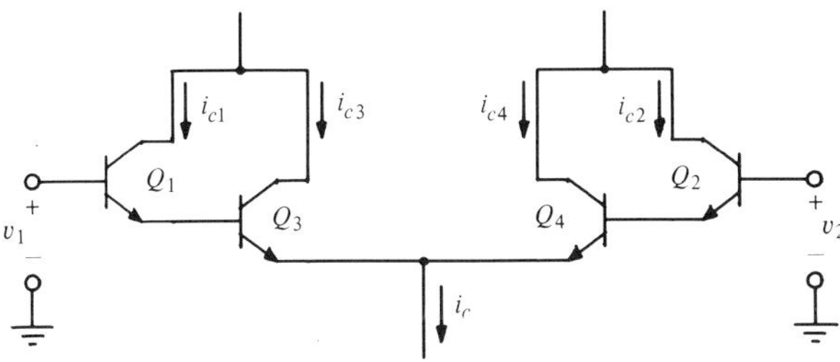

Figure 1.28 Part of the differential amplifier circuit in Fig. 1.24 using compound transistors.

Therefore

$$
\begin{aligned}
r_{be1} &= (\beta_1 + 1)r_{e1} \\
&= (\beta_1 + 1)\frac{V_T}{I_{E1}} \\
&= \frac{(101)(25 \times 10^{-3})}{5.057 \times 10^{-6}} \\
&= 499.3 \text{ k}\Omega
\end{aligned}
$$

Therefore

$$
\begin{aligned}
r_{be} &= 499.3 + (101)(4.944) \\
&= 998.7 \text{ k}\Omega \\
&\simeq 1 \text{ M}\Omega
\end{aligned}
$$

Thus the input impedance of the differential amplifier is 2 MΩ, which is the typical input impedance of an op amp. What is the effect of using compound transistors on the differential-mode and common-mode gains? In the example considered, this effect can be demonstrated by calculating the effective g_m, which is equal to β/r_{be}. Then, for the compound transistor,

$$
\begin{aligned}
g_m &= \frac{10{,}200}{998.7} \\
&= 10.21 \text{ mA/V}
\end{aligned}
$$

This value is approximately half the value we obtained in Example 1.11. This means that both the differential-mode and common-mode gains will be reduced to half of what we obtained in Example 1.11. This is equivalent to a loss of only 6 dB in both gains, and therefore ρ_{cm} will remain almost the same. However, the advantage is that the input impedance is increased by at least 2000 times. The loss in the differential-mode gain can be compensated for in the next state of amplification, which is the subject matter of the next section. In concluding this section, we should mention that the actual value of ρ_{cm} may be lower than what is calculated because of small dissimilarities in the transistors.

1.4 *High-Gain Stage*

The differential-mode gain of an op amp is usually greater than 100 dB. Generally it is almost impossible to achieve this much gain in a single stage of amplification. Therefore most IC op amps have a second stage of amplification so that high-gain requirements can be met. In most op amps the second stage of amplification is accomplished by a common-emitter amplifier with only a single input. Thus, in such amplifiers, the single-ended output is taken from the input stage and further amplified. However, when low drift is required, it is advantageous to connect a second-stage differential amplifier so that differential inputs are available for the second stage. This in turn means that we can use the balanced output from the first stage. Then, no input drift will appear in the balanced output, or at least it will be very small. When a single-ended output is used, only the second stage will amplify this small drift. In this way, the drift in the output can be reduced by 200 times or more. To understand the importance of this input drift, assume that the dc gain of the op amp is 10^6. With a supply voltage of ± 15 V, an input drift as small as 15 μV can either saturate the op amp or drive it to cutoff.

1.5 *Level-Shifting Amplifier*

Since we take the output of the second stage from the collector terminal, the signal voltage at the output of the second stage of amplification will be superimposed onto a dc voltage. Since an op amp is used to amplify dc signals as well, it must be a direct-coupled amplifier. This means that we cannot connect a capacitor to remove this dc voltage. A dc level-shifting circuit is required so that, when $v_1 = v_2 = 0$, the output voltage is also zero and is without a dc component. Basically, it resembles an emitter follower circuit biased by a constant-current source, similar to the one shown in Fig. 1.22. Such a level-shifting circuit is shown in Fig. 1.29. The dc biasing problem must be solved first, so we set $v_i = 0$. Under these conditions, we must have $v_o = 0$. Therefore the dc current through R must be I. Using KVL, we obtain, ignoring the current through R_I,

$$V_{DC} - V_{BE} - IR = 0$$

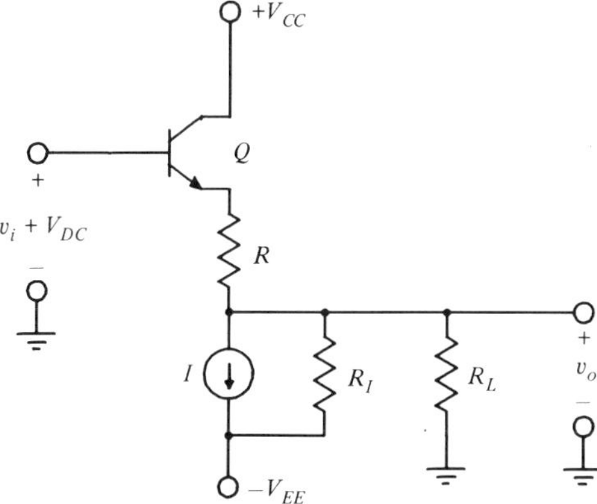

Figure 1.29 Level-shifting circuit.

Solving the above equation, we obtain the required value of R:

$$R = \frac{V_{DC} - V_{BE}}{I}$$

For example, if $V_{DC} = 8$ V and $I = 1$ mA, we will require a resistance value of R equal to 7.3 kΩ. The small-signal voltage gain of this circuit will be

$$\frac{v_o}{v_i} = \frac{R_L \| R_I}{R_L \| R_I + R + r_e} \tag{1.53}$$

where R_I is the equivalent resistance of the current source for small signals and is usually large. Its effect and the effect of r_e, which is very small, can be ignored and the above gain equation approximated as

$$\frac{v_o}{v_i} \simeq \frac{R_L}{R_L + R} \tag{1.54}$$

1.6 *Output Stage*

An additional requirement of an op amp is that the output ideally act as a voltage source. Practically speaking, the output must be able to supply even relatively large amounts of current and the output resistance must be low. The simplest method of achieving this result is to connect an emitter follower circuit, however, this technique is not efficient in terms of power dissipation. Since power dissipation is an important consideration in integrated circuits, this method is not used in such cases. In IC op amps, the output stage is a *push-pull amplifier*. An elementary version of such a circuit is shown in Fig. 1.30. Assume that v_i in Fig. 1.30 is level-shifted and swings about 0 V. Note that Q_1 is an *npn* transistor and that Q_2 is a *pnp* transistor. Also note that the base-emitter junctions of both Q_1 and Q_2 are

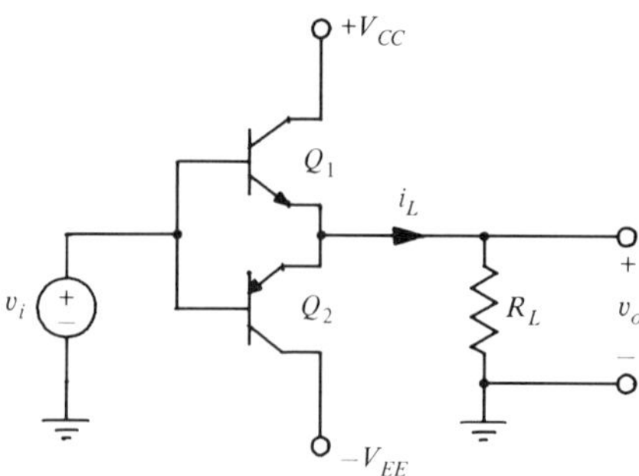

Figure 1.30 Class-B push-pull amplifier.

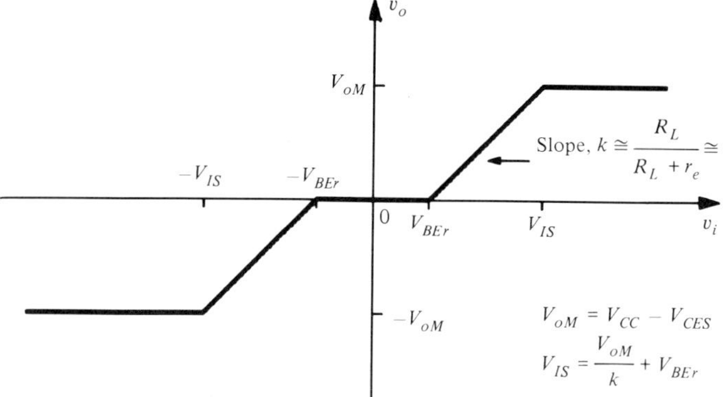

Figure 1.31 Idealized transfer characteristic of the circuit in Fig. 1.30.

not biased. Therefore, when $v_i = 0$, there is no current flowing through either Q_1 or Q_2 and $v_o = 0$; hence it is more efficient than an emitter follower circuit. When v_i increases positively, Q_1 will start conducting if v_i exceeds the cut-in voltage $V_{BE\gamma}$ of Q_1, and therefore i_L will be positive and in turn v_o will also be positive. When v_i increases negatively, Q_2 will conduct if $|v_i|$ exceeds the $|V_{BE\gamma}|$ of Q_2, and therefore i_L will be negative and in turn v_o will also be negative. An idealized transfer characteristic is shown in Fig. 1.31, assuming that both Q_1 and Q_2 have identical characteristics. Note that in Fig. 1.31 there is a dead zone near the origin that causes crossover distortion in the output. This type of crossover distortion can be eliminated by biasing Q_1 and Q_2 properly.

A simple modification of the class-B circuit in Fig. 1.30 is shown in Fig. 1.32. This circuit is a class AB push-pull amplifier. The two diodes D_1 and D_2 provide the biasing for the transistors Q_1 and Q_2, respectively. The

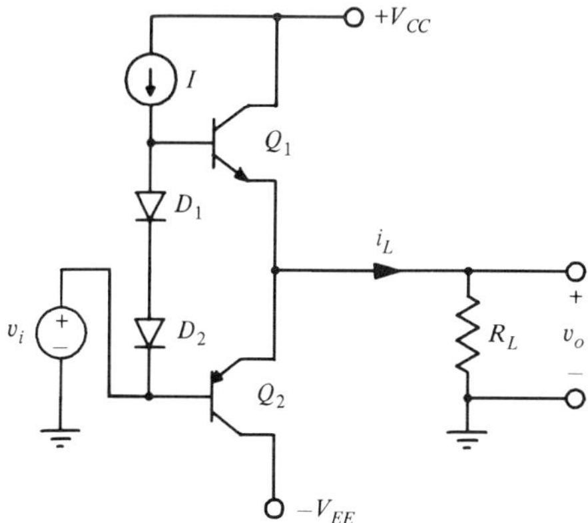

Figure 1.32 Class-AB push-pull amplifier.

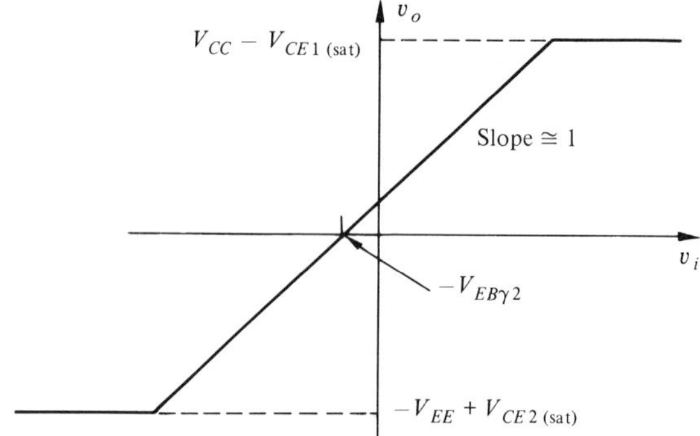

Figure 1.33 Transfer characteristic of a class-AB push-pull amplifier.

working of this circuit is similar to that of the circuit in Fig. 1.30. If $v_i = 0$, the current i_L will still be zero because, with equal biasing, Q_1 and Q_2 carry equal currents and thus $v_o = 0$. However, if v_i increases positively, the current in Q_1 will increase, the current in Q_2 will decrease, and therefore there will be a positive current flow in R_L and v_o will be positive. If v_i increases negatively, the reverse will occur and v_o will be negative. The transfer characteristic will be similar to the one shown in Fig. 1.31 except for the dead zone, and the transfer characteristic of this circuit will be as shown in Fig. 1.33. The output resistance of this amplifier is approximately equal to r_e ohms, which is really a very small value. We should also mention that an accidental short circuit of the output may blow the two transistors Q_1 and Q_2 and that they need to be protected. We shall discuss this further this in a later section.

So far we have dealt with the basic electronic circuits and the modifications required to improve their performance. Before we discuss the complete circuit of an op amp, we must consider another important problem stemming from the need to bias the different stages simultaneously. Basically all the circuits in an op amp are biased using current sources. There has been a need for such current sources in every circuit we have discussed, and therefore we will next consider biasing techniques in IC op amps.

1.7 *Biasing Techniques in Integrated Circuits*

The current source shown in Fig. 1.22 suffers from a major disadvantage in that any change in temperature will cause changes in I_{CO}, β, and V_{BE} and that these changes will in turn affect the value of I_C though, such changes are much reduced because of the negative feedback provided by R_E. A better current source, known as a *Widlar current source*, is shown in Fig. 1.34. In this circuit, Q_1 is a diode-connected transistor. The reference

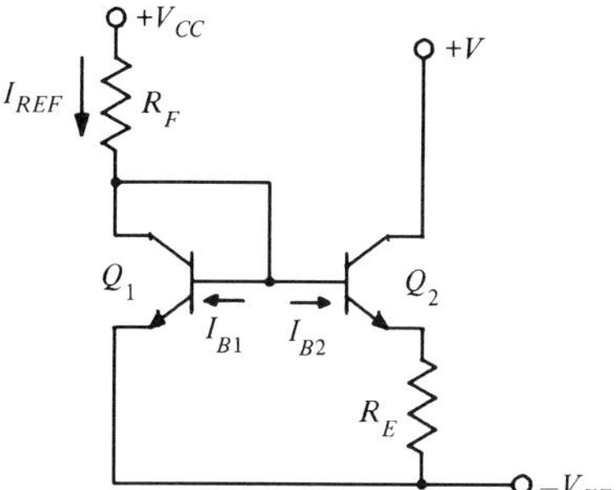

Figure 1.34 Widlar current source.

current I_{ref} can be found as

$$I_{\text{ref}} \simeq \frac{V_{CC} + V_{EE} - V_{BE1}}{R_F} \tag{1.55}$$

Since $V_{CC} + V_{EE} \gg V_{BE1}$, a stable reference current can be obtained. Assuming that Q_1 and Q_2 are identical, we have the collector currents I_{C1} and I_{C2}:

$$I_{C1} = I_{CO}\left(1 + \frac{V_{CE1}}{V_A}\right)\varepsilon^{V_{BE1}/V_T}$$

and

$$I_{C2} = I_{CO}\left(1 + \frac{V_{CE2}}{V_A}\right)\varepsilon^{V_{BE2}/V_T}$$

Therefore

$$\frac{I_{C1}}{I_{C2}} = \frac{1 + V_{CE1}/V_A}{1 + V_{CE2}/V_A}\varepsilon^{(V_{BE1} - V_{BE2})/V_T}$$

Since Q_1 and Q_2 are in close proximity, the changes in V_{BE1} and V_{BE2} will be almost equal to each other, and therefore $V_{BE1} - V_{BE2}$ is likely to remain constant. This means that the ratio I_{C1}/I_{C2} will remain constant independent of the changes in V_{BE} for whatever reason. Now, from the circuit, we have

$$V_{BE1} - V_{BE2} = I_{E2}R_E \simeq I_{C2}R_E$$

Further, if the base currents are ignored, $I_{C1} \simeq I_{\text{ref}}$ and thus we obtain

$$I_{C2} = \frac{V_T}{R_E}\ln\left(\frac{I_{\text{ref}}}{I_{C2}}\,\frac{1 + V_{CE2}/V_A}{1 + V_{CE1}/V_A}\right) \tag{1.56}$$

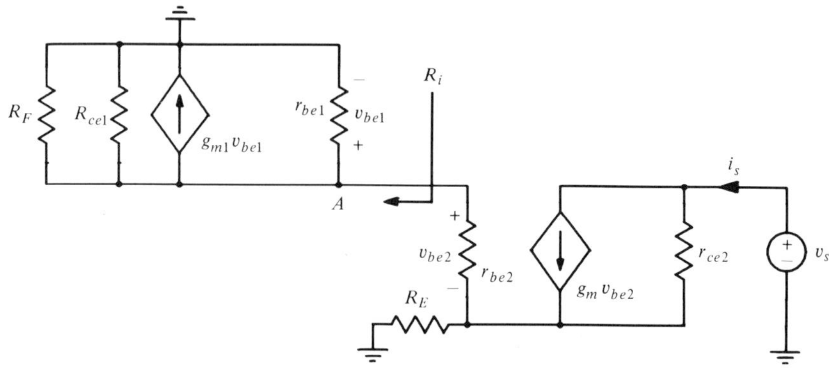

Figure 1.35 Small-signal equivalent circuit for a Widlar current source.

The above design equation is useful in determining the value of I_{C2}. Let us now compute the effective output resistance of this current source. Consider the small-signal equivalent circuit for the current source with a signal source v_s connected to it as shown in Fig. 1.35. The equivalent resistance R_i of the left part of the circuit, looking from node A to ground, can be shown to be approximately equal to r_{e1}. Once we replace this part of the circuit with a resistance equal to r_{e1}, it becomes similar to the circuit in Fig. 1.23, where $R_F = r_{e1}$, and therefore we can use (1.48) to find the output resistance of this circuit. Thus, for the circuit in Fig. 1.35, we obtain

$$\begin{aligned} R_o = \frac{v_s}{i_s} &= \frac{r_{ce2}[(\beta+1)R_E + r_{be1}] + R_E r_{e1}}{R_E + r_{e1}} \\ &\simeq (\beta+1) r_{ce2} \end{aligned} \tag{1.57}$$

This output resistance will be at least on the order of few megohms and therefore will be much higher than the output resistance obtained for the circuit in Fig. 1.22.

Example 1.12: In the Widlar current source in Fig. 1.34, assume that the transistors Q_1 and Q_2 are identical, with $V_A = 120$ V and $\beta = 100$. Given that $V_{CC} = 14.3$ V, $V = 0$ V, $V_{EE} = 15$ V, $R_F = 39$ kΩ, and $R_E = 5$ kΩ, find I_{C2}. Assume that the base bias currents can be ignored. Also, find the output resistance of the current source.

Assuming that $V_{BE1} = 0.7$ V, we first obtain the reference current

$$I_{\text{ref}} = \frac{14.3 + 15 - 0.7}{39} = 0.7333 \text{ mA}$$

Therefore $I_{C1} \simeq 0.7333$ mA. Now we can use (1.56) to find I_{C2}. To do this, we need to know the values of V_{CE1} and V_{CE2}. The value of $V_{CE1} = V_{BE1} \simeq 0.7$ V. Note that $V_{CE2} = 15 - I_{C2}R_E$. Substituting these values of V_{CE1} and V_{CE2} into (1.56), we obtain an equation for I_{C2}:

$$I_{C2} - 5 \times 10^{-3} \ln\left[0.7291\left(\frac{1.125}{I_{C2}} - \frac{1}{24}\right)\right] = 0$$

The above equation for I_{C2} is a nonlinear equation, and it can be solved using Newton's iterative procedure. Thus, solving this equation, we obtain

$$I_{C2} = 18.859\ \mu\text{A}$$

Since the value of I_{C2} is known, r_{ce2} can be calculated:

$$r_{ce2} = \frac{V_A}{I_{C2}}$$
$$= \frac{120}{18.859} = 6.363\ \text{M}\Omega$$

The output resistance is

$$R_o \simeq (\beta + 1) r_{ce2}$$
$$= 642.7\ \text{M}\Omega$$

■

Note that the value of R_o is really very high. For all practical purposes, we can assume that it is an ideal current source. Furthermore, note the strength of the current source I_{C2}, which is on the order of a few tens of microamperes. This is the order of the biasing currents of various transistors in an op amp and mainly serves to keep the power consumption of the device at a low value.

We mentioned earlier that several current sources are needed to bias a complete op amp circuit. In IC technology it is not economical to use large resistances, however, it is possible to produce many transistors cheaply with matched characteristics that track with changes in environmental conditions. Therefore we next consider circuits that reproduce a given reference current or a fraction of a reference current by using as many transistors as required but a minimum number of resistors (if possible, no resistors). Such circuits are called *current mirrors*. A simple current mirror circuit is shown in Fig. 1.36.

Assume that Q_1 and Q_2 are identical transistors and that both dc biasing voltages have been appropriately chosen such that both Q_1 and Q_2 are in the active regions. Then, because of the symmetric way in which the base-emitter junctions of Q_1 and Q_2 have been connected, we conclude that

$$I_{B1} = I_{B2} \qquad \text{and} \qquad I_{E1} = I_{E2}$$

Therefore,

$$I_{C1} = I_{C2} = \beta I_{B1} = \beta I_{B2}$$

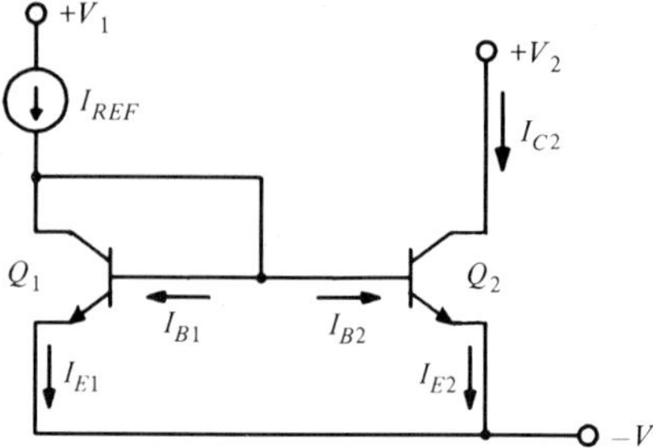

Figure 1.36 Current-mirror circuit.

Using the above equations, we can show that

$$\frac{I_{C2}}{I_{\text{ref}}} = \frac{1}{1 + 2/\beta} \tag{1.58}$$

With large values of β, the ratio $I_{C2}/I_{\text{ref}} \simeq 1$. Also, for large values of β, this current ratio has a very low sensitivity to changes in the β value. For example, when the value of β increases from 50 to 100 (a 100% increase), the ratio changes by only 1.96%. Thus it provides a highly stabilized ratio, and therefore I_{C2} will also be a stabilized current source. The term "current mirror" is based on the fact that I_{ref} is mirrored as I_{C2}. By adding an emitter resistor to Q_2, it is possible to reflect a fraction of I_{ref}. Using a single reference current, it is possible to have as many current sources as we need. As an application of this technique, consider the circuit shown in Fig. 1.37. Assume that the transistors Q_1, Q_2, Q_3, and Q_4 are identical.

In the circuit in Fig. 1.37, if the base bias currents are ignored, then $I_{C1} = I_{C2} = I_{C3} \simeq I_{\text{ref}}$. It is also possible to control the values of I_{C1}, I_{C2}, and I_{C3} to be any fraction of I_{ref}. If we need a current double that of I_{ref}, then by connecting the collectors of Q_2 and Q_3, we can produce a current of $2I_{\text{ref}}$. In fact, we do not need two separate transistors for this purpose. In IC technology, if the area of, for example, Q_2 is twice that of Q_4, then I_{CO2} will be twice I_{CO4}. This implies that $I_{C2} = 2I_{C4}$. Thus the magnitudes of

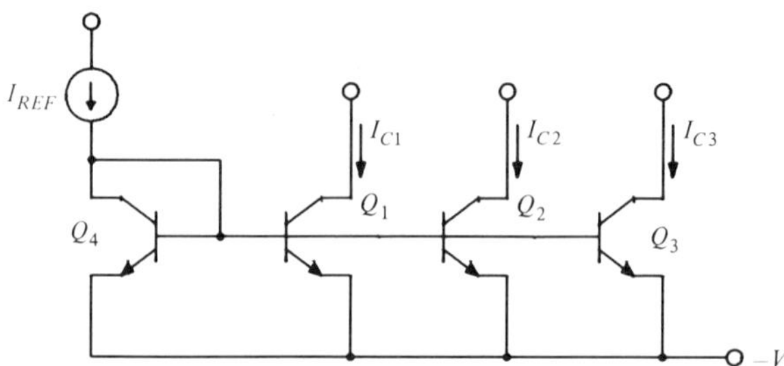

Figure 1.37 Current-mirroring technique for producing several current sources.

the current sources can be controlled by controlling the areas of the transistors. However, if emitter resistors are included, the output resistances of the current sources will be far higher than the output resistances of the same sources without emitter resistors. It should also be mentioned that, for current mirrors with multiple outputs, such as the one shown in Fig. 1.37, errors due to base currents may accumulate and the current transfer ratios may not be nearly equal to unity.

1.8 *Analysis of a Complete Operational Amplifier*

The complete circuit of a 741 type op amp is shown in Fig. 1.38 [3], and it can be partitioned into three parts, namely, the input stage, the intermediate amplifier, and the output stage. In accordance with the requirements of IC technology, the circuit has many transistors, a few resistors, and only one capacitor. The capacitor C_c is used to provide frequency compensation, the details of which will be discussed later. Therefore we can consider the branch of C_c an open circuit for the present.

Biasing

A reference current is formed in the resistor R_1. Then, using a current-mirroring technique, the different parts of the circuit are biased. Here, Q_2 and Q_3 form a current mirror, and Q_4 and Q_5 also form a current mirror. These two current mirrors bias the input stage, and Q_1, Q_{13}, and Q_{14} form another set of current mirrors that bias the intermediate amplifier and the output stage. The transistors Q_{18} and Q_{19} play a role in biasing the output stage.

Input stage: Transistors Q_6 through Q_{12} form the input stage, Q_6 and Q_7 act as emitter-followers to provide a high input impedance, Q_8 and Q_9 form part of a differential common-base amplifier, and Q_{11} and Q_{12} act as active loads. The output of this stage is taken at the collector terminals of Q_9 and Q_{12}.

Intermediate amplifier: The intermediate amplifier is a cascade connection of two amplifiers. The output resistance of the input stage is relatively large, and therefore v_{o1} is first fed to an emitter follower formed by Q_{15}. The output of the emitter is in turn fed to a common-emitter amplifier of Q_{16}. The output of this stage is v_{o2}.

Output stage: The *pnp* transistor Q_{17} acts as an emitter follower so that the gain will be stabilized. Transistors Q_{20} and Q_{21} form a push-pull amplifier biased by Q_{18} and Q_{19}. Transistors Q_{22} through Q_{25} are normally under cutoff conditions and provide short-circuit protection to the amplifier. They conduct and come into action only

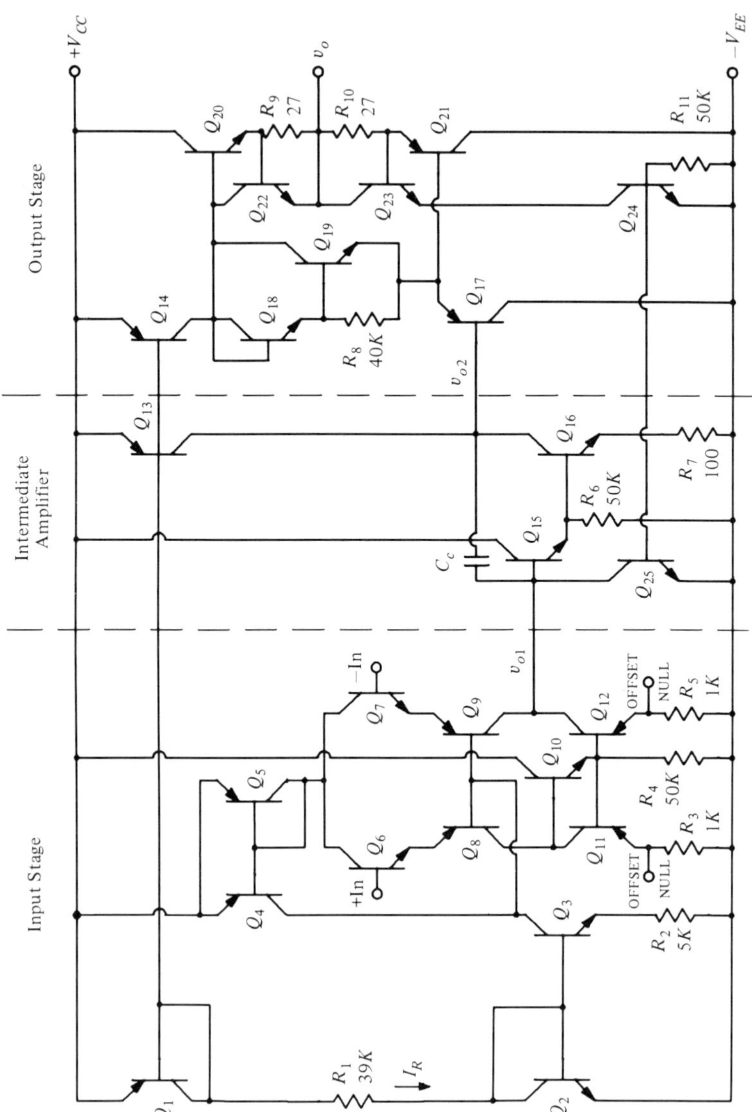

Figure 1.38 A 741 type op amp circuit with internal compensation.

when a large current is drawn at the output v_o. Therefore, when we analyze this circuit, this part of the circuit will not be considered for now.

Before we carry out a small-signal analysis of the circuit, we must first determine the bias currents. Only then can we find their small-signal equivalent circuits. The analysis of such a complex circuit involves certain approximations because there are many unknown quantities and fewer equations to determine. However, by making reasonable assumptions and approximations, the analysis can be made simple. A more accurate analysis requires a graphical approach using device characteristics. Since such characteristics are not available to us, we can only make an approximate analysis. However, this analysis will provide reasonably good answers. For our analysis we require certain basic parameters, which, for all but a few transistors, are:

$$npn\colon I_{CO} = 0.01 \text{ pA} \qquad \beta_n = 200$$

and

$$pnp\colon I_{CO} = 0.01 \text{ pA} \qquad \beta_p = 50$$

The exceptions are Q_{13}, Q_{14}, Q_{20}, and Q_{21}. The areas of transistors Q_{13} and Q_{14} are 0.75 and 0.25 times that of standard transistors, and therefore

$$I_{CO13} = 0.0075 \text{ pA} \qquad \text{and} \qquad I_{CO14} = 0.0025 \text{ pA}$$

Since Q_{20} and Q_{21} are output transistors, they are expected to supply larger currents, their areas are 3 times that of standard transistors, and therefore

$$I_{CO20} = I_{CO21} = 0.03 \text{ pA}$$

With these assumptions for the devices and $V_{CC} = V_{EE} = 15\text{V}$, we proceed with the dc analysis. For the most part, we will use the exponential relationship (1.6) to determine the dc bias quantities. The same relationship can also be written as

$$V_{BE} = V_T \ln \frac{I_C}{I_{CO}} \tag{1.59}$$

We start by finding the reference current I_R. From Fig. 1.38, we note that

$$I_R = \frac{V_{CC} + V_{EE} - V_{EB1} - V_{BE2}}{R_1}$$

Also,

$$V_{EB1} = V_T \ln \frac{I_{C1}}{I_{CO}} \qquad \text{and} \qquad V_{BE2} = V_T \ln \frac{I_{C2}}{I_{CO}}$$

Therefore, in order to find the values of V_{EB1} and V_{BE2}, we first find I_{C1} and I_{C2}. It can be shown that

$$I_{C1} = \frac{I_R}{1 + 2/\beta_p}$$

Ignoring the base bias current of Q_3 (because of large emitter resistance), we obtain

$$I_{C2} = \frac{I_R}{1 + 1/\beta_n}$$

Substituting all of the above into the equation for I_R, we obtain an equation for I_R:

$$I_R - \frac{30}{39k} + \frac{V_T}{39k}\left[\ln\frac{I_R}{I_{CO}(1 + 1/\beta_n)} + \ln\frac{I_R}{I_{CO}(1 + 2/\beta_p)}\right] = 0$$

After substituting the values of $V_T = 25$ mV, $I_{CO} = 0.01$ pA, and of β_p and β_n, we solve the above equation iteratively for I_R and obtain

$$I_R = 0.7372 \text{ mA}$$

The value of V_{BE2} can be found from

$$V_{BE2} = V_T \ln\frac{I_{C2}}{I_{CO}}$$

$$= 25 \ln\frac{I_R}{1.005 I_{CO}}$$

$$= 625.5 \text{ mV}$$

Next we see that Q_2 and Q_3 form a current mirror. For this current mirror, we calculate the value of I_{C3} from the equation

$$I_{C3} \simeq I_{E3} = \frac{V_{BE2} - V_{BE3}}{R_2}$$

$$= \frac{V_{BE2}}{R_2} - \frac{V_T}{R_2}\ln\frac{I_{C3}}{I_{CO}}$$

After substituting values for V_{BE2}, V_T, R_2, and I_{CO} and solving the above equation, we find that

$$I_{C3} = 18.422\ \mu\text{A}$$

Note that Q_4 and Q_5 form a current mirror. Also, note from the symmetry of the circuit that

$$I_{C6} = I_{C7} \simeq I_{C8} = I_{C9}$$

and therefore

$$I_{C4} = 2I_{C6}$$

Now,

$$I_{C3} = I_{C4} + \frac{I_{C8} + I_{C9}}{\beta_p} = 2I_{C6}\left(1 + \frac{1}{\beta_p}\right)$$

Since we know the value of I_{C3}, we can find the value of I_{C6}, hence

$$I_{C6} = I_{C7} \simeq I_{C8} = I_{C9} = 9.03\ \mu\text{A}$$

Because of the large emitter resistances in Q_{10} and Q_{15}, we make the assumption that $I_{B10} = I_{B15} \simeq 0$, which we shall prove momentarily. Under such an assumption, we have

$$I_{C11} = I_{C12} = 9.03\ \mu\text{A}$$

Since Q_1, Q_{13}, and Q_{14} form current mirrors, and using the fact that $I_{CO13} = 0.0075\ p\text{A}$ and $I_{CO14} = 0.0025$ pA, we can show that

$$I_{C13} = \frac{0.75I_R}{1 + 2/\beta_p} = 531.6\ \mu\text{A}$$

and

$$I_{C14} = \frac{0.25I_R}{1 + 2/\beta_p} = 177.2\ \mu\text{A}$$

Since Q_{22} and Q_{23} are cutoff under normal conditions,

$$I_{C20} = I_{C21}$$

Then, using KCL, we find that

$$I_{E17} = I_{C14} = 177.2\ \mu\text{A}$$

Therefore

$$I_{B17} = \frac{-177.2}{50} = -3.54\ \mu\text{A}$$

Thus

$$I_{E16} \simeq I_{C16} = I_{C13} - I_{B17} = 535.16\ \mu\text{A}$$

Using the above value of I_{C16}, we find V_{BE16} to be 617.6 mV. Therefore

$$V_{B16} = V_{BE16} + 100I_{E16} = 671.2\ \text{mV}$$

Now,

$$I_{C15} \simeq I_{E15} = \frac{V_{B16}}{50} + I_{B16} = \frac{671.2}{50} \times 10^{-6} + 2.68 \times 10^{-6}$$
$$= 16.1\ \mu\text{A}$$

Therefore

$$I_{B15} = \frac{16.1}{200} = 0.081\ \mu\text{A}$$

This value of I_{B15} is really small and can be ignored, as mentioned before. Ignoring the base bias currents of Q_{20} and Q_{18}, we can write

$$I_{C14} = I_{C18} + I_{C19}$$

In terms of I_{C19}, we have V_{BE19} as

$$V_{BE19} = V_T \ln \frac{I_{C19}}{I_{CO}}$$

Thus we have

$$I_{C18} \simeq I_{E18} = \frac{I_{C19}}{\beta_n} + \frac{V_{BE19}}{R_8}$$

Substituting for I_{C18} in the equation for I_{C14}, we obtain

$$I_{C19}\left(1 + \frac{1}{\beta_n}\right) + \frac{V_T}{R_8} \ln \frac{I_{C19}}{I_{CO}} = I_{C14}$$

Since we know the value of I_{C14}, we can solve for I_{C19}. Solving for I_{C19} iteratively, we obtain

$$I_{C19} = 161.7\,\mu\text{A} \qquad \text{and} \qquad I_{C18} = 15.51\ \mu\text{A}$$

Now,

$$\begin{aligned} V_{BE20} + V_{EB21} &= V_{BE18} + V_{BE19} \\ &= V_T \ln\left(\frac{I_{C19} I_{C18}}{I_{CO}^2}\right) \\ &= 1.1167\text{ V} \end{aligned}$$

Since $I_{C20} = I_{C21}$, we have the following equation for I_{C20} and I_{C21}:

$$V_{BE20} + V_{EB21} = V_T \ln \frac{I_{C20}^2}{I_{CO}^2} = 1.1167\text{ V}$$

The above equation gives the values of I_{C20} and I_{C21}:

$$I_{C20} = I_{C21} = 150.22\ \mu\text{A}$$

Now we are in a position to find the equivalent circuit parameters of all the active transistors for small-signal analysis.

Small-Signal Analysis

Small-signal analysis can be carried out by replacing each active transistor with its corresponding small-signal equivalent circuit. It would be futile to analyze the entire circuit in one attempt, however, we can analyze the different stages of the op amp individually and replace each part with its two-port equivalent. In an op amp we are essentially interested in finding the overall gain, the input resistance, and the output resistance. Therefore we can replace each stage of the op amp with its input resistance, gain, and output resistance. Toward this end, we attempt to find the equivalent circuit for each stage in terms of its gain, input resistance, and output resistance. The equivalent circuits for the input stage, intermediate amplifier, and output stage are shown in Figs. 1.39, 1.40, and 1.41, respectively.

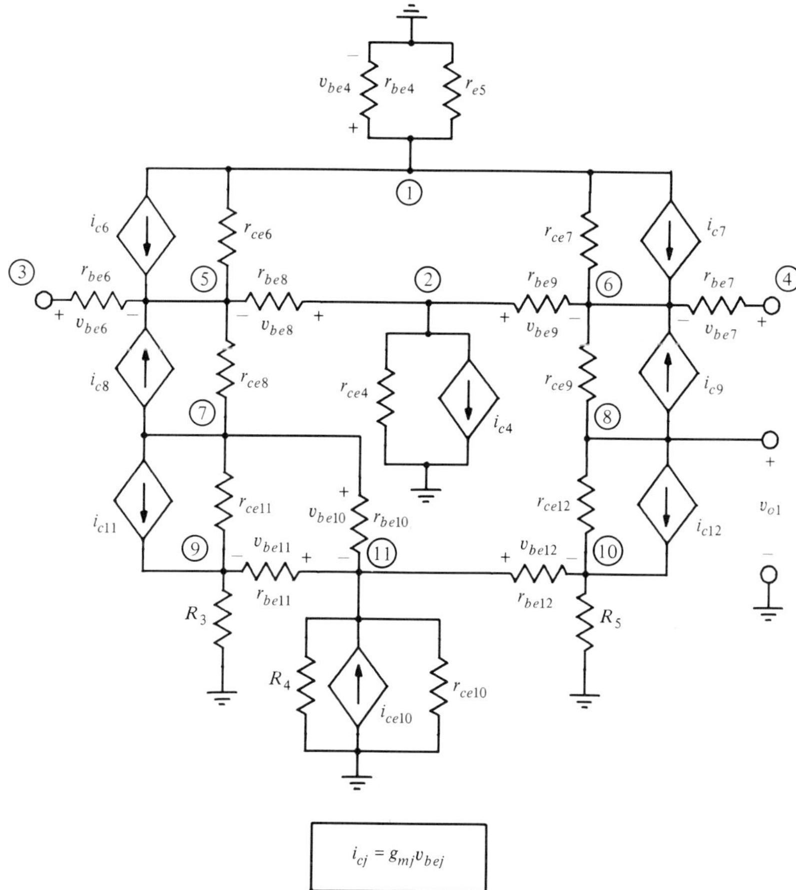

Figure 1.39 Small-signal equivalent circuit for the input stage of the op amp circuit in Fig. 1.38.

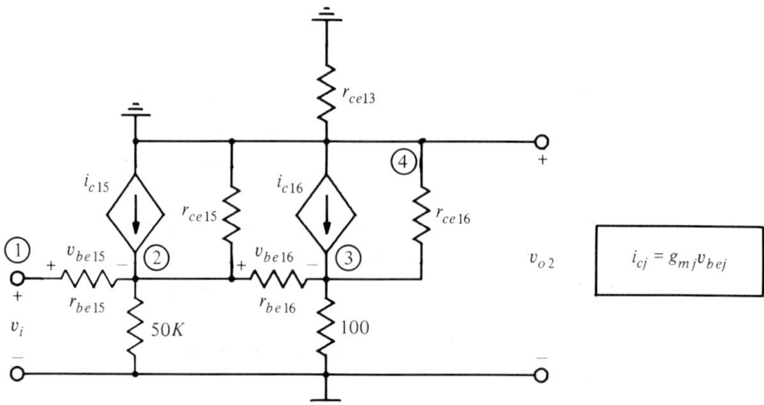

Figure 1.40 Small-signal equivalent circuit for an intermediate amplifier.

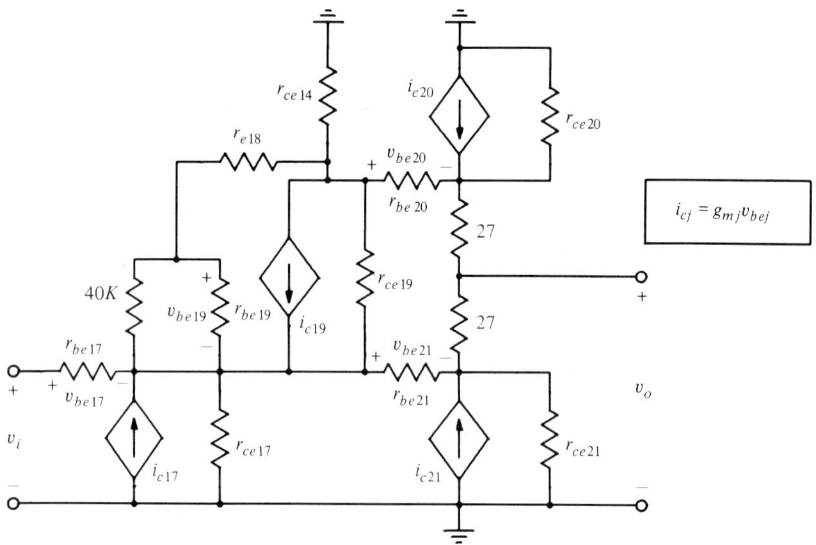

Figure 1.41 Small-signal equivalent circuit for the output stage.

We note from the equivalent circuit in Fig. 1.39 that there are 11 nodes and, even though it is not impossible, it is very difficult to analyze this network with paper and pencil. Therefore we sought the aid of a computer to find the transfer ratio, the input resistance, and the output resistance using a circuit analysis program called SPICE. All three stages were analyzed by the computer, and the input listings for each case are provided in Program 1A at the end of the chapter. These listings also include the element values for the equivalent circuits and their interconnections. For

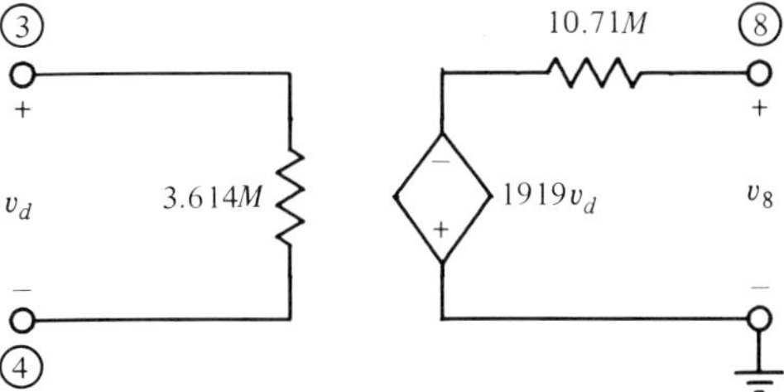

Figure 1.42 Simplified equivalent circuit for the input stage.

example, referring to the listing for the input stage small-signal analysis, we see that resistor RBE7 is connected between nodes 4 and 6 and that its value is 553.7 kΩ. The current source $i_{c6} = g_{m6}v_{be6}$ is specified in terms of the conductance value GM6 = 0.36121 mS, the current source being connected between nodes 1 and 5 (current direction is from 1 to 5) controlled by the potential v_{be6} appearing between the terminals 3 and 5 (v_{be6} is positive at 3 and negative at 5). After feeding in all the data in this way, the output of the analysis is given as small-signal characteristics. The differential-mode gain of the input stage is

$$\frac{v_8}{v_{34}} = -1919$$

The input and output resistances are 3.614 and 10.71 MΩ, respectively. Thus the equivalent circuit for the input stage will be as shown in Fig. 1.42. In a similar way, extracting the information from the small-signal characteristics of the intermediate and output stages, the simplified equivalent circuit for the entire op amp will be as shown in Fig. 1.43.

The overall differential-mode gain v_o/v_d can be obtained as

$$\frac{v_o}{v_i} = \frac{v_o}{v_{o2}}\frac{v_{o2}}{v_{o1}}\frac{v_{o1}}{v_d}$$

From the circuit in Fig. 1.43, we have

$$\frac{v_o}{v_{o2}} = 0.9993$$

$$\frac{v_{o2}}{v_{o1}} = -1044\frac{16{,}880}{16{,}880 + 166.5}$$

$$= -1033.8$$

We also obtain v_{o1}/v_d as

$$\frac{v_{o1}}{v_d} = -1919\frac{3.134}{10.71 + 3.134}$$

$$= -434.4$$

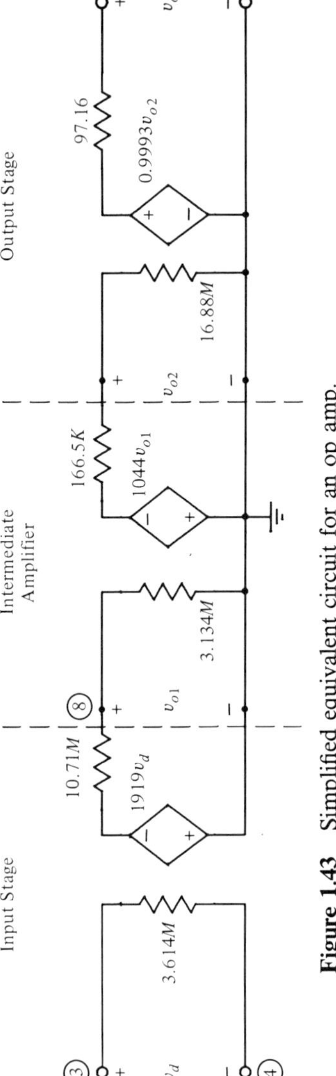

Figure 1.43 Simplified equivalent circuit for an op amp.

Therefore the dc gain of the op amp is

$$\frac{v_o}{v_d} = 448{,}793 = 113.04\ \text{dB}$$

Obviously the input impedance is 3.614 MΩ and the output impedance is 97.16 Ω. The common-mode gain of the differential amplifier was also obtained using the program SPICE and was found to be negligibly small, indicating a very high CMRR. Such a high CMRR cannot be obtained in reality because the transistors will not be exactly identical and resistors R_3 and R_5 will not be equal either. A realistic value of CMRR will be on the order of only 100 dB. Furthermore, the differential-mode gain and input impedance values we have calculated are only typical values and in actual amplifiers, may be lower than those calculated above.

Finally, let us consider the effect of connecting the capacitor C_c to the circuit. This capacitor is used internally for frequency compensation (which will be discussed further in the next section). For the present, let us include this capacitance, whose value is on the order of 30 pF, in our analysis and find the transfer function of the overall op amp circuit. Note that this capacitance is connected between the input and output of the intermediate stage. Therefore, connecting this capacitance C_c across these terminals, we can obtain the transfer function for the entire network:

$$\frac{V_o}{V_d} = \frac{A_3 \dfrac{g_{m1} g_{m2}}{C_c^2}\left(1 - s\dfrac{C_c}{g_{m1}}\right)}{s\left(\dfrac{1}{R_1 C_c} + \dfrac{1}{R_2 C_c} + \dfrac{g_{m2}}{C_c}\right) + \dfrac{1}{R_1 R_2 C_c^2}}$$

where

$$R_1 = 10.71\ \text{k}\Omega \| 3.134\ \text{M}\Omega$$

$$R_2 = 166.5\ \text{k}\Omega \| 16.88\ \text{M}\Omega$$

$$g_{m1} = \frac{1919}{10.71}\ \mu\text{S} \qquad g_{m2} = \frac{1044}{166.5}\ \text{mS}$$

and

$$A_3 = 0.9993$$

Substituting the various values, we have

$$\frac{V_o}{V_d} = \frac{(5.962 \times 10^6)(1 - 167.4 \times 10^{-9} s)}{s + 13.29}$$

The effect of the zero, for low-frequency applications, can be ignored and therefore we have

$$\frac{V_o}{V_d} \simeq \frac{5.962 \times 10^6}{s + 13.29}$$

1.9 *Feedback, Stability, and Compensation*

An op amp is almost never used alone, and it forms part, as a component, in a network along with a passive network included in it. The most general network configuration where an op amp is a component is shown in Fig. 1.44. In most instances, since the output impedance of the op amp is low, the output of the overall network is also the output of the op amp. When an op amp is used in a network, energy is fed to the passive network by both the input source and the op amp, since the op amp is an active element. When an active element is used in a network, the foremost concern of the network designer is its stability. In many cases, the stability of the overall network determines the usefulness of the network itself. It would be futile to attempt to design a network for a specific purpose if the network is unstable, and thus in this section we shall see how we can determine the stability of a network having an op amp as a component. Most of the concepts can be applied in other circuits and systems as well. This stability analysis will illustrate the need for the compensating capacitance C_c in the complete op amp circuit in Fig. 1.38.

Most network designers assume the op amp to be an ideal component, in which case the op amp gain is assumed to be infinite at all frequencies. In practice, the op amp gain is not only finite but also frequency-dependent. When we derived the gain of the op amp in the previous section, we used only the equivalent circuit in Fig. 1.8 for each transistor. Here we make an assumption that, for low-frequency applications, as far as gain calculation is concerned, it is sufficient to use this particular equivalent circuit. However, if we had used the more accurate equivalent circuit in Fig. 1.7, which includes the interelectrode capacitances, we would have found that the gain was also frequency-dependent. The frequency-dependent gain is such that the magnitude of the gain function decreases and the phase of the output with reference to the differential input lags more and more with increasing frequency of the sinusoidal inputs. The typical frequency dependence of an op amp can be described as

$$A_d = \frac{A_o}{(1 + s\tau_1)(1 + s\tau_2)(1 + s\tau_3)} \tag{1.60}$$

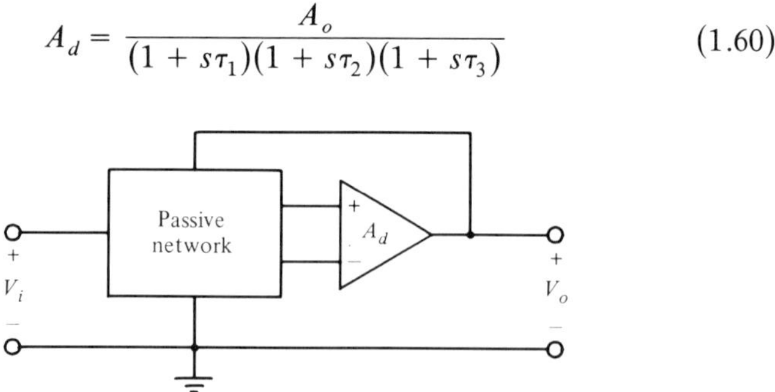

Figure 1.44 General network configuration using an op amp.

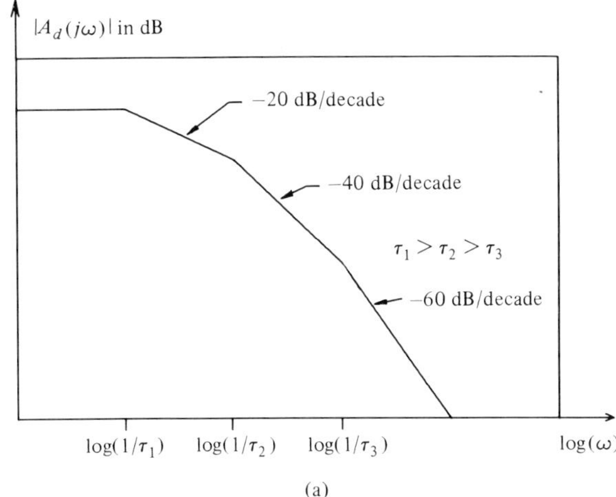

(a)

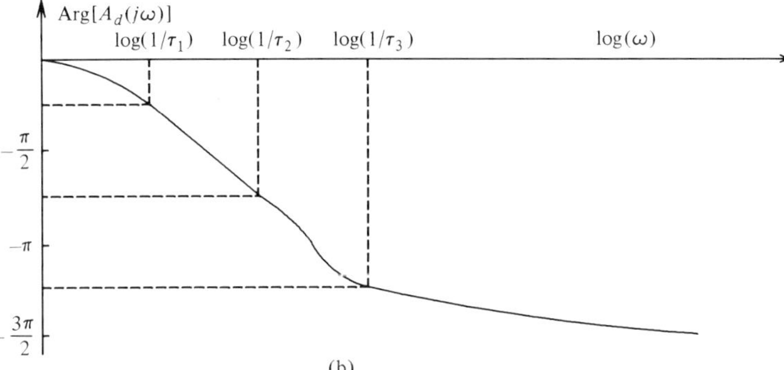

(b)

Figure 1.45 Bode plots of the transfer function in (1.60). (*a*) Magnitude plot. (*b*) Phase plot.

where A_o is the dc gain of the op amp that we calculated in the previous section. The magnitude and phase functions of (1.60) can be obtained for sinusoidal inputs with $s = j\omega$. Given the values of A_o, τ_1, τ_2, and τ_3, we can plot them as functions of ω known as *frequency responses*. However, in most of the applications in this chapter, we can use approximate plots of these responses called *Bode plots*. Bode plots of the magnitude and phase functions of (1.60) are shown in Fig. 1.45, where τ_1, τ_2, and τ_3 are assumed to be real. For more details on Bode plots, the reader can refer to any book on circuits and systems. For a given amplifier, the manufacturer provides the typical magnitude characteristic from which the phase characteristic can be obtained.

The op amp as such is a stable component. This can be seen from the poles of $A_d(s)$ as given by (1.60), because all the poles, which are at

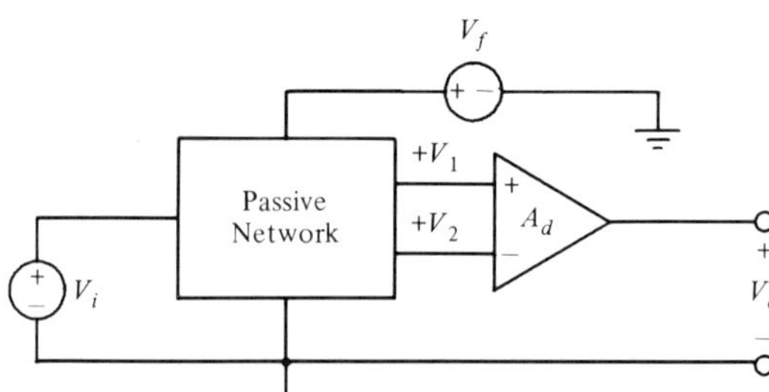

Figure 1.46 Same network as in Fig. 1.44, with the feedback path broken to illustrate feedforward and feedback paths.

$s = -1/\tau_1$, $s = -1/\tau_2$, and $s = -1/\tau_3$, are located on the negative real axis of the s plane. However, the overall stability of the network in Fig. 1.44 is determined by the poles of the transfer function V_o/V_i of the network. This means that the passive network also plays a role in the stability of the network, in addition to the frequency-dependent characteristic of the op amp. Before we determine the stability of the overall network, we must define certain other quantities. The output is not only dependent on V_i but also on V_o itself because V_o is fed back into the passive network. The output signal can be thought of as a composition of two outputs, namely, output due to the input signal alone through what is known as the *forward path*, and output due to the output itself through a *feedback path*. To understand this concept, let us break open the feedback path in the network in Fig. 1.44 and feed a signal V_f to the feedback path as shown in Fig. 1.46. Since the op amp is a linear element and the passive network is a linear network, the output voltage V_o can be found as a linear combination of the two input signals V_i and V_f:

$$V_o = A_f V_i + A_l V_f \tag{1.61}$$

where A_f and A_l are the *forward path gain* and the *loop gain*, respectively. The defining equations for these gains are

$$A_f = \left.\frac{V_o}{V_i}\right|_{V_f=0} \tag{1.62}$$

and

$$A_l = \left.\frac{V_o}{V_f}\right|_{V_i=0} \tag{1.63}$$

The loop gain is usually negative, and the *closed-loop gain* after closing the feedback path, V_o/V_i, can be found in terms of A_f and A_l. Comparing

Figs. 1.44 and 1.46, we find that $V_f = V_o$. Therefore in Fig. 1.44 we have

$$V_o = A_f V_i + A_l V_o$$

or

$$\frac{V_o}{V_i} = \frac{A_f}{1 - A_l} \tag{1.64}$$

The effect of the passive network can be separated from that of the amplifier in the above forward path and loop gains. To do this, we assume that the input impedances of the op amp are infinite and the output impedance of the op amp is zero. That is, the op amp does not load the passive network and the passive network does not load the op amp. This assumption is true in practice and will be discussed further in later chapters. If this assumption holds, we can write

$$A_f = \alpha A_d$$

and

$$A_l = -\beta A_d$$

where α and β are defined by the following equations for the network in Fig. 1.46:

$$\alpha = \left.\frac{V_1 - V_2}{V_i}\right|_{V_f = 0} \tag{1.65}$$

and

$$-\beta = \left.\frac{V_1 - V_2}{V_f}\right|_{V_i = 0} \tag{1.66a}$$

Note that both α and β depend entirely on the passive network under consideration. A negative sign is associated with β, the *feedback factor*, because $(V_1 - V_2)/V_f$ usually results in a negative quantity. The above equation can also be written as

$$\beta = \left.\frac{V_2 - V_1}{V_f}\right|_{V_i = 0} \tag{1.66b}$$

In terms of α and β, the closed-loop transfer function is

$$\frac{V_o}{V_i} = \frac{\alpha A_d}{1 + \beta A_d} \tag{1.67}$$

If the op amp is assumed to be ideal, then $A_d \to \infty$. In this case,

$$\frac{V_o}{V_i} = \frac{\alpha}{\beta} \tag{1.68}$$

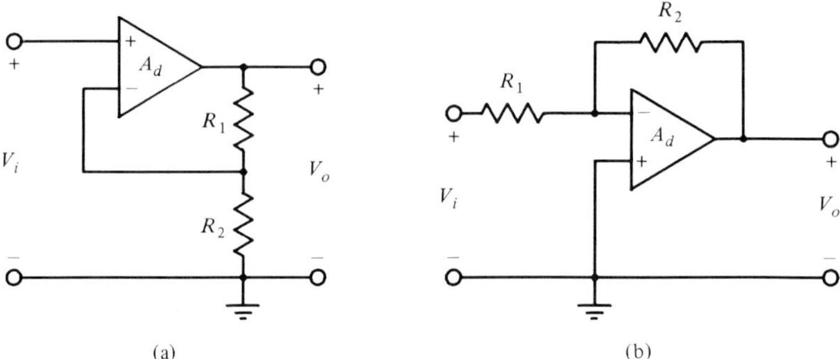

Figure 1.47 Finite gain amplifiers. (*a*) Noninverting amplifier. (*b*) Inverting amplifier.

The closed-loop transfer function depends entirely on the passive network in the ideal case in which the op amp gain tends to ∞. Equation (1.68) is approximately true even if A_d is a finite but very large value. Therefore, in practice, most designs can be carried out using (1.68).

Example 1.13: Shown in Fig. 1.47 are two circuits whose overall gains depend entirely on the two resistances R_1 and R_2 if the op amp is ideal. They are called *finite gain amplifiers*. Find the forward path gain, the loop gain, and the overall gain, all as functions of A_d. Take the limiting condition as $A_d \to \infty$ and find the overall gain.

To find the forward path gain and the loop gain, we have to break open the feedback. With the feedback open, the circuits will be as shown in Fig. 1.48.

Noninverting amplifier:

$$\alpha = \left.\frac{V_1 - V_2}{V_i}\right|_{V_f=0} = 1$$

Applying 1.66*b* to the circuit in Fig. 1.48*a*, we have

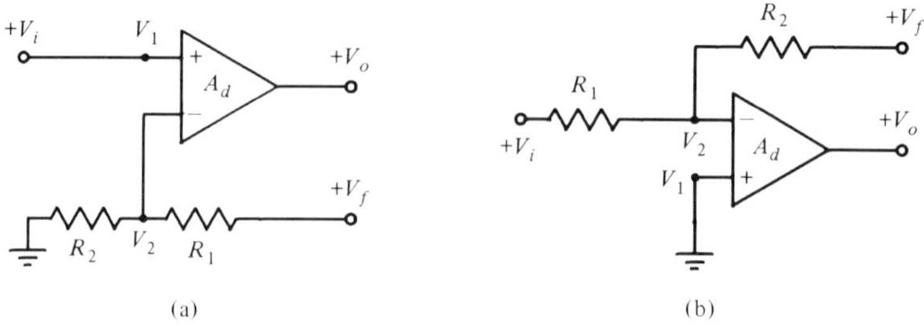

Figure 1.48 Open feedback. (*a*) Noninverting amplifier. (*b*) Inverting amplifier.

$$\beta = \left.\frac{V_2 - V_1}{V_f}\right|_{V_i=0}$$

$$= \frac{V_f R_2/(R_1 + R_2) - 0}{V_f}$$

$$= \frac{R_2}{R_1 + R_2}$$

Therefore

$$A_f = A_d \qquad \text{and} \qquad A_l = -\frac{R_2 A_d}{R_1 + R_2}$$

The overall gain is

$$\frac{V_o}{V_i} = \frac{A_d}{1 + [(R_2/(R_1 + R_2))]A_d}$$

$$= \left(1 + \frac{R_1}{R_2}\right)\frac{1}{1 + (1 + R_1/R_2)/A_d} \tag{1.69}$$

In the limiting case where $A_d \to \infty$, we obtain

$$\frac{V_o}{V_i} = 1 + \frac{R_1}{R_2} \tag{1.70}$$

The above expression is approximately true if A_d is very high. For example, if $R_1/R_2 = 9$ and $A_d = 40{,}000$, for direct current the actual gain will be 9.998, whereas the approximation gives a value of 10. Thus, for dc and low-frequency applications, (1.70) can be used to determine the overall gain of the noninverting amplifier. Sometimes we call this the *nominal value* of the gain, denoted by K. Therefore, for a noninverting amplifier circuit, the nominal value of the gain is

$$K = 1 + \frac{R_1}{R_2} \tag{1.71}$$

However, if we want to take the finite gain of the op amp into consideration, then we should use (1.69).

Inverting amplifier:

$$\alpha = \left.\frac{V_1 - V_2}{V_i}\right|_{V_f=0}$$

$$= \frac{-R_2}{R_1 + R_2}$$

and

$$-\beta = \left.\frac{V_1 - V_2}{V_f}\right|_{V_i=0}$$

$$= \frac{-R_1}{R_1 + R_2}$$

Therefore the forward path gain and the loop gains are

$$A_f = \alpha A_d = \frac{-R_2 A_d}{R_1 + R_2}$$

and

$$A_l = -\beta A_d = \frac{R_1 A_d}{R_1 + R_2}$$

The overall gain is

$$\frac{V_o}{V_i} = \frac{-R_2 A_d/(R_1 + R_2)}{1 + R_1 A_d/(R_1 + R_2)}$$

$$= \frac{-R_2}{R_1} \frac{1}{1 + (1 + R_2/R_1)/A_d} \tag{1.72}$$

In the limiting case where $A_d \to \infty$, the overall gain becomes

$$\frac{V_o}{V_i} = -\frac{R_2}{R_1} = -K \tag{1.73}$$

which is entirely dependent on the ratio of the two resistors. Note the negative sign, which means that there is a phase reversal in the output signal with reference to the input signal. In this case the nominal gain K is given by (1.73). For dc and low-frequency applications, we can use (1.73), however, to include the effect of the finite gain of the op amp, we must use (1.72) for an inverting amplifier. ■

The stability of the overall system can be tested by checking the zeros of the characteristic equation

$$1 + \beta A_d = 0 \tag{1.74}$$

For strictly stable systems, the zeros of the characteristic equation must lie in the left half of the s plane, excluding the $j\omega$ axis. Using a method called *Routh's test* on the polynomial corresponding to the zeros of $1 + \beta A_d$, one can determine the stability of the overall network. Only a "yes" or a "no" answer can be obtained if one uses Routh's test, however, network designers are interested in knowing how close the network is to instability if it is stable. There is an alternative method using the frequency responses of the loop gain that gives one an idea not only of the absolute stability but also of the *relative stability*. The relative stability indicates how close the network is to becoming unstable, and this is also important to engineers. Furthermore, the frequency response of the op amp is already known because the manufacturer provides this information, and so it is easy to find the loop gain. Instead of deriving the condition on the loop gain, we shall derive the condition in terms of βA_d. If there is some frequency ω such that

$$-A_l(j\omega) = \beta(j\omega)A_d(j\omega) = -1$$

then the network will oscillate at this frequency. That is, the network will oscillate at the frequency where the loop gain becomes unity. The most general test for determining network stability using the loop gain or βA_d involves the *Nyquist criterion*, a particular form of which will be used here. The poles of both β and A_d lie in the left half of the s plane, and therefore the poles of the function βA_d also lie in the left half of the s plane. In this case, the Nyquist test for stability using the frequency response of βA_d is simple, and the rule is as follows: When the phase of $\beta A_d(j\omega)$ passes through $\pm 180°$ at only one frequency ω_p, the system is stable if the magnitude is less than unity (0 dB) at this frequency ω_p.

This information can be obtained from Bode plots for $\beta A_d(j\omega)$. We see that

$$20\log|\beta A_d(j\omega)| = 20\log|\beta(j\omega)| + 20\log|A_d(j\omega)|$$

and

$$\arg \beta A_d(j\omega) = \arg \beta(j\omega) + \arg A_d(j\omega)$$

By adding the logarithmic magnitudes of $A_d(j\omega)$ (provided by the manufacturer) and the feedback factor $\beta(j\omega)$, we can obtain the logarithmic magnitude of $\beta A_d(j\omega)$. A similar statement can be made for their phase responses as well. Therefore it is easy to determine the stability based on the frequency responses of $\beta A_d(j\omega)$. Assume that, by carrying out the above test, we find the network to be stable. Using the same frequency responses we can also find the relative stability, which is determined by two quantities, the gain margin and the phase margin. These two quantities are determined at two critical frequencies: The frequency at which the phase of $\beta A_d(j\omega)$ passes through $-180°$ is known as the *phase crossover frequency* ω_p, and the frequency at which the logarithmic magnitude of $\beta A_d(j\omega)$ passes through the 0-dB line is called the *gain crossover frequency* ω_g. Therefore, by definition, we can find these frequencies by using the equations

$$\arg \beta A_d(j\omega_p) = -180° \tag{1.75}$$

and

$$20 \log \left| \beta A_d(j\omega_g) \right| = 0 \tag{1.76}$$

The *phase margin* ϕ_m is defined as

$$\phi_m = 180° + \arg \beta A_d(j\omega_g) \tag{1.77}$$

and is a measure of the additional negative phase at the gain crossover

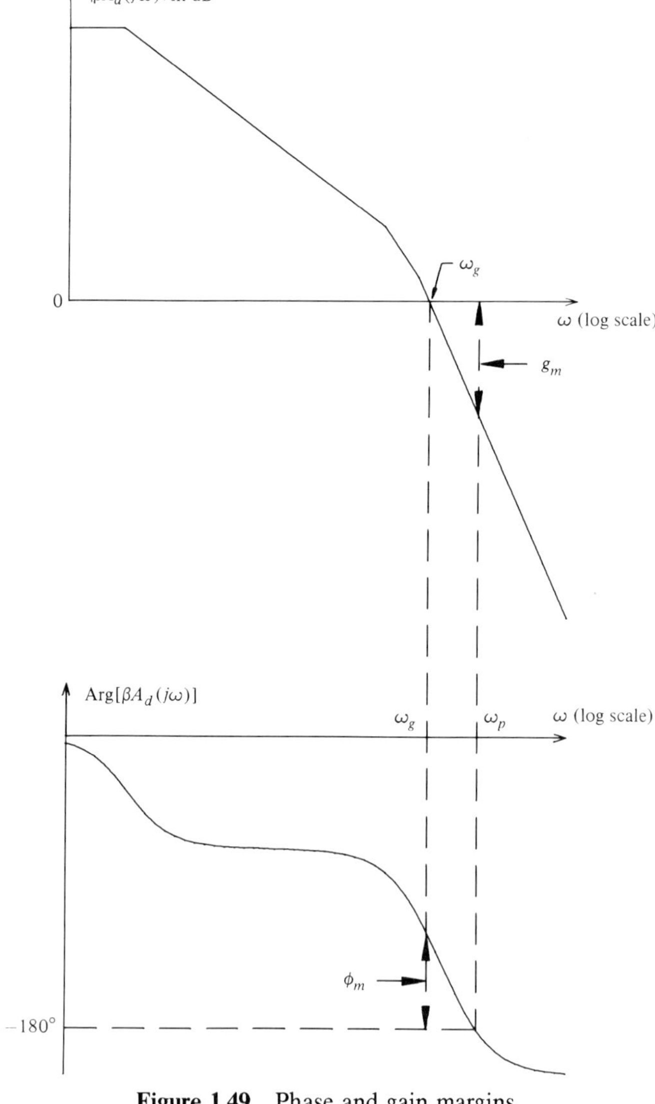

Figure 1.49 Phase and gain margins.

frequency that will cause instability. Similarly, the *gain margin* is the amount of gain increase required by $|\beta A_d(j\omega)|$ to make the gain 0 dB at ω_p (and therefore create instability). The gain margin g_m is

$$g_m = -20 \log \left| \beta A_d(j\omega_p) \right| \tag{1.78}$$

The logarithmic magnitude and the phase plots of $\beta A_d(j\omega)$ for a network are shown in Fig. 1.49. The phase plot passes through the $-180°$ phase at only one frequency, and the gain at this frequency ω_p is less than 0 dB. Therefore the network is absolutely stable. In addition, both the gain and phase margins are positive, and these values are also indicated in Fig. 1.49. A phase margin of 30 to 60° combined with a gain margin of 10 dB is acceptable for good relative stability.

Example 1.14: Investigate the stability problem associated with an inverting amplifier with a nominal closed-loop gain of 9 when it uses an op amp whose differential mode gain is

$$A_d = \frac{10^6}{(1 + 10^{-6}s)(1 + 10^{-8}s)(1 + 10^{-9}s)} \qquad \text{case a}$$

and

$$A_d = \frac{10^6}{(1 + 10^{-1}s)(1 + 10^{-6}s)(1 + 10^{-8}s)} \qquad \text{case b}$$

Also, find the gain and phase margins in each of the above cases.

The logarithmic magnitude and phase plots for $A_d(j\omega)$ for the two amplifiers can be obtained using the given transfer functions as shown in Figs. 1.50*a* and *b*, respectively. In the case of the inverting amplifier,

$$\beta = \frac{1}{K+1} = \frac{1}{9+1} = -20 \text{ dB}$$

The feedback ratio, a constant in this case, does not provide any additional phase shift, and thus arg $\beta A_d(j\omega)$ is the same as arg $A_d(j\omega)$. Further, the magnitude response of $\beta A_d(j\omega)$ can be obtained simply by subtracting a constant value of 20 dB from the magnitude response of $A_d(j\omega)$ at every frequency. This is equivalent to shifting the ω axis of the plot up by 20 dB. This is also illustrated in Fig. 1.50 without having to draw the log plots of $|\beta A_d(j\omega)|$ in the two cases. The gain crossover frequencies for cases a and b are 2.15×10^9 and 10^6 r/s, respectively. At these frequencies, the phase values can be read from the corresponding phase plots and are $-243°$ and $-135°$, respectively. This clearly indicates that an inverting amplifier with a nominal dc gain of 9 employing the first op amp will be unstable, whereas one using the

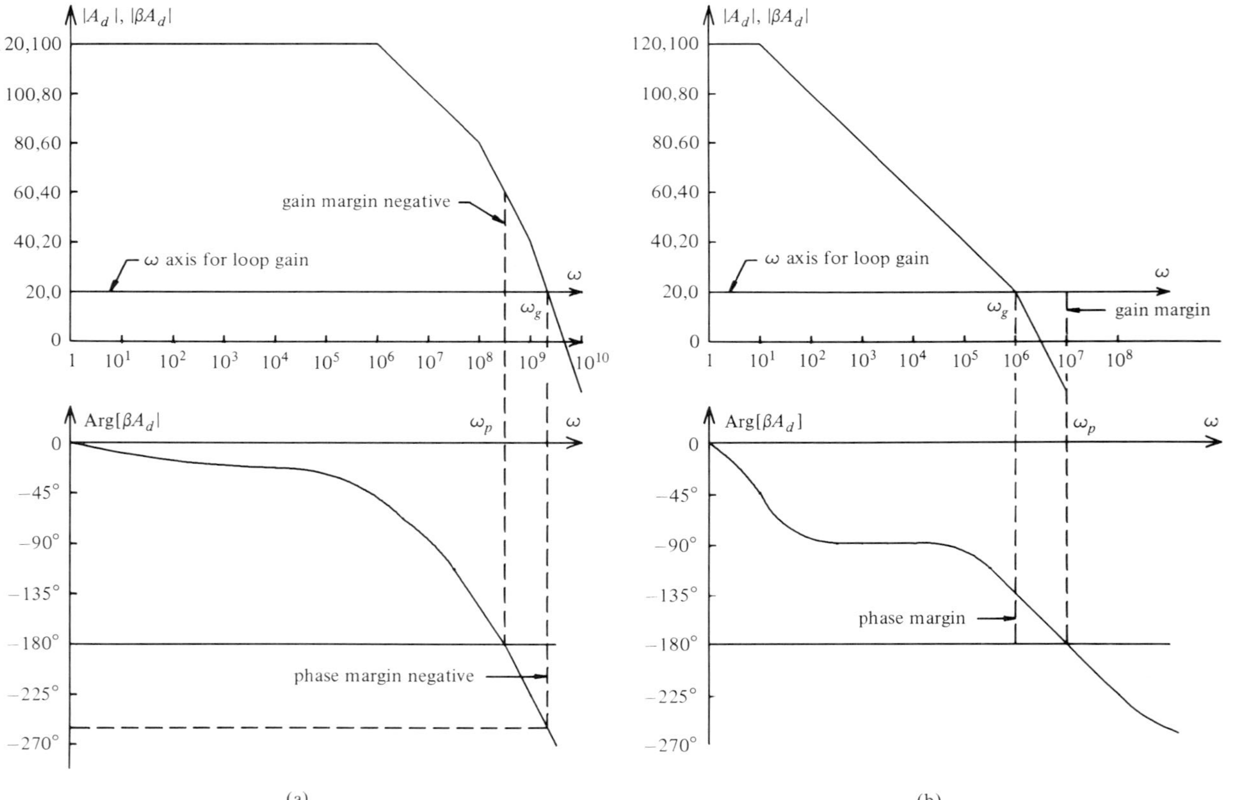

Figure 1.50 Frequency responses of A_d and the loop gain for Example 1.14. (*a*) Amplifier case a. (*b*) Amplifier case b.

second op amp will be absolutely stable. The phase margins are

$$\phi_m = -63° \qquad \text{case a}$$

$$\phi_m = 45° \qquad \text{case b}$$

The phase crossover frequencies are 3.162×10^8 and 10^7 r/s in cases a and b, respectively. Calculating the gain margins, we find that

$$g_m = -40 \text{ dB} \qquad \text{case a}$$

$$g_m = 40 \text{ dB} \qquad \text{case b}$$

■

Example 1.15: Investigate the stability problem of a noninverting unity gain amplifier when it uses the op amp whose gain is given in Example 1.14*b*. Find the phase and gain margins.

The magnitude and phase characteristics of the op amp are shown in Fig. 1.50*b*, and the feedback factor β of a noninverting amplifier is unity. Therefore $\beta A_d(j\omega)$ is the same as $A_d(j\omega)$ in this case. Thus, from the op amp characteristics in Fig. 1.50*b*, we find that the gain crossover frequency ω_g is 3.162×10^6 r/s and $\arg \beta A_d(j\omega_g)$ is $-164°$. This means that the phase margin is only 16°. The phase crossover frequency is 10^7 r/s, and the gain margin is about 20 dB. Though the gain margin is acceptable, the phase margin definitely is not. ■

Compensation

If the calculation of the phase and gain margins indicates poor relative stability, one can improve the stability by connecting external circuits so that these margins are within acceptable limits. This is called *compensation*. For a specific application, one can tailor the compensation to obtain the desired level. In Example 1.14, the op amp in case a has three poles, and each provides a large phase lag before the magnitude characteristic drops to 0 dB. When the op amp in case b is used in the same circuit, the circuit not only is absolutely stable but also has good relative stability. The reason for this is that in case b the phase shift is considerably lower than 180° before its magnitude characteristic passes through the 0-dB line. This implies that, if the op amp has too many poles in the frequency range before its magnitude characteristic passes through the 0-dB line, it is likely that there will be a stability problem. In particular, when the feedback ratio is constant, as in the previous two examples, for absolute stability, the op amp transfer function cannot have more than two poles in the frequency range before its magnitude characteristic passes through the 0-dB line. For good relative stability, even the second pole can be located only below the gain crossover frequency. Otherwise, the phase margin will be considerably lower, as in Example 1.15. This implies that op amp characteristics such as the ones shown in Fig. 1.50*a*, which are typical of uncompensated op amps, must be modified similar to the ones shown in Fig. 1.50*b*, which are typical of compensated op amps. In fact, the capacitance C_c in Fig. 1.38 is used for frequency compensation. In the process of compensation, one should note that the bandwidth of the op amp is reduced and that the performance of the closed-loop circuit is degraded in terms of sensitivity, bandwidth, and transient response. Usually a compromise is achieved.

Dominant Pole Compensation

This type of compensation is also known as *lag compensation*. In some op amps, such as the 741 type, dominant pole compensation is provided internal to the op amp, as shown in Fig. 1.38. Internally compensated op amps are mostly general-purpose devices and are useful in most applications. In uncompensated op amps terminals are available to connect the compensating network.

Example 1.14 clearly shows that, if the loop gain is dominated by a single pole, then for any amount of constant feedback, a stable system results. Further, if the loop gain is such that it is dominated by a single pole in the vicinity of the gain crossover frequency, then an ample phase margin can be obtained. Thus most compensation techniques essentially reduce the gain such that a single pole dominates the magnitude response of the op amp near its unity gain crossover frequency. One brute-force method of obtaining the dominant pole is to connect a capacitance in the signal path to ground or virtual ground. The capacitance value can be chosen to achieve the desired result described earlier; i.e., the single pole caused by the capacitance dominates the magnitude characteristic response of the op amp near its unity gain crossover frequency. This kind of modification can be dealt with in terms of the poles and zeros of the op amp transfer function. If A_d is the gain of an uncompensated op amp, it can be modified by the compensating network so that the compensated op amp will have a gain of

$$A_{dc} = \frac{A_d}{1 + s\tau_p} \tag{1.79}$$

The new pole, $s = -1/\tau_p$, is introduced by the compensating network. Since in this type of compensation the single pole dominates the magnitude and phase characteristics over a wide frequency range, it is called *dominant pole compensation*. This is achieved by the compensating capacitance C_c in the 741 type circuit in Fig. 1.38. Other methods of achieving this compensation are shown in Fig. 1.51. The capacitance C_c in combination with R_o and R_c, respectively, in the two circuits provides the dominant pole. Another method of providing this type of compensation is to connect the

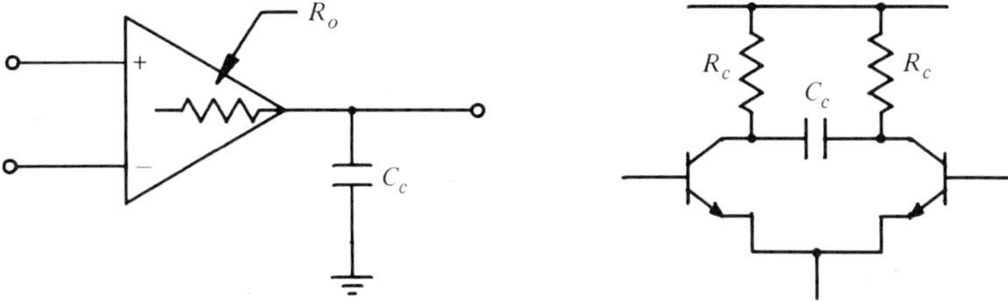

Figure 1.51 Methods for providing dominant pole compensation.

capacitance value across the terminals as suggested by the manufacturer. Manufacturer-suggested compensation usually requires a very low capacitance. The following example illustrates the concept of dominant pole compensation.

Example 1.16: The open-loop gain of an uncompensated op amp is given by

$$A_d = \frac{10^5}{(1 + 10^{-6}s)(1 + 10^{-8}s)(1 + 10^{-9}s)}$$

Use dominant pole compensation so that the compensated amplifier will have a 45° phase margin near its unity gain crossover frequency.

The dominant pole compensation modifies the op amp gain as

$$A_{dc} = \frac{10^5}{(1 + s\tau_p)(1 + 10^{-6}s)(1 + 10^{-8}s)(1 + 10^{-9}s)}$$

We want to choose the value of τ_p such that the required condition is met. First consider the logarithmic magnitude and phase functions of the above transfer function:

$$\begin{aligned} |A_{dc}| &= 20\log(10^5) - 20\log\left(1 + \omega^2\tau_p^2\right)^{1/2} \\ &\quad - 20\log(1 + 10^{-12}\omega^2)^{1/2} - 20\log(1 + 10^{-16}\omega^2)^{1/2} \\ &\quad - 20\log(1 + 10^{-18}\omega^2)^{1/2} \end{aligned} \tag{1.80}$$

and

$$\begin{aligned} \arg A_{dc} &= -\tan^{-1}(\omega\tau_p) - \tan^{-1}(10^{-6}\omega) - \tan^{-1}(10^{-8}\omega) \\ &\quad - \tan^{-1}(10^{-9}\omega) \end{aligned} \tag{1.81}$$

We require a phase margin of 45°. This means that, at the unity gain crossover frequency of ω_g, $\arg A_{dc}$ must be −135°. Also note that, at the same frequency, $|A_{dc}| = 1$. Therefore, using these two conditions in

(1.80) and (1.81), we can find the values of ω_g (though not required) and τ_p. Both (1.80) and (1.81) are highly nonlinear, however, to simplify the solution of the problem, we can make some approximations. Furthermore, the exact solution of such problems is neither required nor necessary. The pole $s = -1/\tau_p$ must be chosen such that the gain falls from a dc gain of 100 dB before the second and other poles provide any substantial lagging phase. Since the magnitude characteristic falls at the rate of 20 dB/decade after the first pole of $s = -1/\tau_p$, the second pole must be at least a few decades away from the first pole so that the gain falls to 0 dB. In such a case $-\tan^{-1}(\omega_g \tau_p)$ can be approximated to $-90°$ for large $\omega_g \gg 1/\tau_p$. Then, arg A_{dc} will be approximately $-135°$ when $\omega \simeq 10^6$ r/s. Therefore ω_g must be close to 10^6 r/s. The dc gain of 100 dB falls to 0 dB at the rate of -20 dB/decade, and this means that $1/\tau_p$ must be on the order of 5 decades lower than ω_g, which confirms our previous assumption. With these ideas in mind, we can approximate (1.81) when $\omega = \omega_g$ and equate it to $-135°$:

$$-135° = -90° - \tan^{-1}\left(10^{-6}\omega_g\right) - \tan^{-1}\left(10^{-8}\omega_g\right)$$

From the above equation, we find that

$$\tan^{-1}\frac{(10^{-6} + 10^{-8})\,\omega_g}{1 - 10^{-14}\omega_g^2} = 45°$$

Solving the above equation, we find that $\omega_g = 0.9806 \times 10^6$ r/s. Substituting this value of ω_g for ω in (1.80) and equating it to 0, we find the value of τ_p to be $0.07281s$. Thus the value of $1/\tau_p = 13.73$ r/s. These values of ω_g and τ_p confirm our assumptions. To verify our approximation further, we find arg A_{dc} at $\omega = 0.9806 \times 10^6$ r/s, and it is equal to $-135.1°$ with negligible error. The magnitude and phase characteristics of the op amp before and after compensation are shown in Fig. 1.52.

■

Dominant pole compensation is a very conservative technique and can be used for many applications. The effect of compensation can be clearly seen in Fig. 1.52. The gain at high frequencies is reduced considerably, and the bandwidth is also reduced. This means that this type of compensation can be used only for low-frequency applications. One can use other types of compensation such as lead, and lead-lag compensation where the dynamics of the op amp can be controlled to some extent. For information on other types of compensation, the reader should consult reference [4].

In addition to the poles of the op amp, which give rise to stability problems, there are other reasons why an op amp circuit may oscillate.

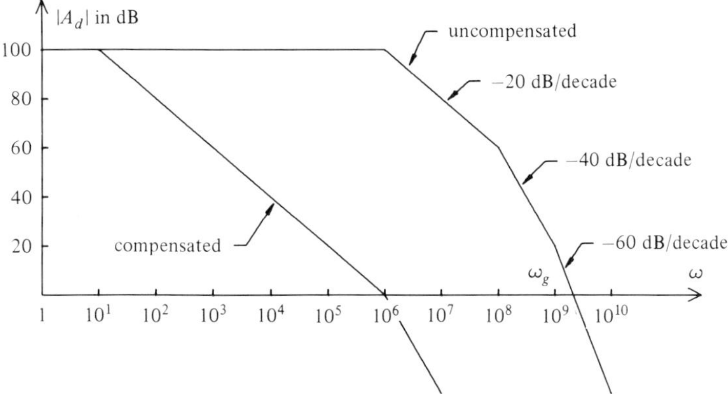

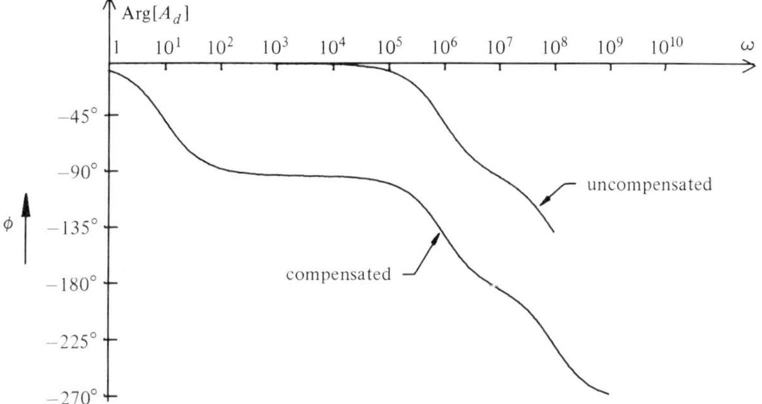

Figure 1.52 Magnitude and phase characteristics of compensated and uncompensated op amps.

Other factors that might affect the loop gain, hence the stability, are

1. Insufficient compensation for a given application
2. Excessive capacitive load
3. Stray capacitances at the input and between the input and output terminals
4. A stability problem caused by the feedback network even when the op amp is sufficiently compensated.
5. Inadequate power supply bypassing.

We next consider an example that shows how stray capacitances, which are on the order of a few picofarads, can affect the stability considerably.

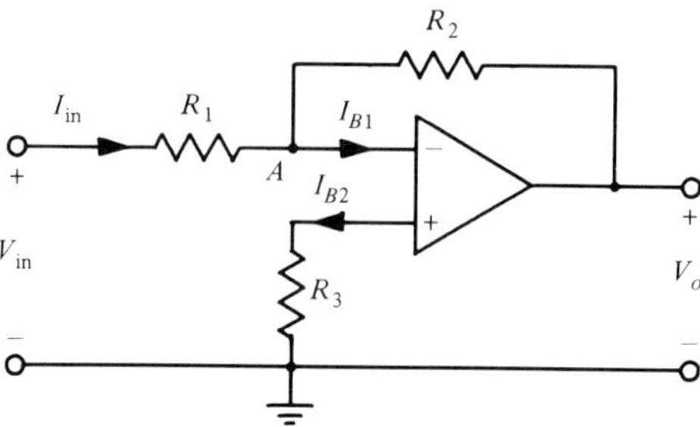

Figure 1.53 The inverting amplifier circuit in Example 1.17.

Example 1.17: An inverting amplifier is designed with a nominal gain of 10 and an input impedance of 100 kΩ using an LM101A op amp with the circuit shown in Fig. 1.53. With the recommended compensation, the logarithmic magnitude characteristic of the op amp is shown in Fig. 1.54*a*.

In the circuit in Fig. 1.53, at the inverting and noninverting input terminals, there will always be dc input bias currents on the order of a few microamperes to a few tenths of a microampere, and $I_{B1} \simeq I_{B2}$. (The actual direction of the flow of these currents may be opposite that indicated in the diagram.) If the resistance $R_3 = 0$ in the circuit, then the dc potential of node A may be raised or lowered (even when the input is ground) by a few millivolts, and this causes an input offset voltage. This in turn will lead to saturation of the op amp. This effect will be particularly pronounced when the actual values of R_1 and R_2 are on the order of many hundreds of kiloohms, but can be avoided to a large extent by connecting resistance R_3 equal to $R_1R_2/(R_1 + R_2)$. Since the input impedance between the inverting and noninverting terminals is on the order of many megohms, the resistance R_3 will not affect calculation of the gain or the feedback factor under normal conditions, and these values can still be obtained using the formulas developed in Example 1.13. For ac operation, node A is at virtual ground, and therefore the input resistance of the amplifier must be equal to R_1. Since we require the input impedance of the amplifier to be 100 kΩ, the value of R_1 must be 100 kΩ, and since the required nominal gain is 10, the value of $R_2 = 1$ MΩ. Of course, the required value of R_3 is 100 kΩ||1 MΩ = 90 kΩ. From the magnitude characteristic of the op amp, we can approximate the gain, for sinusoidal applications, as

$$A_d(jf) \simeq \frac{A_o}{(1 + jf/f_1)(1 + jf/f_2)}$$

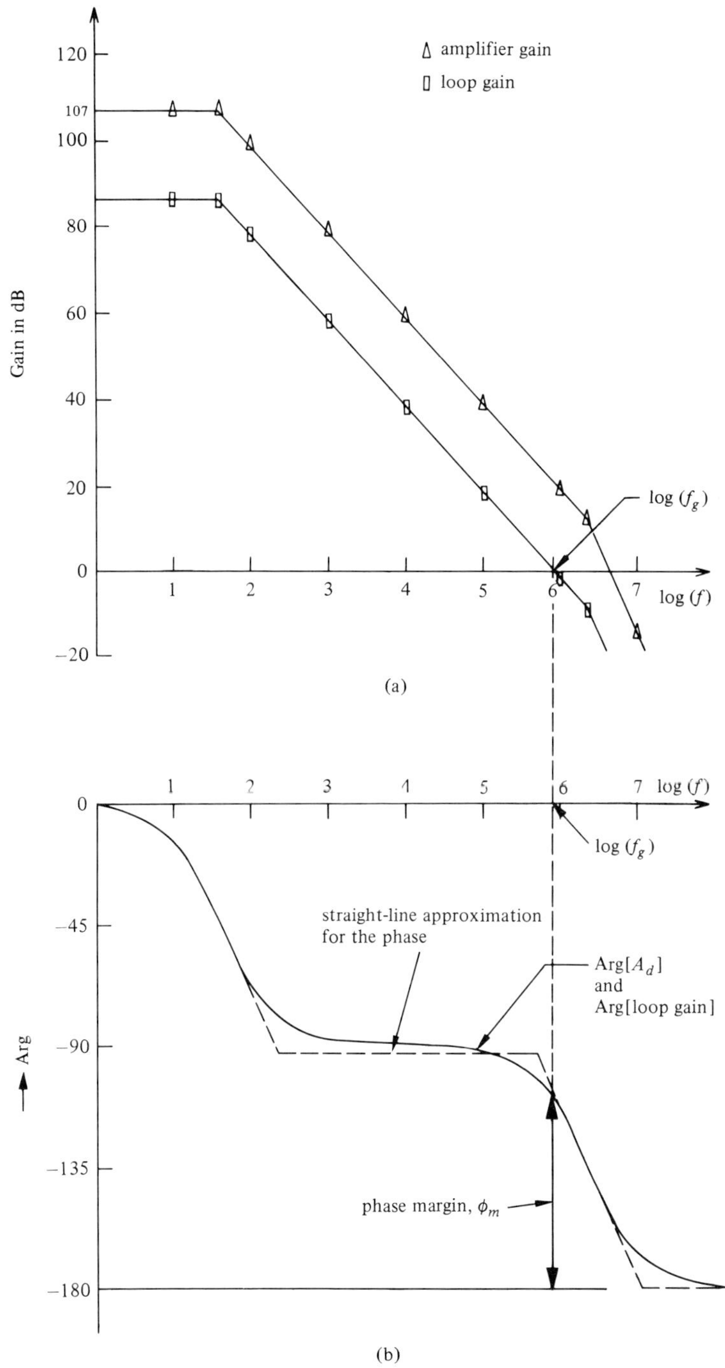

Figure 1.54 Magnitude and phase plots of the op amp gain and the loop gain of an inverting amplifier. (*a*) Magnitude plots. (*b*) Phase plot.

where

$$A_o = 10^{107/20} = 223.9 \times 10^3$$
$$f_1 \simeq 40 \text{ Hz}$$

and

$$f_2 \simeq 2.2 \times 10^6 \text{ Hz}$$

If we do not include any other nonidealness of the op amp, except for the finite gain of the amplifier, we will have

$$\beta A_d = \frac{100}{100 + 1000} A_d$$
$$= \frac{20.35 \times 10^3}{(1 + jf/f_1)(1 + jf/f_2)}$$

Using the above equations, let us calculate the phase margin ϕ_m available in the circuit. To do this we must find f_g such that $|\beta A_d| = 1$ or equivalently

$$20 \log(20.35 \times 10^3) - 10 \log\left[1 + \left(\frac{f_g}{f_1}\right)^2\right] - 10 \log\left[1 + \left(\frac{f_g}{f_1}\right)^2\right] = 0$$

From the above equation, we have

$$\left[1 + \left(\frac{f_g}{f_1}\right)^2\right]\left[1 + \left(\frac{f_g}{f_2}\right)^2\right] = 414.2 \times 10^6$$

Substituting values for f_1 and f_2 and solving for f_g, we obtain

$$f_g = 768.5 \text{ kHz}$$

$$\arg \beta A_d(jf_g) = -\tan^{-1}\frac{f_g}{f_1} - \tan^{-1}\frac{f_g}{f_2}$$
$$= -109.3°$$

Therefore the phase margin is

$$\phi_m = 70.7°$$

Calculation of the phase and gain margins can also be carried out using Bode plots, as illustrated in Fig. 1.54. This indicates that a good phase margin is available in the circuit. However, in practice and under nonideal working conditions, this may not be true. For example, assume that the amplifier is loaded with a 500-pF capacitor at its output. Then,

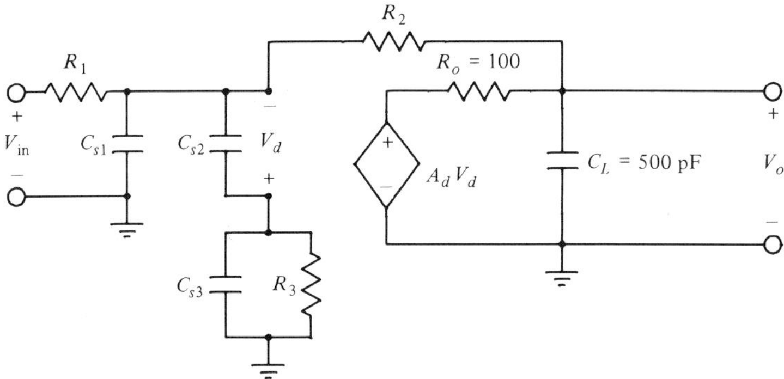

Figure 1.55 Equivalent circuit of an inverting amplifier with stray and load capacitances.

because of the finite output resistance of the op amp, whose typical value is 100 Ω, a pole will be generated by the combination of the load capacitor and the output resistance of the op amp. The input stray capacitances might affect the loop gain seriously, so let us also include these capacitances. The equivalent circuit for the inverting amplifier circuit in Fig. 1.53, including all the effects discussed above, is shown in Fig. 1.55. ■

A conservative estimate of the stray capacitances from the input terminals to the ground is about 3 pF. However, wiring can cause another additional 2-pF capacitance. Thus the values C_{s1} and C_{s3} total about 5 pF. The stray capacitance C_{s2} between the input terminals is very small, hence it will be ignored. This means that this branch can be replaced with an open circuit. Since the output impedance of the op amp is not zero, we will not be able to evaluate β separately. However $A_l = -\beta A_d$ can be evaluated using (1.63). Therefore, after breaking open the feedback path and grounding the input, the circuit for evaluating the loop gain will be as shown in Fig. 1.56. From this circuit, we find that

$$A_l = -A_d \frac{R_1}{R_1 + R_2} \frac{1}{\left(1 + \dfrac{R_1 R_2}{R_1 + R_2} C_{s1} s\right)(1 + R_0 C_L s)}$$

After substituting values for R_1, R_2, R_0, C_{s1}, and C_L and noting that, for sinusoidal inputs $s = j2\pi f$, we have

$$\beta A_d = -A_l = \frac{20.35 \times 10^3}{(1 + jf/f_1)(1 + jf/f_2)(1 + jf/f_3)(1 + jf/f_4)}$$

where f_1 and f_2 are the corner frequencies from the amplifier poles and f_3

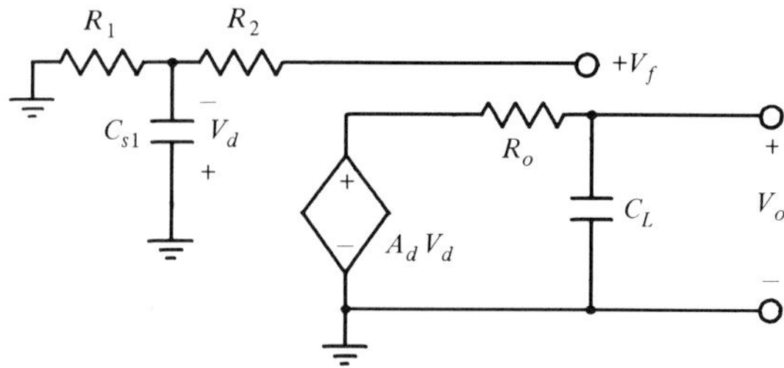

Figure 1.56 Circuit for calculating the loop gain of the circuit in Fig. 1.55.

and f_4 are attributable to the pole frequencies due to $R_1 \| R_2$-C_{s1} and R_0-C_L combinations and their values are

$$f_1 = 40 \text{ Hz} \qquad f_2 = 2.2 \text{ MHz}$$

$$f_3 = 350.1 \text{ kHz} \qquad f_4 = 3.183 \text{ MHz}$$

Note that β is also frequency-dependent and that the loop gain is drastically altered. Specifically, f_3, the pole frequency caused by the stray capacitance, is well before the second pole of the amplifier. Therefore we should expect the unity gain crossover frequency f_g to be considerably lower than the value we calculated earlier. We cannot tell anything about the stability of the circuit until we calculate the stability margins. The new βA_d has more poles, and therefore finding the values of f_p and f_g is a bit difficult. One way is to construct the Bode plots for βA_d and read the values of f_p and f_g from these plots. However, such values will only be approximate, and approximate values can be obtained without plotting the functions. Furthermore, more accurate results can be achieved with calculations a bit more involved. The logarithmic magnitude function is

$$|\beta A_d(jf)| = 86.17 - 10\log\left(1 + \frac{f^2}{f_1^2}\right) - 10\log\left(1 + \frac{f^2}{f_3^2}\right)$$
$$- 10\log\left(1 + \frac{f^2}{f_2^2}\right) - 10\log\left(1 + \frac{f^2}{f_4^2}\right) \text{ dB}$$

The gain drops from 86.17 dB at a rate of -20 dB/decade from 40 Hz and at a rate of about -40 dB/decade well beyond the value of f_3. The approximate value of the gain at 350.1 kHz is

$$|\beta A_d(jf_3)| \simeq 86.17 - 20\log\left(\frac{350.1 \times 10^3}{40}\right) \text{ dB}$$
$$= 7.326 \text{ dB}$$

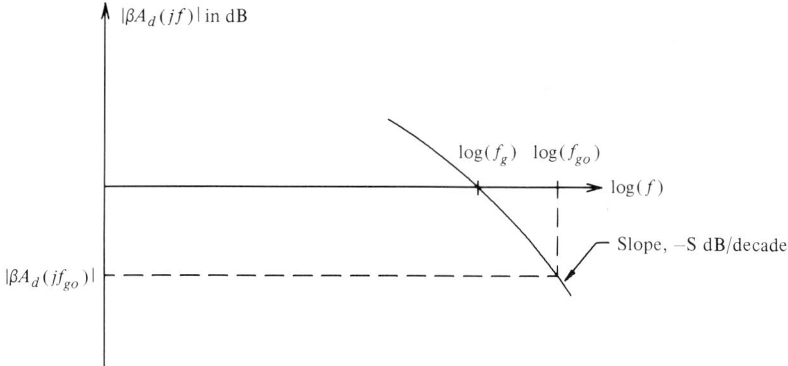

Figure 1.57 Magnitude response of the loop gain around f_g.

The approximate gain at the next pole frequency f_2 must be

$$|\beta A_d(jf_2)| \simeq 7.326 - 10\log\left(1 + \frac{f_2^2}{f_3^2}\right)$$
$$= -8.747 \text{ dB}$$

Therefore f_g must be before f_2 and, approximating the slope to be -40 dB/decade near this frequency, we obtain $f_g = f_{g0}$ to a first-order approximation, as

$$f_{g0} = 350.1 \times 10^{7.326/40} = 533.8 \text{ kHz}$$

This is the value of f_g indicated by the Bode plot, however, a more accurate result can be obtained as follows. Consider the magnitude plot in Fig. 1.57, which is shown near the 0-dB line. Assume that the value of $|\beta A_d(jf_{g0})|$ is not equal to zero (which is more likely than otherwise). Also, let $-S$ decibels/decade be the slope of the characteristic near this frequency. Then, assuming that the logarithmic magnitude characteristic is a straight line, we can find f_g knowing the values of $|\beta A_d(jf_{g0})|$, f_{g0}, and S. This is possible because

$$\frac{|\beta A_d(jf_{g0})| - |\beta A_d(jf_g)|}{\log(f_{g0}) - \log(f_g)} = -S$$

Since $|\beta A_d(jf_g)| = 0$, solving the above equation, we have

$$f_g = f_{g0}10^{|\beta A_d(jf_{g0})|/S}$$

In order to determine whether f_{g0} is really the unity gain crossover frequency, we calculate $|\beta A_d(jf_{g0})|$ anyway. Then, even if we know the approximate value of S, we will be able to evaluate the new value of f_g. In fact, we can use this formula iteratively until sufficient accuracy for f_g is obtained and the formula for iteration becomes

$$f_{g(i+1)} = f_{gi} 10^{|\beta A_d(f_{gi})|/S}$$

For this problem, we can start with $f_{g0} = 533.8$ kHz. Furthermore, we need to know the approximate value of S. At this frequency, the slope is

$$S \simeq 20 + \frac{20\left(f_{g0}^2/f_3^2\right)}{1 + f_{g0}^2/f_3^2}$$

$$= 29.17 \text{ dB/decade}$$

This slope remains almost the same near this frequency, and it is not necessary to update this value. The results of the iterative process are shown in the accompanying table. Iteration is stopped when

$$\left|1 - \frac{f_{g(i+1)}}{f_{gi}}\right| < 0.001$$

i	f_{gi} (kHz)	$\lvert\beta A_d(f_{gi})\rvert$ (dB)	f_{gi+1} (kHz)
0	533.8	-1.923	458.6
1	458.6	0.3686	472.2
2	472.2	-0.0616	469.9
3	469.9	10.6×10^{-3}	470.3

Then, the value of $f_{g(i+1)} = 470.3$ kHz, and thus $f_g \simeq 470.3$ kHz. At this frequency, $|\beta A_d(jf_g)| = -1.85 \times 10^{-3}$, which is really close to zero. Now, the phase of $\beta A_d(jf_g)$ is

$$\arg \beta A_d(jf_g) = -163.9°$$

Therefore the phase margin is only a meager 16.1°, which is definitely *not* acceptable. The phase margin originally calculated to be 70.75° is reduced to such a low value mainly by the finite pole frequency f_3 caused by the stray capacitance, and the circuit is really close to instability. As a matter of interest, let us also find the gain margin. To do this, we have to find the phase crossover frequency f_p using $\arg \beta A_d(jf_p) = -180°$. Since this $f_p \gg f_1$, we also approximate $\tan^{-1}(f_p/f_1) \simeq 90°$. Then, f_p should satisfy

$$\tan^{-1}\frac{f_p}{f_3} + \tan^{-1}\frac{f_p}{f_2} + \tan^{-1}\frac{f_p}{f_4} = 90°$$

or, equivalently,

$$f_p^2\left(\frac{1}{f_2 f_3} + \frac{1}{f_3 f_4} + \frac{1}{f_4 f_2}\right) = 1$$

Solving the above equation, we obtain

$$f_p = 654 \text{ kHz}$$

Calculating the magnitude function, we see that $|\beta A_d(jf_p)| = -5.168$ dB, and thus the gain margin is only 5.168 dB, which is also a small value. Furthermore, the unity gain crossover frequency, which is closely related to the bandwidth of the inverting amplifier, is also reduced to a low value by the stray capacitance. It may be possible to correct such an undesirable situation in many circuits by controlling the dynamics of the circuit externally. For example, in this circuit, by connecting a capacitor C_c across the feedback resistor R_2, we can introduce a zero into the loop gain. Such a zero will reduce the fast drop in the magnitude characteristic and provide a leading phase. This is one reason why C_c is called a lead capacitor. This capacitance value must be chosen so that the zero frequency does not affect the low-frequency performance of the inverting amplifier and at the same time improves the stability of the circuit. Under the same conditions as before, with the introduction of C_c, the feedback factor β will be

$$\beta = \frac{R_1}{R_1 + R_2} \frac{1 + R_2 C_c s}{\left[1 + \dfrac{R_2 R_1}{R_1 + R_2}(C_c + C_{s1})s\right](1 + R_o C_L s)}$$

If we choose $C_c = R_1 C_{s1}/R_2$, then β will reduce to

$$\beta = \frac{R_1}{R_1 + R_2} \frac{1}{1 + R_o C_L s}$$

The influence of the pole frequency at $1/2\pi R_o C_L$ was earlier found to be small, and therefore we will have good stability margins. In this case, we require $C_c = 0.5$ pF, which is really a small value. Wiring may even cause a capacitance of 1 pF. Assuming that $C_c = 1$ pF, let us then find the effect of the zero on the stability margins and unity gain crossover frequency. After substituting the various values, we find that

$$\beta A_d(jf) = \frac{(20.35 \times 10^3)(1 + jf/f_z)}{(1 + jf/f_1)(1 + jf/f_3)(1 + jf/f_2)(1 + jf/f_4)}$$

where

$$f_z = 159.2 \text{ kHz} \qquad \text{and} \qquad f_3 = 291.8 \text{ kHz}$$

and the values of f_1, f_2, and f_4 remain the same as before. Then, the magnitude and phase functions are

$$|\beta A_d(jf)| = 86.17 + 10\log\left(1 + \frac{f^2}{f_z^2}\right) - 10\log\left(1 + \frac{f^2}{f_1^2}\right)$$
$$-10\log\left(1 + \frac{f^2}{f_3^2}\right) - 10\log\left(1 + \frac{f^2}{f_2^2}\right) - 10\log\left(1 + \frac{f^2}{f_4^2}\right)$$

and

$$\arg \beta A_d(jf) = \tan^{-1}\frac{f}{f_z} - \tan^{-1}\frac{f}{f_1} - \tan^{-1}\frac{f}{f_3} - \tan^{-1}\frac{f}{f_2} - \tan^{-1}\frac{f}{f_4}$$

To locate the approximate value of f_g, we note that the gain drops at a rate of -20 dB/decade up to 159.2 kHz starting from 40 Hz. Using a straight-line approximation, we find that the gain must be approximately equal to 14.18 dB at f_z. This gain remains constant up to the frequency f_3 and then drops again at a rate of -20 dB/decade (?) up to f_2. The approximate gain at f_2 must be -3.37 dB. Therefore f_g must lie between f_3 and f_2 and is approximately equal to

$$f_{g0} = 291.8 \times 10^{14.18/20} = 1394 \text{ kHz}$$

But this frequency is close to the next pole frequency f_2, and therefore f_2 has a strong influence on both f_g and the slope at which the magnitude characteristic decreases. To determine the value of f_g accurately, we can use the formula developed earlier and solve for f_g iteratively. Though it is possible to evaluate the slope at f_{g0} exactly, we know that it must be less than -20 dB/decade and more than -40 dB/decade. We shall approximate this value as -30 dB/decade. Starting with $f_{g0} = 1.394$ MHz and $S = 30$, the accompanying table gives the iterations.

i	f_{gi} (MHz)	$\lvert\beta A_d(jf_{gi})\rvert$ (dB)	$f_{g(i+1)}$ (MHz)
0	1.394	-2.628	1.201
1	1.201	-3.487×10^{-3}	1.201

Therefore an accurate estimate of f_g is 1.201 MHz. Here, $\arg \beta A(jf_g)$ is $-133.2°$, and therefore the phase margin is 46.8°. This is a comfortable value of ϕ_m. Also, note that the unity gain crossover frequency is much higher than the value obtained in both of the previous cases. In fact, we have overcompensated for the effects of both stray and load capacitances. Let us check the relative stability of the circuit by calculating the gain margin as well. To do this, we must find the value of f_p using the condition $\arg \beta A_d(jf_p) = -180°$. This frequency f_p must be between f_2 and f_4, and therefore we assume that $\tan^{-1}(f_p/f_1) \simeq 90°$. Then the frequency f_p must satisfy

$$\tan^{-1}\frac{f_p}{f_3} + \tan^{-1}\frac{f_p}{f_2} + \tan^{-1}\frac{f_p}{f_4} - \tan^{-1}\frac{f_p}{f_z} = 90°$$

or, equivalently,

$$f_p^4 - f_p^2[(f_2 f_4 - f_3 f_z) + (f_2 + f_4)(f_3 - f_z)] - f_2 f_3 f_4 f_z = 0$$

Solving the above equation, we find that

$$f_p = 2.777 \text{ MHz}$$

One can verify that indeed this is the phase crossover frequency f_p and that $|\beta A_d(jf_p)| = -12.03$ dB. Thus $g_m = 12.03$ dB. We then have comfortable phase and gain margins, which indicates good relative stability. We also have a higher value for the unity gain crossover frequency, which means that the bandwidth of the inverting amplifier will be higher with the addition of C_c to the circuit.

1.10 *Gain and Bandwidth of Finite Gain Amplifiers*

In most of our applications, we assume that the op amp has been sufficiently compensated and that it is described by its gain equation as

$$A_d = \frac{A_0}{(1 + s\tau_1)(1 + s\tau_2)} \tag{1.82}$$

Without loss of generality, assume that the first pole is at $s = -\omega_1 = -1/\tau_1$. For most compensated op amps, this frequency ranges from a few hertz to a few tens of hertz. The next pole, $s = -1/\tau_2$, is located near the 0-dB line. The value of $f_2 = 1/2\pi\tau_2$ is usually many hundreds of kilohertz to a few megahertz. The higher-order pole frequencies are far beyond these two frequencies, and for all practical purposes the effects of these poles can be ignored. If only these two pole frequencies are taken into account, the magnitude characteristics will be as shown in Fig. 1.58. The frequency ω_1 is the dominant pole and is called the *3-dB frequency*. The typical frequency range in which an op amp is used is such that $\omega_1 \ll \omega \ll \omega_2$. In such cases

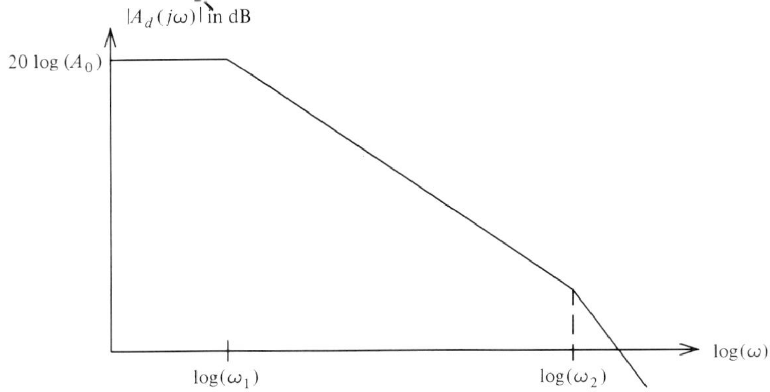

Figure 1.58 Bode plot of magnitude characteristic of a compensated op amp.

the op amp gain can be approximated as

$$A_d \simeq \frac{A_0}{1 + s\tau_1} \simeq \frac{A_0\omega_1}{s}$$

where $\omega_1 = 1/\tau_1$. Since ω_1 is also the bandwidth, the product $A_0\omega_1$ is known as the *gain bandwidth product* (GBP) and is denoted by B. Since we will use only the differential-mode gain of the op amp, we drop the subscript d in A_d and in most of our applications, we shall use the differential-mode gain

$$A = \frac{B}{s} = \frac{1}{s\tau}$$

where

$$\tau = \frac{1}{B} = \frac{1}{A_0\omega_1}$$

The constant τ is known as the *operational amplifier time constant*, and with an ideal gain of $A \to \infty$, it becomes zero. The inverting and noninverting amplifiers are shown in Fig. 1.47. The closed-loop gains of the noninverting and inverting amplifiers were defined in terms of the resistance ratios and A in Sec. 1.9. Since the resistance ratios are fixed by the required nominal value of the gain K in the case of a noninverting amplifier and $-K$ in the case of an inverting amplifier, we can find the overall gains in terms of K:

$$\frac{V_o}{V_i} = \frac{K}{1 + K/A} \qquad \text{noninverting}$$

and

$$\frac{V_o}{V_i} = \frac{-K}{1 + (K+1)/A} \qquad \text{inverting}$$

If we use the finite and frequency-dependent gains of the op amp as approximated earlier, we will obtain

$$\frac{V_o}{V_i} = \frac{K}{1 + Ks\tau} \qquad \text{noninverting}$$

and

$$\frac{V_o}{V_i} = \frac{-K}{1 + (K+1)s\tau} \qquad \text{inverting}$$

The noninverting and inverting amplifiers act as real amplifiers with negligible phase lag only when the frequency of operation ω is such that

$$K\omega\tau \ll 1 \qquad \text{noninverting}$$

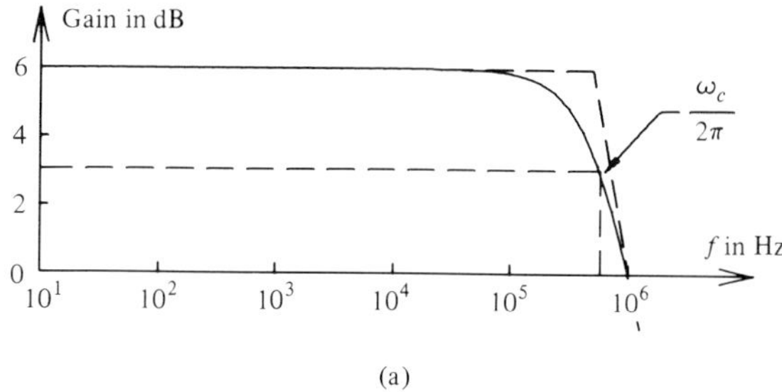

(a)

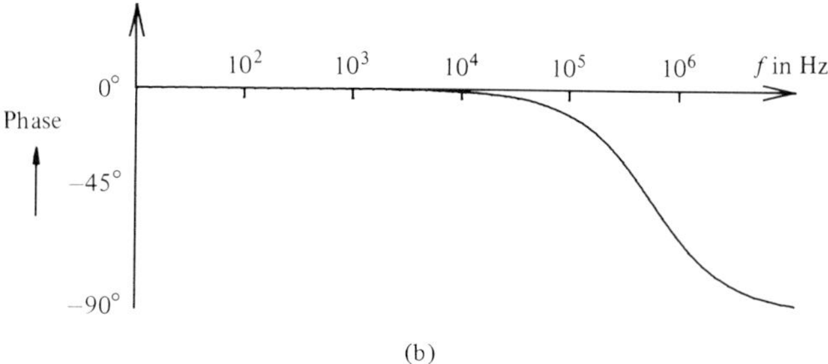

(b)

Figure 1.59 Magnitude (a) and phase (b) characteristics of a noninverting amplifier.

and

$$(K+1)\omega\tau \ll 1 \qquad \text{inverting}$$

When such amplifiers are used as components in other circuits, the frequency-dependent gains of the amplifiers affect the performance of these circuits as well. The frequency response characteristics of a noninverting amplifier with a nominal value of $K = 2$ and those of an inverting amplifier with a nominal value of $K = 100$ are shown in Figs. 1.59 and 1.60, respectively. The assumed value of B, in both cases, is $2\pi \times 10^6$ r/s (typical value of a μA 741 op amp). There is a close relationship between the 3-dB frequency of the amplifiers and the unity gain crossover frequency ω_g of the loop gain of the circuits. For example, for a noninverting amplifier, the 3-dB frequency (or cutoff frequency) is B/K. The loop gain of the circuit is

$$\beta A = \frac{A}{K} = \frac{B}{Ks}$$

$$|\beta A(j\omega_g)| = 1$$

when $\omega_g = B/K$. That is, the unity gain crossover frequency of the loop

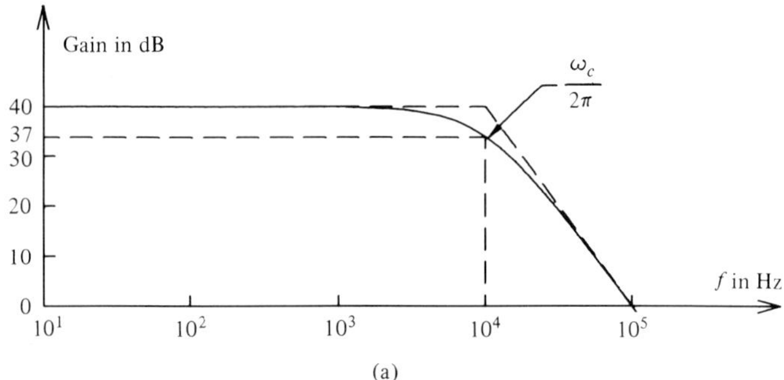

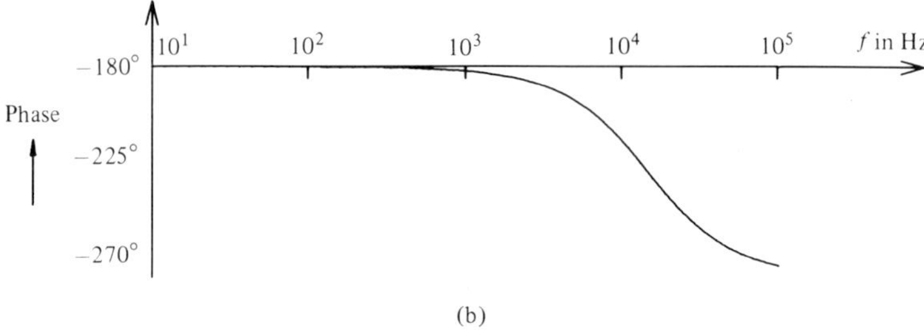

Figure 1.60 The magnitude (a) and phase (b) characteristics of an inverting amplifier.

gain of the circuit is the same as its 3-dB frequency, which is also the bandwidth of the amplifier. The same is true in the case of an inverting amplifier.

1.11 *Unity Gain Crossover Frequency, Full Power Response Frequency, and Slew Rate*

The unity gain crossover frequency of the op amp f_u (note that f_u differs from f_g, which is the unity gain crossover frequency of the loop gain of a circuit) is the frequency at which the op amp gain becomes unity (0 dB). If the op amp gain is approximated by B/s, the unity gain crossover frequency is also B, the gain bandwidth product. However, we cannot use this approximate expression to determine the value of f_u because this frequency is also affected by the second- and higher-order pole frequencies. In high-frequency applications, f_u plays an important role, and its value can be measured by connecting the op amp in the unity gain noninverting amplifier configuration and applying a step input to such an amplifier. The

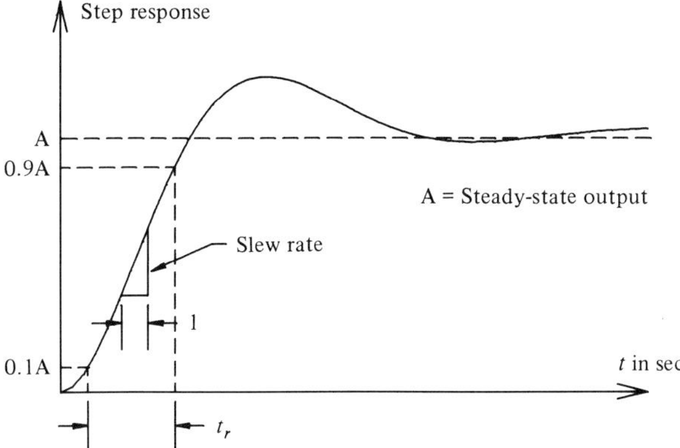

Figure 1.61 Step response of a typical unity gain noninverting amplifier using a compensated op amp.

step response of the unity gain amplifier is typically as shown in Fig. 1.61. The time required for the response to reach 90% of the final steady-state output from a 10% level is called the *rise time* t_r. The unity gain crossover frequency f_u of the op amp is related to t_r as follows:

$$f_u = \frac{0.35}{t_r} \tag{1.83}$$

The closed-loop bandwidth of finite gain amplifiers and other circuits typically depend on this frequency, and thus it plays an important role in small-signal analysis.

Full Power Response Frequency

When an op amp is subjected to a large input signal, the output signal will have large output swings. Under such conditions, the maximum closed-loop bandwidth of the circuits is much less than f_u. The maximum frequency of a sinusoidal output signal of a given amplitude up to which distortionless output can be obtained is called the *full power response frequency*. This full power frequency can be as low as $\frac{1}{100}$ of f_u and depends not only on the output swings expected but also on a variety of factors including the power supply voltage and compensation provided in the op amp. The peak-to-peak output swing that can be obtained without distortion from a 741 type op amp as a function of frequency is shown in Fig. 1.62*a*. Similar characteristics for an externally compensated LM101 op amp with two different compensations are shown in Fig. 1.62*b*. The higher the output swing expected, the lower the full power response frequency.

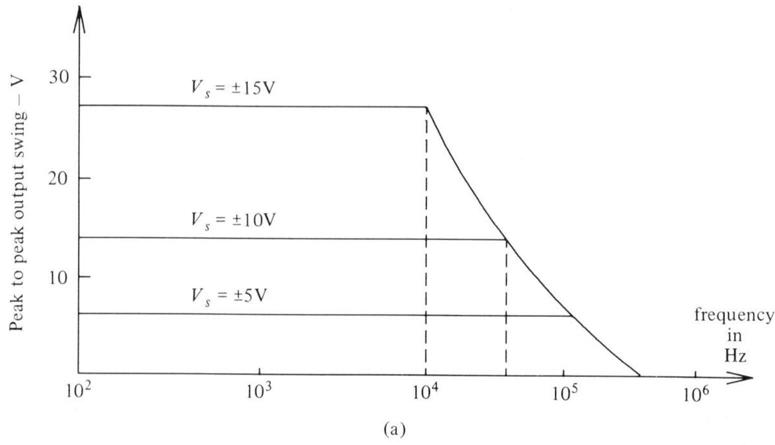

(a)

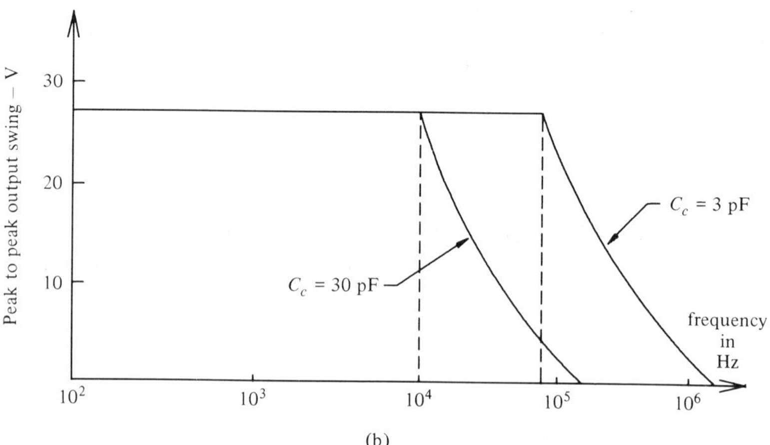

(b)

Figure 1.62 Full power response frequency vs output (pp) swing. (a) Output frequency with various supply voltages to the op amp. (b) Output voltage swing vs frequency with different compensating capacitors.

Slew Rate

There is another quantity of interest that is also closely related to the full power response frequency f_f; it is called the *slew rate*. The slewing of an op amp is essentially a nonlinear phenomenon and depends on almost the same factors as the full power response frequency. Assume that a sinusoidal voltage is applied to a finite gain noninverting amplifier. Then, if we expect a large output swing at some frequency greater than the corresponding full

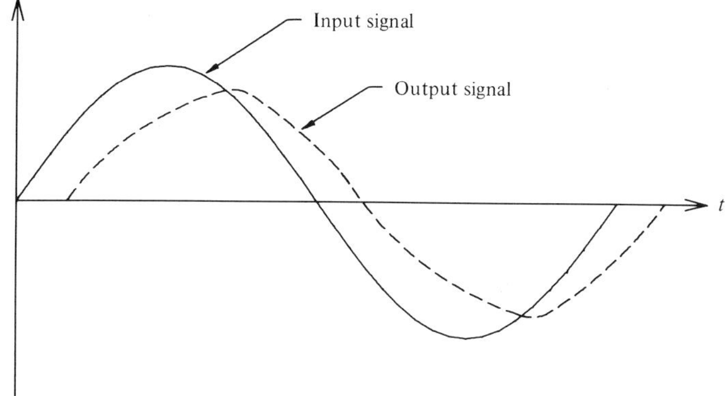

Figure 1.63 Distorted output due to slew rate limitation.

power response frequency, the output signal will be distorted as shown in Fig. 1.63. Obviously the op amp output cannot keep up at the same rate of change as that of the input signal. This phenomenon is called *slewing of the operational amplifier*. Suppose

$$v_o = V_m \sin(\omega t)$$

the maximum rate of change of v_o occurs at $t = 0$ and is

$$\left.\frac{dv_o(t)}{dt}\right|_{\max} = \omega V_m$$

For low frequencies and/or low amplitudes of the output signal, the above rate is small. As ω and/or V_m increase, the maximum rate also increases and the op amp output signal cannot keep up with the input signal. The slew rate is defined as the maximum rate at which the output changes that the op amp can deliver without distortion. One can find this slew rate from the step response of the noninverting unity gain amplifier as illustrated in Fig. 1.61. The slew rate of an op amp is specified by the manufacturer and is usually given in volts per microsecond. The slew rate of an op amp can be related to the full power response frequency f_f as follows. If $\omega = \omega_f$, then the slew rate SR must be equal to $\omega_f V_m$ and therefore

$$f_f = \frac{SR}{2\pi V_m}$$

For an op amp with a slew rate of 0.5 V/μs,

$$f_f V_m = \frac{0.5}{2\pi} \times 10^6 \text{ V/s}$$
$$= 79.58 \times 10^3 \text{ V/s}$$

If the maximum amplitude of the output signal is 1 V, then the full power response frequency will be 79.58 kHz. However, if we expect an output signal with a swing of 20 V peak-to-peak, the full power response frequency will be only 7.958 kHz, and this will be only a small fraction of the f_u of the op amp, as mentioned earlier.

The finite slew rate can also be the cause of an amplitude-dependent phase lag in the loop gain. This means that, when large output swings are expected, the circuit may break into oscillation because of the slew rate limitation. Apart from these limitations and the nonideal properties of the op amps we have discussed, there are other limitations as well. They include finite input impedance, nonzero output impedance, dc offset voltage and current, finite bias currents, and noise. We discussed the effects of some of these quantities in specific problems in earlier sections. Depending on the application, one may have to consider some or all of the nonideal parameters of the op amp, usually only a few important limitations for a given application. (Otherwise, analysis of the circuits using op amps would become unmanageable.) For this information, one has to consult the manufacturer's data book and the applications manual.

1.12 *Conclusions*

We have discussed all the important electronic circuits in an IC op amp and have provided both dc and small-signal analysis of all the circuits in a complete 741 type op amp. In addition, we have included a discussion of some important nonideal properties of op amps and problems arising from them and consider suggested solutions for overcoming such problems. We are now ready to consider some of the applications in which an op amp is used and will concentrate on only one, the most important application, starting with Chap. 4. This particular application is analog signal processing, but before considering the circuits used for this purpose, we will discuss some of the basic concepts.

REFERENCES

1. Ghausi, M. S.: *Electronic Devices and Circuits: Discrete and Integrated*, HRW, New York, 1985.
2. Gray, P.R., and Meyer, R.G.: *Analysis and Design of Analog Integrated Circuits*, 2nd ed., John Wiley, New York, 1984.
3. Fullagar, D.: "A New High Performance Monolithic op amp," Fairchild Semiconductor Applications Brief, May 1968.
4. Roberge, J.K.: *Operational Amplifiers*, John Wiley, New York, 1975.

EXERCISES

1.1. Consider the differential amplifier shown in Fig. E1.1. The transistors are identical except that their current gains are different. Both ac and dc gains are given by $\beta_1 = 20$ and $\beta_2 = 10$. Find the values of R_{E1}, R_{E2}, and R_e that satisfy the following two conditions.
(a) The voltage drop across $R_C = 3\ \text{k}\Omega$ is 6 V.
(b) $R_{E1} + R_{E2} = 100\ \Omega$.

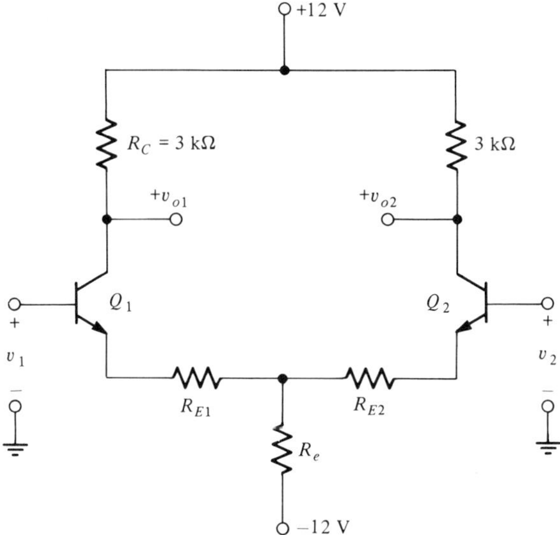

Figure E1.1 Differential amplifier circuit for Exercises 1.1 and 1.2.

1.2. Consider the same circuit in Fig. E1.1. Assume that $\beta_1 = 50$, $\beta_2 = 100$, $R_{E1} = R_{E2} = 50\ \Omega$, and $R_e = 2.7\ \text{k}\Omega$.
(a) Determine all the dc bias currents and voltages in the circuit and whether Q_1 and Q_2 are in the active region.
(b) Assume that the small-signal equivalent circuit for each transistor is in the form of the circuit in Fig. 1.8. Obtain the parameters of the small-signal equivalent circuits for all the transistors. Assume that $V_T = 25$ mV, $V_A = 120$ V, and $I_{CO} = 0.01\ \mu$A.
(c) Use the equivalent circuits obtained in part (b) and determine the following quantities for the differential amplifier assuming that v_{o1} is the output: (i) A_d, (ii) A_{cm}, (iii) ρ_{cm}, (iv) $R_{i1} = v_1/i_1$ when $v_2 = 0$, and (v) $R_{i2} = v_2/i_2$ when $v_1 = 0$.

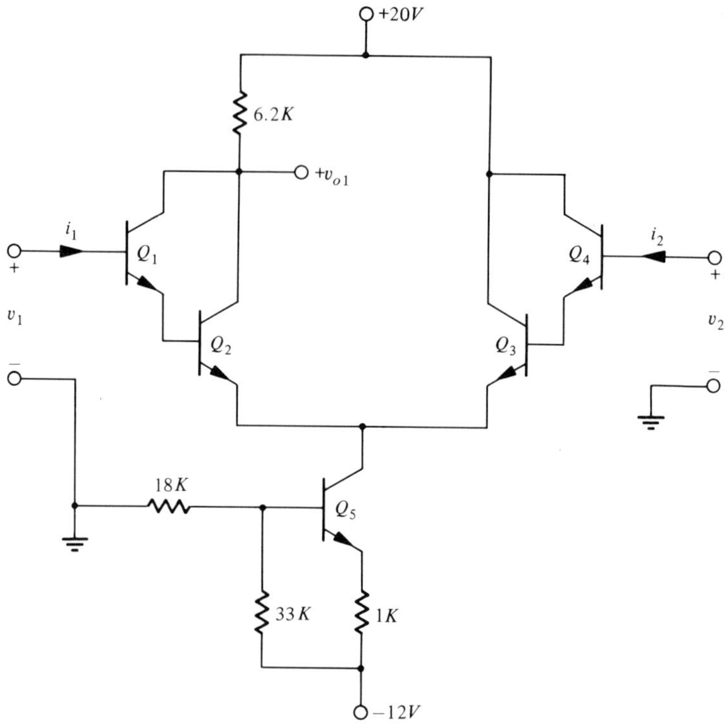

Figure E1.3 Differential amplifier for Exercise 1.3.

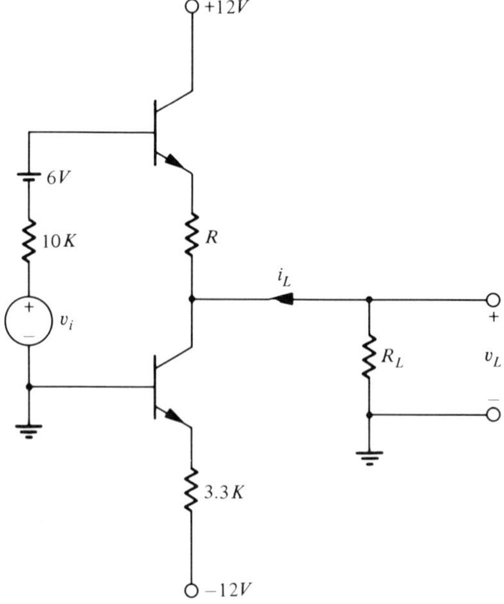

Figure E1.4 Level-shifting circuit for Exercise 1.4.

1.3. A differential amplifier is shown in Fig. E1.3. Assume that all the transistors are identical, $\beta = 50$ for each transistor, v_{o1} is the output, and $V_{BE} = 0.7$ V. Repeat Exercise 1.2.

1.4. In the level-shifting circuit in Fig. E1.4, $\beta_1 = 100$ and $\beta_2 = 50$. Also, assume that $V_{BE1} = V_{BE2} = 0.7$ V.
(a) Find the value of R such that $v_L = 0$ when $v_1 = 0$.
(b) Use this value of R and determine all the bias currents and voltages.
(c) Use the approximate equivalent circuit in Fig. 1.8 for each transistor and determine (i) v_L/v_1 when $i_L = 0$ and (ii) $R_o = v_L/i_L$ when $v_1 = 0$.

1.5. Consider the level-shifting circuit shown in Fig. E1.5. It uses a current mirror to bias Q_1. Assume that $V_{BE} = 0.7$ V for all transistors.
(a) Find the value of R such that $v_o = 0$ when $v_i = 0$.
(b) Find the small-signal voltage gain v_o/v_i when $i_o = 0$.
(c) Find the output resistance R_o of this circuit, i.e., find v_o/i_o when $v_i = 0$.
(d) Find v_o/v_i when $R_L = 15$ kΩ.

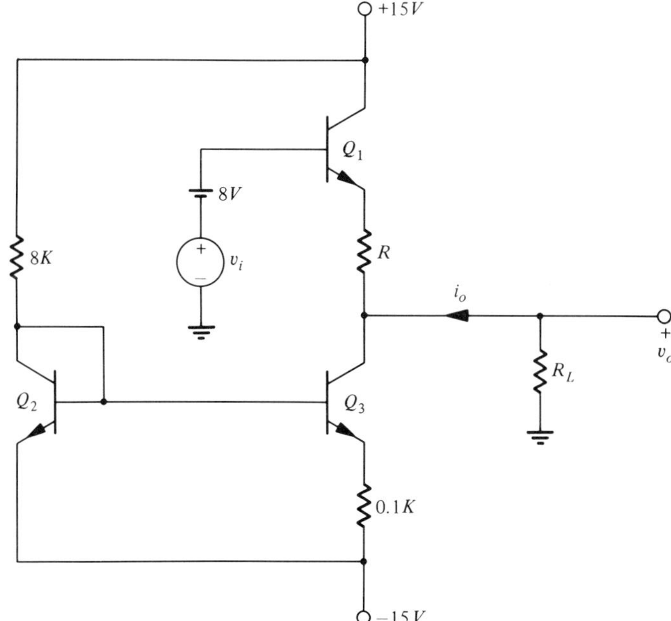

Figure E1.5 Circuit for Exercise 1.5.

1.6. A current source and a current mirror are shown in Fig. E1.6. Assume that $V_{BE} = 0.7$ V and $h_{FE} = 100$ for all the transistors. Find the values of I_{C1}, I_{C2}, and the ratio I_{C1}/I_{C2}.

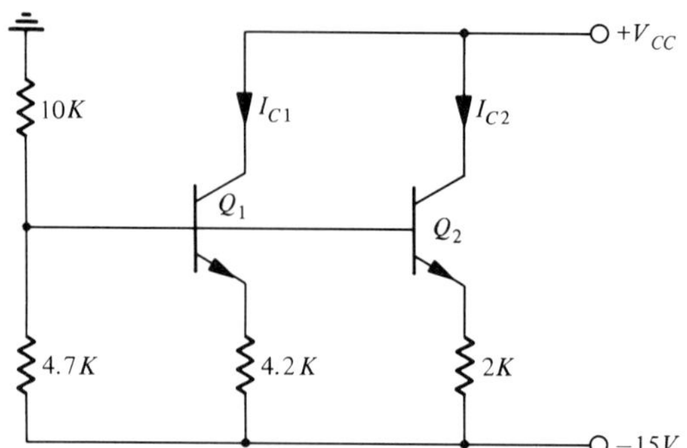

Figure E1.6 Current mirror circuit for Exercise 1.6.

1.7. The circuit shown in Fig. E1.7 is a simple op amp circuit. All the transistors are identical, and $I_C = I_E$ in each transistor ($I_B = 0$). Assume that $V_{BE} = 0.7$ V for each transistor.

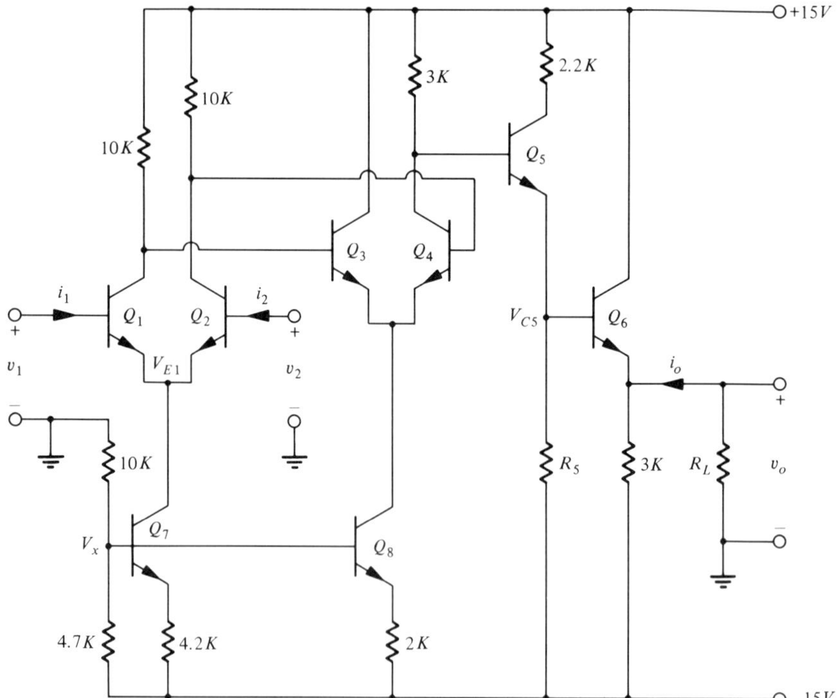

Figure E1.7 Simple op amp circuit for Exercise 1.7.

(a) Perform the dc analysis and determine the value of R_5 such that $v_o = 0$ when $v_1 = v_2 = 0$. Use this value of R_5 to find all dc bias currents and voltages and determine whether all the transistors are in the active region.

(b) Use a value of $\beta = 100$ for all the transistors and bias currents as determined in part (a) to find the small-signal equivalent circuit of each transistor in the form of Fig. 1.8.

(c) Identify the four stages indicated in the circuit. Perform an ac analysis of each stage and find the input and output resistances and the open-circuit gain of each stage. In this way, obtain a simplified equivalent circuit for this amplifier circuit similar to the one shown in Fig. 1.43 (but not the same). Use the simplified equivalent circuit and find the open-circuit differential mode gain and the input and output resistances.

1.8. An actively compensated noninverting finite gain amplifier is shown in Fig. 4.16*b*. Assume that $R_a = (1 - \beta)R$ and $R_b = \beta R$ in this circuit, where R is any arbitrary value. The op amps in this circuit are assumed to be ideal except for their finite gains. The finite gain of each op amp is described by

$$A_1 = A_2 = \frac{10^5}{(1 + s\tau_1)(1 + s\tau_2)}$$

where $\tau_1 = 0.1/2\pi$ s/r and $\tau_2 = 1/2\pi$ μs/r.

(a) Break open the feedback at X and find the forward path and loop gains. Find the overall gain V_o/V_i also.

(b) Obtain Bode plots of the loop gain (both magnitude and phase plots) and determine the phase and gain margins when $\beta = \frac{2}{3}, \frac{1}{2}$, and $\frac{1}{3}$.

(c) Find the closed-loop gains for the values of β when A_1 and A_2 tend to ∞.

1.9. An inverting integrator is shown in Fig. E1.9 with a load resistor. The op amp is ideal except for its finite gain, the nonzero output resistance, and input stray capacitances. The output resistance of the op amp is $R_o = 100\ \Omega$. The input stray capacitance from every input terminal to the ground is 5 pF each.

(a) Find the loop gain and the phase and gain margins when (i) $A = 10^5/(1 + 0.0002s)$ and (ii) $A = 10^5/(1 + 0.01s)$.

(b) Find the frequency at which the closed-loop gain passes through the unity gain in each of the above cases. What is the ratio of this frequency to $1/RC$?

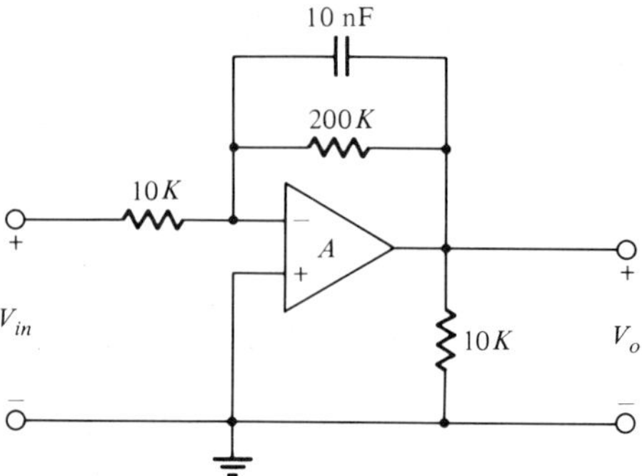

Figure E1.9 Inverting integrator circuit for Exercise 1.9.

1.10. The circuit shown in Fig. E1.10 is an inverting integrator. The op amps are ideal except for their finite gains, and both op amps have the same gain as described in part (ii) of Exercise 1.9a.

(a) Find the phase and gain margins in this circuit.

(b) Also, find the unity gain crossover frequency of the closed-loop gain. Find the ratio of this frequency to $1/RC$.

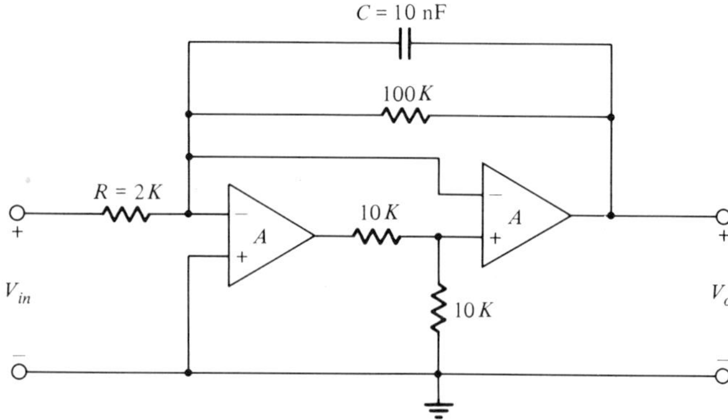

Figure E1.10 Inverting integrator circuit for Exercise 1.10.

1A

Spice Program Listings of Small-Signal Circuits of the Op Amp Circuit

```
*OP. AMP. INPUT STAGE SMALL-SIGNAL ANALYSIS

****      INPUT LISTING       TEMPERATURE = 27.000 DEG C
********************************************************


RE5 1 0 1.3842K
RBE4 1 0 69.212K
RBE6 3 5 553.7K
RBE7 4 6 553.7K
RBE8 2 5 138.42K
RBE9 2 6 138.42K
RBE10 7 11 480.31K
RBE11 11 9 553.7K
RBE12 11 10 553.7K
RCE4 2 0 6.6444MEG
RCE6 1 5 13.289MEG
RCE7 1 6 13.289MEG
RCE8 7 5 13.289MEG
RCE9 8 6 13.289MEG
RCE10 0 11 11.527MEG
RCE11 7 9 13.289MEG
RCE12 8 10 13.289MEG
GM4 2 0 1 0 0.72242MS
GM6 1 5 3 5 0.36121MS
GM7 1 6 4 6 0.36121MS
GM8 7 5 2 5 0.36121MS
GM9 8 6 2 6 0.36121MS
GM10 0 11 7 11 0.4164MS
GM11 7 9 11 9 0.36121MS
GM12 8 10 11 10 0.36121MS
R3 9 0 1K
R4 10 0 1K
R5 11 0 50K
.WIDTH IN=72 OUT=52
VIN 3 4 DC 1
.TF V(8,0) VIN
.END
*******06/22/86******* SPICE 2G.1(15OCT86)    ******* 10.09.03

*OP. AMP. INPUT STAGE SMALL-SIGNAL ANALYSIS

****      SMALL-SIGNAL CHARACTERISTICS

     V(8)/VIN                                  = -1.919D+03
     INPUT RESISTANCE AT VIN                   =  3.614D+06
     OUTPUT RESISTANCE AT V(8)                 =  1.071D+07
```

```
*OP. AMP. INTERMEDIATE STAGE SMALL-SIGNAL ANALYSIS

****       INPUT LISTING       TEMPERATURE = 27.000 DEG C
*********************************************************



RBE15 1 2 310.56K
RE15 0 2 50K
RCE15 2 0 7.453MEG
RBE16 2 3 9.43K
RE16 3 0 100
RCE16 3 4 224.23K
RCE13 4 0 225.73K
GM15 0 2 1 2 0.644MS
GM16 4 3 2 3 21.406MS
.WIDTH IN=72 OUT=52
VIN 1 0 DC 1
.TF V(4,0) VIN
.END
*******06/22/86******* SPICE 2G.1(15OCT86)    ******* 16.10.48

*OP. AMP. INTERMEDIATE STAGE SMALL-SIGNAL ANALYSIS

****      SMALL-SIGNAL CHARACTERISTICS

     V(4)/VIN                                   = -1.044D+03
     INPUT RESISTANCE AT VIN                    =  3.134D+06
     OUTPUT RESISTANCE AT V(4)                  =  1.665D+05
```

```
*OP. AMP. OUTPUT STAGE SMALL-SIGNAL ANALYSIS

****       INPUT LISTING       TEMPERATURE = 27.000 DEG C
*********************************************************



RE18 3 4 1.612K
RBE17 1 2 7.054K
RBE19 3 2 17.44K
RBE20 4 6 33.285K
RBE21 2 5 8.3211K
RCE14 4 0 677.2K
RCE17 2 0 690.74K
RCE19 4 2 742.12K
RCE20 6 0 798.83K
RCE21 5 0 793.83K
GM17 0 2 1 2 6.949MS
GM19 4 2 3 2 6.468MS
GM20 0 6 4 6 6.009MS
GM21 0 5 2 5 6.009MS
R9 6 7 27
R10 5 7 27
.WIDTH IN=72 OUT=52
VIN 1 0 DC 1
.TF V(7,0) VIN
.END
*******06/22/86******* SPICE 2G.1(15OCT86)    ******* 16.10.48

*OP. AMP. OUTPUT STAGE SMALL-SIGNAL ANALYSIS

****      SMALL-SIGNAL CHARACTERISTICS

     V(7)/VIN                                   =  9.993D-03
     INPUT RESISTANCE AT VIN                    =  1.688D+07
     OUTPUT RESISTANCE AT V(7)                  =  9.761D+01
```

2

Approximation Problem and Transformations

THE USE OF MOST NETWORKS in engineering applications is to process electrical signals. Such signal processing networks are also called *filters*. Thus in engineering practice every network can be thought of as a filter network. When a signal is applied to the input of a filter network, a certain band of frequencies is emphasized in the output and another band or bands of frequencies are highly attenuated. Since the input-output characteristic of any network can be described by its transfer function, this function plays an important role. In particular, the magnitude plot of a transfer function with $s = j\omega$ (for the sinusoidal steady state) clearly reveals the input-output response of the network. For example, the magnitude plot of a second-order transfer function is shown in Fig. 2.1. Any network that realizes such a transfer function allows passage of the frequency components centered around $\omega = 1$ r/s, whereas low- and high-frequency components are relatively attenuated in the output. Any such network can be used to select a band of frequencies centered around $\omega = 1$ r/s.

The first phase of a filter design involves the *approximation problem*, which is to obtain an appropriate transfer function satisfying a given set of specifications. Before we consider this problem, we shall consider the major classifications of signal processing networks. In this way, we shall introduce some of the terminology used in discussing this topic.

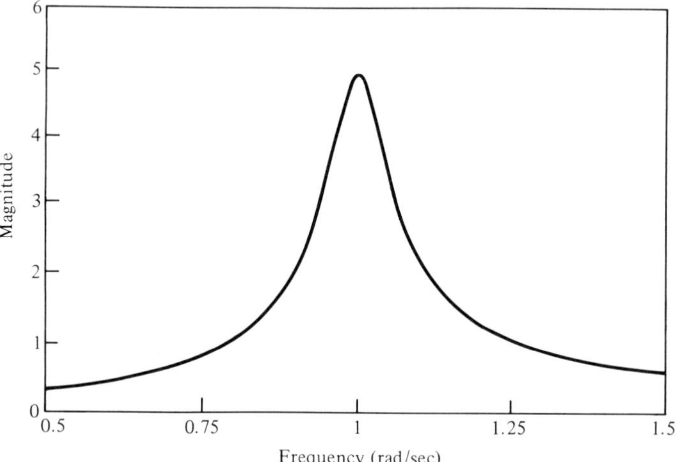

Figure 2.1 Magnitude response of $0.5s/(s^2 + 0.1s + 1)$.

2.1 *Classification of Signal Processing Networks*

Four standard filter characterizations will be described next. The first of these is the *lowpass* (LP) *filter*, whose ideal magnitude characteristic is shown in Fig. 2.2*a*. This type of characteristic is known as the "brick wall" characteristic of a lowpass filter. The ideal lowpass filter, as shown in Fig. 2.2*a*, allows all frequency components from $\omega = 0$ to $\omega = \omega_c$ to pass and attenuates completely all frequency components from $\omega = \omega_c$ to $\omega = \infty$. The range of low frequencies whose components are passed through the network is called the *bandwidth*, and ω_c is the cutoff frequency. A finite network *cannot* realize such a brick wall magnitude response, and physical networks can realize only approximate versions of this ideal characteristic. Three such approximations are shown in Fig. 2.2*b* through *d*, and there are other types of approximations as well.

In Fig. 2.2*b*, the magnitude response decreases smoothly from a maximum value at $\omega = 0$ to zero as $\omega \to \infty$. The frequency at which the magnitude characteristic drops to a specified percentage of the maximum is called the *passband edge frequency* ω_p, and the range of frequencies from 0 to ω_p is called the *passband*. In Fig. 2.2*c* and *d*, there is a kind of ripple in the passband before it goes to zero. The band of frequencies over which the response is below a specified value is called the *attenuation band* or *stopband*. In Fig. 2.2*c*, the response characteristic is monotonically decreasing, whereas there is a ripple in the attenuation band in Fig. 2.2*d*. The lowpass filter characteristic shown in Fig. 2.2*b* and *c* can be realized by networks with network functions having all their zeros at ∞ (called *all-pole functions*),

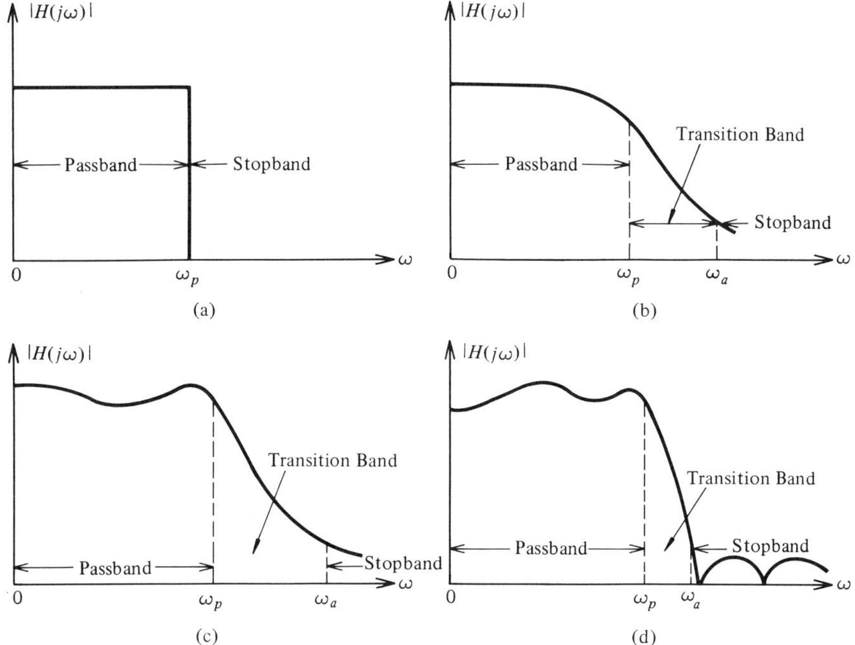

Figure 2.2 Magnitude characteristics of lowpass filters.

which are of the following form:

$$H(s) = \frac{H_0}{D(s)}$$

where H_0 is a constant. For example, a network function of the form

$$H(s) = \frac{1}{s^3 + 2s^2 + 2s + 1} = \frac{1}{(s + 1)(s^2 + s + 1)}$$

is an all-pole function.

In Fig. 2.2*d*, the magnitude characteristic goes to zero at finite frequencies, and the network functions for realizing such a characteristic must have zeros located on the $j\omega$ axis of the s plane and must be of the form

$$H(s) = \frac{N(s)}{D(s)}$$

Finally, the magnitude characteristic shown in Fig. 2.2*b* is described as being *maximally flat*. For the passband magnitude characteristic in Fig. 2.2*c*, all the maximum and minimum values of $|H(j\omega)|$ in the passband are

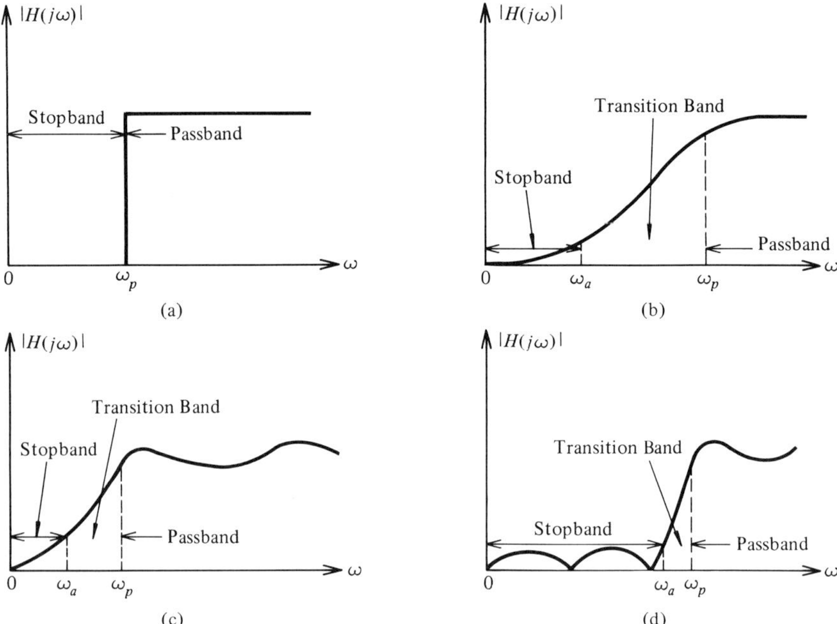

Figure 2.3 Magnitude characteristics of highpass filters.

the same. For this reason such a magnitude characteristic is called *equiripple*. The type of characteristic shown in Fig. 2.2*d* is known as an *elliptic* characteristic for reasons that will become clear later.

The next type of filter is called a *highpass* (HP) *filter*; its ideal and approximate magnitude characteristics are shown in Fig. 2.3. As one can easily see from Fig. 2.3, this filter allows the band of frequencies beyond ω_p up to $\omega = \infty$ (the passband), while low frequencies (the stopband) are highly attenuated. The magnitude response shown in Fig. 2.3*b* is maximally flat, and the one shown in Fig. 2.3*c* is equiripple. Both these responses require network functions of the form

$$H(s) = \frac{H_0 s^n}{D(s)}$$

where H_0 is a constant and n is the degree of the denominator polynomial $D(s)$. All the zeros are located at $s = 0$ (the origin). The network function related to the elliptic characteristic shown in Fig. 2.3*d* should have the form

$$H(s) = \frac{N(s)}{D(s)}$$

where $N(s)$ has all its zeros on the $j\omega$ axis and both $N(s)$ and $D(s)$ are of the same order.

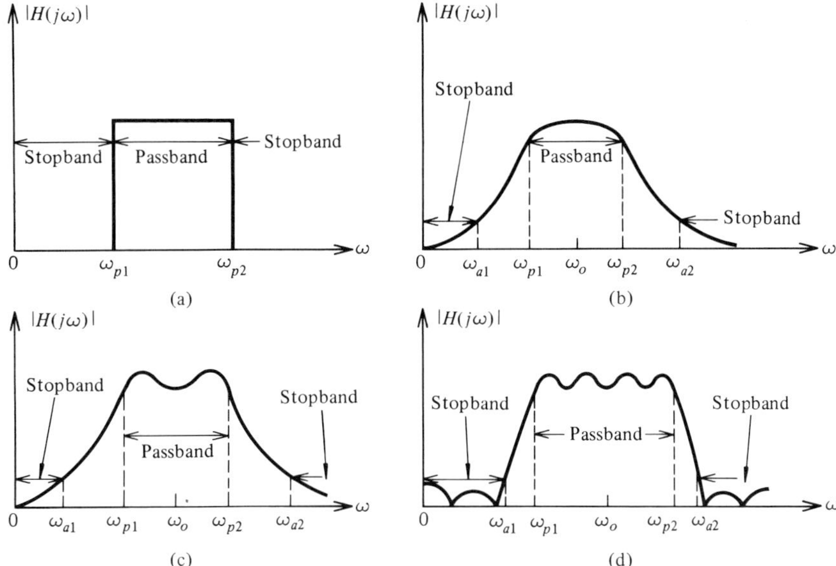

Figure 2.4 Magnitude characteristics of bandpass filters.

The third general type of filter is known as a *bandpass* (BP) *filter*, and its magnitude response characteristics are shown in Fig. 2.4. This type of filter allows a band of frequencies (the passband) to be passed from the input to thc output and attenuates heavily below and above the passband. The band of frequencies passed is called the bandwidth and is defined as the difference between the two passband edge frequencies ω_{p1} and ω_{p2}:

$$B = \omega_{p2} - \omega_{p1}$$

Usually the center frequency ω_0 is the geometric mean of the passband edge frequencies. Thus

$$\omega_0 = \sqrt{\omega_{p1}\omega_{p2}}$$

The maximally flat and equiripple characteristics shown in Fig. 2.4*b* and *c* have network functions of the form

$$H(s) = \frac{H_0 s^{n/2}}{D(s)}$$

where n is the order of $D(s)$. This means that half of the zeros are located at the origin and the other half at $s = \infty$. However, the elliptic characteristic shown in Fig. 2.4*d* requires $j\omega$-axis zeros, as in the previous cases.

The last type of filter is known as a *bandstop* (BS) *filter*. The attenuation characteristics of this filter are complementary to those of the bandpass

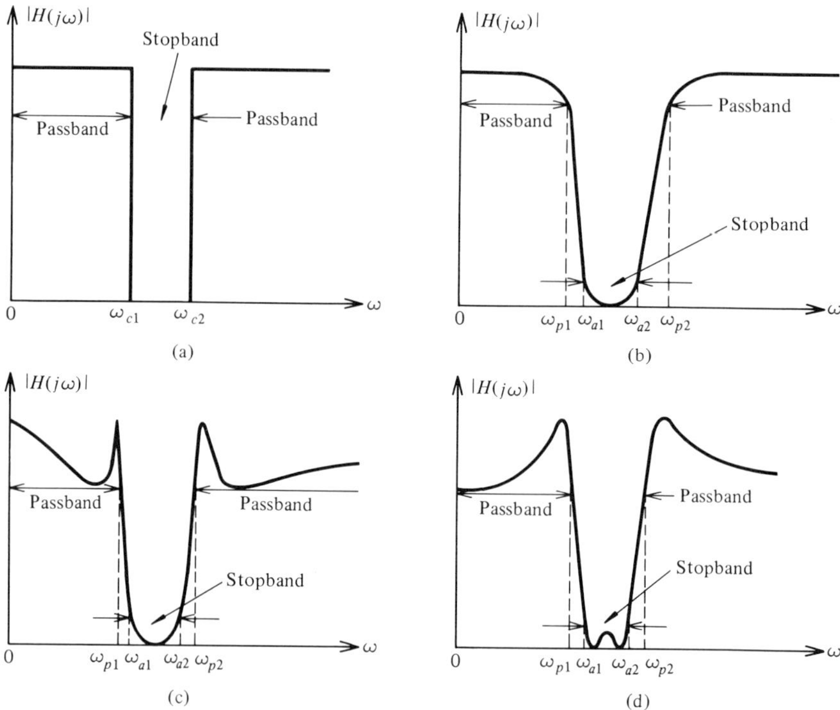

Figure 2.5 Magnitude characteristics of bandstop filters.

filter, and its magnitude characteristics are shown in Fig. 2.5. A bandstop filter allows passage of both low- and high-frequency ends without attenuation. The bandwidth $(\omega_{a2} - \omega_{a1})$ is the stopband width at which the gain is below a specified value, which is usually small. The ideal stopband gain is zero. The degrees of the numerator and denominator polynomials of the network function of a bandstop filter are equal, and the network function has as many zeros on the $j\omega$ axis as the number of poles.

2.2 *Approximation Problem*

In the first phase of a filter design a network function is obtained that satisfies a given set of specifications in terms of frequency response. In most filter design problems, a given set of requirements must be satisfied in terms of magnitude response, but the specifications are usually given in terms of loss in decibels rather than in magnitude. Specifications may also be given for satisfying a given phase characteristic. This will be a phase approximation problem, however, we will concentrate only on the magnitude approximation problem here.

The specifications given for lowpass, highpass, bandpass, and bandstop filters can be translated into another set of requirements for an equivalent *normalized* lowpass filter characteristic by using frequency transformations. Once we obtain a network function satisfying a set of specifications for a normalized lowpass filter characteristic, by using frequency transformations, we can alter the network function to meet other types of filter characteristics and specifications. This implies that basically we need to know how to obtain a network function satisfying the specifications for a normalized lowpass filter. Therefore we next consider the approximation problem of a normalized, or *prototype*, lowpass filter. We consider three different classic approximation techniques, namely, maximally flat, Chebyshev, and elliptic. But before we describe the three individual approximations, we will outline the basic steps common to all three types.

An analog network function $H(s)$ can be represented as a ratio of two polynomials:

$$H(s) = \frac{N(s)}{D(s)} \tag{2.1}$$

Such a function is called a rational function. In the normalized frequency domain, for sinusoidal steady-state operation, the magnitude function can be obtained as $|H(j\Omega)|$, where Ω is the real frequency. The loss or attenuation of the network function is defined as

$$A(\Omega) = -20\log|H(j\Omega)| \tag{2.2}$$

Plots of the loss function $A(\Omega)$ of a prototype lowpass filter function are shown in Fig. 2.6. Here we have assumed the loss to be 0 dB at $\omega = 0$ (reference loss) in all three cases. The first, shown in Fig. 2.6*a*, is the maximally flat response. The second and third, shown in Fig. 2.6*b* and *c* are, respectively, the equiripple and elliptic characteristics. Specifications are usually given for the filter characteristic to be satisfied in terms of loss in certain frequency ranges. For a lowpass filter, the specifications are

1. Maximum allowed passband loss A_p in the frequency range $0 \le \Omega \le \Omega_p$
2. Minimum required stopband loss A_a in the frequency range $\Omega \ge \Omega_a$.

These quantities are also indicated in Fig. 2.6. The approximation problem is essentially to find a rational function $H(s)$ that satisfies the above specifications given in terms of A_p, Ω_p, A_a, and Ω_a using the three classic approximations. In these types of problems, it is convenient to work with the squared magnitude function $|H(j\Omega)|^2$. First we note that the loss function $A(\Omega)$ is

$$A(\Omega) = -10\log|H(j\Omega)|^2 \tag{2.3}$$

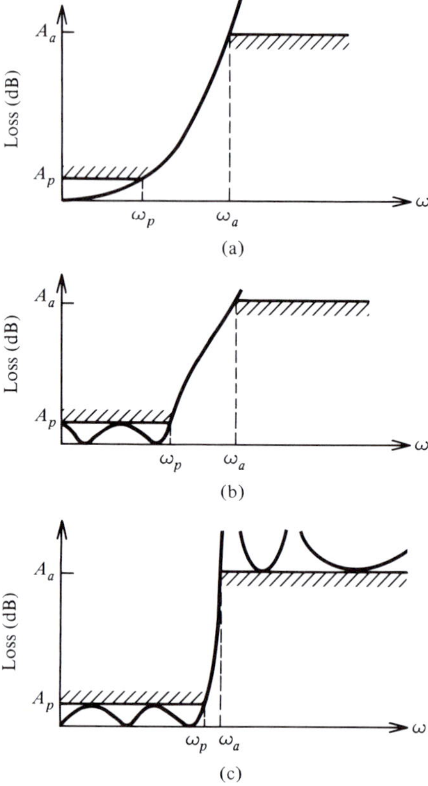

Figure 2.6 Three classical approximations of a lowpass filter and the specifications.

We can write $|H(j\Omega)|^2$ as

$$\begin{aligned} |H(j\Omega)|^2 &= H(j\Omega)H^*(j\Omega) \\ &= H(j\Omega)H(-j\Omega) \end{aligned} \tag{2.4}$$

If $H(s)$ is a rational function, using (2.3) we can show that $|H(j\Omega)|^2$ is a rational function in Ω^2 and an even function in Ω. This information will be used later in selecting the functions for $|H(j\Omega)|^2$. In order to see how $H(s)$ can be obtained from $|H(j\Omega)|^2$, assume that $|H(j\Omega)|^2 = T(\Omega^2)$:

$$T(\Omega^2)|_{\Omega=s/j} = T(-s^2) \tag{2.5}$$

and, by employing analytic continuation, we obtain

$$T(-s^2) = H(s)H(-s) \tag{2.6}$$

Assume that $T(\Omega^2)$, hence $T(-s^2)$, is known. How do we find $H(s)$? Since

$T(-s^2)$ is known, we can find the poles and zeros of the function $T(-s^2)$, and thus it can be written as

$$T(-s^2) = T_o \frac{(s+z_1)(s-z_1)(s+z_2)(s-z_2)\cdots}{(s+p_1)(s-p_1)(s+p_2)(s-p_2)\cdots} \tag{2.7}$$

Note that in (2.7), for a pole $s = -p_i$ in $T(-s^2)$, there is a corresponding pole $s = p_i$. The same is true with respect to the zeros. This is due to the fact that $T(-s^2)$ is an even function of s. Therefore, if $s = -p_i(-z_i)$ is assigned to $H(s)$, $s = p_i(z_i)$ can automatically be assigned to $H(-s)$, and this is suggested by (2.6). Thus half of the poles and zeros of $T(-s^2)$ are assigned to $H(s)$, and the other half are automatically assigned to $H(-s)$. Of course, in this process, in order for $H(s)$ to have real numerator and denominator coefficients, we assign complex conjugate pairs of poles and zeros to the same group. Further, for stability reasons, poles with negative real parts are assigned to $H(s)$. As far as zeros are concerned, we can assign zeros with positive or negative real parts to $H(s)$. However, as we shall see later, the zero assignment problem does not arise in all three classic approximations of the lowpass filter characteristic.

2.3 *Maximally Flat Approximation*

A lowpass magnitude squared function $T(\Omega^2)$ has a maximally flat characteristic when as many derivatives of $T(\Omega^2)$ at $\Omega = 0$ as possible are zeros. We start with such a squared magnitude function of the form

$$|H(j\Omega)|^2 = T(\Omega^2) = \frac{H_o^2}{1 + \varepsilon^2\Omega^{2N}} \tag{2.8}$$

The first $2N - 1$ derivatives of $|H(j\Omega)|^2$ at $\Omega = 0$ are zeros. Also, note that $|H(j\Omega)|^2$ is an even function in Ω as required and that $|H(j\Omega)|$ reaches its maximum value of H_o at $\Omega = 0$. The quantity ε is called the *ripple factor* and, as will be seen later, N is the order of the network function $H(s)$. These two quantities, ε and N, control the passband loss and the rate at which the loss characteristic rises with the frequency Ω, and the values of these two quantities must be determined before we can use (2.8) to determine the poles and zeros of the network function $H(s)$. For this purpose, we use the given set of specifications to determine the values of ε and N. First, we find the loss $A(\Omega)$ as

$$A(\Omega) = 10\log(1 + \varepsilon^2\Omega^{2N}) - 20\log(H_o) \tag{2.9}$$

The value of H_o can be used to fix the gain or loss requirement at a given frequency. Usually, for a lowpass filter, the gain requirement is specified at

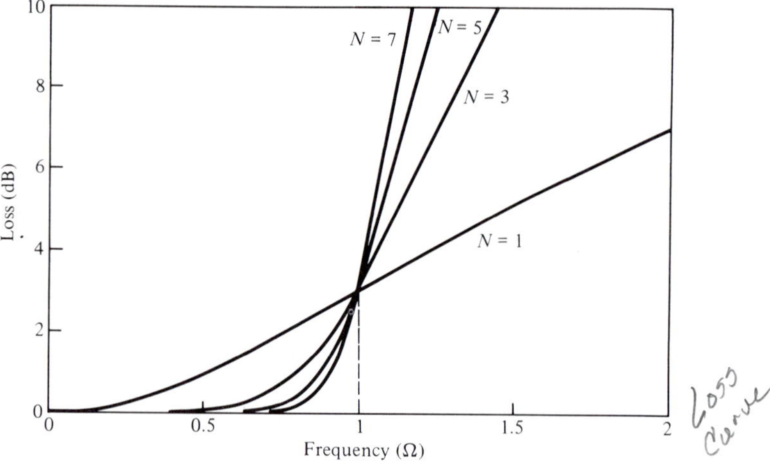

Figure 2.7 Plots of (2.10) with $\varepsilon = 1$ and $N = 1, 3, 5, 7$.

Loss Curve

$\Omega = 0$. This requirement can be satisfied even after finding the network function $H(s)$, which is also the most convenient procedure. Furthermore, in most applications we are interested only in the relative loss, and therefore, for now, we shall choose $H_o = 1$. Thus $A(\Omega)$ becomes

$$A(\Omega) = 10 \log(1 + \varepsilon^2 \Omega^{2N}) \tag{2.10}$$

Plots of (2.10) with $\varepsilon = 1$ and various values of N are shown in Fig. 2.7. From these plots, one can easily see that (1) the loss increases monotonically with Ω, and (2) the loss characteristic becomes a better approximation of the ideal lowpass filter with increasing values of N. Using the first observation, we conclude that the maximum passband loss occurs at $\Omega = \Omega_p$. Further, in a normalized filter design, we assume that $\Omega_p = 1$ r/s. Then, using (2.10), we obtain the value of ε as

$$\varepsilon = \sqrt{10^{0.1A_p} - 1} \tag{2.11}$$

If a minimum stopband loss of A_a decibels is expected at and above the frequency Ω_a, again using (2.10), we have

$$A_a \leq 10 \log\left(1 + \varepsilon^2 \Omega_a^{2N}\right) \tag{2.12}$$

Substituting (2.11) into (2.12) for ε and solving the inequality, we obtain the following inequality for N:

$$N \geq \frac{\log\left[(10^{0.1A_a} - 1)/(10^{0.1A_p} - 1)\right]}{2 \log(\Omega_a)} \tag{2.13}$$

Since the order of the network function must be an integer value, we choose the least integer value for N that satisfies (2.13).

Maximally flat approximation with $\varepsilon = 1$ is called *Butterworth approximation*, and the corresponding filter functions are called *Butterworth functions*. After determining the values of ε and N using (2.11) and (2.13), $T(\Omega^2)$ as given by (2.8) is completely defined. The next step is to determine the poles and zeros of $T(-s^2)$ from which we can find the poles and zeros of $H(s)$. Since the numerator is only a constant, all the zeros are located at $s = \infty$ and we need only determine the poles of $T(-s^2)$ and pick out those in the left half of the s plane. Thus substituting $\Omega = s/j$ into the denominator of (2.8) and setting it equal to zero, we obtain

$$s^{2N} + \frac{(-1)^N}{\varepsilon^2} = 0 \tag{2.14}$$

The roots of the above characteristic equation must be the poles of $T(-s^2)$:

$$p_k = R \exp\left[j\left(\frac{\pi}{2} + \frac{2k-1}{2N}\pi\right)\right], \qquad k = 1, 2, 3, \ldots, 2N$$

where

$$R = \varepsilon^{-1/N}$$

The poles $s = p_k$, $k = 1, 2, \ldots, N$, have negative real parts, which are the ones that should be assigned to $H(s)$. Thus the transfer function $H(s)$ is

$$H(s) = \frac{H_o}{\prod_{k=1}^{N} (s - p_k)} \tag{2.15}$$

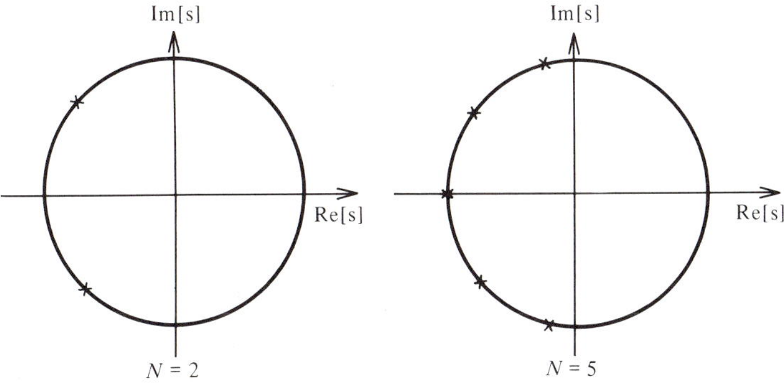

Figure 2.8 Pole positions of maximally flat network functions.

where the values of p_k are

$$p_k = R\exp\left[j\left(1+\frac{2k-1}{N}\right)\frac{\pi}{2}\right], \qquad k=1,2,\ldots,N$$

where

$$R = \varepsilon^{-1/N} \tag{2.16}$$

These roots lie equiangularly in a circle, as shown in Fig. 2.8, for $N = 2$ and $N = 5$.

Example 2.1: Find the network function $H(s)$ using maximally flat approximation to satisfy the following requirements for a lowpass filter:

$$f_p = 10\text{ kHz} \quad\text{and}\quad f_a = 20\text{ kHz}$$
$$A_p \le 0.5\text{ dB} \quad\text{and}\quad A_a \ge 30\text{ dB}$$

Although we only know how to design a normalized lowpass filter function with $\Omega_p = 1$ r/s, we can solve this problem with a simple frequency scaling. Let us use the frequency scaling $s = K_f p$, where s and p are two complex frequencies in the normalized and denormalized frequency domains, respectively, such that $s = j\Omega$ and $p = j\omega$. Then the real frequencies Ω and ω are also related through K_f such that $\Omega = K_f\omega$. Thus we have

$$\Omega_p = K_f\omega_p \quad\text{and}\quad \Omega_a = K_f\omega_a$$

From the above equations, since we know the values of ω_p and ω_a, we can find the values of K_f and Ω_a by imposing the condition $\Omega_p = 1$ r/s. Doing this for the problem, we obtain $K_f = 10^{-4}/2\pi$ and $\Omega_a = 2$ r/s. In fact, this frequency scaling is one of the frequency transformations called LP-LP transformation, which will be discussed later. Now, proceeding with the problem, we find

$$\varepsilon = \sqrt{10^{0.1A_p} - 1} = \sqrt{10^{0.05} - 1} = 0.349311$$

Using (2.13), we find the required order N:

$$N \ge \frac{\log\left[(10^3 - 1)/(10^{0.05} - 1)\right]}{2\log 2} = 6.5$$

Thus we choose the value $N = 7$. The value of R can immediately be found to be 1.162132. Then, using (2.16), we find the values of p_k:

$$p_7^* = p_1 = -0.258599 + j1.132995$$
$$p_6^* = p_2 = -0.724577 + j0.908591$$
$$p_5^* = p_3 = -1.047044 + j0.504230$$

and

$$p_4 = -1.162132$$

Then the normalized network function can be obtained using (2.15). From the realization point of view, we have to combine factors corresponding to complex conjugate poles, and the network function must be found in the form

$$H(s) = \frac{H_o}{D_o(s)\prod_{i=1}^{N_1}\left(s^2 + A_i s + B_i\right)} \tag{2.17}$$

where

$$D_o(s) = \begin{cases} s + \sigma_o & N \text{ odd} \\ 1 & N \text{ even} \end{cases}$$

and

$$N_1 = \begin{cases} (N-1)/2 & N \text{ odd} \\ N/2 & N \text{ even} \end{cases}$$

For our problem, $H(s)$ can be expressed in the form of (2.17), and the values of σ_o, A_i, and B_i are given in Table 2.1.

To find the lowpass network function satisfying the original specifications, we need only apply the frequency transformation $s = K_f p$ to (2.17), where K_f, for this example, is $10^{-4}/2\pi$, and $H(p)$ will also be in the form of (2.17) (where p is the variable):

$$H(p) = \frac{H_o}{D_o(p)\prod_{i=1}^{N_1}\left(p^2 + A_i p + B_i\right)} \tag{2.18}$$

Table 2.1 Coefficients of the factors for the normalized lowpass filter in Example 2.1[a]

i	A_i	B_i
1	0.517197	1.35055
2	1.44915	1.35055
3	2.09409	1.35055

[a]Ripple factor = 0.349311; σ_o = 1.16213. H_o = 2.86278 (for 0 dB maximum gain).

Table 2.2 Coefficients of the factors in Example 2.1[a]

i	A_i	B_i
1	3.24965E + 04	5.33176E + 09
2	9.10531E + 04	5.33176E + 09
3	1.31575E + 05	5.33176E + 09

[a] Ripple factor = 0.349311; σ_o = 7.30189E + 04. H_o = 11.06743 × 10^{33}.

However, the values of σ_o, A_i, and B_i will be different because of denormalization, and these coefficients are given in Table 2.2.

The value of H_o can be chosen to fix the dc gain. For example, if the dc gain is to be 1 (0 dB), then

$$H_o = \begin{cases} \sigma_o \prod_{i=1}^{N_1} B_i & N \text{ odd} \\ \prod_{i=1}^{N_1} B_i & N \text{ even} \end{cases}$$

The value of H_o is also given in Table 2.2 for this example. Furthermore, to verify the fact that this transfer function satisfies the given specifications, the plots of the loss characteristics are shown in Fig. 2.9 from which one can see that the specifications are indeed satisfied. ■

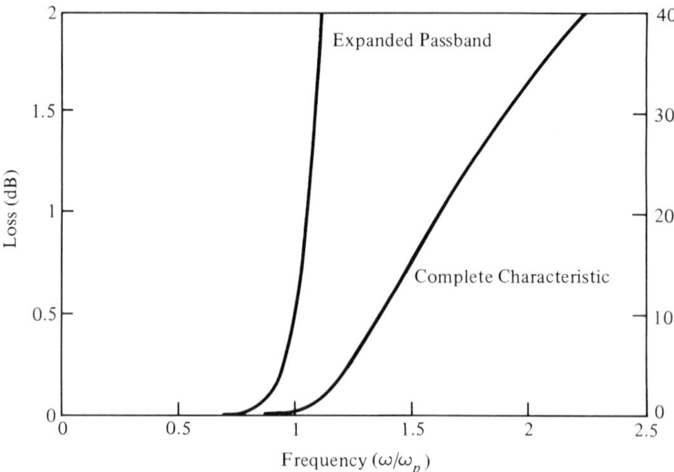

Figure 2.9 Loss characteristics of the network function designed in Example 2.1.

2.4 *Equiripple Approximation*

The *Chebyshev* approximation or what is called *equiripple* approximation uses a magnitude squared function of the form

$$|H(j\Omega)|^2 = T(\Omega^2) = \frac{H_o^2}{1 + \varepsilon^2 C_N^2(\Omega)} \tag{2.19}$$

where ε, as in the case of maximally flat approximation, is used to control the ripple in the passband and $C_N(\Omega)$ is the Chebyshev polynomial of order N. These polynomials have the property

$$0 \le C_N^2(\Omega) \le 1 \qquad 0 \le \Omega \le 1$$

and

$$C_N^2(\Omega) > 1 \qquad \Omega > 1$$

Starting with $C_0(\Omega) = 1$ and $C_1(\Omega) = \Omega$, we can obtain any higher-order Chebyshev polynomial using the following recursive relation:

$$C_N(\Omega) = 2\Omega C_{N-1}(\Omega) - C_{N-2}(\Omega)$$

with

$$C_0(\Omega) = 1 \qquad \text{and} \qquad C_1(\Omega) = \Omega$$

For example,

$$\begin{aligned} C_2(\Omega) &= 2\Omega C_1(\Omega) - C_0(\Omega) \\ &= 2\Omega^2 - 1 \end{aligned}$$

Another way of describing these polynomials is

$$C_N(\Omega) = \begin{cases} \cos[N\cos^{-1}(\Omega)] & 0 \le \Omega \le 1 \\ \cosh[N\cosh^{-1}(\Omega)] & \Omega \ge 1 \end{cases} \tag{2.20}$$

The plots of the loss characteristics with $\varepsilon = 1$ and $N = 2$ and 5 are shown in Fig. 2.10. The stopband loss near and above $\Omega = 1$ r/s is higher in this case compared to the filter functions of the same order in maximally flat approximation.

For a normalized filter design, we assume that $\Omega_p = 1$ r/s. If the maximum passband loss allowed is A_p (in decibels), substituting (2.20) into (2.19) and evaluating (2.19) at $\Omega = 1$ r/s, we have

$$A_p = 10\log(1 + \varepsilon^2)$$

where H_o is assumed to be unity for convenience, as in the previous case. Solving the above equation for ε, we obtain thc samc cquation as (2.11).

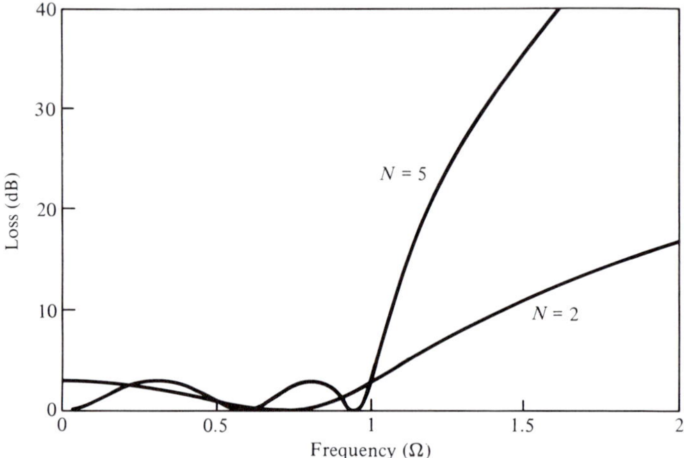

Figure 2.10 Equiripple loss characteristics with $\varepsilon = 1$, $N = 2$ and 5.

Therefore, to find the value of ε, we can still use (2.11). If the minimum stopband loss required is A_a at and above Ω_a r/s ($\Omega_a > 1$), again using (2.19), we have

$$A_a \leq 10 \log\left[1 + \varepsilon^2 \cosh^2\left(N \cosh^{-1}\Omega_a\right)\right]$$

Using (2.11) for ε and solving the above inequality for N, we have

$$N \geq \frac{\cosh^{-1}\left[(10^{0.1A_a} - 1)/(10^{0.1A_p} - 1)\right]^{0.5}}{\cosh^{-1}\Omega_a} \tag{2.21}$$

Since the value of N gives the order of $H(s)$, it must be an integer and therefore we choose the least integer value for N greater than or equal to the right side of (2.21).

Example 2.2: Determine the ripple factor and the order N of the network function that uses Chebyshev approximation to satisfy the following requirements:

$$A_p \leq 0.5 \text{ dB} \qquad \Omega_p = 1 \text{ r/s}$$

and

$$A_a \geq 30 \text{ dB} \qquad \Omega_a = 2 \text{ r/s}$$

The above specifications are the same as those in Example 2.1. The ripple factor is still determined by (2.11), and therefore $\varepsilon = 0.349311$. The order of the filter function can be chosen after evaluating the right side of (2.21):

$$N \geq 3.95$$

Therefore we choose the value of N as 4. Comparing this value with that of 7 in the previous case, one can easily see that the Chebyshev filter function is efficient in filtering out frequency components, and thus equiripple approximation requires a function of a lower order than that required by maximally flat approximation to satisfy a given set of specifications. ■

Once ε and N are known in (2.19), what we need to show next is how to find the poles of the corresponding network function. This can be done by setting the denominator of (2.19) equal to zero after substituting $\Omega = s/j$. After substituting (2.20) into (2.19) for $C_N(\Omega)$ and replacing Ω with s/j, we have

$$T(-s^2) = H(s)H(-s) = \frac{H_o^2}{1 + \varepsilon^2 \cos^2[N \cos^{-1}(s/j)]}$$

We can find the poles of $H(s)$ from the values of s that make the denominator zero and have negative real parts. Therefore, setting the denominator of the above to zero, we have

$$\cos\left[N \cos^{-1}\left(\frac{s}{j}\right)\right] = \frac{\pm j}{\varepsilon}$$

To solve the above equation, let us first define a new complex variable z such that

$$z = u + jv = \cos^{-1}\left(\frac{s}{j}\right) \tag{2.22}$$

Then,

$$\cos(Nz) = \cos(Nu + jNv)$$

and thus

$$\cos(Nu)\cosh(Nv) - j\sin(Nu)\sinh(Nv) = \frac{\pm j}{\varepsilon}$$

Since $\cosh(Nv) \geq 1$ for any v, equating the real and imaginary parts of the above equation, we have

$$\cos(Nu) = 0 \tag{2.23}$$

and, for any value of u satisfying (2.23),

$$v = \frac{1}{N}\sinh^{-1}\left(\frac{1}{\varepsilon}\right) \tag{2.24}$$

Solving for u from (2.23), we obtain

$$u_i = \frac{2i-1}{2N}\pi \qquad i = 1, 2, 3, \ldots, 2N \tag{2.25}$$

The poles of $H(s)$ are the values of $s = p_i$ satisfying (2.22). Therefore, using (2.22), it follows that

$$p_i = j\cos(u_i + jv) \tag{2.26}$$

where u_i and v are given by (2.25) and (2.24), respectively. Expanding the right side of (2.26), we obtain

$$p_i = \sin(u_i)\sinh(v) + j\cos(u_i)\cosh(v) \qquad i = 1, 2, 3, \ldots, 2N$$

where the values of u_i and v can be substituted from (2.25) and (2.24). We are interested only in one half of the poles that have negative real parts [the other half belong to $H(-s)$], and they are obtained from values $i = N + 1$ through $2N$ in the above equation. Thus retaining only these poles and substituting $i = N + k$, we find the required poles to be

$$p_k = -\sin\left[\frac{(2k-1)\pi}{2N}\right]\sinh(v) + j\cos\left[\frac{(2k-1)\pi}{2N}\right]\cosh(v)$$

$$k = 1, 2, \ldots, N \tag{2.27}$$

The poles of the network function obtained using Chebyshev approximation lie on an ellipse, as shown in Fig. 2.11 for $N = 2$ and $N = 5$. Once we find the poles of the required network function, then it is a simple matter to put the network function in the form of (2.17) by combining complex conjugate poles.

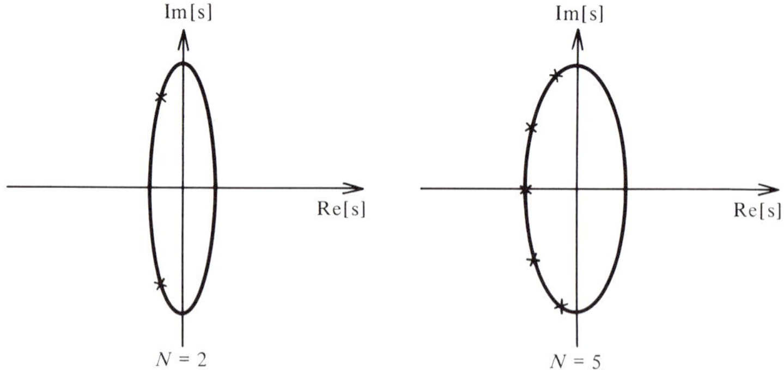

Figure 2.11 Pole positions of the Chebyshev network functions.

Table 2.3 Coefficients of the network function in Example 2.3[a]

i	A_i	B_i
1	0.350706	1.06352
2	0.84668	0.35641

[a] $H = 0.357847$ is required to give a maximum passband gain of 0 dB; ripple factor = 0.349311; order of filter function = 4.

Example 2.3: Using Chebyshev approximation, find the network function $H(s)$ that satisfies the specifications in Example 2.2.

From the solution for Example 2.2, we find that

$$\varepsilon = 0.349311 \qquad \text{and} \qquad N = 4$$

Then, using (2.24), we find that $v = 0.44354$. Using this value of v in (2.27), we have

$$p_4^* = p_1 = -0.17535 + j1.01625$$

and

$$p_3^* = p_2 = -0.42334 + j0.42095$$

For this example, $H(s)$ can be found in the form of (2.17), and the coefficients are given in Table 2.3.

To verify that the above network function satisfies the requirements, the loss characteristics of this network function are shown in Fig. 2.12, where one can see that the specifications are met. ■

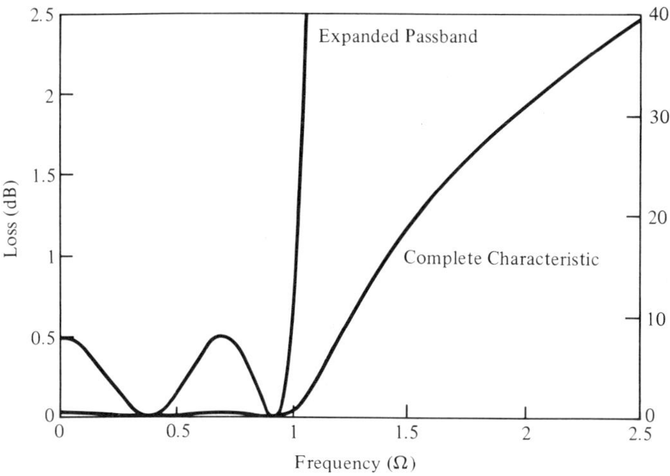

Figure 2.12 Loss characteristics of the network function designed in Example 2.3.

The design of network functions satisfying a prescribed set of specifications as in Examples 2.1 through 2.3 can be done using the computer program called MAXCHY in App. 2A.

2.5 *Elliptic Approximation*

Although Chebyshev approximation is more efficient than maximally flat approximation, an even more efficient magnitude approximation uses elliptic functions. The development of formulas for finding the poles and zeros for such approximations is more involved and therefore is not dealt with here. Interested readers can find the details in reference [1]. However, the steps involved in obtaining network functions using elliptic approximation will be given here. We first consider the design of a prototype normalized lowpass filter. The network function that results from using elliptic approximation is of the form

$$H_N(s) = \frac{H_o}{D_o(s)} \prod_{i=1}^{N_1} \frac{s^2 + C_i}{s^2 + A_i s + B_i} \tag{2.28}$$

where

$$N_1 = \begin{cases} (N-1)/2 & N \text{ odd} \\ (N/2) & N \text{ even} \end{cases}$$

and

$$D_o(s) = \begin{cases} s + \sigma_o & N \text{ odd} \\ 1 & N \text{ even} \end{cases}$$

and N is the order of the network function. Assume that Ω_p and Ω_a are the passband and stopband edge frequencies, respectively. Let the maximum passband loss allowed and the minimum stopband loss required be A_p and A_a, respectively. Then, the following sequence of design steps can be used to find the order and the coefficients of the network function [1]:

1. Calculate the selectivity factor k as

$$k = \frac{\Omega_p}{\Omega_a}$$

2. Use the following sequence of formulas to calculate q and D:

$$k' = \sqrt{1 - k^2}$$

$$q_o = \frac{1}{2}\,\frac{1 - \sqrt{k'}}{1 + \sqrt{k'}}$$

$$q = q_o + 2q_o^5 + 15q_o^9 + 150q_o^{13}$$

$$D = \frac{10^{0.1A_a} - 1}{10^{0.1A_p} - 1}$$

3. Choose the order of the network function N such that it is an integer and

$$N \geq \frac{\log 16D}{\log(1/q)}$$

4. Calculate the following auxiliary quantities in sequence:

$$D_1 = \frac{10^{0.05A_p} + 1}{10^{0.05A_p} - 1}$$

$$\lambda = \frac{\ln D_1}{2N}$$

$$\sigma_o = \frac{2q^{0.25} \sum_{m=0}^{\infty} (-1)^m q^{m(m+1)} \sinh[(2m+1)\lambda]}{1 + 2 \sum_{m=1}^{\infty} (-1)^m q^{m^2} \cosh(2m\lambda)}$$

$$W = \left(1 + k\sigma_o^2\right)\left(1 + \frac{\sigma_o^2}{k}\right)$$

5. Calculate A_i, B_i, and C_i using the following formulas:

$$\Omega_i = \frac{2q^{0.25} \sum_{m=1}^{\infty} (-1)^m q^{m(m+1)} \sin[(2m+1)\mu\pi/N]}{1 + 2 \sum_{m=0}^{\infty} (-1)^m q^{m^2} \cos(2m\mu\pi/N)}$$

$$\mu = \begin{cases} i & N \text{ odd} \\ i - 0.5 & N \text{ even} \end{cases} \qquad i = 1, 2, 3, \ldots, N_1$$

$$V_i = \left(1 - k\Omega_i^2\right)\left(1 - \frac{\Omega_i^2}{k}\right)$$

$$C_i = \frac{1}{\Omega_i^2}$$

$$B_i = \frac{V_i\sigma_o^2 + W\Omega_i^2}{\left(1 + \sigma_o^2\Omega_i^2\right)^2}$$

$$A_i = \frac{2\sigma_o\sqrt{V_i}}{1 + \sigma_o^2\Omega_i^2}$$

6. If a minimum loss of 0 dB is required, then the gain constant H_o can be chosen as

$$H_o = \begin{cases} \sigma_o \prod_{i=1}^{N_1} (B_i/C_i) & N \text{ odd} \\ 10^{-0.05A_p} \prod_{i=1}^{N_1} (B_i/C_i) & N \text{ even} \end{cases}$$

The infinite series involved in the calculation of σ_o and Ω_i converge rapidly, and three or four terms are sufficient for most purposes. For a given set of values for N, k, and A_p, the actual stopband loss will be

$$A_a = 10 \log\left(1 + \frac{10^{0.1A_p} - 1}{16q^N}\right) \tag{2.29}$$

The coefficients of the network function of (2.28) can be easily determined using the program called ELIPFT in App. 2B.

Example 2.4: Using elliptic approximation, design a lowpass network function that satisfies the specifications in Example 2.3.

Table 2.4 Parameters of the network function designs in Example 2.4[a]

i	A_i	B_i	C_i
1	0.537932	1.14858	5.15321

[a] $H = 0.154286$ is required to give a maximum passband gain of 0 dB; ripple factor = 0.349311; order of filter function = 3; $\sigma_o = 0.692218$.

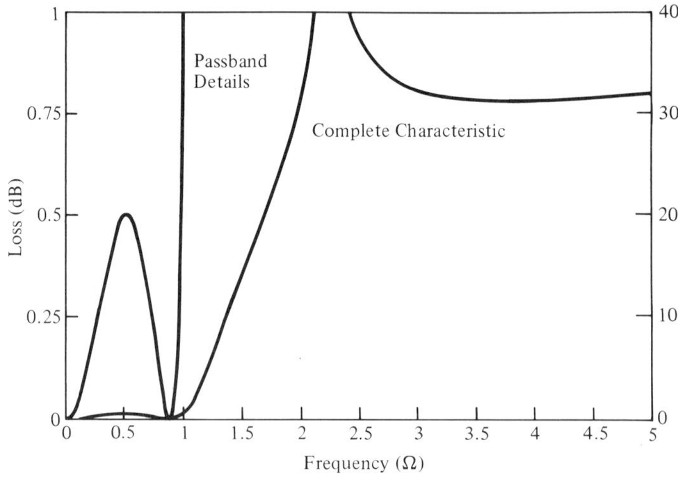

Figure 2.13 Loss characteristics of the network function designed in Example 2.4.

The network function has been designed using the program in App. 2B, and the results are given in Table 2.4. Note that we require an order of only 3, indicating that elliptic approximation is even more efficient than equiripple approximation.

The actual stopband loss is 31.19 dB, greater than that required. The loss characteristics of the network function are shown in Fig. 2.13 to prove that this network function indeed satisfies the requirements. ■

2.6 Frequency Transformations

In the preceding sections of this chapter, we considered only the design of lowpass filter functions and, in particular, the approximation of (prototype) normalized lowpass filter functions where $\Omega_p = 1$ r/s. Other types of filter functions with arbitrary frequency and loss specifications can be designed by using *frequency transformations*. We have already seen one form of frequency transformation, LP-LP transformation, in Example 2.1. This transformation is called *frequency scaling* because the shape of the loss characteristic remains the same except for the change in scale of the frequency axis.

In all future discussions of frequency transformation, we shall use the complex frequency variable s in the prototype LP network function and the complex frequency variable p in the network function that results after applying frequency transformation to the prototype LP network function. In addition, we shall use $s = j\Omega$ and $p = j\omega$, where Ω and ω are the real frequency variables in the above two cases, respectively.

Lowpass-Lowpass Transformation

Though we discussed this transformation in Example 2.1, for the sake of clarity and completeness we consider it again here. The LP-LP complex frequency transformation is of the form

$$s = \frac{p}{\omega_o} \tag{2.30}$$

where ω_o is a constant.

For a sinusoidal steady state with $s = j\Omega$ and $p = j\omega$, the above transformation implies that

$$\Omega = \frac{\omega}{\omega_o} \quad \text{or} \quad \omega = \omega_o \Omega \tag{2.31}$$

When we apply the transformation of (2.30) to any network function $H_N(s)$, the new network function is

$$H_N(s) = H_N\left(\frac{p}{\omega_o}\right)$$

Then, we have

$$|H_N(j\Omega)| = \left| H_N\left(\frac{j\omega}{\omega_o}\right) \right| = |H(j\omega)|$$

which means that $|H(j\omega)|$ at $\omega = \omega_o\Omega$ r/s will be the same $|H_N(j\Omega)|$ at Ω r/s. This implies that the loss value of $|H(j\omega)|$ at $\omega = \omega_o$ r/s will be the same as that of $|H(j\Omega)|$ at $\Omega = \Omega_p = 1$ r/s. Similarly, the loss value of $|H(j\omega)|$ at $\omega = \omega_o\Omega_a$ r/s will be the same as that of $|H_N(j\Omega)|$ at $\Omega = \Omega_a$r/s. That is, the passband edge of 1 r/s in $|H_N(j\Omega)|$ will become ω_o r/s in $|H(j\omega)|$ and, similarly, the stopband edge of Ω_a r/s in $|H_N(j\Omega)|$ becomes $\omega_o\Omega_a$ in $|H(j\omega)|$. The following network function $H_N(s)$ represents an LP prototype filter with a maximum passband loss of 3 dB and a minimum stopband loss of 30 dB:

$$H_N(s) = \frac{H_o(s^2 + 2.80601)}{(s + 0.348788)(s^2 + 0.2387668s + 0.885128)} \tag{2.32}$$

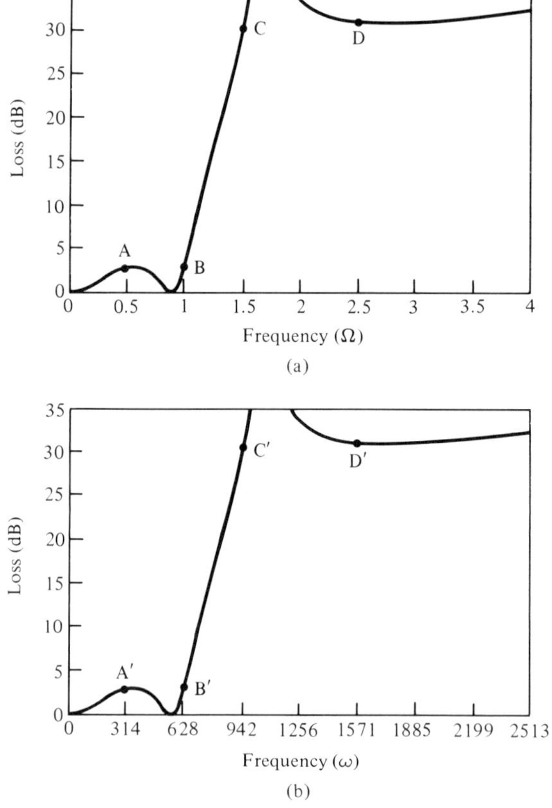

Figure 2.14 Loss characteristics of (a) $|H_N(j\Omega)|$ and (b) $|H(j\omega)|$.

where

$$H_o = 0.110022$$

The passband and stopband edge frequencies are 1 and 1.5 r/s, respectively, and a plot of the loss characteristic is shown in Fig. 2.14*a*. We substitute $s = p/200\pi$ into the above network function and obtain

$$H(p) = \frac{H_{01}(p^2 + 1.1078 \times 10^6)}{(p + 219.15)(p^2 + 150.02p + 349.435 \times 10^3)}$$

where

$$H_{01} = 69.129$$

The loss characteristic of $|H(j\omega)|$ as a function of ω is shown in Fig. 2.14*b*. Note that the loss characteristics are identical except for the change in the frequency scaling by a factor of 200π.

How do we use this frequency transformation to design a network function satisfying an arbitrary set of specifications for a lowpass filter function? Assume that the specifications for a lowpass filter function are given as illustrated in Fig. 2.15*a*. The following are the design steps.

1. Apply LP-LP transformations to the original set of specifications to obtain a set of specifications that can be satisfied by a normalized prototype lowpass network function as shown in Fig. 2.15*b*. This requires that

   $$\Omega_p = 1$$

   Then from the transformation $\Omega = \omega/\omega_0$, we get

   $$\omega_o = \frac{\omega_p}{\Omega_p} = \omega_p \tag{2.33}$$

   and

   $$\omega_a = \omega_o \Omega_a \qquad \text{or} \qquad \Omega_a = \frac{\omega_a}{\omega_o} = \frac{\omega_a}{\omega_p} \tag{2.34}$$

   Note that we know the values of ω_o and ω_a.
2. Design a normalized LP network function, $H_N(s)$ with the values of A_a, A_p, and Ω_a using any one of the approximations discussed earlier.
3. Obtain $H(p)$ by substituting $s = p/\omega_o$ with the value of ω_o calculated in step 1. The network function $H(p)$ will satisfy the original set of specifications given in Fig. 2.15*a*. The form of $H(p)$ will be the same as the form of $H_N(s)$ from which $H(p)$ is derived.

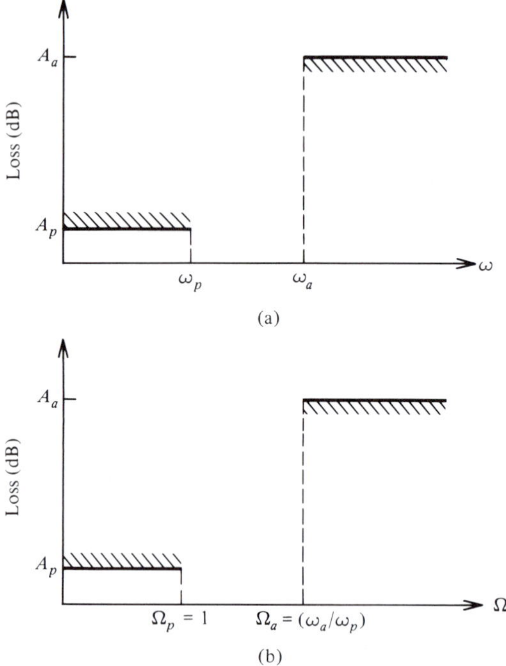

Figure 2.15 (*a*) Arbitrary specifications for a lowpass filter. (*b*) Normalized specifications for a prototype lowpass filter.

Lowpass-Highpass

The LP-HP complex frequency transformation requires that

$$s = \frac{\omega_o}{p} \tag{2.35}$$

where ω_o is again a constant.

For sinusoidal steady-state responses with $s = j\Omega$ and $p = j\omega$, the above transformation implies that

$$\Omega = -\frac{\omega_o}{\omega} \quad \text{or} \quad \omega = -\frac{\omega_o}{\Omega} \tag{2.36}$$

When we use (2.35) in a network function $H_N(s)$, the new network function is then

$$H_N(s) = H_N\left(\frac{\omega_o}{p}\right) = H(p)$$

That is,

$$|H_N(j\Omega)| = \left|H_N\left(\frac{\omega_o}{j\omega}\right)\right| = |H(j\omega)|$$

The above implies that the loss value of $|H(j\omega)|$ at $\omega = \omega_o$ will be same as that of $|H_N(j\Omega)|$ at $\Omega = -1$ r/s. However, note that both $|H(j\omega)|$ and $|H_N(j\Omega)|$ are even functions. Therefore the value of $|H(j\omega)|$ at $\omega = \omega_o$ will be the same as that of $|H_N(j\Omega)|$ at $\Omega = 1$ r/s. Similarly, the loss value of $|H(j\omega)|$ at $\omega = \omega_a$ will be the same as that of $|H_N(j\Omega)|$ at $\Omega_a = \omega_o/\omega_a$. Also note that the roles of $\Omega = 0$ and $\Omega = \infty$ in $|H_N(j\Omega)|$ are changed to the roles of $\omega = \infty$ and $\omega = 0$ in $|H(j\omega)|$, respectively. This means that the loss characteristic of $|H_N(j\Omega)|$ in the frequency range $\Omega = 0$ to $\Omega = 1$ r/s will be similar to $|H(j\omega)|$ in the frequency range $\omega = \omega_o$ to $\omega = \infty$, and the loss characteristic of $|H_N(j\Omega)|$ in the frequency range $\Omega = \Omega_a$ to $\Omega = \infty$ will be the similar to that of $|H(j\omega)|$ in the frequency range $\omega = \omega_a$ to $\omega = 0$. Thus, by using the transformation in (2.35), a lowpass network function can be converted to a highpass network function. To show this pictorially, consider the lowpass prototype network function in (2.32) whose plot is shown in Fig. 2.14*a*. When we apply the transformation of $s = 300\pi/p$ to this transfer function, the network function becomes

$$H(p) = \frac{p(p^2 + 316.557 \times 10^3)}{(p + 2.70215 \times 10^3)(p^2 + 254.2p + 1.00354 \times 10^6)}$$

The loss characteristics of $|H(j\omega)|$ as a function of ω are shown in Fig. 2.16. The points *A*, *B*, *C*, and *D* in Fig. 2.14*a* are located at *A*′, *B*′, *C*′, and *D*′, respectively, in Fig. 2.16. Obviously, from the plots in Fig. 2.16, one can easily tell that the above network function acts as a highpass function with $\omega_p = 300\pi$ r/s and $\omega_a = 200\pi$ r/s.

We have to ask a question similar to that raised in the previous case. How do we obtain a network function satisfying a given set of arbitrary

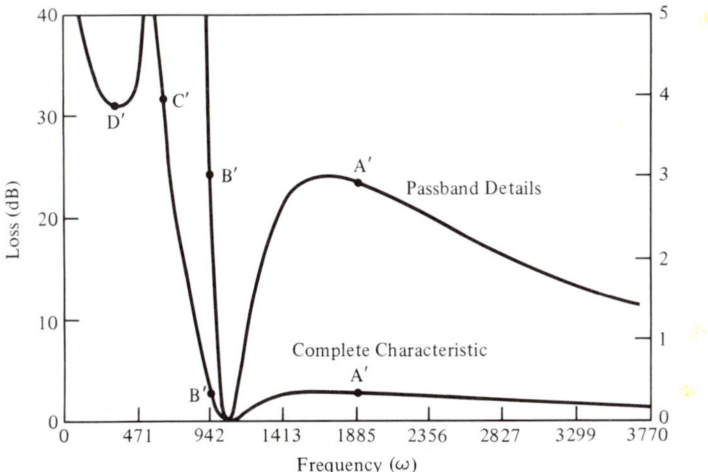

Figure 2.16 Loss characteristics of (2.32) when $s = 300\pi/p$ and $p = j\omega$.

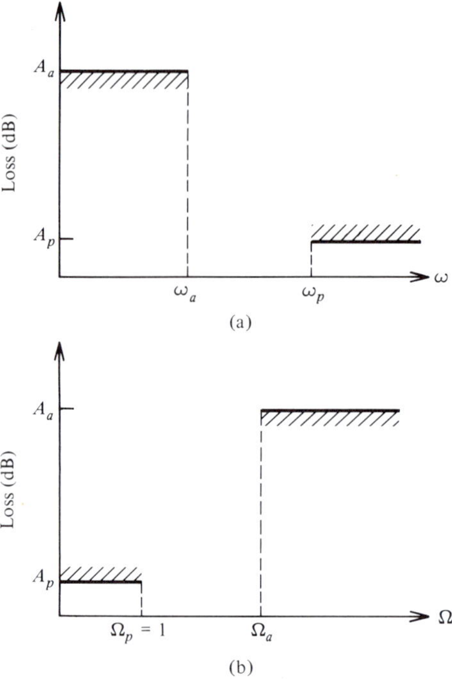

Figure 2.17 (*a*) Arbitrary specifications for a highpass filter. (*b*) Normalized specifications for a prototype lowpass filter.

specifications for a highpass filter as illustrated in Fig. 2.17*a*? The following steps outline the procedure for achieving these results.

1. Apply HP-LP transformation first (the reverse of LP-HP transformation) such that $p = \omega_o/s$, so that $\omega = \omega_o/\Omega$, where ω_o is a constant to be determined. This implies that

$$\omega_o = \omega_p$$

(since we want $\Omega_p = 1$ r/s) and

$$\omega_a = \frac{\omega_o}{\Omega_a} \qquad \text{or} \qquad \Omega_a = \frac{\omega_a}{\omega_o}$$

2. Use the above value of Ω_a and the values of A_p and A_a to obtain a prototype LP function that satisfies the LP specifications in Fig. 2.17*b*.
3. Apply the transformation $s = \omega_o/p$ to the network function obtained in step 2 with the value of ω_o obtained in step 1. The new function will satisfy the requirements of the original highpass filter requirement.

The form of $H(p)$ will differ from that of $H_N(s)$ when LP-HP transformation is used. It will also depend on the original form of $H_N(s)$. When maximally flat or Chebyshev approximation is used, $H_N(s)$ is an all-pole function, whereas with elliptic approximation $H_N(s)$ will have some finite zeros also. If the network function $H_N(s)$ was obtained using the first two approximations, then $H_N(s)$ will have the form of (2.17). Then $H(p)$ obtained using LP-HP transformation will have the form

$$H(p) = \frac{H_o p^N}{D_o(p) \prod_{i=1}^{N_1} (p^2 + a_i p + b_i)} \tag{2.37}$$

where N is the order of $H_N(s)$ as well as that of $H(p)$,

$$D_o(p) = \begin{cases} p + \sigma_o & N \text{ odd} \\ 1 & N \text{ even} \end{cases}$$

and

$$N_1 = \begin{cases} (N-1)/2 & N \text{ odd} \\ N/2 & N \text{ even} \end{cases}$$

If $H_N(s)$ is obtained using elliptic approximation, then it will have the form of (2.28). When LP-HP transformation is applied to (2.28), the resultant $H(p)$ will have the form

$$H(p) = \frac{H_o p^j}{D_o(p)} \prod_{i=1}^{N_1} \frac{p^2 + c_i}{p^2 + a_i p + b_i} \tag{2.38}$$

where

$$N_1 = \begin{cases} (N-1)/2 & N \text{ odd} \\ N/2 & N \text{ even} \end{cases}$$

$$D_o(p) = \begin{cases} p + \sigma_o & N \text{ odd} \\ 1 & N \text{ even} \end{cases}$$

and

$$j = \begin{cases} 1 & N \text{ odd} \\ 0 & N \text{ even} \end{cases}$$

Lowpass-Bandpass Transformation

The third frequency transformation we consider is the LP-BP transformation

$$s = \frac{1}{B} \frac{p^2 + \omega_o^2}{p} \tag{2.39}$$

where s is the lowpass filter frequency variable and p is the bandpass filter frequency variable. Here ω_o is the center frequency and B is the bandwidth of the bandpass network function. This bandwidth B is just the difference between the passband edge frequencies, as our next discussion will prove.

When this frequency transformation is used, the frequency $s = 0$ in the lowpass function transforms to $p = \pm j\omega_o$ (two frequencies) in the bandpass. For any frequency $s = j\Omega$ at which the loss is $A(\Omega)$ in the lowpass, there will be two frequencies in the bandpass network function at which the same loss will occur. Note that, at $s = -j\Omega$ also, the loss will be $A(\Omega)$, and therefore there will be two more frequencies in the bandpass network function at which the loss will again be $A(\Omega)$. Thus there will be a total of four different frequencies at which the bandpass network function will have the same loss $A(\Omega)$. Selecting only the two positive frequencies (the other two are negatives of the positive ones), we have

$$\omega_2 = \frac{B\Omega}{2} + \sqrt{\omega_o^2 + \left(\frac{B\Omega}{2}\right)^2}$$

and (2.40)

$$\omega_1 = -\frac{B\Omega}{2} + \sqrt{\omega_o^2 + \left(\frac{B\Omega}{2}\right)^2}$$

That is, $s = \pm j\Omega$ in the lowpass function transform to two positive frequencies $p = j\omega_2$ and $p = j\omega_1$ (and two negative ones also). The LP-BP transformation is illustrated in Fig. 2.18. The two *positive* frequencies given by (2.40) are related by

$$\omega_2\omega_1 = \omega_o^2 \tag{2.41}$$

and

$$\omega_2 - \omega_1 = B\Omega \tag{2.42}$$

In particular, the passband edge frequencies $\Omega = 1$ r/s and $\Omega = -1$ r/s in the lowpass are transformed to two *positive* frequencies ω_{p2} and ω_{p1}, respectively. These two frequencies are then related by

$$\omega_{p2}\omega_{p1} = \omega_o^2 \tag{2.43}$$

and

$$\omega_{p2} - \omega_{p1} = B \tag{2.44}$$

In a similar way, the two stopband edge frequencies $\Omega = \Omega_a$ and $\Omega = -\Omega_a$ in the lowpass are transformed to two positive frequencies ω_{a2} and ω_{a1}, and these two frequencies are related by

$$\omega_{a2}\omega_{a1} = \omega_o^2 \tag{2.45}$$

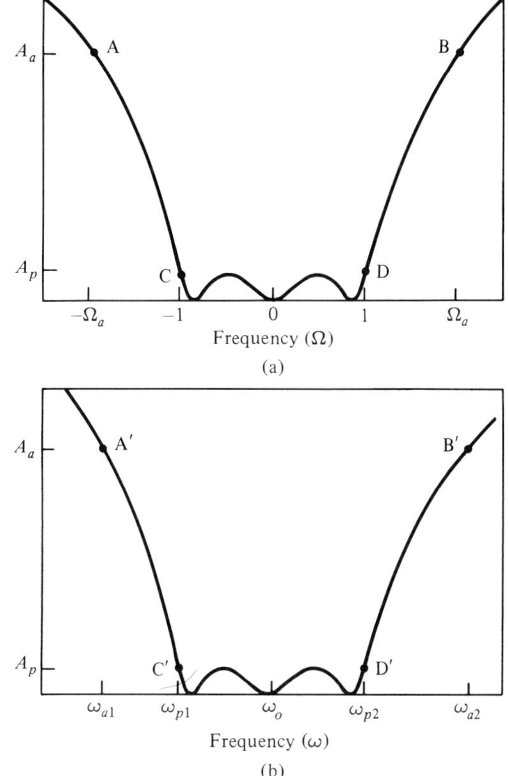

Figure 2.18 LP-BP transformation. (*a*) Lowpass characteristic. (*b*) Bandpass characteristic.

and

$$\omega_{a2} - \omega_{a1} = B\Omega_a \tag{2.46}$$

As is obvious, ω_{p2} and ω_{p1} are the edge frequencies of the passband, and $\omega_{p2} - \omega_{p1} = B$ is the bandwidth (selected) of the bandpass network function. There are two stopbands in the frequency ranges $0 \leq \omega \leq \omega_{a1}$ and $\omega \geq \omega_{a2}$. Now, using (2.44) and (2.46), we obtain

$$\Omega_a = \frac{\omega_{a2} - \omega_{a1}}{\omega_{p2} - \omega_{p1}} \tag{2.47}$$

Next, we shall see how to obtain a network function that satisfies a set of specifications for a bandpass filter as illustrated in Fig. 2.19.

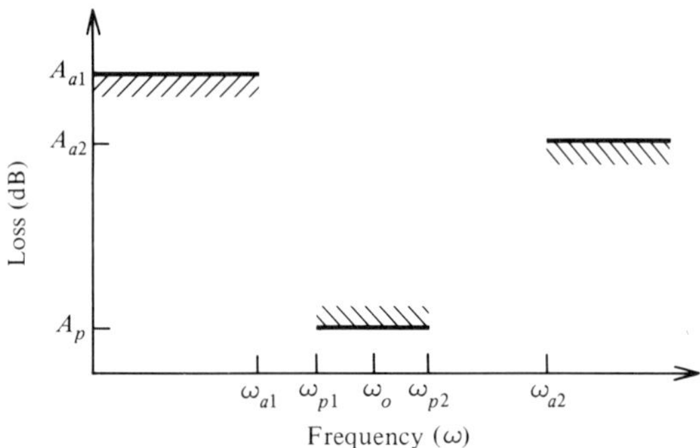

Figure 2.19 Arbitrary specifications for a bandpass filter. It is possible that $A_{a2} \geqq A_{a1}$.

One conservative but effective method of obtaining the required network function is given in the following steps.

1. Select $A_a = \max[A_{a1}, A_{a2}]$ and $A_p = A_p$.
2. Determine $\omega_o = \sqrt{\omega_{p2}\omega_{p1}}$ and $B = \omega_{p2} - \omega_{p1}$.
3. Determine $\omega_a = \omega_o^2/\omega_{a1}$.

4a. If $\omega_a \leq \omega_{a2}$, choose $\omega_{a1} = \omega_{a1}$ and $\omega_{a2} = \omega_a$, or

4b. If $\omega_a > \omega_{a2}$, choose $\omega_{a2} = \omega_{a2}$ and $\omega_{a1} = \omega_o^2/\omega_{a2}$.

5. With the values of ω_{a1} and ω_{a2} selected in step 4a or 4b use (2.47) to find the value of Ω_a.
6. Since the values of A_p (originally given), A_a (from step 1), and Ω_a (from step 5) are known, use any (or a given) approximation procedure to find a prototype lowpass network function $H_N(s)$.
7. Knowing the values of ω_o and B (step 2), use (2.39) to obtain $H(p)$ that satisfies the original set of specifications.

Example 2.5: Using Chebyshev approximation, find a network function that satisfies the specifications in Fig. 2.19: $A_p \leq 0.5$ dB, $A_{a1} \geq 20$ dB, $A_{a2} \geq 17$ dB. The frequencies are $f_{a1} = 500$ Hz, $f_{a2} = 5$ kHz, $f_{p1} =$ 1 kHz, and $f_{p2} = 4$ kHz.

1. Select $A_a = \max[20, 17] = 20$ dB
2. Determine

$$\omega_o = 2\pi\sqrt{(1)(4 \times 10^6)} = 4\pi \times 10^3 \text{ r/s}$$

$$B = 2\pi(4 - 1) \times 10^3 = 6\pi \times 10^3 \text{ r/s}$$

3. Determine

$$\omega_a = \frac{(4\pi \times 10^3)^2}{2\pi \times 500} = 16\pi \times 10^3 \text{ r/s}$$

4. Since $\omega_a > \omega_{a2}$, we choose $\omega_{a2} = 2\pi \times 5 \times 10^3$ r/s and

$$\omega_{a1} = \frac{16\pi^2 \times 10^6}{10\pi \times 10^3} = 1.6\pi \times 10^3 \text{ r/s}$$

5. Find

$$\Omega_a = \frac{\omega_{a2} - \omega_{a1}}{B} = \frac{8.4\pi \times 10^3}{6\pi \times 10^3} = 1.4$$

6. Using the values $A_p = 0.5$ dB, $A_a = 20$ dB, and $\Omega_a = 1.4$ r/s (of course, $\Omega_p = 1$ r/s), equiripple approximation gives the following prototype lowpass function:

$$H_N(s) = \frac{H_o}{(s + 0.36232)(s^2 + 0.223926s + 1.0358)(s^2 + 0.58624s + 0.47677)}$$

where H_o can be chosen to satisfy any gain requirement.

7. In this step, we substitute

$$s = \frac{p^2 + \omega_o^2}{Bp} = \frac{p^2 + 16\pi^2 \times 10^6}{6\pi \times 10^3 p}$$

into $H_N(s)$ to find $H(p)$. Consider a factor of the form $s + \sigma_o$ in the denominator of $H_N(s)$. Such a factor becomes

$$s + \sigma_o \to \frac{p^2 + \omega_o^2}{Bp} + \sigma_o = \frac{p^2 + (B\sigma_o)p + \omega_o^2}{Bp}$$

In our example $s + 0.36232$ is transformed to

$$s + 0.36232 \to \frac{(p^2 + 6.82957 \times 10^3 p + 157.914 \times 10^6)}{18.8496 \times 10^3 p}$$

Next consider a factor of the form $s^2 + a_i s + b_i$ in $H_N(s)$. When we substitute (2.39) into such a factor, we get

$$s^2 + a_i s + b_i \to \left(\frac{p^2 + \omega_o^2}{Bp}\right)^2 + a_i \frac{p^2 + \omega_o^2}{Bp} + b_i = \frac{p^4 + (a_i B)p^3 + (2\omega_o^2 + b_i B^2)p^2 + (a_i B\omega_o^2)p + \omega_o^4}{(Bp)^2}$$

The above equations mean that a first-order factor in the denominator of $H_N(s)$ will result in a second-order factor, and that a second-order factor in $H_N(s)$ will result in a fourth-order factor in the denominator of $H(p)$. Thus an *all-pole function* $H_N(s)$ of the form of (2.17), repeated here for convenience,

$$H_N(s) = \frac{H_o}{D_o(s)\prod_{i=1}^{N_1}(s^2 + a_i s + b_i)} \tag{2.17}$$

where

$$D_o(s) = \begin{cases} s + \sigma_o & N \text{ odd} \\ 1 & N \text{ even} \end{cases}$$

and

$$N_1 = \begin{cases} (N-1)/2 & N \text{ odd} \\ N/2 & N \text{ even} \end{cases}$$

is transformed to $H(p)$, where $H(p)$ will be of the form

$$H(p) = \frac{H_{oB}p^N}{D_{BP}(p)\prod_{i=1}^{N_1}(p^4 + A_{3i}p^3 + A_{2i}p^2 + A_{1i}p + A_{oi})} \tag{2.48}$$

where

$$\begin{aligned} A_{3i} &= a_i B \qquad A_{2i} = 2\omega_o^2 + b_i B^2 \\ A_{1i} &= a_i B\omega_o^2 \qquad \text{and} \qquad A_{oi} = \omega_o^4 \end{aligned} \tag{2.49}$$

and

$$D_{BP}(p) = \begin{cases} p^2 + (B\sigma_o)p + \omega_o^2 & N \text{ odd} \\ 1 & N \text{ even} \end{cases}$$

In our example, $N = 5$ (odd) and $N_1 = 2$.
Also,

$$D_{BP}(p) = p^2 + 6.82957 \times 10^3 p + 157.914 \times 10^6$$

The coefficients of the other factors are

$$\begin{aligned} A_{31} &= 6.61073 \times 10^3 & A_{32} &= 15.9595 \times 10^3 \\ A_{21} &= 693.702 \times 10^6 & A_{22} &= 485.226 \times 10^6 \\ A_{11} &= 1.04392 \times 10^{12} & A_{12} &= 2.52023 \times 10^{12} \\ A_{01} &= 24.9367 \times 10^{15} & A_{02} &= 24.9367 \times 10^{15} \end{aligned}$$

The value of H_o can be chosen for a given gain requirement usually at ω_o. ■

Next we shall see how a lowpass network function obtained through elliptic approximation, such as the one given by (2.28), is transformed when LP-BP transformation is used. When such a transformation is used on (2.28), the resulting function $H(p)$ is of the form

$$H(p) = \frac{H_{oB}p^j}{D_{BP}(p)} \prod_{i=1}^{N_1} \frac{p^4 + C_i B^2 p^2 + \omega_o^4}{p^4 + A_{3i}p^3 + A_{2i}p^2 + A_{1i}p + A_{oi}} \tag{2.50}$$

where A's and $D_{BP}(p)$ are exactly the same as calculated in (2.49). Here, in addition, we must define the value of j:

$$j = \begin{cases} 1 & N \text{ odd} \\ 0 & N \text{ even} \end{cases}$$

We still have to consider another problem involving the transfer functions in (2.48) and (2.50). This problem arises from the point of view of realization and construction of these filters. Most realizations of network functions are based on second-order filter building blocks. In such cases, we have to express the fourth-order factors as second-order factors. The numerator of (2.50) can be split into two second-order factors quite easily. Splitting the fourth-order factors of the denominator of both (2.48) and (2.50) requires a little more work, but can be achieved by using the formulas developed next. In principle, we want to express $p^4 + A_{3i}p^3 + A_{2i}p^2 + A_{1i}p + A_{oi}$ in the form of $(p^2 + \alpha_i p + \beta_i)(p^2 + \gamma_i p + \delta_i)$. Another way of writing a second-order factor such as $p^2 + \alpha_i p + \beta_i$ is in the form of $p^2 + (\omega_{oi}/Q_i)p + \omega_{oi}^2$, where ω_{oi} and Q_i are called the *pole frequency* and *pole Q factor*, respectively, and will be considered further in the next chapter. Using the second-order factors in terms of the pole frequency and the pole Q factor, let

$$p^4 + A_{3i}p^3 + A_{2i}p^2 + A_{1i}p + A_{oi}$$

$$= \left(p^2 + \frac{\omega_{oi}}{Q_i}p + \omega_{oi}^2\right)\left(p^2 + \frac{\omega_{oj}}{Q_j}p + \omega_{oj}^2\right)$$

Our next step is to develop formulas for finding ω_{oi}, ω_{oj}, Q_i, and Q_j in terms of the given quantities, which are a_i, b_i, ω_o, and B. To do this, we obtain the following equations by comparing the coefficients of p on both sides of the above equation:

$$\omega_{oi}^2\omega_{oj}^2 = \omega_o^4 \tag{2.51}$$

$$\frac{\omega_{oi}^2\omega_{oj}}{Q_j} + \frac{\omega_{oj}^2\omega_{oi}}{Q_i} = a_i B\omega_o^2 \tag{2.52}$$

$$\omega_{oi}^2 + \omega_{oj}^2 + \frac{\omega_{oj}\omega_{oi}}{Q_j Q_i} = 2\omega_o^2 + b_i B^2 \tag{2.53}$$

and

$$\frac{\omega_{oi}}{Q_i} + \frac{\omega_{oj}}{Q_j} = a_i B \tag{2.54}$$

where (2.49) has been used. After using (2.51) in (2.52), and from (2.52) and (2.54), we have

$$\frac{\omega_{oi}}{Q_j} + \frac{\omega_{oj}}{Q_i} = a_i B = \frac{\omega_{oj}}{Q_j} + \frac{\omega_{oi}}{Q_i}$$

or

$$\omega_{oi}\left(\frac{1}{Q_j} - \frac{1}{Q_i}\right) - \omega_{oj}\left(\frac{1}{Q_j} - \frac{1}{Q_i}\right) = 0$$

Irrespective of the values of ω_{oi} and ω_{oj}, as long as $\omega_{oi} \neq \omega_{oj}$ the solution of the above equation is $Q_j = Q_i = Q$ and

$$Q = \frac{\omega_{oi} + \omega_{oj}}{a_i B} \tag{2.55}$$

Substituting the above value of Q for Q_i and Q_j in (2.53), and after some mathematical manipulation, we have

$$(\omega_{oi} + \omega_{oj})^4 - X(\omega_{oi} + \omega_{oj})^2 + Y = 0$$

where

$$X = 4\omega_o^2 + b_i B^2$$

and

$$Y = (a_i B\omega_o)^2 \tag{2.56}$$

We are interested in only the positive values of ω_{oi} and ω_{oj}, and therefore, of the two possible solutions for $(\omega_{oi} + \omega_{oj})^2$, we pick the one that will yield positive solutions for ω_{oi} and ω_{oj}:

$$(\omega_{oi} + \omega_{oj})^2 = \frac{X}{2} + \sqrt{\left(\frac{X}{2}\right)^2 - Y}$$

$$\triangleq Z^2 \tag{2.57}$$

Note that Z^2 is a known quantity as soon as we find X and Y using (2.56). Therefore, we have

$$\omega_{oi} + \omega_{oj} = Z$$

and, using $\omega_{oj} = \omega_o^2/\omega_{oi}$ (from 2.51), the following results:

$$\omega_{oi}^2 - Z\omega_{oi} + \omega_o^2 = 0$$

Using one of the solutions of the above equation for ω_{oi} (the other solution gives ω_{oj}), we obtain

$$\omega_{oi} = \frac{Z}{2} + \sqrt{\left(\frac{Z}{2}\right)^2 - \omega_o^2} \tag{2.58}$$

An easy way to find ω_{oj} is to use

$$\omega_{oj} = \frac{\omega_o^2}{\omega_{oi}} \tag{2.59}$$

The steps are obvious. Use (2.56), (2.57), and (2.58) in that order to find ω_{oi}. Then use (2.59) and (2.55), respectively, to calculate ω_{oj} and Q in that order. Now, we have found the two quadratic factors. Finally, the network function $H(p)$ can be expressed either as

$$H(p) = \frac{H_o p^N}{\prod_{i=1}^{N} (p^2 + A_i p + B_i)} \tag{2.60}$$

in the case of maximally flat or Chebyshev approximation or, in the case of elliptic approximation, as

$$H(p) = \frac{H_o p^j \prod_{i=1}^{2N_1} (p^2 + c_i)}{\prod_{i=1}^{N} (p^2 + A_i p + B_i)} \tag{2.61}$$

where

$$N_1 = \begin{cases} (N-1)/2 & N \text{ odd} \\ N/2 & N \text{ even} \end{cases}$$

$$j = \begin{cases} 1 & N \text{ odd} \\ 0 & N \text{ even} \end{cases}$$

and N is the order of the network function $H_N(s)$. Also note that the order of $H(p)$ is equal to $2N$.

Lowpass-Bandstop Transformation

The last transformation, we consider, is a lowpass-to-bandstop transformation. This complex frequency transformation, which converts a prototype lowpass function $H_N(s)$ to a bandstop function $H(p)$ is

$$s = \frac{Bp}{(p^2 + \omega_o^2)} \tag{2.62}$$

where B is the difference between the two passband edge frequencies and ω_o is the geometric mean of the two stopband edge frequencies.

The above transformation can be considered a combination of two transformations, first

$$s = \frac{1}{z} \tag{2.63}$$

and then

$$z = \frac{p^2 + \omega_o^2}{Bp} \tag{2.64}$$

The first transformation (2.63) is a LP-HP transformation, where z is a complex frequency variable, and the second is a LP-BP transformation. Consider a prototype lowpass function

$$H_N(s) = \frac{H_o(s^2 + 2.80601)}{(s + 0.348788)(s^2 + 0.238766s + 0.885128)}$$

The loss characteristic of the above lowpass function is shown in Fig. 2.20*a*, with $s = j\Omega$. After we apply the transformation (2.63) to this lowpass function, with $z = j\Omega'$, the loss characteristic is obtained. This characteristic must clearly be a highpass characteristic, as shown in Fig. 2.20*b*. The transformation (2.64) is applied to this highpass network function, which is now a function of z, and the resultant loss characteristic with $p = j\omega$ is shown in Fig. 2.20*c*. Any frequency $\pm\Omega$ in the lowpass characteristic at which the loss is $A(\Omega)$ will be transformed to two positive frequencies in the bandstop characteristic at which the loss will be the same. These two positive frequencies are

$$\omega_1 = \frac{-B}{2\Omega} + \sqrt{\omega_o^2 + \left(\frac{B}{2\Omega}\right)^2}$$

and (2.65)

$$\omega_2 = \frac{B}{2\Omega} + \sqrt{\omega_o^2 + \left(\frac{B}{2\Omega}\right)^2}$$

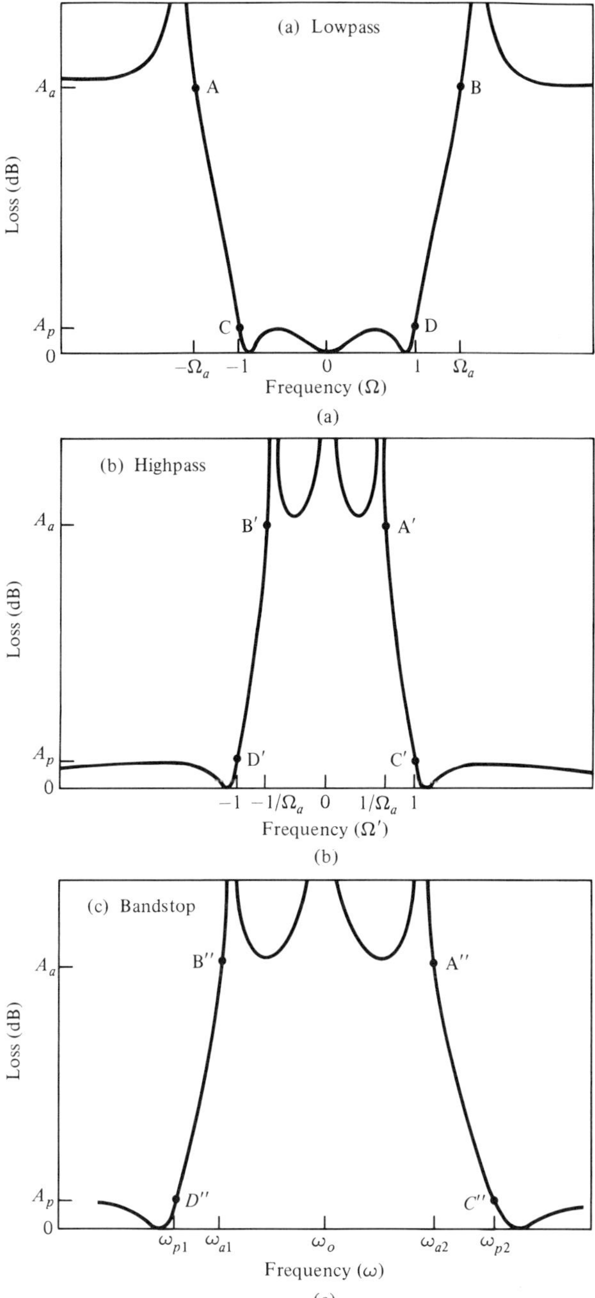

Figure 2.20 An LP-BS transformation. (*a*) Lowpass characteristic. (*b*) Highpass characteristic. (*c*) Bandstop characteristic.

In particular, the two frequencies $\pm\Omega_p = \pm 1$ are transformed to ω_{p1} and ω_{p2}, respectively, and thus

$$\omega_{p1} = -\frac{B}{2} + \sqrt{\omega_o^2 + \left(\frac{B}{2}\right)^2}$$

and

$$\omega_{p2} = \frac{B}{2} + \sqrt{\omega_o^2 + \left(\frac{B}{2}\right)^2}$$

Therefore the difference between the passband edge frequencies is $\omega_{p2} - \omega_{p1}$ and is equal to B. That is,

$$B = \omega_{p2} - \omega_{p1} \tag{2.66}$$

Similarly, the frequencies $\pm\Omega_a$ in the lowpass characteristic become ω_{a1} and ω_{a2}, respectively. Therefore, using (2.65) again, we have

$$\omega_{a1} = -\frac{B}{2\Omega_a} + \sqrt{\omega_o^2 + \left(\frac{B}{2\Omega_a}\right)^2}$$

and (2.67)

$$\omega_{a2} = \frac{B}{2\Omega_a} + \sqrt{\omega_o^2 + \left(\frac{B}{2\Omega_a}\right)^2}$$

Using (2.67), we find that

$$\omega_o = \sqrt{\omega_{a1}\omega_{a2}} \tag{2.68}$$

Also,

$$\Omega_a = \frac{B}{\omega_{a2} - \omega_{a1}} = \frac{\omega_{p2} - \omega_{p1}}{\omega_{a2} - \omega_{a1}} \tag{2.69}$$

Given the set of specifications for a bandstop filter requirement graphically illustrated in Fig. 2.21, we next proceed to show how the required network function can be obtained. The steps are analogous to those for the bandpass case:

1. Select $A_p = \min\{A_{p1}, A_{p2}\}$ and $A_a = A_a$.
2. Determine $\omega_o = \sqrt{\omega_{a1}\omega_{a2}}$.
3. Determine $\omega_p = \omega_o^2/\omega_{p1}$.

4a. If $\omega_p \le \omega_{p2}$, choose $\omega_{p1} = \omega_{p1}$ and $\omega_{p2} = \omega_p$, or

4b. If $\omega_p > \omega_{p2}$, choose $\omega_{p2} = \omega_{p2}$ and $\omega_{p1} = \omega_o^2/\omega_{p2}$.

5. Find $B = \omega_{p2} - \omega_{p1}$ and $\Omega_a = B/(\omega_{a2} - \omega_{a1})$.

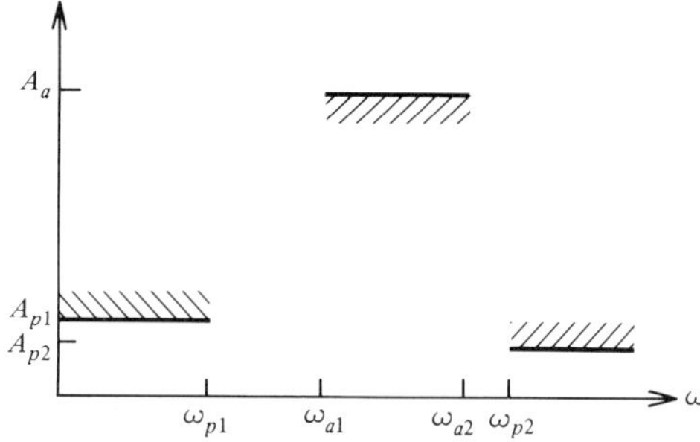

Figure 2.21 Arbitrary specifications for a bandstop filter. It is possible that $A_{p2} \gtreqqless A_{p1}$.

6. With the selected values of A_p and A_a and with $\Omega_p = 1$ r/s and Ω_a as found in step 5, use any (or a given) approximation procedure to find a prototype lowpass function $H_N(s)$.
7. Apply the frequency transformation (2.62) to obtain $H(p)$ that satisfies the original set of specifications.

An all-pole function $H_N(s)$ of the form of (2.17) will be transformed to

$$H(p) = \frac{\left(p^2 + \omega_o^2\right)^N}{\prod_{i=1}^{N}\left(p^2 + A_i p + B_i\right)} \tag{2.70}$$

where N is the order of the prototype lowpass function.

If elliptic approximation is used, the prototype function $H_N(s)$ will be of the form given by (2.28). In this case, $H(p)$ will be of the form

$$H(p) = \prod_{i=1}^{N} \frac{p^2 + C_i}{p^2 + A_i p + B_i} \tag{2.71}$$

Note that the order of $H(p)$ in both (2.70) and (2.71) is equal to $2N$.

2.7 *Conclusions*

In this chapter, we have described the use of important approximation procedures in designing a lowpass network function satisfying a set of prescribed specifications. A discussion of frequency transformations was included to illustrate how the other three types of filter functions, namely, highpass, bandpass, and bandstop, can also be obtained for a set of prescribed specifications. Appendixes 2A and 2B include two programs called MAXCHY and ELIPFT, mainly written in GW BASIC language for use in

IBM and its compatibles, that can be used for to obtain the parameters for four types of network functions. The program MAXCHY can be used for maximally flat or Chebyshev approximation, and the program ELIPFT uses elliptic approximation.

REFERENCE

1. Antoniou, A.: *Digital Filters: Analysis and Design*, McGraw-Hill, New York, 1979.

EXERCISES

2.1. Show that, when $s = j\Omega$, the magnitude function of a rational function is an even function of Ω.

2.2. The specifications of a lowpass filter with a maximally flat passband are $A_p \leq 0.5$ dB, $f_p = 1$ kHz, $A_a \geq 20$ dB, $f_a = 1.5$ kHz. Use (2.11), (2.13), and (2.16) to find the ripple factor, the order, and the pole positions of the network function. Verify your answers using the program MAXCHY.

2.3. Find the Chebyshev polynomial $C_5(\Omega)$. Assume that $H_o = 1$ and $\varepsilon = 1$. Use this polynomial in (2.19) and plot the function $|H(j\Omega)|$ as a function of Ω over the range 0 to 2 r/s. Also, use (2.20) in (2.19) with the same values of H_o and ε and verify that these two curves are identical.

2.4. The specifications of a Chebyshev lowpass filter are $A_p \leq 1$ dB, $f_p =$ 3.2 kHz, $A_a \geq 30$ dB, $f_a = 4$ kHz.
(a) Use (2.11) and (2.21) to find the ripple factor and the required order N. Also, find the pole positions.
(b) Verify your answers using the program MAXCHY.
(c) Plot the attenuation characteristics of the designed network function and determine whether it satisfies the specifications.

2.5. A lowpass filter has to satisfy the specifications of Prob. 2.4. Use the elliptic approximation and find the required network function. Verify your answers using the program ELIPFT.

2.6. A bandpass filter is required to satisfy the following specifications: lower passband edge frequency $f_{p1} = 16$ kHz, upper passband edge frequency $f_{p2} = 20$ kHz, passband ripple not to exceed 1 dB, lower stopband edge frequency $f_{a1} = 14$ kHz, upper stopband edge frequency $f_{a2} = 22$ kHz, minimum stopband loss required is 35 dB.

(a) Use Chebyshev approximation and find the required network function.
(b) Verify your answers using the program MAXCHY.

2.7. The specifications of a bandstop filter are as follows: stopband, f_{a1} = 8 kHz, f_{a2} = 12 kHz, $A_a \geq 30$ dB; passband edge frequencies, f_{p1} = 7 kHz and f_{p2} = 14 kHz; maximum passband loss, A_p = 0.5 dB.
(a) Use elliptic approximation and find the required network function that meets the above specifications.
(b) Verify your answers using the program ELIPFT.
(c) Plot the attenuation characteristic of the network function and determine whether the specifications are satisfied.

2A

Program MAXCHY

THE PROGRAMS LISTED in Apps. 2A and 2B are mainly written for use on a HP85. Since these programs are written in HP BASIC, they can be modified for other systems by the user. After loading the programs, simply run. The specifications for any type of filter are inputted by answering a few questions, and the following inputs are necessary for the different types of filters.

1. Lowpass and highpass filters
   a. Maximum passband loss and minimum stopband loss (in decibels)
   b. Passband and stopband edge frequencies.
2. Bandpass and bandstop filters
   a. Maximum passband loss and minimum stopband loss (in decibels)
   b. The difference between the passband edge frequencies and the difference between the stopband edge frequencies
   c. The geometric mean of the passband or stopband edge frequencies.

Using the program listed in this appendix, one can obtain a network function using either maximally flat or Chebyshev approximation.

The output is the order and the set of coefficients of the appropriate network function. Depending on the filter function desired, the parameters must be associated with an appropriate network function. The mathematical forms of those network functions were dealt with in Chap. 2. One may refer to the appropriate equation as indicated in the accompanying table.

Network function	Equation no.	Output quantities
Lowpass	(2.17)	N, σ_o (if any), A_i and B_i
Highpass	(2.37)	N, σ_o (if any), A_i and B_i
Bandpass	(2.60)	$N_2 = 2N$, A_i, and B_i
Bandstop	(2.70)	$N_2 = 2N$, A_i, and B_i

```
10 '
20 '
30 '
40 '                        PROGRAM  "MAXCHY"
50 '
60 '
70 'PROGRAM TO DESIGN ALL FOUR TYPES OF FILTERS USING MAX.FLAT OR CHEBYSHEV APPR
OXIMATION.
80 DIM A(20),B(20)
90 PRINT" THIS PROGRAM CAN BE USED TO DESIGN LOWPASS,HIGHPASS,BANDPASS AND BANDS
TOP  FILTERS USING EITHER MAXIMALLY FLAT OR CHEBYSHEV APPROXIMATION."
100 PRINT"ENTER THE TYPE OF FILTER FUNCTION TO BE DESIGNED.ENTER 'LP'(for lowpas
s),HP(for highpass),BP(for bandpass) AND BS(for bandstop)";:INPUT T$
110 IF T$="LP" OR T$="HP" OR T$="BP" OR T$="BS" THEN 130 ELSE 120
120 IF T$="lp" OR T$="hp" OR T$="bp" OR T$="bs" THEN 130 ELSE 100
130 IF T$="lp" OR T$="LP" THEN K=1:LPRINT" DESIGN OF LOWPASS FILTER":GOTO 170
140 IF T$="hp" OR T$="HP" THEN K=2:LPRINT" DESIGN OF HIGHPASS FILTER":GOTO 170
150 IF T$="bp" OR T$="BP" THEN K=3:LPRINT" DESIGN OF BANDPASS FILTER":GOTO 170
160 IF T$="bs" OR T$="BS" THEN K=4:LPRINT" DESIGN OF BANDSTOP FILTER"
170 LPRINT STRING$(80,"_")
180 PI2=8!*ATN(1)
190 LPRINT:LPRINT" THE SPECIFICATIONS ARE":LPRINT
200 'Inputs are given in the next few lines.
210 PRINT"INPUT THE MAXIMUM PASSBAND LOSS ALLOWED  IN dB";:INPUT A1
220 PRINT"INPUT THE MINIMUM STOPBAND LOSS REQUIRED IN dB";:INPUT A2
230 LPRINT" THE MAXIMUM PASSBAND LOSS ALLOWED IN dB=";USING"##.#####^^^^";A1
240 LPRINT" THE MINIMUM STOPBAND LOSS REQUIRED IN dB=";USING"##.#####^^^^";A2
250 IF K=3 OR K=4 THEN 300
260 PRINT"INPUT THE PASSBAND EDGE FREQUENCY IN Hz";:INPUT W1:W1=PI2*W1
270 PRINT"INPUT THE STOPBAND EDGE FREQUENCY IN Hz";:INPUT W2:W2=PI2*W2
280 LPRINT" THE PASSBAND EDGE FREQUENCY IN RADS./SEC.=";USING"##.######^^^^";W1
290 LPRINT" THE STOPBAND EDGE FREQUENCY IN RADS./SEC.=";USING"##.######^^^^";W2:
GOTO 360
300 PRINT"INPUT THE DIFFERENCE OF PASSBAND(s) EDGE FREQUENCIES IN Hz";:INPUT W1:
W1=PI2*W1
310 PRINT"INPUT THE DIFFERENCE OF STOPBAND(s) EDGE FREQUENCIES IN Hz";:INPUT W2:
W2=PI2*W2
320 PRINT"INPUT THE GEOMETRIC MEAN OF THE PASSBAND(for bandpass) OR THE STOPBAND
(for bandstop) EDGE FREQUENCIES IN Hz";:INPUT W0:W0=PI2*W0
330 LPRINT" THE DIFFERENCE OF PASSBAND(s) EDGE FREQUENCIES IN RADS./SEC.=";USING
"##.######^^^^";W1
340 LPRINT" THE DIFFERENCE OF STOPBAND(s) EDGE FREQUENCIES IN RADS./SEC.=";USING
"##.######^^^^";W2
350 LPRINT" THE GEOMETRIC MEAN OF THE PASSBAND(for bandpass) OR THE STOPBAND(for
 bandstop)  EDGE FREQUENCIES IN RADS./SEC.=";USING"##.######^^^^";W0
360 W3=W2/W1:IF K=2 OR K=4 THEN W3=1/W3
370 PRINT"IF YOU WANT TO USE MAX. FLAT APPROXIMATION ENTER 'MAX'. IF YOU WANT TO
USE CHEBYSHEV APPROXIMATION ENTER 'CHY'.":INPUT T$
380 IF T$="MAX" OR T$="max" OR T$="CHY" OR T$="chy" THEN 390 ELSE 370
390 IF T$="MAX" OR T$="max" THEN LPRINT:LPRINT STIRNG$(80,"_"):LPRINT" MAXIMALLY
 FLAT APPROXIMATION":I=1
400 IF T$="CHY" OR T$="chy" THEN LPRINT:LPRINT STRING$(80,"_"):LPRINT" CHEBYSHEV
 APPROXIMATION":I=2
```

```
410 LPRINT STRING$(80,"_"):LPRINT:LPRINT" DESIGN PARAMETERS ARE":LPRINT
420 PRINT"DO YOU WANT TO KNOW THE DESIGN PARAMETERS OF PROTOTYPE LOWPASS FILTER?
 Y/N";:INPUT A$
430 IF A$="Y" OR A$="y" OR A$="N" OR A$="n" THEN 440 ELSE 420
440 IF A$="Y" OR A$="y" THEN K1=1 ELSE K1=0
450 'calculation of ripple factor and the order of the prototype filter.
460 E=SQR(10^(.1*A1)-1):E1=SQR(10^(.1*A2)-1)
470 IF I=2 THEN 490
480 X=LOG(E1/E)/LOG(W3):GOTO 500
490 Y=E1/E:X=LOG(Y+SQR(Y*Y-1))/LOG(W3+SQR(W3*W3-1))
500 N=INT(X)+1:N1=INT(N/2):F=N-2*N1:P1=PI2/4:A0=0:N3=N:N4=N1
510 'calculation of the coefficients of the prototype transfer function.
520 IF I=2 THEN 560
530 R=1/E^(1/N):X1=R*R:H=1
540 FOR J=1 TO N1:P2=P1*((2*J-1)/N+1):A(J)=-2*R*COS(P2):B(J)=X1:H=H*X1:NEXT J
550 GOTO 600
560 E1=1/E:V=LOG(E1+SQR(E1*E1+1))/N:V1=EXP(V):V2=EXP(-V):R=(V1-V2)/2:C=(V1+V2)/2

570 IF F=0 THEN H=10^(-A1/20) ELSE H=1
580 FOR J=1 TO N1:P2=P1*(2*J-1)/N:R1=R*SIN(P2):A(J)=2*R1:R3=C*COS(P2)
590 B(J)=R1*R1+R3*R3:H=H*B(J):NEXT J
600 IF F=0 THEN 610 ELSE A0=R:H=H*R
610 IF K1=0 THEN 630
620 LPRINT" PARAMETERS OF THE PROTOTYPE LOWPASS FILTER":LPRINT:GOSUB 900
630 LPRINT:LPRINT" PARAMETERS OF THE DENORMALIZED FILTER":LPRINT STRING$(80,"_")
:LPRINT
640 IF K=3 OR K=4 THEN 750
650 'LP to LP or LP to HP transformation.
660 W4=W1*W1:IF F=0 THEN 680
670 IF K=1 THEN A0=A0*W1 ELSE A0=W1/A0
680 FOR J=1 TO N1
690 IF K=2 THEN 700 ELSE A(J)=W1*A(J):B(J)=B(J)*W4:GOTO 710
700 A(J)=W1*A(J)/B(J):B(J)=W4/B(J)
710 NEXT J
720 IF K=2 THEN 730 ELSE H=H*W1^N:GOTO 740
730 H=1:IF I=2 AND F=0 THEN H=10^(-A1/20)
740 GOTO 880
750 IF K=3 THEN 780
760 'LP to HP transformation for bandstop filter. Then LP to BP transformation
770 FOR J=1 TO N1:A(J)=A(J)/B(J):B(J)=1/B(J):NEXT J:IF A0>0 THEN A0=1/A0
780 N2=2*N:N3=N2:N4=N
790 'LP to BP transformation. For bandstop also this transformation is used.
800 W4=W0*W0:IF F=0 THEN 820
810 A(N)=W1*A0:B(N)=W4
820 FOR J=1 TO N1:Y=W4*(W1*A(J))^2:B3=(4*W4+B(J)*W1*W1)/2
830 X=B3+SQR(B3*B3-Y):Z=SQR(X)/2:B1=Z+SQR(Z*Z-W4):B2=W4/B1
840 Q=(B1+B2)/(A(J)*W1):A(J)=B1/Q:A(J+N1)=B2/Q:B(J)=B1*B1:B(J+N1)=B2*B2
850 NEXT J
860 IF K=3 THEN H=H*W1^N
870 F=0:IF K=4 THEN 730
880 GOSUB 900
890 END
900 'printing subroutine
910 LPRINT"         ";CHR$(238);"=";USING"##.######^^^^";E:LPRINT
920 LPRINT" THE ORDER OF THE FILTER=";N3:LPRINT STRING$(80,"_"):LPRINT
930 IF F=0 THEN 960
940 LPRINT" THERE IS A REAL POLE IN THIS FUNCTION":LPRINT
950 LPRINT"          ";CHR$(229);"0=";USING"##.######^^^^";A0:LPRINT
960 LPRINT" THE COEFFICIENTS OF THE SECOND ORDER FACTORS ARE AS FOLLOWS:":LPRINT

970 LPRINT"   I","A(I)","   B(I)":LPRINT STRING$(80,"_"):LPRINT
980 FOR J=1 TO N4
990 LPRINT"  "; J;USING"    ##.######^^^^";A(J);B(J):LPRINT
1000 NEXT J
1010 LPRINT STRING$(80,"_")
1020 LPRINT" THE VALUE OF H REQUIRED TO GIVE A MAXIMUM GAIN OF 0 dB IS";H
1030 RETURN
```

2B

Program ELIPFT

THE PROGRAM LISTED in this appendix, called ELIPFT, uses an elliptic approximation procedure. It is also interactive, and the inputs can be given through the terminal while it is being run. The necessary inputs for different types of network functions are the same as those listed in App. 2A.

The output is the set of coefficients and the order of an appropriate network function. The different mathematical forms of such network functions are in the forms of equations in Chap. 2 as indicated in the accompanying table.

Network function	Equation no.	Output quantities
Low	(2.28)	N, σ_o (if any), A_i and B_i
Highpass	(2.38)	N, σ_o (if any), A_i, B_i, and C_i
Bandpass	(2.61)	$N_2 = 2N$, A_i, B_i, and C_i
Bandstop	(2.71)	$N_2 = 2N$, A_i, B_i, and C_i

```
10 '
20 '
30 '            PROGRAM LISTING OF 'ELIPFT'
40 '
50 '
60 'PROGRAM TO DESIGN ALL FOUR TYPES OF FILTERS USING ELLIPTIC APPROXIMATION.
70 DIM A(20),B(20),C(20)
80 PRINT" THIS PROGRAM CAN BE USED TO DESIGN LOWPASS,HIGHPASS,BANDPASS AND BANDS
TOP  FILTERS USING ELLIPTIC APPROXIMATION."
90 PRINT"ENTER THE TYPE OF FILTER FUNCTION TO BE DESIGNED.ENTER 'LP'(for lowpass
),HP(for highpass),BP(for bandpass) AND BS(for bandstop)";:INPUT T$
100 IF T$="LP" OR T$="HP" OR T$="BP" OR T$="BS" THEN 120 ELSE 110
110 IF T$="lp" OR T$="hp" OR T$="bp" OR T$="bs" THEN 120 ELSE 90
120 IF T$="lp" OR T$="LP" THEN K=1:LPRINT" DESIGN OF LOWPASS FILTER":GOTO 160
```

```
130 IF T$="hp" OR T$="HP" THEN K=2:LPRINT" DESIGN OF HIGHPASS FILTER":GOTO 160
140 IF T$="bp" OR T$="BP" THEN K=3:LPRINT" DESIGN OF BANDPASS FILTER":GOTO 160
150 IF T$="bs" OR T$="BS" THEN K=4:LPRINT" DESIGN OF BANDSTOP FILTER"
160 LPRINT STRING$(80,"_")
170 PI2=8!*ATN(1)
180 LPRINT:LPRINT" THE SPECIFICATIONS ARE":LPRINT
190 'Inputs are given in the next few lines.
200 PRINT"INPUT THE MAXIMUM PASSBAND LOSS ALLOWED  IN dB";:INPUT A1
210 PRINT"INPUT THE MINIMUM STOPBAND LOSS REQUIRED IN dB";:INPUT A2
220 LPRINT" THE MAXIMUM PASSBAND LOSS ALLOWED IN dB=";USING"##.#####^^^^";A1
230 LPRINT" THE MINIMUM STOPBAND LOSS REQUIRED IN dB=";USING"##.#####^^^^";A2
240 IF K=3 OR K=4 THEN 300
250 PRINT"INPUT THE PASSBAND EDGE FREQUENCY IN Hz";:INPUT W1:W1=PI2*W1
260 PRINT"INPUT THE STOPBAND EDGE FREQUENCY IN Hz";:INPUT W2:W2=PI2*W2
270 LPRINT" THE PASSBAND EDGE FREQUENCY IN RADS./SEC.=";USING"##.######^^^^";W1
280 LPRINT" THE STOPBAND EDGE FREQUENCY IN RADS./SEC.=";USING"##.######^^^^";W2:
W3=W2/W1:IF K=2 THEN W3=1/W3
290 GOTO 380
300 PRINT"INPUT THE DIFFERENCE OF PASSBAND(s) EDGE FREQUENCIES IN Hz";:INPUT B1:
B1=PI2*B1
310 PRINT"INPUT THE DIFFERENCE OF STOPBAND(s) EDGE FREQUENCIES IN Hz";:INPUT B2:
B2=PI2*B2
320 PRINT"INPUT THE GEOMETRIC MEAN OF THE PASSBAND(for bandpass) OR THE STOPBAND
(for bandstop) EDGE FREQUENCIES IN Hz";:INPUT W0:W0=PI2*W0
330 LPRINT" THE DIFFERENCE OF PASSBAND(s) EDGE FREQUENCIES IN RADS./SEC.=";USING
"##.######^^^^";B1
340 LPRINT" THE DIFFERENCE OF STOPBAND(s) EDGE FREQUENCIES IN RADS./SEC.=";USING
"##.######^^^^";B2
350 LPRINT" THE GEOMETRIC MEAN OF THE PASSBAND(for bandpass) OR THE STOPBAND(for
 bandstop)  EDGE FREQUENCIES IN RADS./SEC.=";USING"##.######^^^^";W0
360 W3=B2/B1:IF K=4 THEN W3=1/W3
370 W1=1:W2=W3
380 LPRINT:LPRINT STRING$(80,"_"):LPRINT" ELLIPTIC APPROXIMATION"
390 LPRINT STRING$(80,"_"):LPRINT:LPRINT" DESIGN PARAMETERS ARE":LPRINT
400 IF K=1 OR K=2 THEN 450
410 PRINT"DO YOU WANT TO KNOW THE DESIGN PARAMETERS OF PROTOTYPE LOWPASS FILTER?
 Y/N";:INPUT A$
420 IF A$="Y" OR A$="y" OR A$="N" OR A$="n" THEN 430 ELSE 410
430 IF A$="Y" OR A$="y" THEN K1=1 ELSE K1=0
440 'calculation of ripple factor and the order of the prototype filter.
450 E=SQR(10^(.1*A1)-1):E1=10^(.1*A2)-1
460 S=1/W3:S1=SQR(SQR(1-S*S)):Q=.5*(1-S1)/(1+S1):Q1=Q^4
470 Q=Q*(1+Q1*(2+Q1*(15+150*Q1))):X=LOG(16*E1/E^2)/LOG(1/Q)
480 N=INT(X)+1:N1=INT(N/2):F=N-2*N1:A0=0:L=10^(A1/20):L=LOG((L+1)/(L-1))/(2*N)
490 D=1:L1=EXP(L):L2=1/L1:N2=L1-L2:L3=L1*L1:L4=L2*L2
500 Y1=1:Y2=1:X2=1:Q0=1:M=0
510 M=M+1:X=1:IF (M-2*INT(M/2))>0 THEN X=-1
520 Q0=Q0*Q:X1=X2*Q0:X2=X1*Q0:Y1=Y1*L3:Y2=Y2*L4
530 D1=X1*X*(Y1+Y2):D=D+D1:N3=X2*X*(Y1*L1-L2*Y2):N2=N2+N3
540 IF ABS(D1)>=.000001 OR ABS(N3/N2)>=.000001 THEN 510
550 Q4=SQR(SQR(Q)):S0=ABS(Q4*N2/D):IF F>0 THEN A0=S0
560 'Calculation of second order factors.
570 S2=S0*S0:P1=PI2/(2*N):H=S0:IF F=0 THEN H=10^(-A1/20)
580 FOR J=1 TO N1:U=J:IF F=0 THEN U=J-.5
590 U=U*P1:U1=2*U:U2=0:D=1:X2=1:Q0=1:N2=SIN(U):M=0
600 M=M+1:X=1:IF(M-2*INT(M/2))>0 THEN X=-1
610 Q0=Q0*Q:X1=X2*Q0:X2=X1*Q0:U2=U2+U1
620 D1=2*X*X1*COS(U2):D=D+D1:N3=X*X2*SIN(U2+U):N2=N2+N3
630 IF ABS(D1)>=.000001 OR ABS(N3/N2)>=.000001 THEN 600
640 O=2*N2*Q4/D:W=(1+S*S2)*(1+S2/S):O2=O*O
650 V2=(1-S*O2)*(1-O2/S):V=SQR(V2):C(J)=1/O2
660 D=1+S2*O2:B(J)=(S2*V2+O2*W)/(D*D)
670 A(J)=2*S0*V/D:H=H*B(J)/C(J):NEXT J
680 ' LP to LP and LP to HP transformation
690 W4=W1*W2:W1=SQR(W4):IF A0>0 THEN A0=A0*W1
700 IF K=2 THEN A0=W4/A0
710 FOR J=1 TO N1:IF K=2 THEN 730
720 A(J)=W1*A(J):B(J)=W4*B(J):C(J)=W4*C(J):GOTO 740
730 A(J)=W1*A(J)/B(J):B(J)=W4/B(J):C(J)=W4/C(J)
740 NEXT J
750 IF K=3 OR K=4 THEN IP=0 ELSE IP=1
760 IF K=1 OR K=2 THEN 790
770 IF K1=0 THEN 840
780 LPRINT:LPRINT" PARAMETERS OF THE PROTOTYPE LOWPASS FILTER":LPRINT
790 Z=0:IF K=2 THEN 810
800 IF A0>0 THEN H=H*W1:GOTO 990
810 H=1:IF F=0 THEN H=10^(-A1/20)
820 IF F>0 THEN Z=1 ELSE Z=0
```

```
830 GOTO 990
840 IF K=3 THEN 880
850 'LP to HP transformation for bandstop filter.
860 IF A0>0 THEN A0=1/A0
870 FOR J=1 TO N1:A(J)=A(J)/B(J):B(J)=1/B(J):C(J)=1/C(J):NEXT J
880 W4=W0*W0:IF A0=0 THEN 900 ELSE A(N)=B1*A0:B(N)=W4
890 IF K=4 AND F>0 THEN C(N)=W4:IF K=3 AND F>0 THEN C(N)=0
900 FOR J=1 TO N1:Y=W4*(B1*A(J))^2:B3=(4*W4+B(J)*B1*B1)/2
910 X=B3+SQR(B3*B3-Y):Z=SQR(X)/2:W1=Z+SQR(Z*Z-W4):W2=W4/W1
920 Q=(W1+W2)/(A(J)*B1):A(J)=W1/Q:A(J+N1)=W2/Q:B(J)=W1*W1:B(J+N1)=W2*W2:NEXT J
930 W7=B1*B1/W4
940 FOR J=1 TO N1:G=1+C(J)*W7/2:P=SQR(G*G-1)
950 C(J)=W4*(G-P):C(J+N1)=W4*(G+P):NEXT J
960 N1=N:N=N*2:Z=0:IP=1
970 IF K=4 THEN 1010
980 IF F>0 THEN H=H*B1:Z=1
990 GOSUB 1040
1000 IF IP=0 THEN 840 ELSE END
1010 H=1:IF F=0 THEN H=10^(-A1/20)
1020 GOTO 990
1030 'printing subroutine
1040 LPRINT"       ";CHR$(238);"=";USING"##.######^^^^";E:LPRINT
1050 LPRINT" THE ORDER OF THE FILTER=";N:LPRINT STRING$(80,"_"):LPRINT
1060 IF (K=3 OR K=4) AND IP=1 THEN 1100
1070 IF F=0 THEN 1100
1080 LPRINT" THERE IS A REAL POLE IN THIS FUNCTION":LPRINT
1090 LPRINT"          ";CHR$(229);"0=";USING"##.######^^^^";A0:LPRINT
1100 IF Z=0 THEN 1110 ELSE LPRINT"THERE IS A ZERO AT THE ORIGIN.IF IT IS A BANDP
ASS FILTER, THEN C(";N1;") IS SET TO ZERO AND THERE ARE ONLY";N1-1;"SECOND ORDER
 FACTORS IN THE NUMERATOR":LPRINT
1110 LPRINT" THE COEFFICIENTS OF THE SECOND ORDER FACTORS ARE AS FOLLOWS:":LPRIN
T
1120 LPRINT"   I","A(I)","   B(I)","     C(I)":LPRINT STRING$(80,"_"):LPRINT
1130 FOR J=1 TO N1
1140 LPRINT"  "; J;USING"    ##.######^^^^";A(J);B(J);C(J):LPRINT
1150 NEXT J
1160 LPRINT STRING$(80,"_")
1170 LPRINT" THE VALUE OF H REQUIRED TO GIVE A MAXIMUM GAIN OF 0 dB IS";H
1180 RETURN
```

3

Sensitivity Analysis

A GIVEN NETWORK FUNCTION $H(s)$ can be realized using a wide variety of circuits. One important problem faced by the designer is the evaluation of a design in comparison with many other designs that meet the same specifications. The values of components with which the electronic circuits are built are subject to change due to various factors. For example, the op amp dc gain decreases by about 0.25% per degree Celsius increase in the operating temperature. As another example, carbon film resistors have a temperature coefficient of $-0.2\%/°C$ and, in addition, their values have tolerances and drift during many hours of use. Ceramic capacitors have temperature coefficients of as much as $2\%/°C$. Thus changes in the working conditions involving temperature, humidity, and aging can alter the performance of these circuits. In order to evaluate a particular design, changes in the performance due to changes occurring in the various elements of each circuit must be calculated. In order to measure such changes in the electronic circuits, the concept of sensitivity has been introduced, and we will lay the mathematical foundation for calculating this value. *Sensitivity* is the measure of change in some network performance characteristic (or network function) resulting from a change in one or several elements in the circuit. A particular design may be attractive based on many other considerations, but a high sensitivity may make it useless in practice. Thus, in this chapter, we will introduce various sensitivity concepts that will later aid us in choosing realizations that have minimum sensitivity. These concepts will also help us in seeking methods for minimizing the sensitivities of the realizations we desire to use.

3.1 *Definition of Sensitivity*

The symbol S is used to denote sensitivity. Further, a superscript is used to indicate the performance characteristic being evaluated and a subscript to indicate the network element that causes the change. For example, if a network parameter x causes a change in a network function y, the sensitivity of y with respect to x is denoted by the symbol S_x^y. The *classic logarithmic sensitivity* of y with respect to x is defined as

$$S_x^y = \frac{\partial(\ln y)}{\partial(\ln x)} = \frac{\partial y/y}{\partial x/x} = \frac{\partial y}{\partial x}\frac{x}{y}\bigg|_{\text{nominal values}} \tag{3.1}$$

The above definition of sensitivity can be interpreted as the ratio of the per-unit or percentage change in y to the per-unit or percentage change in x, as long as the changes that take place are incremental (that is, small). In the above expression, y is evaluated with the nominal value of x, and this value is called the nominal value of y. In some cases, the nominal value of y is zero, but it is possible for the change in y to be nonzero. If the above definition of sensitivity is used in these cases, the sensitivity will tend to ∞, which, however, does not mean that the changes are infinite. Thus, in these cases, the definition of sensitivity is changed to that of *unnormalized sensitivity*, defined as

$$US_x^y = \frac{\partial y}{\partial \ln x} = \frac{\partial y}{\partial x}x \tag{3.2}$$

The use of sensitivity can be seen simply by rewriting (3.1) and (3.2). For small changes in x, the per-unit change in x is $\Delta x/x$. If the sensitivities are known, the per-unit change in y can be found (with a nonzero nominal value of y) simply as

$$\frac{\Delta y}{y} \simeq S_x^y \frac{\Delta x}{x} \tag{3.3}$$

If the nominal value of y is zero, one can use the unnormalized sensitivity to find the change in y as

$$\Delta y \simeq US_x^y \frac{\Delta x}{x} \tag{3.4}$$

Expressions such as (3.3) and (3.4) can be used not only to find future changes that will take place after realization of the network but also to find the deviation of y from the nominal value of y due to the tolerance of the network element x represents. Because of the various uses of sensitivity, considerable attention has been given to this concept in the recent past.

Since we will be making use of classic sensitivity extensively, it is worthwhile to remember a number of sensitivity identities, which are listed

Table 3.1 Properties of the logarithmic sensitivity function[a]

Number	Identity or property
1	$S_x^{ky} = S_{kx}^y = S_x^y$
2	$S_{1/x}^y = S_x^{1/y} = -S_x^y$
3	$S_x^{yu} = S_x^y + S_x^u$
4	$S_x^{y/u} = S_x^y - S_x^u$
5	$S_x^{y^n} = nS_x^y$
6	$S_{x^n}^y = \frac{1}{n}S_x^y$
7	$S_{x_1}^y = S_{x_2}^y S_{x_1}^{x_2}$
8[a]	$S_x^y = S_x^{\lvert y\rvert} + jUS_x^{\arg y}$
9[a]	$S_x^{\lvert y\rvert} = \mathrm{Re}\, S_x^y$
10[a]	$US_x^{\arg y} = \mathrm{Im}\, S_x^y$
11	$S_x^{y+u} = \frac{1}{y+u}(yS_x^y + uS_x^u)$
12	$S_x^{\ln y} = \frac{1}{\ln y}S_x^y$

[a] In this relationship, y is complex and k is a constant.

in Table 3.1. These identities can be derived using well-known familiar rules for partial derivatives. Here y and u are single-valued differentiable functions of x, and k and n are real constants.

3.2 *Function Sensitivity*

The sensitivity of the network function $H(s)$, which is realized by a circuit, due to any passive or active element in the same circuit is the *function sensitivity*:

$$S_x^{H(s)} = \frac{\partial H(s)}{\partial x}\frac{x}{H(s)} \tag{3.5}$$

The network function $H(s)$ can be written as a ratio of two polynomials:

$$H(s) = \frac{N(s)}{D(s)} \tag{3.6}$$

Using property 4 in Table 3.1, we can write the function sensitivity $S_x^{H(s)}$

$$S_x^{H(s)} = S_x^{N(s)} - S_x^{D(s)} \tag{3.7}$$

For sinusoidal steady-state applications, the function sensitivity, using

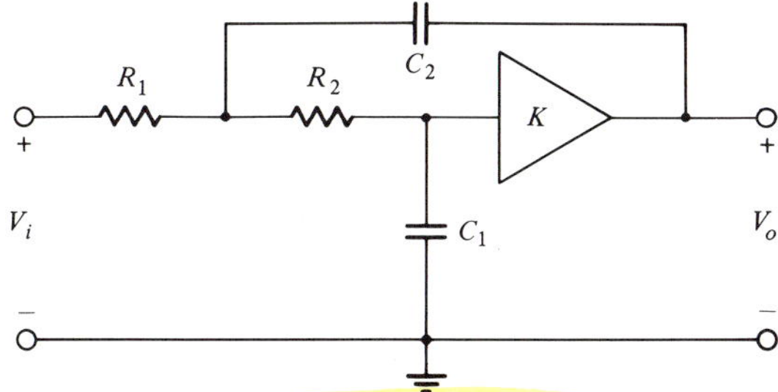

Figure 3.1 Sallen and Key lowpass filter.

property 8, is

$$S_x^{H(j\omega)} = S_x^{|H(j\omega)|} + jUS_x^{\arg H(j\omega)}$$

Example 3.1: Find the function sensitivities $S_x^{H(s)}$ and $S_x^{|H(j\omega)|}$ at $\omega = 1$ r/s, where $H(s)$ is V_o/V_i for the circuit in Fig. 3.1 and x represents G_1, G_2, C_1, C_2, and K. The nominal values are $G_1 = G_2 = C_1 = C_2 = 1$, and $K = 2.8$.

The triangle in Fig. 3.1 represents a voltage-controlled voltage source (VCVS) of gain K. The voltage transfer function $H(s)$ of the above network can be found by routine circuit analysis:

$$H(s) = \frac{K(G_1G_2/C_1C_2)}{s^2 + s\left[\dfrac{G_2 + G_1}{C_2} + \dfrac{G_2}{C_1}(1 - K)\right] + \dfrac{G_1G_2}{C_1C_2}}$$

For convenience, let $1/C_2 = S_2$ and $1/C_1 = S_1$. Then,

$$H(s) = \frac{KG_1G_2S_1S_2}{s^2 + s[S_2(G_2 + G_1) + G_2S_1(1 - K)] + G_1G_2S_1S_2}$$

$$= \frac{N(s)}{D(s)}$$

Since $S_x^{H(s)} = S_x^{N(s)} - S_x^{D(s)}$, we have

$$S_{G_1}^{H(s)} = \frac{KG_2S_1S_2G_1}{KG_1G_2S_1S_2} - \frac{G_1(S_2s + G_2S_1S_2)}{D(s)}$$

Substituting the nominal values of the different parameters, we have

$$S_{G_1}^{H(s)} = 1 - \frac{s+1}{s^2 + 0.2s + 1}$$
$$= \frac{s(s-0.8)}{s^2 + 0.2s + 1}$$

Similarly,

$$S_{G_2}^{H(s)} = \frac{s(s+1)}{s^2 + 0.2s + 1}$$

$$S_{C_1}^{H(s)} = -S_{S_1}^{H(s)} = -\frac{s(s+2)}{s^2 + 0.2s + 1}$$

$$S_{C_2}^{H(s)} = -S_{S_2}^{H(s)} = -\frac{s(s-1.8)}{s^2 + 0.2s + 1}$$

and

$$S_K^{H(s)} = \frac{s^2 + 3s + 1}{s^2 + 0.2s + 1}$$

Since $S_x^{|H(j\omega)|} = \operatorname{Re} S_x^{H(j\omega)}$, we have the following at $s = j1$:

$$S_{G_1}^{|H(j\omega)|} = -4 \qquad S_{G_2}^{|H(j\omega)|} = 5$$

$$S_{C_1}^{|H(j\omega)|} = -10 \qquad S_{C_2}^{|H(j\omega)|} = 9$$

and

$$S_K^{|H(j\omega)|} = 15$$

■

Plots of $S_x^{|H(j\omega)|}$ for different elements in the network are shown in Fig. 3.2. They clearly show that the magnitude of the network function is most sensitive to all the elements in the frequency range near $\omega = 1$ r/s. We will see later that this is the frequency range that is important for this network function.

Now we will find the effect on the network function for variations in more than one parameter. The per-unit change in $H(s)$ will be

$$\frac{\Delta H(s)}{H(s)} = d[\ln H(s)] = \sum_{i=1}^{n} S_{x_i}^{H(s)} \frac{dx_i}{x_i}$$

where n is the number of elements being considered. From the above equation the per-unit change in the magnitude function for the sinusoidal steady state can be obtained:

$$\frac{\Delta|H(j\omega)|}{|H(j\omega)|} = \sum_{i=1}^{n} S_{x_i}^{|H(j\omega)|} \frac{\Delta x_i}{x_i}$$
$$= \sum_{i=1}^{n} \operatorname{Re} S_{x_i}^{H(j\omega)} \frac{\Delta x_i}{x_i} \qquad (3.8)$$

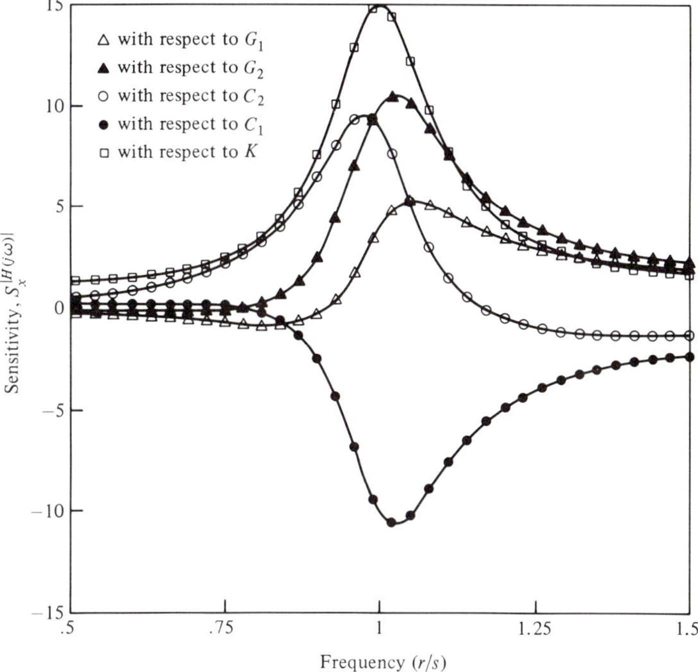

Figure 3.2 Plots of the sensitivities of the magnitude function of the network in Fig. 3.1.

Similarly, the phase change can be found to be

$$\Delta[\arg H(j\omega)] = \sum_{i=1}^{n} \operatorname{Im} S_{x_i}^{H(j\omega)} \frac{\Delta x_i}{x_i} \tag{3.9}$$

The worst-case deviation in (3.8) and (3.9) will occur when all the terms in (3.8) and (3.9) add to each other. Therefore, in particular, the worst-case deviation in the magnitude is

$$\left.\frac{\Delta|H(j\omega)|}{|H(j\omega)|}\right|_{\text{worst case}} = \sum_{i=1}^{n} |\operatorname{Re} S_{x_i}^{H(j\omega)}| \left|\frac{\Delta x_i}{x_i}\right| \tag{3.10}$$

Suppose the tolerance limit for each element is ε, then $|dx_i/x_i| = \varepsilon$. Using this expression,

$$\left.\frac{\Delta|H(j\omega)|}{|H(j\omega)|}\right|_{\text{worst case}} = \varepsilon \sum_{i=1}^{n} |\operatorname{Re} S_{x_i}^{H(j\omega)}|$$

and the worst-case multiparameter sensitivity is defined as

$$W^{|H(j\omega)|} = \sum_{i=1}^{n} |\mathrm{Re}\, S_{x_i}^{H(j\omega)}| \tag{3.11}$$

Example 3.2: Consider the lowpass network shown in Fig. 3.1. The percentage deviation in the resistors is 1%, and the percentage deviation in the capacitors is 0.5%. The expected percentage deviation in the gain of VCVS is 0.01%. Find the worst-case deviation in $|H(j\omega)|$ at $\omega = 1$ r/s. Using (3.10), we have

$$\left.\frac{\Delta|H(j1)|}{|H(j1)|}\right|_{\text{worst case}} = \sum_{i=1}^{2} |S_{G_i}^{|H(j1)|}|\left|\frac{\Delta G_i}{G_i}\right| + \sum_{i=1}^{2} |S_{C_i}^{|H(j1)|}|\left|\frac{\Delta C_i}{C_i}\right| + |S_K^{|H(j1)|}|\frac{|\Delta K|}{|K|}$$

Using the results from Example 3.1, we have

$$\left.\frac{\Delta|H(j1)|}{|H(j1)|}\right|_{\text{worst case}} = 1(5+4) + 0.5(10+9) + 0.01(15) = 18.65\%$$

■

This example clearly shows the necessity for having tighter tolerance on the circuit elements. From (3.8), one can also observe that it is possible to make the temperature-dependent changes zero or a minimum by adopting appropriate technology such that the resistances and capacitances have positive and negative temperature coefficients.

For some classes of networks, the worst-case magnitude sensitivity can be shown to have a lower bound. In networks having resistors and capacitors as passive elements and with commonly used active elements such as generalized immittance converters (GICs), voltage-controlled voltage sources (VCVSs), current-controlled voltage sources (CCVSs), and current-controlled current sources (CCCSs), it may be shown [1] that, for any dimensionless network function $T(s)$,

$$\sum_i S_{x_i}^{T(s)} = 2S_s^{T(s)} \tag{3.12}$$

where x_i is taken to include all passive elements and the active elements of CCVSs and gyrators. The worst-case sensitivity is then

$$W^{|T(j\omega)|} = \sum_i |S_{x_i}^{|T(j\omega)|}| \geq \left|\sum S_{x_i}^{|T(j\omega)|}\right| = |2S_\omega^{|T(j\omega)|}| \tag{3.13}$$

Comparing the first and last terms in (3.13), we can define a lower bound on the worst-case magnitude sensitivity as

$$LW^{|T(j\omega)|} = 2|S_{\omega}^{|T(j\omega)|}| \tag{3.14}$$

The right side of (3.14) indicates that the lower bound of the worst-case sensitivity depends on the network function $T(s)$ and is independent of the network used to realize $T(s)$.

When optimization techniques are used to select the element values so as to minimize the sensitivities of a network to changes in all the elements, a frequent choice is the sum of the squares of the magnitude sensitivities. If this sensitivity is defined as $Q^{|H(j\omega)|}$, then

$$Q^{|H(j\omega)|} = \sum_{i=1}^{n} \left(S_{x_i}^{|H(j\omega)|}\right)^2 \tag{3.15}$$

If the sensitivities of both the magnitude and the phase are to be minimized, one can minimize

$$\sum_{i=1}^{n} S_{x_i}^{H(j\omega)}\left[S_{x_i}^{H(j\omega)}\right]^* = \sum_{i=1}^{n} |S_{x_i}^{H(j\omega)}|^2 \tag{3.16}$$

The above sensitivity function is known as *Schoffler sensitivity.*

3.3 *Coefficient Sensitivity*

The network function $H(s)$ can be expressed in a general form as

$$H(s) = \frac{N(s)}{D(s)} = \frac{n_0 + n_1 s + \cdots + n_m s^m}{d_0 + d_1 s + d_2 s^2 + \cdots + d_n s^n} \tag{3.17}$$

The coefficients n_i and d_i are real functions of the variable x_i. The sensitivities of the coefficients can be easily calculated where the coefficient sensitivities are defined as

$$S_x^{n_i} = \frac{\partial n_i}{\partial x}\frac{x}{n_i} \quad \text{and} \quad S_x^{d_i} = \frac{\partial d_i}{\partial x}\frac{x}{d_i} \tag{3.18}$$

The transfer function of the network shown in Fig. 3.1 was

$$H(s) = \frac{KG_1G_2S_1S_2}{s^2 + s[S_2(G_1 + G_2) + G_2S_1(1 - K)] + G_1G_2S_1S_2}$$

The coefficient sensitivities with respect to one element, say G_1, are

$$S_{G_1}^{n_0} = 1 \qquad S_{G_1}^{d_0} = 1 \qquad \text{and} \qquad S_{G_1}^{d_1} = \frac{S_2G_1}{S_2(G_1 + G_2) + G_2S_1(1 - K)}$$

and $S_{G_1}^{d_2} - 0$.

One can find the network function sensitivity $S_{x_i}^{H(s)}$ using the coefficient sensitivities defined in (3.18). Using (3.17), we have

$$S_{x_i}^{H(s)} = S_{x_i}^{N(s)} - S_{x_i}^{D(s)}$$
$$= \frac{\sum S_{x_i}^{n_i} n_i s^i}{N(s)} - \frac{\sum S_{x_i}^{d_i} d_i s^i}{D(s)} \tag{3.19}$$

For example, $S_{G_1}^{H(s)}$ in the network function $H(s)$ in Example 3.1 can be obtained as

$$S_{G_1}^{H(s)} = 1 - \frac{1 + 0.2s(5)}{s^2 + 0.2s + 1} = \frac{s(s - 0.8)}{s^2 + 0.2s + 1}$$

This is the same result as that obtained for $S_{G_1}^{H(s)}$ in Example 3.1. Thus one use of coefficient sensitivities is in terms of (3.19). The computer evaluation of $S_x^{H(s)}$ is easy through (3.19).

3.4 *Root Sensitivity*

One meaningful criterion used in determining network properties is how the changes in elemental values affect the network poles and zeros of the network function. This can be determined in terms of pole and zero sensitivities. The unnormalized root sensitivity is defined as

$$US_x^{p_i} = \frac{\partial p_i}{\partial x} x \qquad \text{for a pole} \tag{3.20a}$$

and

$$US_x^{z_i} = \frac{\partial z_i}{\partial x} x \qquad \text{for a zero} \tag{3.20b}$$

where p_i and z_i are the poles and zeros, respectively, of a network function. Since the treatment of zero sensitivity is identical to that of pole sensitivity, we shall consider only pole sensitivity here.

If we are interested in pole sensitivity with respect to an element x whose nominal value is x_0, then we must consider the denominator polynomial as a function of two variables s and x. Assume that this denominator polynomial is $D(s, x)$. Because we are considering linear networks, this polynomial $D(s, x)$ has a first-order dependence on any network element. Thus $D(s, x)$ can always be expressed as

$$D(s, x) = D_0(s) + xD_1(s) \tag{3.21}$$

Let $s = p$ be the nominal value of the pole with the nominal value for x as x_0. Then,

$$D(p, x_0) = 0 \tag{3.22}$$

Now, if x increases from x_0 to $x_0 + \Delta x$, the new pole position will be different. Let this pole position be $p' = p + \Delta p$, where Δp is the shift in the pole position. Then, the following must also be true:

$$D(p + \Delta p, x_0 + \Delta x) = 0 \tag{3.23}$$

Using Taylor series approximation, we can write

$$D(p + \Delta p, x_0 + \Delta x) \cong D(p, x_0) + \left.\frac{\partial D(s, x)}{\partial s}\right|_{s=p,\ x=x_0} \Delta p + \left.\frac{\partial D(s, x)}{\partial x}\right|_{s=p,\ x=x_0} \Delta x$$

Using (3.22) and (3.23) in the above equation, we obtain

$$\frac{\Delta p}{\Delta x} \cong \left.\frac{-\partial D(s, x)/\partial x}{\partial D(s, x)/\partial s}\right|_{s=p,\ x=x_0}$$

In the limiting case $\Delta x \to 0$, the above becomes an exact relationship:

$$\frac{\partial p}{\partial x} = \left.\frac{-\partial D(s, x)/\partial x}{\partial D(s, x)/\partial s}\right|_{s=p,\ x=x_0} \tag{3.24}$$

Using (3.21) in (3.24), we obtain

$$\frac{\partial p}{\partial x} = \frac{-D_1(p)}{D_0'(p) + x_0 D_1'(p)} \tag{3.25}$$

where the prime indicates differentiation with respect to s. Once $\partial p/\partial x$ is found using (3.24) or (3.25), we can substitute its value in (3.20a) to find the unnormalized pole sensitivity evaluated with $x = x_0$. We shall now illustrate the use of the above expressions in finding the pole sensitivity with respect to K, the gain of the VCVS in Fig. 3.1, with the same nominal values as in Example 3.1.

Substituting the nominal values of all the elements except the one under consideration, the denominator polynomial can be written in the form of (3.21):

$$D(s, K) = s^2 + 3s + 1 - Ks \tag{3.26}$$

Verify that

$$D_0(s) = s^2 + 3s + 1$$

and

$$D_1(s) = -s$$

The nominal value of one of the complex poles can be found by substituting the nominal value $K = 2.8$ in (3.26):

$$p = -0.1 + j0.995$$

Then, using (3.25) and (3.20b), we obtain

$$US_K^p = \frac{Kp}{2p + 0.2} = \frac{2.8(-0.1 + j0.995)}{-0.2 + j1.99 + 0.2}$$
$$= 1.4 + j0.1407$$

Assume that the gain of the VCVS increases from 2.8 to 2.94, so that the per-unit change is $\Delta K/K = 0.05$. Then, the shift in the pole position can be calculated:

$$\Delta p = US_K^p \frac{\Delta K}{K} = 0.07 + j0.0074$$

The approximate position of the new pole is

$$p' = p + \Delta p = -0.03 + j1.0024$$

The exact value of the new pole position can also be found directly from (3.26) in this simple case by substituting the new value $K = 2.94$ and obtaining $p' = -0.03 + j0.9995$. This value is very close to the approximate value calculated before. Thus for small changes we can predict the shifts in the pole positions quite accurately using root sensitivities as indicated in the above case.

3.5 ω_p *and* Q_p *Sensitivities*

In signal processing, second-order circuits (where the name is derived from the order of the network function realized by the circuit) play an important role. The most important second-order filters are lowpass, highpass, and bandpass filters. One main reason for a sensitivity study is to compare the sensitivity performance of various second-order circuit realizations of the same network function and choose the best one for a given application. Another reason for such a study is to choose free design parameters available to the network designer of a given circuit such that the realized network will have the minimum possible sensitivity. For these reasons, we shall develop criteria based on the sensitivities of this important class of filters. The magnitude and phase functions of a second-order network function are mainly controlled by the pole positions. Usually the two poles form a pair of complex conjugate poles. The function sensitivity or the sensitivities of the magnitude and phase functions are too complex to be used because they are specified over an entire frequency range of use. For purposes of comparison, it is better to use a performance criterion that emphasizes the selective characteristics of these network functions. The ω_p and Q_p sensitivities are two such criteria. The treatment starts with the definition of the second-order network functions and of the pole frequency

ω_p and pole Q factor Q_p. The second-order lowpass, bandpass, and highpass network functions are

$$H_{LP}(s) = \frac{H_0}{s^2 + d_1 s + d_0} \tag{3.27a}$$

$$H_{BP}(s) = \frac{H_0 s}{s^2 + d_1 s + d_0} \tag{3.27b}$$

and

$$H_{HP}(s) = \frac{H_0 s^2}{s^2 + d_1 s + d_0} \tag{3.27c}$$

Consider $H_{BP}(s)$. The maximum value of $|H_{BP}(j\omega)|$ occurs at $\omega_p = \sqrt{d_0}$, the *resonant frequency* (also called the *undamped natural frequency*), and we shall consider this frequency the pole frequency because it controls the pole positions of $H_{BP}(s)$.

Also, a quality factor Q_p can be defined for this transfer function:

$$Q_p = \frac{\omega_p}{BW} = \frac{\sqrt{d_0}}{d_1}$$

where BW is the bandwidth defined as the difference between the 3-dB frequencies in $|H_{BP}(j\omega)|$ and can be shown to be equal to d_1. In terms of ω_p and Q_p, $H_{BP}(s)$ becomes

$$H_{BP}(s) = \frac{H_0 s}{s^2 + (\omega_p/Q_p)s + \omega_p^2}$$

where $d_0 = \omega_p^2$ and $d_1 = \omega_p/Q_p$.

We shall call Q_p the pole Q factor because it also controls the pole positions. In fact, the poles are (for $Q_p > 0.5$)

$$p_{1,2} = \sigma \pm j\omega = -\frac{d_1}{2} \pm j\sqrt{d_0 - \left(\frac{d_1}{2}\right)^2}$$

$$= -\frac{\omega_p}{2Q_p} \pm j\frac{\omega_p}{2Q_p}\sqrt{4Q_p^2 - 1}$$

In terms of the coefficients of the network function.

$$\omega_p = \sqrt{d_0} \qquad \text{and} \qquad Q_p = \frac{\sqrt{d_0}}{d_1} \tag{3.28}$$

In cases when $Q_p > 5$ (which are of practical interest), all three types of network functions have resonant peaks near the pole frequency ω_p. When

$Q_p > 5$, the magnitude characteristics near ω_p in all three types of filters are controlled by ω_p and Q_p. Thus, though the results we obtain now are exactly true only for bandpass networks, they are approximately true for other types as well when $Q_p > 5$. The sensitivities of ω_p and Q_p are now defined:

$$S_x^{\omega_p} = \frac{\partial \omega_p}{\partial x}\frac{x}{\omega_p} \qquad \text{and} \qquad S_x^{Q_p} = \frac{\partial Q_p}{\partial x}\frac{x}{Q_p} \tag{3.29}$$

Using (3.28) and (3.29), we can find $S_x^{\omega_p}$ and $S_x^{Q_p}$ in terms of coefficient sensitivities:

$$S_x^{\omega_p} = \tfrac{1}{2}S_x^{d_0} \qquad \text{and} \qquad S_x^{Q_p} = \tfrac{1}{2}S_x^{d_0} - S_x^{d_1} \tag{3.30}$$

The ω_p and Q_p sensitivities can be obtained easily using the coefficient sensitivities as suggested by (3.30). The following example illustrates this procedure.

Example 3.3: Consider the network in Fig. 3.1. Find the ω_p and Q_p sensitivities with respect to the passive elements R_1, R_2, C_1, and C_2 and the active parameter K. If 1% resistors and 0.5% capacitors are used in the network, find the worst-case per-unit change in ω_p and Q_p. Assume that K is realized to an exact value.

The transfer function of the network is given in the solution for Example 3.1. From this transfer function, we have

$$d_1 = \frac{\omega_p}{Q_p} = S_2(G_2 + G_1) + G_2S_1(1 - K)$$

and

$$d_0 = \omega_p^2 = G_1G_2S_1S_2$$

where $S_i = 1/C_i$ and $G_i = 1/R_i$.

In order to find the sensitivities of d_1 and d_0, we will need the nominal values of d_1 and d_0. Given that $G_1 = G_2 = C_1 = C_2 = 1$ and $K = 2.8$, we find that

$$d_1 = \frac{\omega_p}{Q_p} = 0.2 \qquad \text{and} \qquad d_0 = 1$$

By inspection, we see that

$$S_x^{d_0} = 1$$

where $x = G_1, G_2, S_1, S_2$. And thus

$$S_y^{d_0} = -1$$

where $y = R_1, R_2, C_1, C_2$. Therefore

$$S_y^{\omega_p} = -\tfrac{1}{2}$$

where $y = R_1, R_2, C_1, C_2$. Also,

$$S_K^{\omega_p} = 0$$

since d_0 and ω_p do not depend on K.

Let us find the derivative of d_1 with respect to G_1 first. We have

$$\frac{\partial d_1}{\partial G_1} = S_2 = 1$$

Therefore

$$S_{R_1}^{d_1} = -S_{G_1}^{d_1} = -\frac{\partial d_1}{\partial G_1}\frac{G_1}{d_1} = -5$$

Next we find that

$$S_{R_1}^{Q_p} = \tfrac{1}{2}S_{R_1}^{d_0} - S_{R_1}^{d_1} = -\tfrac{1}{2} + 5 = 4.5$$

Following similar steps, we find the other Q_p sensitivities:

$$S_{R_2}^{Q_p} = -4.5 \qquad S_{C_1}^{Q_p} = -9.5$$

$$S_{C_2}^{Q_p} = 9.5 \qquad \text{and} \qquad S_K^{Q_p} = 5$$

The per-unit change in ω_p can be written as

$$\frac{\Delta\omega_p}{\omega_p} = \sum_{x_i} S_{x_i}^{\omega_p}\frac{\Delta x_i}{x_i}$$

and, in the worst case,

$$\left.\frac{\Delta\omega_p}{\omega_p}\right|_{WC} = \sum_{x_i} |S_{x_i}^{\omega_p}|\left|\frac{\Delta x_i}{x_i}\right|$$

Since all the resistors have a 1% tolerance and all the capacitors have a 0.5% tolerance, we can write the sum as two groups of terms:

$$\left.\frac{\Delta\omega_p}{\omega_p}\right|_{WC} = \left|\frac{\Delta R}{R}\right|\sum_{R_i} |S_{R_i}^{\omega_p}| + \left|\frac{\Delta C}{C}\right|\sum_{C_i} |S_{C_i}^{\omega_p}|$$

$$= 0.01 \times 1 + 0.005 \times 1$$

$$- 0.015 = 1.5\%$$

Similarly,

$$\left.\frac{\Delta Q_p}{Q_p}\right|_{WC} = \left|\frac{\Delta R}{R}\right| \sum_{R_i} |S_{R_i}^{Q_p}| + \left|\frac{\Delta C}{C}\right| \sum_{C_i} |S_{C_i}^{Q_p}| + \left|\frac{\Delta K}{K}\right| S_K^{Q_p}$$

Since K is realized exactly, $\Delta K/K = 0$. Then, substituting the different values, we have

$$\left.\frac{\Delta Q_p}{Q_p}\right|_{WC} = 18.5\% \qquad ■$$

The above per-unit changes, due to tolerance effects, can be tuned out by adjusting the resistance values. (This is the usual case, however, capacitors are usually not amenable to adjustment.) More importantly, we are interested in predicting the per-unit changes that will result from future changes in resistances, capacitances, and other amplifier parameters after tuning out the tolerance effects. For example, assume that resistance values have been adjusted to realize a given ω_p and Q_p at a nominal room temperature of 20°C. However, the actual application requires that the filter be used at an operating temperature of 80°C. The increase in operating temperature will cause changes in resistances, capacitances, and amplifier parameters. Any such changes will in turn cause the actual values of ω_p and Q_p to differ from the realized values. Such future changes can also be predicted using the sensitivities, as illustrated in the following example.

Example 3.4: In the network in Fig. 3.1, the resistors and capacitors have temperature coefficients of $5 \times 10^{-3}\%/°\text{C}$ and $-10^{-2}\%/°\text{C}$, respectively. The gain K has a temperature coefficient of $10^{-5}\%/°\text{C}$. Find the per-unit change in ω_p and Q_p due to an increase of 40°C in the operating temperature.

An increase of 40°C in the operating temperature will result in

$$\frac{\Delta R}{R} = 0.2\%$$

$$\frac{\Delta C}{C} = -0.4\%$$

and

$$\frac{\Delta K}{K} = 4 \times 10^{-4}\%$$

In order to find the per-unit changes, we must find the algebraic sum of the changes caused by each element. Thus

$$\begin{aligned}\frac{\Delta\omega_p}{\omega_p} &= \sum_{R_i} S_{R_i}^{\omega_p}\frac{\Delta R_i}{R_i} + \sum_{C_i} S_{C_i}^{\omega_p}\frac{\Delta C_i}{C_i}\\ &= \frac{\Delta R}{R}(-1) + \frac{\Delta C}{C}(-1)\\ &= -0.2 + 0.4 = 0.2\%\end{aligned}$$

■

Note that, if we had calculated the worst-case change, it would have been 0.6%. However, the changes caused by the increase in the resistance values are partially compensated for by the decrease in the capacitance values, and thus it should be taken into account. Further, also note that, if $\Delta R/R = -\Delta C/C$, then $\Delta\omega_p/\omega_p = 0$ in this circuit, which means that, by choosing a technology of fabrication such that $\Delta R/R = -\Delta C/C$, it is possible to eliminate future changes in ω_p. This is possible not only in this network but in all networks. This is due to the reason that

$$\sum_{R_i} S_{R_i}^{\omega_p} = \sum_{C_i} S_{C_i}^{\omega_p}$$

and the assumption that all the resistance and capacitance values track each other, which is satisfied in all the networks considered in later chapters. Let us now calculate the percentage change in Q_p.

$$\begin{aligned}\frac{\Delta Q_p}{Q_p} &= \sum_{R_i} S_{R_i}^{Q_p}\frac{\Delta R_i}{R_i} + \sum_{C_i} S_{C_i}^{Q_p}\frac{\Delta C_i}{C_i} + S_K^{Q_p}\frac{\Delta K}{K}\\ &= \frac{\Delta R}{R}\sum_{R_i} S_{R_i}^{Q_p} + \frac{\Delta C}{C}\sum S_{C_i}^{Q_p} + S_K^{Q_p}\frac{\Delta K}{K}\\ &= 0 + 0 + 5 \times 4 \times 10^{-4}\%\\ &= 2 \times 10^{-3}\%\end{aligned}$$

Note that

$$\sum_{R_i} S_{R_i}^{Q_p} = \sum_{C_i} S_{C_i}^{Q_p} = 0$$

and that this is possible because Q_p is a function of the ratios of the resistances and ratios of the capacitances. This is true not only for this network but for every network considered in later chapters. Let us verify this fact for this network. Since $d_0 = \omega_p^2 = G_1G_2S_1S_2$, we have

$$\omega_p = \sqrt{G_1G_2S_1S_2}$$

We use the above relation in the equation for d_1, which is also equal to

ω_p/Q_p, to find $1/Q_p$:

$$\frac{1}{Q_p} = \sqrt{\frac{S_2}{S_1}}\left(\sqrt{\frac{G_2}{G_1}} + \sqrt{\frac{G_1}{G_2}}\right) + \sqrt{\frac{G_2 S_1}{G_1 S_2}}(1 - K)$$

$$= \sqrt{\frac{C_1}{C_2}}\left(\sqrt{\frac{R_1}{R_2}} + \sqrt{\frac{R_2}{R_1}}\right) + \sqrt{\frac{R_1 C_2}{R_2 C_1}}(1 - K)$$

From the above discussion, it can be seen that $1/Q_p$, hence Q_p, is a function of the ratio between the resistance and capacitance values.

In conclusion, by realizing the active filters appropriately, it is possible to make the per-unit change in ω_p, caused by the change in the values of passive elements, zero. In addition, since Q_p is dependent only on the ratio of the resistance and capacitance values, changes in passive elements do not affect Q_p, at least to a first-order effect. The same statements can be made about all the active filters considered in later chapters. However, the effect of a change in the active element characteristic cannot be ignored and a discussion of this subject is postponed until a later section.

3.6 *Function and Pole Sensitivities in Terms of ω_p and Q_p Sensitivities*

One method of comparing different realizations is based on their ω_p and Q_p sensitivities, as developed in the previous section. However, in some cases ω_p sensitivities are more critical than Q_p sensitivities, and in other cases Q_p sensitivities are more critical than ω_p sensitivities. In order to compare networks in two such different groups, a common strategy must be established. For the purpose of comparing all types of networks, we can use a comparison based on the function sensitivity or the per-unit change in the function magnitude. Toward this end, let us find the per-unit change in the function magnitude $|H(j\omega)|$ in terms of the per-unit changes in ω_p and Q_p. Since $|H(j\omega)|$, second-order network function, is dependent on of both ω_p and Q_p, we can write

$$\frac{\Delta|H(j\omega)|}{|H(j\omega)|} = S_{\omega_p}^{|H(j\omega)|}\frac{\Delta\omega_p}{\omega_p} + S_{Q_p}^{|H(j\omega)|}\frac{\Delta Q_p}{Q_p} \tag{3.31}$$

Toward finding $S_{\omega_p}^{|H(j\omega)|}$ and $S_{Q_p}^{|H(j\omega)|}$, using any one type of transfer function, we can show that

$$S_{\omega_p}^{H(s)} = \frac{-\omega_p(s/Q_p + 2\omega_p)}{s^2 + (\omega_p/Q_p)s + \omega_p^2} \tag{3.32}$$

and

$$S_{Q_p}^{H(s)} = \frac{(\omega_p/Q_p)s}{s^2 + (\omega_p/Q_p)s + \omega_p^2} \tag{3.33}$$

With $s = j\omega$ in (3.32) and (3.33) and using identity 9 in Table 3.1, we can show that

$$S_{\omega_p}^{|H(j\omega)|} = \frac{(2 - 1/Q_p^2)x^2 - 2}{(1 - x^2)^2 + x^2/Q_p^2} \tag{3.34}$$

and

$$S_{Q_p}^{|H(j\omega)|} = \frac{x^2/Q_p^2}{(1 - x^2)^2 + x^2/Q_p^2} \tag{3.35}$$

where $x = \omega/\omega_p$.

Now, using (3.34) and (3.35) in (3.31), we can obtain the per-unit change in the magnitude function $|H(j\omega)|$, and this will be a function of the frequency, as shown in Fig. 3.3. For network functions of practical use, the per-unit change will have ripples in the frequency band of interest—

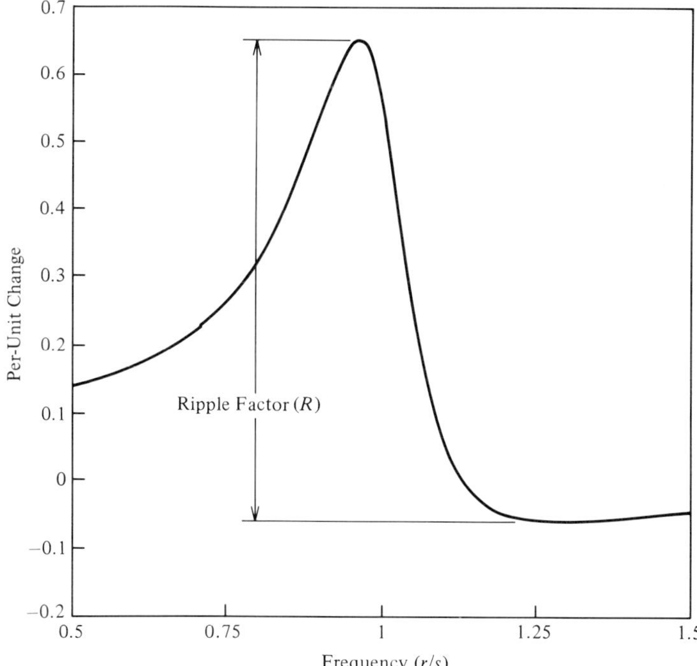

Figure 3.3 Per-unit change in the magnitude function as a function of frequency.

the passband. Therefore, if we want to guarantee an accuracy of $\pm n$ percent for the filter characteristic, the ripple R should be less than or equal to $2n/100$. Note that this ripple R is

$$R = \left.\frac{\Delta|H(j\omega)|}{|H(j\omega)|}\right|_{\max} - \left.\frac{\Delta|H(j\omega)|}{|H(j\omega)|}\right|_{\min}$$

and we can use this quantity as a figure of merit for comparison purposes because it includes both maximum and minimum per-unit changes [2]. Furthermore, it will not be a function of the frequency. This is the kind of figure of merit we have been looking for. Therefore we shall proceed to find R in terms of ω_p, Q_p, and the per-unit changes in ω_p and Q_p. We first define

$$r \triangleq \frac{\Delta Q_p/Q_p}{\Delta\omega_p/\omega_p}$$

Then, the per-unit change in magnitude function can be found as

$$\frac{\Delta|H(j\omega)|}{|H(j\omega)|} = \frac{Ex^2 - 2}{(1-x^2)^2 + x^2/Q_p^2}\,\frac{\Delta\omega_p}{\omega_p} \tag{3.36}$$

where

$$E = 2 + \frac{r-1}{Q_p^2}$$

The frequencies at which the above per-unit change reaches its maximum and minimum are given by

$$x_{\max} = \sqrt{\frac{2(1+C)}{E}} \tag{3.37a}$$

and

$$x_{min} = \sqrt{\frac{2(1-C)}{E}} \tag{3.37b}$$

where

$$C = \frac{\sqrt{4Q_p^2 + r^2 - 1}}{2Q_p^2}$$

Using (3.37) in (3.36), we can find the maximum and minimum per-unit changes and, in turn, the ripple R. Thus R can be calculated to be

$$R = \frac{4Q_p^2\sqrt{4Q_p^2 + r^2 - 1}}{4Q_p^2 - 1}\frac{\Delta\omega_p}{\omega_p}$$

$$= \frac{4Q_p^2}{4Q_p^2 - 1}\sqrt{(4Q_p^2 - 1)\left(\frac{\Delta\omega_p}{\omega_p}\right)^2 + \left(\frac{\Delta Q_p}{Q_p}\right)^2} \tag{3.38}$$

For high values of Q_p (> 5), one can approximate the above equation as

$$R \cong \sqrt{\left(2Q_p\frac{\Delta\omega_p}{\omega_p}\right)^2 + \left(\frac{\Delta Q_p}{Q_p}\right)^2} \tag{3.39}$$

Equation (3.39) clearly shows that the effect of a change in ω_p is $2Q_p$ times greater than that of a comparable change in Q_p. Any comparison of filter networks should take into consideration this weighting factor for the per-unit change in ω_p. In other words, for comparable changes in ω_p and Q_p, a change in ω_p is $2Q_p$ times more effective in changing the magnitude characteristic, and this fact should be taken into account when making comparisons of networks based on ω_p and Q_p sensitivities. Particularly, when minimizing sensitivities of networks, we should try to minimize the ripple factor R as given by (3.38) or (3.39) rather than minimizing the ω_p or Q_p sensitivity only. This is because R provides information about the maximum change that will occur in the magnitude characteristic and, in the final analysis, this is the most important consideration. Thus, the ripple factor R as given by (3.38) or (3.39) can be used as a figure of merit for comparisons as well as for minimizing network sensitivity. However, using this factor as a figure of merit creates a problem. In order to calculate R, one has to find both the ω_p and Q_p sensitivities and/or the per-unit changes in ω_p and Q_p. At this juncture, a question arises whether there is a single quantity that gives the same information about the network that R does. The next logical step is to find the pole sensitivity or the per-unit change in the pole position. After all, the pole position is fixed by both ω_p and Q_p, and thus pole sensitivity must reveal some information about the ω_p and Q_p sensitivities. Let us next calculate the per-unit change in the pole position.

$$p = \frac{-\omega_p}{2Q_p} + j\omega_p\sqrt{1 - \frac{1}{4Q_p^2}} \tag{3.40}$$

Using (3.40), we can show that

$$\frac{\Delta p}{p} = \frac{\Delta \omega_p}{\omega_p} - \frac{j}{\left(4Q_p^2 - 1\right)^{1/2}} \frac{\Delta Q_p}{Q_p} \tag{3.41}$$

and

$$\left|\frac{\Delta p}{p}\right| = \frac{1}{\left(4Q_p^2 - 1\right)^{1/2}} \sqrt{\left(4Q_p^2 - 1\right)\left(\frac{\Delta \omega_p}{\omega_p}\right)^2 + \left(\frac{\Delta Q_p}{Q_p}\right)^2}$$

Using the above equation in (3.38), we have

$$R = \frac{4Q_p^2}{\left(4Q_p^2 - 1\right)^{1/2}} \left|\frac{\Delta p}{p}\right| \tag{3.42}$$

That is,

$$R = \frac{4Q_p^2}{\left(4Q_p^2 - 1\right)^{1/2}} |S_x^p| \frac{\Delta x}{x} \tag{3.43}$$

For a given pole position the nominal value of Q_p is fixed, and for a particular technology $\Delta x/x$ is also given. Therefore, from (3.43), it can be clearly seen that $|S_x^p|$ can be used as a figure of merit, either for comparison or for minimization. The advantage of this approach is that there is only one sensitivity to be calculated, and at the same time it provides the same information that the ripple factor R does about the maximum change in the magnitude function. Thus, in the remaining chapters, we will use $|S_x^p|$ or $|\Delta p/p|$ as a figure of merit. Further, if one is interested in the per-unit changes in ω_p and Q_p or the ω_p and Q_p sensitivities, one can use the following relationships, which are direct results of (3.41):

$$\frac{\Delta \omega_p}{\omega_p} = \text{Re}\frac{\Delta p}{p}$$

and (3.44)

$$S_x^{\omega_p} = \text{Re}\, S_x^p$$

$$\frac{\Delta Q_p}{Q_p} = -\left(4Q_p^2 - 1\right)^{1/2} \text{Im}\, \frac{\Delta p}{p}$$

and (3.45)

$$S_x^{Q_p} = -\left(4Q_p^2 - 1\right)^{1/2} \text{Im}\, S_x^p$$

In both (3.44) and (3.45), the parameter x for which the sensitivities are calculated is assumed to be real. Equations (3.44) and (3.45) provide an alternative way of calculating the ω_p and Q_p sensitivities in conjunction with (3.25). In order to verify the fact that these equations can also be used to obtain the ω_p and Q_p sensitivities, one can rework Example 3.2. For example, assume that we want to obtain $S_{G_1}^{\omega_p}$ and $S_{G_1}^{Q_p}$. First we calculate $S_{G_1}^{p}$:

$$S_{G_1}^{p} = \left.\frac{-G_1[\partial D(s, G_1)/\partial G_1]}{\partial D(s, G_1)/\partial s}\right|_{s=p,\ G_1=1} \frac{1}{p}$$

The nominal values of ω_p, Q_p, and p are

$$\omega_p = 1 \text{ r/s} \qquad Q_p = 5 \qquad p = -0.1 + j\sqrt{0.99}$$

Now,

$$S_{G_1}^{p} = \frac{-G_1[\partial D(s, G_1)/\partial G_1]|_{s=p,\ G_1=1}}{p(2p + \omega_p/Q_p)}$$

Note that the numerator depends on the value of G_1 and the derivative of $D(s, G_1)$ with respect to G_1. However, the denominator is independent of G_1 and depends only on the nominal values of ω_p and Q_p. This means that, irrespective of the element for which we calculate the sensitivity, the denominator has to be calculated only once and that the same denominator can be used for calculation of the pole sensitivities with respect to other elements as well. Also note that the denominator of the above is

$$p\left(2p + \frac{\omega_p}{Q_p}\right) = p^2 - \omega_p^2$$

Proceeding further, we first find

$$-G_1\frac{\partial D(s, G_1)}{\partial G_1} = -(G_1 S_2 s + G_1 G_2 S_1 S_2)$$

Therefore, after substituting the nominal values of the elements, we have

$$S_{G_1}^{p} = \frac{-(p+1)}{p^2 - 1}$$

Now, substituting the value of p, we obtain

$$S_{G_1}^{p} = 0.5 + j0.4523$$

Therefore

$$S_{G_1}^{\omega_p} = 0.5 \qquad \text{and} \qquad S_{G_1}^{Q_p} = -4.5$$

which are in fact the values calculated in Example 3.2.

3.7 *Active Sensitivity*

Apart from the resistances and capacitances used in signal processing networks, the operational amplifier is also an important component. In this section, we discuss the effects of the nonidealness of this component. Usually the nominal designs of operational amplifier circuits are based on the assumption that the op amps used are ideal. In such cases, these components are assumed to possess infinite gain, infinite input impedance, and zero output impedance. However, in reality, we know that this is not true, as indicated in Chap. 1. Most active filters are designed such that the finite (but high) input impedances and nonzero (but very low) output impedances will have a negligible effect on the performance of the circuits. This is done by appropriately scaling the passive network. In most cases, we will be interested only in voltage transfer ratios. Remember that transfer ratios are not affected by the magnitude scaling of the impedances. In such cases, we can choose the magnitude scaling such that the resulting impedance levels of the passive part of the network can be adjusted so that the high input impedances and the low output impedance of the op amps will not have a substantial effect on the overall circuit performance. However, the frequency-dependent gains of the op amps affect the performance of the circuit. Sometimes when the operating frequency is high enough, the circuits may even become oscillatory, which we will consider in later chapters. In most of our applications, we will use only frequency-compensated op amps. For most cases, the gain of an op amp can be described by

$$A(s) = \frac{A_0}{(1 + s\tau_1)(1 + s\tau_2)(1 + s\tau_3)}$$

The frequency range of our application is such that

$$|s| \ll \frac{1}{\tau_2}, \frac{1}{\tau_3}$$

and

$$|s| \gg \frac{1}{\tau_1}$$

Therefore the above gain equation can be approximated as

$$A(s) \simeq \frac{A_0}{s\tau_1} = \frac{1}{s\tau} \tag{3.46}$$

where

$$\tau = \frac{\tau_1}{A_0}$$

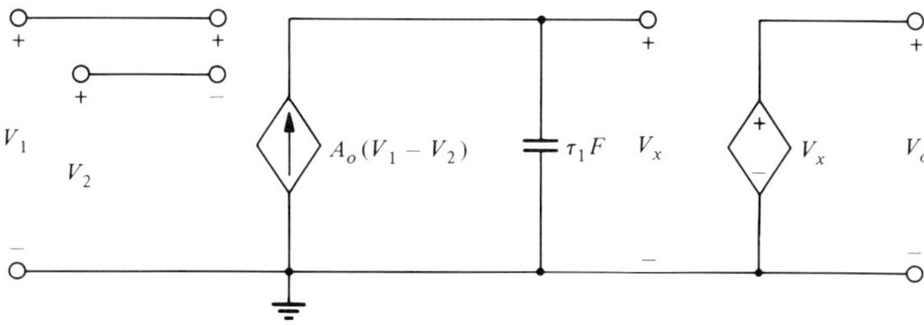

Figure 3.4 Equivalent circuit for the nonideal model of an op amp.

The quantity τ is the op amp time constant and is equal to the inverse of the gain bandwidth product of the op amp B. The ideal value of τ is zero, which corresponds to the case of infinite gain of the op amp. However, in practice, τ has a nonzero value. For example, for a 741 type op amp, data books give the value of B as $2\pi \times 10^6$ r/s, and therefore, for such an op amp, $\tau = 0.5/\pi$ μs. Thus we assume that the nonideal model used in all our applications will be ideal except for the finite frequency-dependent gain as given by (3.46). In this section and in all later chapters, this is the nonideal model that will be used. This means that the input currents to the op amp are zeros and that the op amp can supply any amount of current as required by the rest of the circuit. These facts must be kept in mind whenever we refer to the nonideal model of the op amp. This crude nonideal model can be represented by the equivalent circuit as shown in Fig. 3.4.

The nonzero value of τ gives rise to two main problems. Since we are interested in realizing only second-order transfer functions, we emphasize such networks only in our discussions. However, most of our results are more general. Networks for realizing second-order transfer functions are designed with the assumption that $\tau = 0$ in (3.46). However, if the nonideal model of the op amp is included in the analysis, then the transfer function will be a higher-order one whose denominator polynomial has a degree greater than 2, depending on the number of op amps used in the circuit. To be precise, the degree of the denominator polynomial of the transfer function, when the crude nonideal model of the op amp is used, will be 2 plus the number of op amps used in the circuit. This means that, in addition to the two desired poles corresponding to the original second-order transfer function, there will be other poles caused by the nonideal model of the op amps. The number of additional poles is exactly equal to the number of op amps used. The second problem is that the design poles, derived from the given transfer function,

$$p_{1,2} = -\omega_p\left[\frac{1}{2Q_p} \pm j\sqrt{1 - \left(\frac{1}{2Q_p}\right)^2}\right] \tag{3.47}$$

will shift to new positions by the nonzero values of τ, which can be described by

$$p'_{1,2} = -\omega'_p \left[\frac{1}{2Q'_p} \pm j\sqrt{1 - \left(\frac{1}{2Q'_p} \right)^2} \right] \tag{3.48}$$

where p'_1 and p'_2 are the actual pole positions.

In addition to the excess poles, there may also be additional zeros other than those intended. These excess poles and zeros do not substantially affect the frequency response of the circuit in the frequency range of interest. Their magnitudes are usually more than a few tenths of the gain bandwidth products of the op amps, which means that they are in excess of a few hundreds kilohertz for most practical op amps. Such poles and zeros do not substantially affect the frequency response of circuits whose f_p values are not in excess of a few tens of kilohertz. This is one of the major limitations of active filters—that their applicable frequency range cannot exceed 100 kHz. Now, returning to our problem, the shift in the pole positions is the most serious problem in almost all active filters. This shift, caused by the nonzero value of τ, may be so dangerous that the circuit may even become oscillatory in some cases. Even when the nonzero value of τ is taken into account and such shifts are compensated for (this is possible in most circuits using a technique called predistortion), such that ω'_p and Q'_p are the desired pole frequency and pole Q factor, respectively, both are sensitive to variations in A_0 and τ. Furthermore, the changes that occur in A_0 and τ are not correlated with the changes that take place in resistances and capacitances. The sensitivity of quantities such as pole position p, pole frequency ω_p, and pole Q factor Q_p with respect to τ is called the *active sensitivity*. The logarithmic sensitivity of a function y can be defined as

$$S_\tau^y = \frac{\partial y}{\partial \tau} \frac{\tau}{y} \tag{3.49}$$

Since the nominal value of τ is zero, the above sensitivity goes to zero as $\tau \to 0$. In such cases, where the nominal value of the element goes to zero (called the zero element), an alternative definition of sensitivity is given. This sensitivity, called *semilogarithmic sensitivity*, is defined as

$$S_\tau^y = \frac{\partial y}{\partial \tau} \frac{1}{y} \tag{3.50}$$

and we will use it in all later chapters. Furthermore, when more than one op amp is used in a circuit, we shall assume that all the op amps have the same nominal values of τ. This is not unreasonable, because dual and quad op amps are now available. However, a stricter analysis requires that each op amp time constant be treated individually. In the author's experience, measurements of the gain bandwidth products of individual op amps in dual and quad packages have shown that they have different gain bandwidth products, but however these differences are small and therefore the assumption of equal time constants for all the op amps will simplify our analysis.

Earlier we concluded that $|\Delta p/p|$ and $|S_x^p|$ can be used as figures of merit. There is a strong indication in the literature that the active pole sensitivity is the most important of all, because $|\Delta p/p|$ due to all passive elements can be made zero by choosing the appropriate technology, as discussed earlier. Therefore $|\Delta p/p|$ due to a nonzero value of τ and $|S_\tau^p|$ will be used as figures of merit in all later chapters, and we shall develop techniques for finding these quantities.

The transfer function of a circuit can be found using the finite gains of the op amps as variables, and then we can substitute $1/s\tau$ for the gain of each op amp. The denominator polynomial of the transfer function will be a function of two variables, s and τ, therefore let it be $D(s, \tau)$. Let $D_0(s)$ be the denominator polynomial of the transfer function when all the op amps are ideal, with $\tau = 0$. If we are interested only in the active sensitivity, we can also substitute the nominal values of the passive elements into the transfer function. After normalizing the coefficients of the transfer function such that the coefficient of the s^2 term in $D_0(s)$ is unity, we obtain

$$D_0(s) = s^2 + \frac{\omega_p}{Q_p}s + \omega_p^2 \tag{3.51}$$

where ω_p and Q_p are the design values. The denominator polynomial $D(s, \tau)$ can always be expressed in the form of a power series as

$$D(s, \tau) = D_0(s) + s\tau D_1(s) + (s\tau)^2 D_2(s) + \cdots \tag{3.52}$$

In (3.52), $D_0(s)$, $D_1(s)$, $D_2(s)$, etc., are usually second-degree polynomials, and the actual number of terms depends on the number of amplifiers used in the circuit. For example, in a single-amplifier circuit, $D(s, \tau)$ is of the form

$$D(s, \tau) = D_0(s) + s\tau D_1(s)$$

It is sufficient to calculate the shift in one member of the complex pair of poles because the shift in the other pole must be the complex conjugate of the first one. Therefore we shall consider only one of the poles, whose nominal position is given by

$$p_1 = -\omega_p\left[\frac{1}{2Q_p} - j\sqrt{1 - \left(\frac{1}{2Q_p}\right)^2}\,\right] \tag{3.53}$$

There are two ways to find the shift in the pole position Δp. The first is to use some form of root finder technique and a computer. (A programmable calculator can also be used to do this.) This method works as follows.

Since we know the nominal value of τ, we can substitute its value into (3.52) and find the polynomial in the form

$$D(s,\tau) = d_0 + d_1 s + d_2 s^2 + d_3 s^3 + \cdots \tag{3.54}$$

where the coefficients d_0, d_1, d_2, etc., are known. This polynomial must have a complex pair of poles close to the ideal pole positions p_1 and p_2, respectively. These are the dominant poles, denoted as p_1' and p_2' and given by (3.48). Using root finder techniques, we can always find the values of p_1' and p_2'. A computer program in BASIC is given at the end of this chapter in Appendix 3A. There will be other parasitic poles as well, but we will not be interested in them. Once we find the desired poles, we can write $D(s,\tau)$ as

$$D(s,\tau) = K(s - p_1')(s - p_2')D_p(s) \tag{3.55}$$

where

$$D_p(s) = 1 + d_{1p}s + d_{2p}s^2 + \cdots$$

and contains the parasitic poles. Note that the constant term in $D_p(s)$ has been made to be unity by extracting an appropriate constant K from the polynomial $D(s,\tau)$. The reason for this is that the coefficients d_{1p}, d_{2p}, etc., will be very small for small values of τ, such that in the frequency range of application their contribution will be small. In fact, in most practical cases, we will be able to approximate $D(s,\tau)$ as

$$D(s,\tau) \cong K(s - p_1')(s - p_2') \tag{3.56}$$

If we are interested only in the per-unit change in the pole position, then we can find it as

$$\frac{\Delta p}{p_1} = \frac{p_1'}{p_1} - 1 \tag{3.57}$$

Once $\Delta p/p_1$ is calculated using (3.57), we can obtain the values of $\Delta\omega_p/\omega_p$ and $\Delta Q_p/Q_p$ using (3.44) and (3.45), respectively. In this procedure, it is not necessary to make approximations, the results obtained are almost exact (except for computer accuracy), and large as well as small changes are obtained irrespective of the value of τ. The large-scale pole sensitivity can then be found to be

$$S_\tau^p = \frac{1}{\tau}\frac{\Delta p}{p_1} \tag{3.58}$$

In fact, we do not need to find S_τ^p in any of our applications, and it is sufficient to find $|\Delta p/p_1|$.

The second method of obtaining S_τ^p and $\Delta p/p_1$ will be explained now. Note that the pole position is a function of τ, hence $p(\tau)$ can be expanded in a Taylor series as

$$p(\tau) = p(0) + \left.\frac{\partial p}{\partial \tau}\right|_{\tau=0} \tau + \left.\frac{\partial^2 p}{\partial \tau^2}\right|_{\tau=0} \frac{\tau^2}{2} + \cdots \tag{3.59}$$

where $p(0)$ is the nominal value of the pole position.

For small values of τ, to a first-order approximation, we can write

$$\Delta p = p(\tau) - p(0) \cong \left.\frac{\partial p}{\partial \tau}\right|_{\tau=0} \tau \tag{3.60}$$

Once we calculate Δp, we can find other quantities of interest immediately. To do this we need to evaluate $\partial p/\partial \tau$, and (3.24) can be used:

$$\left.\frac{\partial p}{\partial \tau}\right|_{\tau=0} = \left.\frac{-\partial D(s,\tau)/\partial \tau}{\partial D(s,\tau)/\partial s}\right|_{s=p(0),\ \tau=0} \tag{3.61}$$

because the nominal value of τ is zero. Picking the nominal pole position $p(0)$ as p_1, we find the shift in this pole position. Hence we need to find $\partial p/\partial \tau$ for this pole position. Applying (3.61) to (3.52), we have

$$\left.\frac{\partial p}{\partial \tau}\right|_{\tau=0} = \frac{-p_1 D_1(p_1)}{2p_1 + \omega_p/Q_p}$$

Using the above in (3.60) results in

$$\frac{\Delta p}{p_1} \cong \frac{-D_1(p_1)}{2p_1 + \omega_p/Q_p}\tau \tag{3.62}$$

The semilogarithmic sensitivity of p_1 is then

$$S_\tau^{p_1} = \frac{-D_1(p_1)}{2p_1 + \omega_p/Q_p} \tag{3.63}$$

Note that $\Delta p/p_1$ and $S_\tau^{p_1}$ depend only on $D_0(s)$ and $D_1(s)$, and that the higher-order terms associated with $\tau^2, \tau^3, \ldots$ [even if they are present in $D(s,\tau)$] do not figure in these calculations. The denominators of both (3.62) and (3.63) can be simplified further by using (3.53):

$$2p_1 + \frac{\omega_p}{Q_p} = 2j\omega_p\sqrt{1 - \left(\frac{1}{2Q_p}\right)^2} \tag{3.64}$$

The calculation of $D_1(p_1)$ can also be simplified. Usually, $D_1(s)$ is a second-degree polynomial, so let this polynomial be of the form

$$D_1(s) = a_2 s^2 + a_1 s + a_0 \tag{3.65}$$

Then we can write $D_1(s)$ as

$$\begin{aligned} D_1(s) &= a_2\left(s^2 + \frac{a_1}{a_2}s + \frac{a_0}{a_2}\right) \\ &= a_2\left[s^2 + \frac{\omega_p}{Q_p}s + \omega_p^2 + \left(\frac{a_1}{a_2} - \frac{\omega_p}{Q_p}\right)s + \left(\frac{a_0}{a_2} - \omega_p^2\right)\right] \end{aligned}$$

Therefore $D_1(p_1)$ becomes

$$D_1(p_1) = \left(a_1 - a_2\frac{\omega_p}{Q_p}\right)p_1 + \left(a_0 - a_2\omega_p^2\right) \tag{3.66}$$

It is easy to evaluate (3.66) compared to the direct evaluation of (3.65) with $s = p_1$, and such simplification may be used whenever possible. In some networks, $a_1 = a_2(\omega_p/Q_p)$ and $a_0 = a_2\omega_p^2$, and in such cases $D_1(p_1) \equiv 0$. This means that the active pole sensitivity becomes zero, however, it does not mean that $\Delta p/p_1$ is zero. This term becomes approximately zero only to a first-order approximation. Note that the ripple factor R is directly related to $\Delta p/p_1$ and not to $S_\tau^{p_1}$. In cases where the first-order effect of τ is zero, the second- and higher-order effects of τ must be taken into account. Next we shall develop a formula for Δp that includes both first- and second-order effects of τ, where third- and high-order effects are ignored. This is sufficient for most of our applications. We will use the Taylor series expansion of a two-variable function. Specifically,

$$\begin{aligned} &D(p + \Delta p, \tau) \\ &\quad \simeq D(p,0) + \left.\frac{\partial D(s,\tau)}{\partial s}\right|_{s=p,\,\tau=0}\Delta p + \left.\frac{\partial D(s,\tau)}{\partial \tau}\right|_{s=p,\,\tau=0}\tau \\ &\qquad + \frac{1}{2}\left[\left.\frac{\partial^2 D(s,\tau)}{\partial s^2}\right|_{s=p,\,\tau=0}(\Delta p)^2 + 2\left.\frac{\partial^2 D(s,\tau)}{\partial s\,\partial \tau}\right|_{s=p,\,\tau=0}\Delta p\,\tau \right. \\ &\qquad \left. + \left.\frac{\partial^2 D(s,\tau)}{\partial \tau^2}\right|_{s=p,\,\tau=0}\tau^2\right] \end{aligned}$$

Choosing $p = p_1$ and applying the above expansion to (3.52), we have

$$\begin{aligned} &D(p_1 + \Delta p, \tau) \\ &\quad \simeq D_o(p_1) + \left(2p_1 + \frac{\omega_p}{Q_p}\right)\Delta p + (p_1\tau)D_1(p_1) + (\Delta p)^2 \\ &\qquad + \left[D_1(p_1) + p_1 D_1'(p_1)\right]\Delta p\,\tau + D_2(p_1)(p_1\tau)^2 \end{aligned}$$

where

$$D_1'(p_1) = \frac{\partial D_1(s)}{\partial s}\bigg|_{s=p_1}$$

Since

$$D(p_1 + \Delta p, \tau) = 0 \qquad \text{and} \qquad D_o(p_1) = 0$$

we obtain a quadratic equation for Δp. However, using the fact that when Δp is small $(\Delta p)^2$ can also be ignored in the above equation and solving for Δp, one has

$$\frac{\Delta p}{p_1} \simeq \frac{-[D_1(p_1) + (p_1\tau)D_2(p_1)]\tau}{2p_1 + \dfrac{\omega_p}{Q_p} + \tau[D_1(p_1) + p_1 D_1'(p_1)]} \tag{3.67}$$

If τ is small, or in the limiting case where $\tau \to 0$, the above expression reduces to

$$\frac{\Delta p}{p_1} \simeq \frac{-D_1(p_1)}{2p_1 + \omega_p/Q_p}\tau$$

which is same as (3.62). Thus (3.67) is more general and can be used even if $D_1(p_1)$ is zero.

Example 3.4: The transfer function of a bandpass filter network that uses a single op amp has the following denominator polynomial:

$$D(s) = s^2 + \frac{\omega_p}{Q_p}s + \omega_p^2 + 1.55A^{-1}\left(s^2 + 4.75\omega_p s + \omega_p^2\right)$$

where A is the op amp gain and is given as

$$A = \frac{2\pi \times 10^6}{s}$$

The nominal values of ω_p and Q_p are

$$\omega_p = 2\pi \times 10^4 \text{ r/s}$$

and

$$Q_p = 10$$

Find (a) the per-unit changes in ω_p and Q_p due to the finite gain of the op amp, (b) the per-unit changes in $|H(j\omega)|$ due to the finite gain of the op amp at the 3-dB frequencies and at the center frequency, and (c) the ripple factor R.

(a) For the given op amp,

$$\tau = \frac{10^{-6}}{2\pi} \text{ s}$$

The denominator polynomial in terms of τ (using this as a variable) is

$$D(s,\tau) = s^2 + \frac{\omega_p}{Q_p}s + \omega_p^2 + 1.55s\tau\left(s^2 + 4.75\omega_p s + \omega_p^2\right)$$

With $\tau = 0$, we find that

$$D_0(s) = D(s,0) = s^2 + \frac{\omega_p}{Q_p}s + \omega_p^2 = s^2 + 0.1\omega_p s + \omega_p^2$$

and

$$D_1(s) = 1.55\left(s^2 + 4.75\omega_p s + \omega_p^2\right)$$

Also, the nominal value of one of the poles is

$$p_1 = \omega_p(-0.05 + j0.99875)$$

As much as possible, we shall avoid using a number for ω_p. The first step is to find $\Delta p/p_1$ due to the nonzero value of τ. We first use the mathematical formulas developed in the previous section. Here again, we have two choices. We can use either (3.62) or (3.67), but since (3.67) is more accurate, we shall use it to find $\Delta p/p_1$. But first, we evaluate the different quantities needed:

$$\begin{aligned}
D_1(p_1) &= 1.55\left(s^2 + 4.75\omega_p s + \omega_p^2\right)\Big|_{s=p_1} \\
&= 1.55\left(4.75 - \frac{1}{Q_p}\right)\omega_p p_1 \\
&= 7.208\omega_p p_1 = \omega_p^2(-360.38 \times 10^{-3} + j7.1985) \\
D_1(p_1) + p_1 D_1'(p_1) &= D_1(p_1) + p_1(2p_1 + 4.75\omega_p) \\
&= 2p_1^2 + 11.958\omega_p p_1 \\
&= 11.758\omega_p p_1 - 2\omega_p^2 \\
&= \omega_p^2(-2.5879 + j11.743)
\end{aligned}$$

Using (3.64), we have

$$\begin{aligned}
2p_1 + \frac{\omega_p}{Q_p} &= 2j\omega_p\sqrt{1 - \left(\frac{1}{2Q_p}\right)^2} \\
&= j1.9975\omega_p
\end{aligned}$$

Substituting all these numbers in (3.67), we obtain

$$\frac{\Delta p}{p_1} = \frac{-(-0.3604 + j7.1985)(\omega_p\tau)}{j1.9975 + (\omega_p\tau)(-2.5879 + j11.743)}$$

The value of $\omega_p \tau$ can be found to be 0.01. Substituting this value in the above equation and reducing it, we have

$$\frac{\Delta p}{p_1} = -34.052 \times 10^{-3} - j1.2873 \times 10^{-3}$$

Using (3.44) and (3.45), $\Delta\omega_p/\omega_p$ and $\Delta Q_p/Q_p$ can be calculated:

$$\frac{\Delta\omega_p}{\omega_p} = \mathrm{Re}\frac{\Delta p}{p_1} = -34.052 \times 10^{-3}$$

and

$$\frac{\Delta Q_p}{Q_p} = -\left(4Q_p^2 - 1\right)^{1/2}\mathrm{Im}\frac{\Delta p}{p_1} = 25.714 \times 10^{-3}$$

To find the exact values of $\Delta\omega_p/\omega_p$ and $\Delta Q_p/Q_p$, we evaluate the polynomial $D(s, \tau)$ by substituting the actual values of τ and then find the dominant roots from this polynomial. The polynomial $D(s, \tau)$ is

$$D(s, \tau) = 246.69 \times 10^{-9}s^3 + 1.07363s^2 + 0.1155\omega_p s + \omega_p^2$$

Note that the value of ω_p is known and that we can substitute this value into the above equation and find the roots. After extracting the dominant poles, we find that

$$D(s, \tau) = \left(s^2 + 0.09424\omega_p s + 0.93269\omega_p^2\right)(1.07216 + 246.69 \times 10^{-9}s)$$

There is one parasitic pole at

$$s = -4.3462 \times 10^6 = -69.172\omega_p$$

Near the operating frequency of ω_p, the influence of this pole on the transfer function is very small. Also, one can see that

$$246.69 \times 10^{-9}\omega_p = 15.688 \times 10^{-3} \ll 1.07216$$

In any case, considering only the dominant poles, we have

$$(\omega_p')^2 = 0.93269\omega_p^2$$

and

$$\frac{\omega_p'}{Q_p'} = 0.09424\omega_p$$

Solving for ω_p' and Q_p', we obtain

$$\omega_p' = 0.965761\omega_p$$

and

$$Q_p' = 10.248$$

Thus

$$\frac{\Delta\omega_p}{\omega_p} = \frac{\omega_p'}{\omega_p} - 1 = -34.239 \times 10^{-3}$$

and

$$\frac{\Delta Q_p}{Q_p} = \frac{Q_p'}{Q_p} - 1 = 24.763 \times 10^{-3}$$

These so-called exact values are quite close to the values calculated using the formula, which shows the accuracy of (3.67) and how closely it can predict the per-unit changes in the frequency ω_p and the pole Q factor Q_p. In fact, (3.62) can also predict these values with moderate accuracy.

(b) To find the per-unit changes in $|H(j\omega)|$, we use (3.36). The value of E can be found to be

$$E = 1.9824$$

The lower and upper 3-dB frequencies can be shown to be

$$\omega_l \simeq \omega_p - \frac{\omega_p}{2Q_p} = 0.95\omega_p$$

and

$$\omega_h \simeq \omega_p + \frac{\omega_p}{2Q_p} = 1.05\omega_p$$

Therefore

$$x_l = 0.95 \qquad \text{and} \qquad x_h = 1.05$$

Using all these values and the value of $\Delta\omega_p/\omega_p$ obtained in (3.36) in part (a), we have

$$\left.\frac{\Delta|H(j\omega)|}{|H(j\omega)|}\right|_{\omega_l} = 0.3874$$

$$\left.\frac{\Delta|H(j\omega)|}{|H(j\omega)|}\right|_{\omega_h} = -0.2936$$

and

$$\left.\frac{\Delta|H(j\omega)|}{|H(j\omega)|}\right|_{\omega_p} = 0.0598$$

Note that the highest percentage change in magnitude function is

38.74%. This high percentage is mainly due to $\Delta\omega_p/\omega_p$. A small change in ω_p on the order of 3.4% is magnified by approximately Q_p times to yield this change in magnitude function at the lower and upper 3-dB frequencies. In fact, this can be shown to be true using (3.36). In other words, if the allowed change in the transfer function magnitude is x percent, then the change in the pole frequency ω_p must not exceed x/Q_p percent. For high Q_p values, this value must be very small.

(c) The ripple factor can be calculated using (3.38) or (3.42) as

$$R = 0.681$$

Also note that

$$\left.\frac{\Delta|H(j\omega)|}{|H(j\omega)|}\right|_{\omega_l} - \left.\frac{\Delta|H(j\omega)|}{|H(j\omega)|}\right|_{\omega_h} = 0.3874 - (-0.2936) = 0.681$$

It is only a coincidence that these numbers are almost exactly equal, however, in most cases, this is true. ■

3.8 *Predistortion*

We discussed the effect of the nonzero value of τ of the op amp and developed techniques for finding the shift in the pole positions from their ideal values. This shift, as a first-order effect, can be made zero by making $D_1(p_1) \equiv 0$. In this way, the active sensitivity as given by (3.63) can also be made zero. This method of making the shifts zero is known as active compensation, and we will consider some of these techniques with reference to specific types of circuits in later chapters. Another method of realizing the *required pole positions*, which includes the effect of the nonzero τ values, is called *predistortion*. To understand this procedure, assume that we obtain the denominator polynomial, which includes the nominal value of τ, as given by (3.54). This polynomial in turn can be written in the form of (3.55), and for small values of τ we can approximate it as given by (3.56), reproduced here for convenience:

$$D(s,\tau) \simeq K(s - p_1')(s - p_2') \tag{3.56}$$

Usually, given pole positions such as p_1 and p_2 as expressed by (3.47) or the ω_p and Q_p values, the nominal design is carried with the assumption of $\tau = 0$ so that the ideal polynomial $D_0(s)$ is exactly given by (3.51). This means that the passive elements in the network are chosen such that we obtain $D_0(s)$ as given by (3.51). The question is, Why not change the passive elements so that $D(s,\tau)$ has p_1 and p_2 as its dominant poles? Yes, it is possible to do this in active filters where the nominal values of the op amp time constants are included in the analysis and design of the network.

Such a design procedure is known as predistortion because the passive element values obtained in the ideal case are predistorted to compensate for the effects of nonzero values of τ. Before we discuss a technique for achieving this, we will see how it is possible to do at all. Assume that we have designed the network to realize the nominal pole positions p_1 and p_2 with the assumption that all the τ are equal to 0. Let $x_1, x_2, \ldots, x_n$ be the element values of the passive components. Assume that we alter these values to $x_1', x_2', \ldots, x_n'$, where the increments are small. Now, including the effects of τ, for small changes, we write

$$\Delta p = \sum_{i=1}^{n} \frac{\partial p}{\partial x_i} \Delta x_i + \frac{\partial p}{\partial \tau} \tau \tag{3.68}$$

where

$$\Delta x_i = x_i' - x_i \qquad i = 1, 2, \ldots, n$$

All the derivatives of p can be found using (3.24). If the value of τ is known, then the right side of (3.68) is already known except for Δx_i, $i = 1, 2, \ldots, n$. Note that we want $\Delta p = 0$. Therefore suitable choices for Δx_i, $i = 1, 2, \ldots, n$, can be made such that $\Delta p = 0$, which does not mean that $s_\tau^p = 0$. If $\Delta p = 0$,

$$\sum_{i=1}^{n} \frac{\partial p}{\partial x_i} \Delta x_i = -\frac{\partial p}{\partial \tau} \tau \tag{3.69}$$

Using (3.24) in (3.69), we have

$$\sum_{i=1}^{n} \left. \frac{\partial D(s, x_i)}{\partial x_i} \right|_{\text{nominal values, } s=p} \Delta x_i = -\left. \frac{\partial D(s, \tau)}{\partial \tau} \right|_{\text{nominal values, } s=p} \tau \tag{3.70}$$

Usually, all the terms in (3.70) are complex. Therefore, by equating real and imaginary parts, we obtain two equations from (3.70). However, unique choices for Δx_i, $i = 1, 2, \ldots, n$, exist only when $n = 2$, since we have only two equations. Therefore we choose any two elements in the circuit (usually resistances) and keep all the other element values fixed at the x_i values obtained for the ideal case. We treat these two elements and τ as variables in the denominator polynomial and obtain the derivatives as required by (3.70). Then we obtain the two equations and solve for the Δx_i values of these two elements, from which the new x_i values can be obtained. Consider the following example.

Example 3.5: The denominator polynomial of a filter circuit was found for a given set of capacitance values, and all other parameters were treated as variables. This denominator polynomial is

$$D(s, \tau) = s^2 + (2G_3 + G_1 - \delta G_2)s + G_3(G_1 + G_2) \\ + (1 + \delta)s\tau\left[s^2 + (2G_3 + G_1 + G_2)s + G_3(G_1 + G_2)\right]$$

The required pole Q factor Q_p and pole frequency ω_p are 5 and 1 r/s, respectively, and the actual value of $\tau = 0.005$. The ideal design with an assumption of $\tau = 0$ was carried out, and the following element values obtained:

$$G_1 = 0.5 \qquad G_2 = 2 \qquad G_3 = 0.4 \qquad \delta = 0.55$$

It was decided to change the values of G_2 and G_3 so that the dominant poles of $D(s, \tau)$ would be the required poles. One of the required poles is

$$p_1 = -0.1 + j0.9950$$

This can be found by substituting the values of ω_p and Q_p in (3.53). Repeat this problem with $\tau = 0.05$.

The solution of the problem requires the following derivatives evaluated at the nominal values.

$$\left.\frac{\partial D(s,\tau)}{\partial G_2}\right|_{\tau=0} = -\delta s + G_3$$

$$\left.\frac{\partial D(s,\tau)}{\partial G_3}\right|_{\tau=0} = 2s + G_1$$

$$\left.\frac{\partial D(s,\tau)}{\partial \tau}\right|_{\tau=0} = (1+\delta)s\left[s^2 + (2G_3 + G_1 + G_2)s + G_3(G_1 + G_2)\right]$$

Substituting the different nominal values and $s = p_1$ into the above derivatives and using (3.70), we have

$$(0.455 - j0.5472)\,\Delta G_2 + (2.3 + j1.99)\,\Delta G_3$$
$$= 23.54 \times 10^{-3} + j4.781 \times 10^{-3}$$

Equating the real and imaginary parts, we have

$$0.455\,\Delta G_2 + 2.3\,\Delta G_3 = 23.54 \times 10^{-3}$$
$$-0.5472\,\Delta G_2 + 1.99\,\Delta G_3 = 4.781 \times 10^{-3}$$

Solving for ΔG_2 and ΔG_3, we have

$$\Delta G_2 = 16.57 \times 10^{-3} \qquad \text{and} \qquad \Delta G_3 = 6.959 \times 10^{-3}$$

Thus the new values of G_2 and G_3 are

$$G_2 = 2.0166 \qquad \text{and} \qquad G_3 = 0.4070$$

Before concluding this problem, let us verify whether, with the above values for G_2 and G_3 and the values $G_1 = 0.5$, $\delta = 0.55$, and $\tau = 0.005$, the dominant poles are the required poles. To find the dominant poles, we obtain $D(s, \tau)$:

$$D(s, \tau) = 7.75 \times 10^{-3} s^3 + 1.026 s^2 + 0.2128 s + 1.024$$

For the dominant poles, we have

$$p_1' = -0.10010 + j0.99498$$

Comparing the values of p_1' and p_1, we find that we have been successful in shifting the pole to the required position by changing the values of G_2 and G_3 from 2 and 0.4 to 2.0166 and 0.4070, respectively.

Let us now repeat the calculations with $\tau = 0.05$. Since the derivatives are evaluated with the nominal value $\tau = 0$, the left sides of the equations for ΔG_2 and ΔG_3 remain the same. Increasing the value of τ by 10 times causes the right sides of the equations to become 10 times greater. Since the values for ΔG_2 and ΔG_3 are to be 10 times the values we obtained before, we have

$$\Delta G_2 = 0.1657 \qquad \text{and} \qquad \Delta G_3 = 0.06959$$

Thus the new values of G_2 and G_3 are

$$G_2 = 2.1657 \qquad \text{and} \qquad G_3 = 0.4696$$

The next question is whether we will have as much success as we had earlier in shifting the pole to the required position with the above values of G_2 and G_3. Let us find the dominant pole positions again. We first calculate the coefficients of the polynomial $D(s, \tau)$ with $G_1 = 0.5$, $\delta = 0.55$, $\tau = 0.05$, $G_2 = 2.1657$, and $G_3 = 0.4696$, and the resulting polynomial is

$$D(s, \tau) = 77.5 \times 10^{-3} s^3 + 1.2794 s^2 + 0.3451 s + 1.2518$$

From the above expression, the dominant pole is

$$p_1' = -0.1062 + j0.9899$$

Therefore we find that

$$|\Delta p| = 7.998 \times 10^{-3}$$ ■

The value of $|\Delta p|$ is nearly 80 times the value of $|\Delta p|$ in the previous case, though we increased the value of τ by only 10 times. Though this method works quite efficiently for very small values of τ, it fails in cases where the value of τ is relatively large. Therefore we next develop a method where the actual value of τ is used in the calculations. The second method uses polynomial division to obtain the error terms as well as the derivatives.

Consider the denominator polynomial $D(s, \tau)$ after the value of τ has been substituted. We still need two parameters x_1 and x_2 to shift the poles to the required positions. Therefore we can substitute the values of all the passive elements, except x_1 and x_2, and let

$$D(s, x_1, x_2) = d_0 + d_1 s + d_2 s^2 + d_3 s^3 + \cdots \tag{3.71}$$

Obviously, the coefficients d_0, d_1, d_2, etc., are functions of x_1 and x_2. For arbitrary values of x_1 and x_2, using polynomial division, we can write $D(s, \tau)$ as

$$D(s, x_1, x_2) = \left(s^2 + \frac{\omega_p}{Q_p}s + \omega_p^2\right)P(s, x_1, x_2) + r_1(x_1, x_2)s + r_0(x_1, x_2) \tag{3.72}$$

where r_1 and r_0 are the remainder terms and the polynomial $P(s)$ contains the parasitic poles. Note that r_1 and r_0 will be functions of x_1 and x_2 and we are looking for values of x_1 and x_2 such that $r_1 = r_0 = 0$.

Assume that the initial values of the elements x_1 and x_2 are x_{10} and x_{20} (obtained with the nominal design assuming that all τ values are equal to zero). Let us increase the values of x_{10} and x_{20} to $x_{10} + \Delta x_1$ and $x_{20} + \Delta x_2$, respectively. Then, using a Taylor series, we have

$$r_1(x_{10} + \Delta x_1, x_{20} + \Delta x_2) \simeq r_1(x_{10}, x_{20}) + \left.\frac{\partial r_1}{\partial x_1}\right|_{x_1 = x_{10},\, x_2 = x_{20}} \Delta x_1 + \left.\frac{\partial r_1}{\partial x_2}\right|_{x_1 = x_{10},\, x_2 = x_{20}} \Delta x_2 \tag{3.73}$$

and

$$r_0(x_{10} + \Delta x_1, x_{20} + \Delta x_2) \simeq r_0(x_{10}, x_{20}) + \left.\frac{\partial r_0}{\partial x_1}\right|_{x_1 = x_{10},\, x_2 = x_{20}} \Delta x_1 + \left.\frac{\partial r_0}{\partial x_2}\right|_{x_1 = x_{10},\, x_2 = x_{20}} \Delta x_2 \tag{3.74}$$

Since we want to have

$$r_1(x_{10} + \Delta x_1, x_{20} + \Delta x_2) = r_0(x_{10} + \Delta x_1, x_{20} + \Delta x_2) = 0$$

we can write (3.73) and (3.74) as

$$\begin{aligned} A\,\Delta x_1 + B\,\Delta x_2 &= -r_1(x_{10}, x_{20}) \\ C\Delta x_1 + D\Delta x_2 &= -r_0(x_{10}, x_{20}) \end{aligned} \tag{3.75}$$

where A, B, C, and D are used to simplify the notation:

$$A = \left.\frac{\partial r_1}{\partial x_1}\right|_{x_1=x_{10},\, x_2=x_{20}} \qquad B = \left.\frac{\partial r_1}{\partial x_2}\right|_{x_1=x_{10},\, x_2=x_{20}}$$

$$C = \left.\frac{\partial r_0}{\partial x_1}\right|_{x_1=x_{10},\, x_2=x_{20}} \qquad D = \left.\frac{\partial r_0}{\partial x_2}\right|_{x_1=x_{10},\, x_2=x_{20}} \tag{3.76}$$

The values of $r_1(x_{10}, x_{20})$ and $r_0(x_{10}, x_{20})$ can be calculated by substituting $x_1 = x_{10}$ and $x_2 = x_{20}$ into (3.71) and using polynomial division to find the error terms r_1 and r_0 as suggested by (3.72), and they must be the values of r_1 and r_0 we are looking for. What is more interesting is the fact that the derivatives of r_1 and r_0 with respect to x_1 and x_2 can also be obtained by using polynomial division, as we will now show. Substituting (3.71) into (3.72) for $D(s, x_1, x_2)$ in terms of the coefficients $d_0, d_1, d_2, \ldots,$ and then differentiating with respect to x_1 and x_2, we obtain two equations:

$$\sum_i \frac{\partial d_i}{\partial x_1} s^i = D_1(s, x_1, x_2) = \left(s^2 + \frac{\omega_p}{Q_p} s + \omega_p^2\right) \frac{\partial P}{\partial x_1} + \frac{\partial r_1}{\partial x_1} s + \frac{\partial r_o}{\partial x_1} \tag{3.78}$$

and

$$\sum_i \frac{\partial d_i}{\partial x_2} s^i = D_2(s, x_1, x_2) = \left(s^2 + \frac{\omega_p}{Q_p} s + \omega_p^2\right) \frac{\partial P}{\partial x_2} + \frac{\partial r_1}{\partial x_2} s + \frac{\partial r_o}{\partial x_2} \tag{3.79}$$

Evaluating (3.78) and (3.79) with $x_1 = x_{10}$ and $x_2 = x_{20}$, we obtain

$$\begin{aligned}\sum_i \left.\frac{\partial d_i}{\partial x_1}\right|_{x_1=x_{10},\, x_2=x_{20}} s^i &= D_1(s, x_{10}, x_{20}) \\ &= \left(s^2 + \frac{\omega p}{Q_p} s + \omega_p^2\right) \left.\frac{\partial P}{\partial x_1}\right|_{x_1=x_{10},\, x_2=x_{20}} + As + C\end{aligned} \tag{3.80}$$

and

$$\begin{aligned}\sum_i \left.\frac{\partial d_i}{\partial x_2}\right|_{x_1=x_{10},\, x_2=x_{20}} s^i &= D_2(s, x_{10}, x_{20}) \\ &= \left(s^2 + \frac{\omega_p}{Q_p} s + \omega_p^2\right) \left.\frac{\partial P}{\partial x_2}\right|_{x_1=x_{10},\, x_2=x_{20}} + Bs + D\end{aligned} \tag{3.81}$$

By carrying out two more polynomial divisions as suggested by (3.80) and (3.81), we can obtain the constants A, B, C, and D. Furthermore, we wish to point out that, because we are considering linear networks, the coefficients have a linear dependence on x_1 and x_2 and thus finding the polynomials $D_1(s, x_1, x_2)$ and $D_2(s, x_1, x_2)$ is not difficult. Thus, after determining all the quantities in (3.75) except Δx_1 and Δx_2, we can solve for these values using (3.75). Note that we do not have to substitute $s = p_1$ as in the first case, which involves complex number notation. This procedure can be programmed into a programmable calculator. After calculating Δx_1 and Δx_2, the required values of x_1 and x_2 can be found to be

$$x_1 = x_{10} + \Delta x_1 \qquad \text{and} \qquad x_2 = x_{20} + \Delta x_2$$

In this procedure also, there is *no* guarantee that we will achieve $r_1 = r_0 = 0$ after making corrections for x_1 and x_2 in one step, because of the approximations involved in (3.73) and (3.74). In all such cases, one has to iterate the procedure by updating the element values so that the final values obtained at the end of one iteration can become the initial values for the next iteration. Thus we can reduce the gap between p_1 and p_1' and, eventually, make $\Delta p \simeq 0$. In this procedure, we hope to get $r_1 \simeq 0$ and $r_0 \simeq 0$ or, at least, r_1 and r_0 are very small. Then the next question is, How small should the values of r_1 and r_0 be before we stop the iterations? It is difficult to place terminating conditions on r_1 and r_0, because we do not know how small the values of r_1 and r_0 should be. However, in practical engineering problems such as this one, we can impose conditions on Δx_1 and Δx_2. Remember that x_1 and x_2 are practical elements and therefore that it is easy to impose conditions on Δx_1 and Δx_2. For example, we can stop the iterative procedure when both $|\Delta x_1/x_1|$ and $|\Delta x_2/x_2|$ are less than ε, where ε may be 1, 0.1, or 0.01% depending on the accuracy we need. Just to be on the safe side, we shall assume in most of our cases that $|\Delta x_1/x_1|$ and $|\Delta x_2/x_2|$ should be less than or equal to 0.0001, corresponding to 0.01%. Consider the following example.

Example 3.6: Repeat Example 3.5 with $\tau = 0.05$ and find the values of G_2 and G_3 so that the dominant poles are the required poles.

First, we substitute the values of G_1, δ, and τ, which are fixed, and thus $D(s, G_2, G_3)$ is

$$D(s, G_2, G_3) = d_0 + d_1 s + d_2 s^2 + d_3 s^3$$

where

$$d_0 = G_3(0.5 + G_2)$$

$$d_1 = (0.5 + 2G_3 - 0.55G_2) + 77.5 \times 10^{-3} G_3(0.5 + G_2)$$

$$d_2 = 1 + 77.5 \times 10^{-3}(0.5 + 2G_3 + G_2)$$

and

$$d_3 = 77.5 \times 10^{-3}$$

Choosing $x_1 = G_2$ and $x_2 = G_3$, we obtain

$$D_1(s, G_2, G_3) = G_3 + (0.0775G_3 - 0.55)s + 0.0775s^2$$

and

$$D_2(s, G_2, G_3) = (G_2 + 0.5) + \left[2 + (77.5 \times 10^{-3})(G_2 + 0.5)\right]s + 0.155s^2$$

Since $\omega_p = 1$ r/s and $Q_p = 5$, we have

$$s^2 + \frac{\omega_p}{Q_p}s + \omega_p^2 = s^2 + 0.2s + 1$$

For this simple problem, using polynomial division, we can easily find r_1, r_0, A, B, C, and D:

$$r_1 = d_1 - d_3 - 0.2d_2 + 0.04d_3$$
$$r_0 = d_0 - d_2 + 0.2d_3$$
$$A = 77.5 \times 10^{-3}G_3 - 0.55 - 15.5 \times 10^{-3}$$
$$B = 2 + (77.5 \times 10^{-3})(G_2 + 0.5)$$
$$C = G_3 - 77.5 \times 10^{-3}$$
$$D = G_2 + 0.5 - 0.155$$

The salient calculations are shown in Table 3.2. At the end of the first iteration, we find that

$$\frac{\Delta G_2}{G_2} = 10.02\% \qquad \text{and} \qquad \frac{\Delta G_3}{G_3} = 17.68\%$$

Therefore, using the new values of G_2 and G_3, we proceed with the second iteration, and at the end of the second iteration, we find that

$$\frac{\Delta G_2}{G_2} = -0.247\% \qquad \text{and} \qquad \frac{\Delta G_3}{G_3} = -0.171\%$$

which are very small. However, we proceed and at the end of the third iteration we find that

$$\frac{\Delta G_2}{G_2} = -0.0038\% \qquad \text{and} \qquad \frac{\Delta G_3}{G_3} = -0.0067\%$$

Therefore, for all practical purposes, the values of G_2' and G_3' at the end of the third iteration can be accepted as the required values of G_2 and G_3. Indeed, with these values given as

$$G_2 = 2.1948 \qquad \text{and} \qquad G_3 = 0.4699$$

Table 3.2 Calculations for Example 3.6

Iteration	G_2 (x_{10})	G_3 (x_{20})	r_1	r_0	ΔG_2	ΔG_3	G_2' (x_1)	G_3' (x_2)
1	2.0000	0.4000	-48.05×10^{-3}	-240.3×10^{-3}	200.3×10^{-3}	70.7×10^{-3}	2.2003	0.4707
2	2.2003	0.4707	-1.094×10^{-3}	4.298×10^{-3}	-5.43×10^{-3}	-805.2×10^{-6}	2.1949	0.4699
3	2.1949	0.4699	25.30×10^{-6}	116.7×10^{-6}	-83.17×10^{-6}	-31.38×10^{-6}	2.1948	0.4699

the polynomial $D(s, \tau)$ has its dominant poles as the required poles, which can be verified by using any root finder computer routine. ■

Note that, at the end of the first iteration, the new values of G_2 and G_3 are within about 0.25% of the final values. This accuracy may be sufficient for most practical purposes. Thus it looks as if the second method is more accurate and powerful compared to the first, which can be confirmed by evaluating $|\Delta p|$ after predistortion with only one step in each case. We have already calculated $|\Delta p|$ in the previous case. Let us find the same quantity with $G_2 = 2.2003$ and $G_3 = 0.4707$, which are the values obtained at the end of first iteration in the second case. We find that

$$p_1' = -0.9945 + j0.99673$$

Therefore

$$|\Delta p| = |p_1' - p_1| = 1.83 \times 10^{-3}$$

which is less than a quarter of the value obtained with the first method. This calculation clearly shows that the second method is more efficient, and there are several reasons for this. The second method uses the actual value of τ and only changes are calculated, whereas in the first method, when the changes and the derivatives are calculated it is assumed that $\tau = 0$, which is not accurate. For this reason and because we do not need complex arithmetic, the second method is preferable even for one-step calculations.

In conclusion, we have presented two methods for predistorting an active filter design so that the amplifier time constants can become part of the design. Note that, in each case, we obtain the initial values of the parameters with $\tau = 0$. For very small values of τ, predistortion may not be necessary. Of the two methods, the second has been found to be better than the first. We should also mention that the same procedure can be applied to the numerator polynomial of the transfer function to place the zeros of the transfer function at the required positions. However, if an attempt is made to place the poles and zeros of the second-order transfer functions at the required positions, we will need 4 independent parameters for adjustment and 16 derivatives. Though conceptually it requires the same procedure that we have used, computationally it is too much involved even in the case of second-order transfer functions. However, the computation part can be done in a computer or a calculator.

3.9 *Conclusions*

In this chapter we have discussed the important concepts related to sensitivity analysis and the methods for calculating various sensitivity measures.

The various sensitivity measures may be used for evaluating different network designs, as we shall see in the remaining chapters. This also helps in selecting networks with minimum sensitivity. The sensitivity measures may also be used for the deterministic tuning of electronic circuits, which is exactly what we did in the last section of this chapter.

REFERENCES

1. Blostein, M. L.: Some Bounds on the Sensitivity in *RLC* Networks, *Proc. 1st Allerton Conference on Circuits and Systems Theory*, pp. 488–501, 1963.
2. Weyten, L.: A Useful Sensitivity Measure for Second-Order *RC* Active Filter Design, *IEEE Trans. Circuits. Systs.*, Vol. CAS-23, pp. 506–508, Aug. 1936.

EXERCISES

3.1. Derive properties 2, 3, 5, and 9 in Table 3.1.

3.2. Consider the network shown in Fig. E3.2.

(a) Obtain the transfer function $H(s)$.

(b) Obtain the sensitivities of the magnitude function as a function of ω with respect to all the elements in the network.

(c) Plot these sensitivities as functions of ω over the range $0 \le \omega \le 2$ r/s.

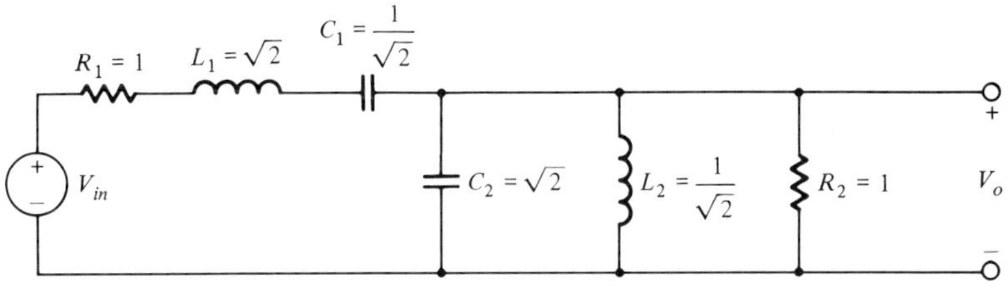

Figure E3.2 Circuit for calculation of the sensitivities in Exercises 3.2 and 3.3.

3.3. For the network in Fig. E3.2, find the worst-case sensitivity using (3.11) at the following frequencies:

(a) $\omega = 0.5$ r/s.

(b) $\omega = 0.75$ r/s.

(c) $\omega = 1$ r/s.

(d) $\omega = 2$ r/s.

3.4. Consider the network in Fig. 3.1. Assume that $R_1 = R_2 = 15.915$ kΩ and $C_1 = C_2 = 10$ nF. Also, $K = 2.8$. The temperature coefficients of the resistors and capacitors are 100 ppm/°C and −40 ppm/°C, respectively, and K does not change with temperature.
(a) Find the per-unit changes in ω_p and Q_p.
(b) Find the worst-case per unit changes in ω_p and Q_p.

3.5. Prove (3.36) through (3.38).

3.6. In the network in Fig. 4.4, $C_1 = C_2 = 4.7$ nF, $R_1 = 56.438$ kΩ, and $R_2 = 2.258$ kΩ.
(a) Find the passive ω_p and Q_p sensitivities.
(b) Assume that both K_1 and K_2 are realized using op amps having a finite gain-bandwidth product of 1 MHz. Find the values of $\Delta\omega_p/\omega_p$, $\Delta Q_p/Q_p$, and R (ripple factor) due to the nonzero value of τ.

3.7. The denominator polynomial of the transfer function of an active filter, which includes the effect of the nonzero values of τ of the op amps used, is

$$D(s,\tau) = s^2 + s\left[(G_1+G_2+G_3)S_1 - \frac{\beta G_3 S_2}{1-\beta}\right] + G_3\left(G_2 - \frac{\beta G_1}{1-\beta}\right)$$
$$+\left(\frac{s\tau}{1-\beta}\right)\left[s^2 + s\left[(G_1+G_2+G_3)S_1 + G_3S_2\right]\right.$$
$$\left. + G_3(G_1+G_2)S_1S_2\right]$$

where $S_i = 1/C_i$ and $G_i = 1/R_i$, for all i. A nominal design was obtained with the assumption of $\tau = 0$, and the element values were $\beta = 0.3$, $C_1 = 5$ nF, $C_2 = 1$ nF, $R_1 = 6.247$ kΩ, $R_2 = 3.364$ kΩ, and $R_3 = 2.897$ kΩ.
(a) Assume that $\tau = 0.5/\pi$ μs/r. Find the values of $\Delta\omega_p/\omega_p$, $\Delta Q_p/Q_p$, and the ripple factor R due to the nonzero value of τ.
(b) It is decided to use predistortion to account for the nonzero value of τ so that the ω_p and Q_p values of the dominant poles are the desired ones. Use the parameters G_2 and G_3 and the method of Sec. 3.8 to find the new values of R_2 and R_3. Use the iteration procedure until successive values of R_2 and R_3 are within 0.1%.
(c) Evaluate the pole positions with the new values of R_2 and R_3 as obtained in part (b) using a root-finding routine and determine whether they are the desired poles.

3A

Program ROOT FINDER

THIS APPENDIX PROVIDES the details of the program ROOT FINDER written in GW BASIC language that can be run in IBM PC or its compatibles. This program can be used to find the zeros of a polynomial. The procedure to use this program is simple. Just enter the program and run it.

Inputs Necessary for the Program

(a) The order, N of the polynomial
(b) Maximum number of iterations allowed (usually this number is in tens)
(c) Coefficients of the polynomial, $A_0, A_1, \ldots, A_N$.

After the above inputs are given, the program prints the coefficients on the screen and you have the option to change some or all of them, if needed. Once the coefficients are correctly entered through the keyboard, the program finds the zeros of the polynomial and prints them.

Outputs from the Program

(a) The coefficients of the polynomial
(b) The zeros of the polynomial.

```
10 '
20 '
30 '                    PROGRAM LISTING OF 'ROOT FINDER'
40 '                    ------------------------------
50 '
60 'program to find the roots of a polynomial
70 PRINT" THIS PROGRAM FINDS THE ZEROS OF A POLYNOMIAL"
80 DEFINT N,I,J,K,L,X
90 PRINT"INPUT THE ORDER OF THE POLYNOMIAL";:INPUT N
100 DIM A(N),B(N),C(N)
110 E=.0000001
120 OPTION BASE 0
130 PRINT"MAXIMUM NUMBER OF ITERATIONS FOR CONVERGENCE, YOU WANT TO ALLOW"
140 PRINT"TYPICALLY, THIS NUMBER IS IN TENS";:INPUT N1
150 LPRINT STRING$(40,"_"):LPRINT
160 LPRINT"THE COEFFICIENTS OF THE POLYNOMIAL ARE"
170 LPRINT STRING$(40,"_")
180 PRINT"INPUT THE COEFFICIENTS OF THE POLYNOMIAL DIRECTED BY THE PROMPT"
190 FOR I=N TO 0 STEP -1
200 PRINT "A(";I;")=";:INPUT A(I)
210 NEXT I
220 FOR I=N TO 0 STEP -1
230 PRINT
240 PRINT"A(";I;")=";A(I)
250 NEXT I
260 PRINT" DO YOU WANT TO CHANGE ANY ONE OR MORE OF THE COEFFICIENTS";
270 PRINT" IF YOU DO, ANSWER Y ELSE ANSWER N";
280 INPUT Y$
290 IF  Y$="Y" OR  Y$="y" OR Y$="N" OR Y$= "n" THEN 310
300 PRINT" Y OR N EXPECTED": GOTO 280
310 IF Y$="N" OR Y$="n" THEN 360
320 PRINT" INDICATE THE ORDER OF S WHOSE COEFFICIENT, YOU WANT TO CHANGE";
330 INPUT I
340 PRINT" A(";I;")";:INPUT A(I)
350 GOTO 260
360 FOR I=N TO 0 STEP -1
370 LPRINT "        A(";I;")=";USING"##.#####^^^^";A(I):LPRINT
380 NEXT I
390 LPRINT"                ROOTS ARE":LPRINT STRING$(40,"_")
400 LPRINT"     REAL PART         IMAG.PART"
410 IF N<=3 THEN 720
420 Q=A(0)/A(2):P=A(1)/A(2)-A(0)*A(3)/(A(2)*A(2))
430 I1=0:I2=0
440 I1=I1+1
450 X=N
460 FOR I=0 TO N :B(I)=A(I):NEXT I
470 GOSUB 1070
480 R3=R1:R2=R0
490 IF R3=0 AND R2=0 THEN 600
500 IF I2>0 THEN 600
510 X=N-2
520 FOR I=0 TO X :B(I)=C(I):NEXT I
530 GOSUB 1070
540 D0=R0-P*R1:D1=Q*R1:D2=R0*D0+R1*D1:D3=(R2*D0+D1*R3)/D2:D4=(R3*R0-R2*R1)/D2
550 P=P+D4:Q=Q+D3
560 IF ABS(D3)<E AND ABS(D4)<E THEN 590
570 IF I1>N1 THEN 870
580 GOTO 440
590 I2=I2+1:GOTO 440
600 FOR I=0 TO N-2 :A(I)=C(I):NEXT I
610 N=N-2:R1=-P/2:R2=R1*R1-Q
620 IF R2<0 THEN 680
630 R2=SQR(R2)
640 R3=R1+R2:R4=R1-R2
650 LPRINT USING"    ##.#####^^^^";R3;0:LPRINT
660 LPRINT USING"    ##.#####^^^^";R4;0:LPRINT
670 GOTO 710
680 R3=SQR(-R2)
690 LPRINT USING"    ##.#####^^^^";R1;R3:LPRINT
700 LPRINT USING"    ##.#####^^^^";R1;-R3:LPRINT
710 IF N=0 THEN 1050  ELSE 410
720 IF N<=2 THEN 860
730 P=-A(0)/A(1)
740 I1=0
750 I1=I1+1
760 F1=((A(3)*P+A(2))*P+A(1))*P+A(0)
770 F2=(3*A(3)*P+2*A(2))*P+A(1)
780 P1=-F1/F2:P=P+P1
```

```
790 IF ABS(P1)<E THEN 810
800 IF I1>N1 THEN 1020 ELSE 750
810 N=N-1:LPRINT
820 LPRINT USING"    ##.#####^^^^";P;0:LPRINT
830 R1=(A(2)+A(3)*P)/A(3):Q=A(1)/A(3)+R1*P
840 P=R1
850 GOTO 610
860 P=A(1)/A(2):Q=A(0)/A(2):GOTO 610
870 PRINT"THERE MAY BE A QUADRATIC FACTOR OF 'S^2+P*S+Q' WITH THE FOLLOWING APPR
OXIMATE VALUES OF P&Q":I3=2
880 PRINT "P=";USING"     ##.####^^^^";P
890 PRINT "Q=";USING"     ##.####^^^^";Q
900 PRINT"THIS MAY BE A REPEATED ROOT OR YOUR MAX. # OF ITERATIONS MAY BE INSUFF
ICIENT"
910 PRINT"YOU MAY WANT TO SUGGEST INITIAL VALUES FOR P(AND Q) AND START ALLOVER
AGAIN"
920 PRINT"ENTER 'Y' TO CONTINUE. TO STOP ENTER 'N' FOR THE PROMPT"
930 INPUT Y$
940 IF Y$="Y" OR Y$="y" OR Y$="N" OR Y$="n" THEN 960 ELSE 950
950 PRINT"Y OR N EXPECTED FOR THE PROMPT":GOTO 920
960 IF Y$="N" OR Y$="n" THEN 1050
970 PRINT"INPUT THE VALUE OF P FOR THE PROMPT";
980 INPUT P
990 IF I3=1 THEN 740
1000 PRINT"INPUT THE VALUE OF Q FOR THE PROMPT";
1010 INPUT Q:GOTO 430
1020 PRINT"THERE MAY BE A LINEAR FACTOR OF 'S-P' WITH THE FOLLOWING APPROXIMATE
VALUE OF P"
1030 PRINT "P=";USING"##.#####^^^^    ";P
1040 I3=1:GOTO 900
1050 END
1060 ' subroutine to divide a polynomial with a quadratic factor of s^2+p*s+q
1070 C(X)=0:C(X-1)=0
1080 FOR I=(X-2) TO 0 STEP -1
1090 C(I)=B(I+2)-P*C(I+1)-Q*C(I+2)
1100 NEXT I
1110 R1=B(1)-C(0)*P-C(1)*Q:R0=B(0)-C(0)*Q
1120 RETURN
```

4

Active Filters with Finite Gain Amplifiers

IN THIS AND THE FOLLOWING CHAPTERS, we will consider the design of *active filters*. The term "active" refers to the use of active devices in filter circuits. The filter circuits considered use only resistors and capacitors as passive elements. Active *RC* filters usually weigh less and require less space than passive ones. More importantly, they can be fabricated in microminiature form using integrated circuit technology. For these and many other reasons, the present communications industry uses active filters more widely compared to any other type of filter.

Given a network function, one can use direct methods to realize it. There are other methods, such as cascade, primary resonator, and multiple feedback topology, that use second-order active filters as the basic building blocks. The latter methods are the ones used exclusively in practice. It has been proven in the literature that network realizations using direct methods of synthesis are more sensitive to component variations compared to networks obtained using these other methods. In addition, the latter methods, which use second-order building blocks, possess other important advantages. As an example, consider the cascade method. In this method, the higher-order transfer function is factored into a product of second-order terms. Then, each second-order transfer function is individually realized by an active *RC* circuit, and when all the individual circuits are connected in cascade, the entire combination can realize the given higher-order transfer function. In addition to the fact that such cascade combinations possess excellent sensitivity properties compared to those of networks using direct methods of synthesis, they also have other advantages. Since each network is expected to realize only a second-order transfer function, the design procedure is relatively simple. Furthermore, the design and fabrication procedures can be standardized. Usually the output of each such second-order network is taken at the output of an op amp, thus providing the necessary isolation from one another. This means that each circuit acts independently and further implies that each circuit can be tuned independently to make certain that every network satisfies its own requirements.

4.1 *Amplifier Filters*

Among many second-order active *RC* networks, the most popular ones are *Sallen and Key* type filters. These filters use a voltage-controlled voltage source (VCVS) of gain K. In practice, this active device has proven to be the preferred one. The common name for this device is *voltage amplifier* or simply *amplifier*. The circuit symbol and its equivalent circuit are shown in Fig. 4.1. If the gain K is positive, then the amplifier is called a *noninverting amplifier*. Op amp realizations of noninverting and inverting amplifiers are shown in Fig. 4.2. A unity gain noninverting amplifier, also known as a *voltage follower*, requires no resistors, as shown in Fig. 4.2*b* and is simple to realize using a single op amp. A noninverting VCVS of gain $K = 2$ can be realized using any two equivalued resistors. These are some important design considerations that one has to keep in mind when designing active filters using a VCVS as the active element. Though there are active *RC* filters using inverting amplifiers as active elements, they are not popular for two reasons: (1) The op amp realization of an inverting amplifier possesses a finite input impedance (equal to R in Fig. 4.2*c*), and (2) the gain requirement of such amplifiers is large and thus they can be used only in the low-frequency range. For these reasons, active filters using noninverting amplifiers are more widely used.

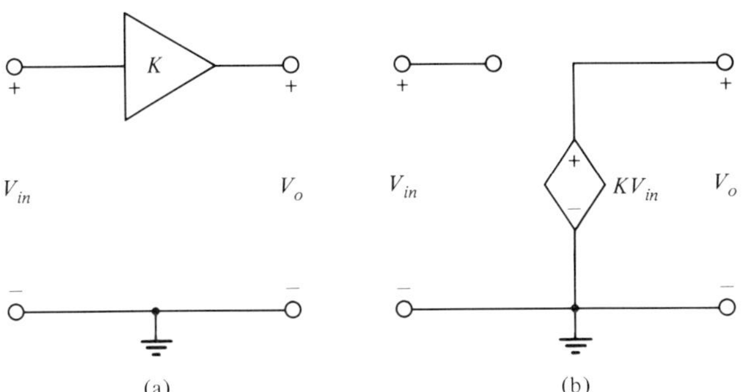

Figure 4.1 Circuit symbol and the equivalent circuit for a VCVS.

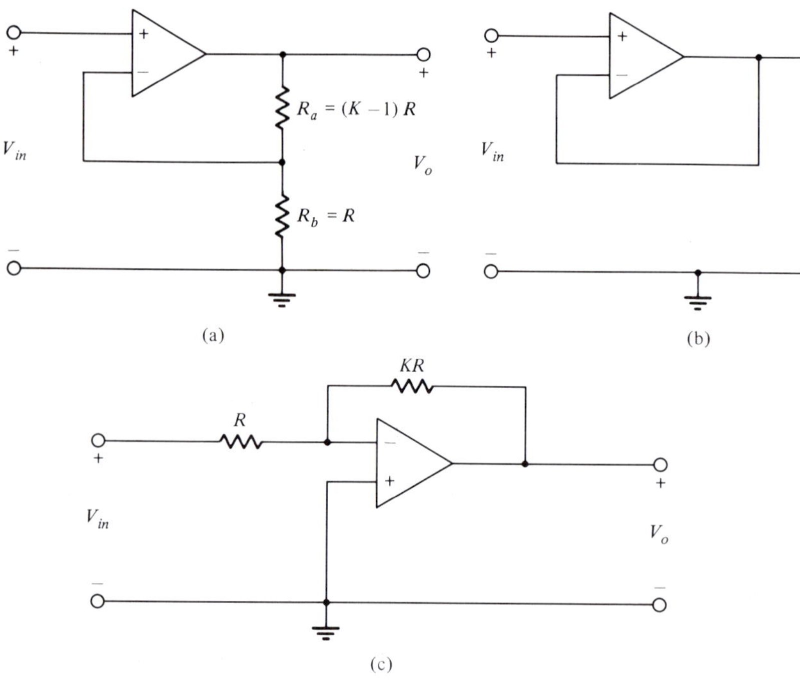

Figure 4.2 Operational amplifier realizations of finite-gain amplifiers. (*a*) Noninverting. (*b*) Unity gain. (*c*) Inverting.

Table 4.1 Types of second-order filter functions

Type of filter	Form of network function
Lowpass	$\dfrac{H_0\omega_p^2}{D(s)}$
Highpass	$\dfrac{H_0 s^2}{D(s)}$
Bandpass	$\dfrac{H_0(\omega_p/Q_p)s}{D(s)}$
Lowpass notch	$\dfrac{H_0(s^2+\omega_z^2)}{D(s)}, \ \omega_z > \omega_p$
Highpass notch	Same as lowpass notch except that $\omega_z < \omega_p$
All-pass	$H_0\dfrac{s^2-(\omega_p/Q_p)s+\omega_p^2}{D(s)}$

In all cases, the denominator polynomial $D(s)$ is

$$D(s) = s^2 + \frac{\omega_p}{Q_p}s + \omega_p^2$$

The most general form of the second-order transfer function is

$$H(s) = H_0 \frac{s^2 + n_1 s + n_0}{s^2 + d_1 s + d_0} \tag{4.1}$$

The above transfer function is usually expressed as

$$H(s) = H_0 \frac{s^2 + (\omega_z/Q_z)s + \omega_z^2}{s^2 + (\omega_p/Q_p)s + \omega_p^2} \tag{4.2}$$

where ω_z and Q_z describe the zeros of the network function and ω_p and Q_p describe the poles. The constant H_0 is known as the gain constant. Table 4.1 gives the specific names of the various filter functions and their descriptions. Finally, it should be mentioned that the transfer functions in which we are interested are voltage transfer ratios.

4.2 *Active Lowpass Filters Using Noninverting Finite Gain Amplifiers*

Figure 4.3 shows a lowpass filter using a single noninverting finite gain amplifier, which is called a Sallen and Key lowpass filter. Assuming that the op amp is ideal, the voltage transfer function of this circuit can be derived as

$$\frac{V_o}{V_i} = \frac{KG_1G_2S_1S_2}{s^2 + s[S_2(G_1 + G_2) + S_1G_2(1 - K)] + G_1G_2S_1S_2} \tag{4.3}$$

where $G_i = 1/R_i$, $S_i = 1/C_i$, $i = 1, 2$, and $K = 1 + R_a/R_b$. Note that, when the op amp is ideal,

$$K = 1 + \frac{R_a}{R_b} = \frac{V_o}{V_X}$$

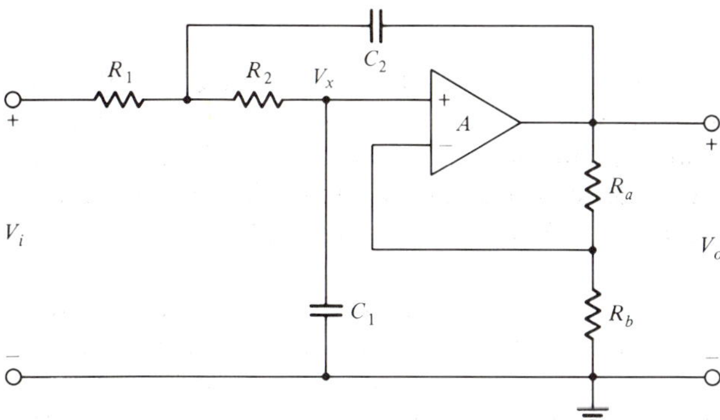

Figure 4.3 Sallen and Key lowpass filter using a noninverting finite amplifier.

For a given set of specifications for ω_p, Q_p, and H_0, we need to find the values of the elements. To do this, we compare the transfer function in (4.3) with the lowpass transfer function given in Table 4.1 and obtain the basic design equations, which, for this network, are

$$\omega_p^2 = G_1 G_2 S_1 S_2 \tag{4.4a}$$

$$\frac{\omega_p}{Q_p} = S_2(G_1 + G_2) + S_1 G_2 (1 - K) \tag{4.4b}$$

and

$$H_0 = K \tag{4.4c}$$

There are five parameters to choose from, satisfying only three equations. We can choose two of these parameters arbitrarily, however, we can impose certain design considerations in our selection. Before we impose any such considerations, since $K = H_0$, we have practically no control over K. But, in this network, we usually do not impose any restriction on K to satisfy a given H_0. The value of K is generally chosen to fulfill conditions other than satisfying H_0. In such cases (4.4c) is no longer a constraint to be used, and H_0 in turn is constrained by the value of K. Then, we are left with three arbitrary parameters. To fix this arbitrariness, one possible design consideration we can impose is that we want to have equivalued capacitances and resistances in the network. This requires that

$$R_1 = R_2 = R \qquad \text{and} \qquad C_1 = C_2 = C$$

When we use the above constraints in (4.4a) and (4.4b), we have

$$RC = \frac{1}{\omega_p} \tag{4.5a}$$

and

$$K = 3 - \frac{1}{Q_p} \tag{4.5b}$$

For a given value of ω_p, we usually select a standard value for capacitance and calculate the required value of R using (4.5a). Then, for a given value of Q_p, we calculate the value of K using (4.5b). The choices for R_a and R_b can then be made to realize the required gain. For example, if we select a standard value $R_b = R$, then $R_a = (K - 1)R$.

Example 4.1: Design a lowpass filter having equivalued resistors and capacitors. The design specifications are $f_p = 1$ kHz and $Q_p = 5$.

Choosing a standard capacitance value for C_1 and C_2 as $C_1 = C_2 = 0.01$ μF and using (4.5a), we have $R_1 = R_2 = 15.915$ kΩ. Then, using

(4.5*b*), we find that $K = 2.8$. This value of K can then be realized using $R_b = 10\ \text{k}\Omega$ and $R_a = 18\ \text{k}\Omega$. ■

Before we consider other possible design procedures, we shall simplify the basic design equations in (4.4) by letting

$$R_2 = m^2 R_1 = m^2 R \tag{4.6a}$$

and

$$C_2 = n^2 C_1 = n^2 C \tag{4.6b}$$

Using (4.6) in (4.4), we obtain

$$\omega_p = \frac{1}{mnRC} \tag{4.7a}$$

and

$$\frac{1}{Q_p} = \frac{m}{n} + \frac{1}{mn} + \frac{n}{m}(1 - K) \tag{4.7b}$$

The RC product can be chosen according to the required value of ω_p, which is in effect equivalent to frequency scaling. Then, of the three parameters m, n, and K, there are still two parameters that can be chosen arbitrarily. This is because we now need to satisfy only (4.7*b*) for a given value of Q_p. In the previous design procedure, we fixed the values of both m and n at 1. Then, solving for K from (4.7*b*), we do indeed obtain (4.5*b*). Next, instead of fixing the values of m and n, we shall develop a design procedure in which we fix the values of K and n and choose the value of m to satisfy (4.7*b*). Note that fixing the value of n (and therefore n^2) allows us to choose standard values for the capacitances C_1 and C_2. Similarly, fixing the value of K allows us to choose standard values for the resistances R_a and R_b. Now, solving for m from (4.7*a*) in terms of n and K, we have

$$m = \frac{n}{2Q_p} \pm \sqrt{\left[\frac{n}{2Q_p}\right]^2 + (K - 1)n^2 - 1} \tag{4.8}$$

In order for m to be real, K and n must satisfy the following inequality:

$$K \geq 1 + \frac{1}{n^2} - \frac{1}{4Q_p^2} \tag{4.9}$$

We may choose either sign in (4.8) to obtain the value of m. However, if we choose n as a positive (negative) number, we also have to choose m as a positive (negative) number.

We shall consider two possible values for K in particular. A good choice is $K = 1$ because it does not require the two resistors R_a and R_b. However, choosing $K = 1$ then requires $n^2 \geq 4Q_p^2$. This means that, to avoid excessive capacitor spread, the choice of $K = 1$ is limited to only low-Q_p designs.

Example 4.2: Design a Sallen and Key lowpass filter to satisfy the specifications $f_p = 1$ kHz and $Q_p = \sqrt{2}$. Choose $K = 1$.

We require that $n^2 \geq 4Q_p^2 = 8$ and we choose $n^2 = 10$. With $K = 1$ and $n^2 = 10$, we have, from (4.8), $m = 1.618$ or 0.618. Since one value is the reciprocal of the other, we can choose either one. Therefore, selecting the value of m as 0.618 and $C = C_1 = 0.01\ \mu$F, we can find the values of the other elements using (4.7a) and (4.6). They are $C_2 = 0.1$ μF, $R_1 = R = 8.143$ kΩ, and $R_2 = 3.111$ kΩ. Then, $R_a = 0$ and $R_b = \infty$ is all that is necessary to realize $K = 1$. ■

Another possible design procedure would be to choose $K = 2$, since such a choice requires R_a and R_b to be equal. This means that *any two equivalued resistors* can be used for R_a and R_b. With $K > 1$, the resistor and capacitor spreads will be considerably reduced. In particular, for $K > 1$, the capacitor spread does not depend on Q_p (or Q_p^2) but only on $K - 1$. Rewriting (4.9), we find that

$$n^2 \geq \frac{1}{K - 1 + 1/4Q_p^2}$$

Thus, if $n^2 \geq 1/(K - 1)$, the above condition can be satisfied. Therefore, even if Q_p is large, the capacitance spread does not have to be large.

Example 4.3: Design a Sallen and Key lowpass filter with $K = 2$ to satisfy the specifications $f_p = 10$ kHz and $Q_p = 10$. Choose (a) $n^2 = 1$ and (b) $n^2 = 2$.

(a) With $K = 2$, $n = 1$, and $Q_p = 10$, the possible value for m is 0.1. Then, choosing $C_1 = C_2 = 0.01\ \mu$F, we have $R_1 = R = 15.92$ kΩ and $R_2 = m^2 R_1 = 159.2\ \Omega$. The possible values for R_a and R_b are any two equivalued resistors, such as $R_a = R_b = 10$ kΩ. Note that the resistor ratio is $m^2 = 1/Q_p^2$, with $K = 2$, and $n = 1$. This again puts a limit on the value of Q_p when $K = 2$ and $n = 1$ are required simultaneously.

(b) With $n = \sqrt{2}$, $K = 2$, and $Q_p = 10$, the possible value for m is 1.0732. Then, choosing $C_1 = 1$ nF, we find that $R_1 = 10.486$ kΩ and $R_2 = 12.078$ kΩ. Note that, by allowing a small capacitance spread, we are able to bring the resistor ratio close to unity. ■

The fourth possible design consideration that can be imposed is to choose any two free parameters such that the magnitude of the semilogarithmic active pole sensitivity due to the nonzero time constant of the op amp τ is minimized. To develop this procedure, we have to first find the active pole sensitivity as a function of the two parameters we choose. To find the active pole sensitivity, we first obtain the denominator polynomial of the transfer function assuming that $A^{-1} = s\tau$. This requires only the substitution of

$$K \to \frac{K}{1 + Ks\tau}$$

into the transfer function in (4.3). After replacing K with $K/(1 + Ks\tau)$ in (4.3) and using (4.6), the denominator polynomial of the transfer function is

$$D(s, \tau) = s^2 + sG_2S_2(m^2 + n^2 + 1 - Kn^2) + m^2n^2G_2^2S_2^2 \\ + Ks\tau\left[s^2 + sG_2S_2(m^2 + n^2 + 1) + m^2n^2G_2^2S_2^2\right]$$

Since in the basic design $\tau = 0$, the basic design equations must then be

$$\omega_p = mnG_2S_2 \tag{4.10a}$$

and

$$\frac{1}{Q_p} = \frac{m^2 + n^2 + 1 - Kn^2}{mn} \tag{4.10b}$$

In fact, (4.10) is the same as (4.7). To realize a given set of specifications for ω_p and Q_p, one can choose the product G_2S_2 and K to fix the values of ω_p and Q_p, respectively. Therefore the active pole sensitivity can be minimized with respect to the remaining two parameters, m and n. Solving for G_2S_2 and K from (4.10) and eliminating these quantities in the denominator polynomial, we have

$$D(s, \tau) = s^2 + \frac{\omega_p}{Q_p}s + \omega_p^2 + Ks\tau\left(s^2 + a\omega_p s + \omega_p^2\right) \tag{4.11}$$

where

$$a = \frac{m^2 + n^2 + 1}{mn} \tag{4.12a}$$

and

$$K = \frac{m^2 + n^2 + 1 - mn/Q_p}{n^2} \tag{4.12b}$$

Note that we can eliminate K in (4.11) using (4.12b). However, we have retained K in (4.11) for ease of writing. Next, applying (3.63) to (4.11), we obtain the semilogarithmic sensitivity of the pole, $p = -\omega_p/(2Q_p) + j\omega_p$:

$$S_\tau^p = j0.5K\left(a - \frac{1}{Q_p}\right)p \tag{4.13}$$

Using (4.12) in (4.13), we find that

$$|S_{\tau}^{p}| = \frac{0.5\left(m^2 + n^2 + 1 - mn/Q_p\right)^2}{mn^3}\omega_p \tag{4.14}$$

We can use (4.13) and the value of τ to obtain

$$\frac{\Delta p}{p} = j0.5K\left(a - \frac{1}{Q_p}\right)\left(-\frac{1}{2Q_p} + j\right)(\omega_p\tau) \tag{4.15}$$

To minimize the effect of the nonzero value of τ on the pole frequency, we have to minimize the function

$$F = \frac{0.5\left(m^2 + n^2 + 1 - mn/Q_p\right)^2}{mn^3}$$

by choosing the appropriate values of m and n. The above function F can also be written as

$$F = 0.5\left[x^{1.5} + x^{-0.5}(1 + n^{-2}) - \frac{x^{0.5}}{Q_p}\right]^2 \tag{4.16}$$

where it is assumed that both m and n are positive together and $x = m/n$. Instead of minimizing F with respect to m and n, we can minimize (4.16) with respect to x and n. By inspection, it can easily be seen that, for any choice of x, F is a minimum when $n \to \infty$. Further, it can be shown that, for a given choice of n, the minimum of F occurs when

$$x = \frac{1}{6Q_p} + \sqrt{\left(\frac{1}{6Q_p}\right)^2 + \frac{1 + n^{-2}}{3}}$$

or

$$m = \frac{n}{6Q_p} + \sqrt{\left(\frac{n}{6Q_p}\right)^2 + \frac{n^2 + 1}{3}} \tag{4.17}$$

The above discussion reveals that the absolute minimum of F occurs when both m and $n \to \infty$ such that $m/n \approx 1/\sqrt{3} + 1/6Q_p$. Since this is a physical problem, neither the capacitance ratio nor the resistance ratio can be infinite. Therefore we can only hope to obtain a suboptimal solution by choosing a convenient capacitance ratio n^2 (for practical reasons) and then choosing m to minimize F. Such a choice can be made using (4.17). Before leaving this topic, we must answer another question: What range of n^2 should be selected to give us a value of F that will approach the absolute minimum? The larger the value of n^2, the better. However, for practical reasons, it is not wise to choose a large capacitance ratio. Even with $n^2 = 10$, we will be very close to the absolute minimum. Consider the following example.

Example 4.4: Design a Sallen and Key lowpass filter with minimum active pole sensitivity that satisfies the specifications $f_p = 10$ kHz and $Q_p = 10$. Choose (a) $n^2 = 10$ and (b) $n^2 = 5$. Calculate the magnitude of the active pole sensitivity in each case and compare the values.

(a) The choice of m should be made using (4.17) to minimize the active pole sensitivity. This value of m is 1.9683. Since the values of m and n are known for a given Q_p, the value of K must be chosen using (4.12*b*), and it is 1.4252. With $m = 1.9683$ and $K = 1.4252$, we will need nonstandard values for all the resistors in the circuit. Furthermore, K must be realized exactly from the viewpoint of passive Q_p sensitivities. Therefore a wise choice is to fix the value of K near the value we calculated earlier and recalculate the value of m using (4.8) to fix the requirement of Q_p. If the choice of K is 1.5, then we will require that $m = 2.1644$. To find the effect of the changes we have made in the values of m and K (for the sake of convenience) on F, we can calculate F in the two cases: (i) $m = 1.9683$ and $K = 1.4252$; (ii) $m = 2.1644$ and $K = 1.5$. Then, $F = 1.6316$ and 1.6437, respectively. With the choice $K = 1.5$ and $m = 2.1644$, there is no question that the active pole sensitivity will increase. This increase is very small (less than 1%) but provides an advantage in that we will be able to use standard resistance values for R_a and R_b.

Let us now complete the design with $K = 1.5$ and $m = 2.1644$. Choosing $C_1 = 1$ nF, we have $C_2 = 10$ nF, $R_1 = 2.3254$ kΩ, and $R_2 = 10.893$ kΩ. The possible choices for R_a and R_b are $R_a = 10$ kΩ and $R_b = 20$ kΩ.

(b) With $n^2 = 5$, the value of m required to minimize the active pole sensitivity is 1.4520. The value of K required to fix the value of Q_p will then be 1.5567. Again, the value of K is inconvenient, and therefore we use $K = 1.55$ and recalculate the required value of m for the given Q_p using (4.8). Thus, with $K = 1.55$ and $n = \sqrt{5}$, we have $m = 1.4394$. To complete the design, we choose $C_1 = 1$ nF and $C_2 = 5$ nF and calculate the values of R_1 and R_2, obtaining $R_1 = 4.945$ kΩ and $R_2 = 10.245$ kΩ. The possible values for R_a and R_b are 11 and 20 kΩ, respectively.

From (4.14), we find that the magnitude of the active pole sensitivity is just $F\omega_p$, and therefore this magnitude is $1.6437\omega_p$ when $K = 1.5$, $n^2 = 10$, and $m = 2.1644$.

In part (b) we can find the value of F to be 1.866, and thus the magnitude of the semilogarithmic active pole sensitivity will be $1.866\omega_p$. We emphasize the fact that this magnitude is proportional to the ripple factor R developed in Chap. 3 for a given ω_p, and thus this comparison is meaningful in the sense that it reflects the changes that will take place in the magnitude function of the transfer function because of the nonzero value of τ. Now, returning to the comparison, we find that there is not much reduction in the magnitude of the active pole sensitivity when we increase the capacitance ratio from 5 to 10. This indicates that, even with a capacitance ratio of 5, we are able to obtain good results. For practical reasons, it is not good to have a large capacitor ratio. For example, it can be easily seen that a larger capacitor spread also means a larger resistor spread. Thus we can keep the resistor spread and the capacitor spread to a low value by choosing $n^2 = 5$ without losing much in terms of the effect on the magnitude of the active pole sensitivity. ■

The design procedure developed takes care of only one of the effects of the nonzero value of τ, namely, that on the active pole sensitivity. However, the magnitude characteristic of the realized filter circuit will differ from the required one because of the changes in ω_p and Q_p due to the nonzero value of τ. The changes can be compensated for by using the predistortion technique developed in Chap. 3. When the value of $\omega_p\tau = \tau_n$ is known, R_1 and R_2 can be predistorted to obtain the required values of ω_p and Q_p for a given value of n^2, and K can be used to minimize the magnitude of the active pole sensitivity. This can be implemented with the program SKLPN in App. 4A. With this program, when $\tau_n = 0.01$, $n^2 = 5$, and $Q_p = 10$, for a normalized value of $\omega_p = 1$ r/s, the element values required to achieve minimum active pole sensitivity after predistortion is applied are

$\Omega_p = 1$
$K = 1.557$
$R_1 = 0.301869$ kΩ
$R_2 = 0.638265$ kΩ
$C_1 = 0.001$ F
$C_2 = 0.005$ F
Possible value of $R_a = 0.557$ kΩ
Possible value of $R_b = 1.000$ kΩ
Minimum logarithmic active pole sensitivity = 1.83418E − 2

The same program can be used to obtain practical element values for a given set of values for ω_p and C_1. For example, if $\omega_p = 2\pi \times 10^4$ r/s and $C_1 = 1$ nF, then the denormalized values for the above design will be

Desired $f_p = 10$ kHz
$R_1 = 4.80439$ kΩ
$R_2 = 10.1583$ kΩ
$C_1 = 1$ nF
$C_2 = 5$ nF
Possible value of $R_a = 11.14$ kΩ
Possible value of $R_b = 20.00$ kΩ

Before we conclude our discussion of this filter circuit, we must consider the passive ω_p and Q_p sensitivities. The passive ω_p sensitivities can be found by inspection:

$$S_{R_1}^{\omega_p} = S_{R_2}^{\omega_p} = S_{C_1}^{\omega_p} = S_{C_2}^{\omega_p} = -0.5 \tag{4.18a}$$

and

$$S_{R_a}^{\omega_p} = S_{R_b}^{\omega_p} = 0 \tag{4.18b}$$

The passive Q_p sensitivities can be derived as

$$S_{R_1}^{Q_p} = -S_{R_2}^{Q_p} = -0.5 + Q_p\frac{m}{n} \tag{4.18c}$$

$$S_{C_2}^{Q_p} = -S_{C_1}^{Q_p} = -0.5 + Q_p\left(\frac{m}{n} + \frac{1}{mn}\right) \tag{4.18d}$$

$$S_{R_a}^{Q_p} = -S_{R_b}^{Q_p} = \frac{(K-1)nQ_p}{m} \tag{4.18e}$$

Example 4.5: Calculate the passive ω_p and Q_p sensitivities of the filter designed in Example 4.3 with $n^2 = 1$ and $K = 2$ and those of the one designed in Example 4.4 with $n^2 = 5$ and $K = 1.55$. Assume that the resistances have a 1% tolerance and the capacitances have a 0.5% tolerance. Find the worst-case per-unit change that can be expected in each of these designs.

The ω_p sensitivities are constants, as is evident from (4.18a). We need only to evaluate the Q_p sensitivities.

For Example 4.3 with $n = 1$, $K = 2$, and $m = 0.1$:

$$S_{R_1}^{Q_p} = -S_{R_2}^{Q_p} = 0.5$$

$$S_{C_2}^{Q_p} = -S_{C_1}^{Q_p} = 100.5$$

$$S_{R_a}^{Q_p} = -S_{R_b}^{Q_p} = 100$$

$$\left.\frac{\Delta Q_p}{Q_p}\right|_{WC} = 301.5\%$$

For Example 4.4 with $n = \sqrt{5}$, $K = 1.55$, and $m = 1.4391$:

$$S_{R_1}^{Q_p} = -S_{R_2}^{Q_p} = 5.94$$

$$S_{C_2}^{Q_p} = -S_{C_1}^{Q_p} = 9.044$$

$$S_{R_a}^{Q_p} = -S_{R_b}^{Q_p} = 8.54$$

$$\left.\frac{\Delta Q_p}{Q_p}\right|_{WC} = 38\%$$

The worst-case percentage changes that can be expected in these two filter designs clearly indicate that the filter designed in Example 4.4 is superior. This is because of the capacitor spread allowed in this case. Also note that, when such a capacitor spread is allowed, the active pole sensitivity is also minimized. These passive sensitivities are also calculated in the program SKLPN for a predistorted design of this filter. ■

Bach Lowpass Filter

Another lowpass filter we shall consider is the *Bach lowpass filter*. This circuit is attractive for two reasons:

1. It uses only unity gain amplifiers. Operational amplifier realizations of unity gain amplifiers do not require additional resistors, and the nominal unity gain can be realized exactly.
2. The passive ω_p and Q_p sensitivities are at their theoretical minimum values.

A Bach lowpass filter is shown in Fig. 4.4. The transfer function of this circuit can be derived in terms of the passive element values and the gains of the amplifiers:

$$\frac{V_0}{V_i} = \frac{K_1K_2G_1G_2S_1S_2}{s^2 + s[G_1S_1 + (1 - K_1K_2)G_2S_2] + G_1G_2S_1S_2} \tag{4.19}$$

where $G_i = 1/R_i$ and $S_i = 1/C_i$, $i = 1, 2$.

With the nominal values $K_1 = K_2 = 1$, the basic design equations are

$$\omega_p^2 = \frac{1}{R_1R_2C_1C_2} \tag{4.20a}$$

and

$$Q_p = \sqrt{\frac{R_1C_1}{R_2C_2}} \tag{4.20b}$$

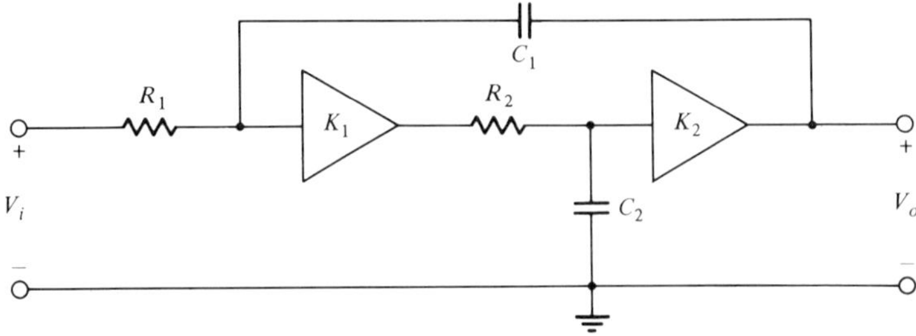

Figure 4.4 Bach lowpass filter ($K_1 = K_2 = 1$).

We will consider two possible design procedures. The first is developed with the assumption that $C_1 = C_2 = C$. Then, the solution of (4.20) gives

$$R_2 = \frac{1}{\omega_p Q_p C} \tag{4.21a}$$

and

$$R_1 = Q_p^2 R_2 \tag{4.21b}$$

Note that, with equivalued capacitors, the resistor spread is on the order of Q_p^2. Therefore this design procedure can be used only to realize low-Q_p filters.

Example 4.6: Design a Bach lowpass filter that satisfies the specifications $f_p = 3$ kHz and $Q_p = 5$.

We choose a convenient capacitance value for C_1 and C_2, namely, 4.7 nF. Then, using (4.21), we have $R_2 = 2.258$ kΩ and $R_1 = 56.438$ kΩ. ■

A second design procedure allows some capacitor spread so that the resistor spread can be reduced. A simple design consideration can be imposed: $C_1/C_2 = Q_p = R_1/R_2$. This might lead to two problems: (1) nonstandard capacitance values for C_1 and C_2, and (2) a large capacitance spread. So a wise step is to choose a convenient capacitor spread of $C_1/C_2 = n^2$ (> 1 for practical cases of interest). Then, the solution of (4.20) leads to

$$R_2 = \frac{1}{Q_p \omega_p C_2} \tag{4.22a}$$

and

$$R_1 = \frac{Q_p^2 R_2}{n^2} \tag{4.22b}$$

Though such a capacitor spread reduces the resistor spread, it is not possible in practice to allow a large capacitor spread. Therefore this design procedure makes it possible to increase the Q_p value that can be realized in this circuit, but to a limited degree.

Example 4.7: Design a Bach lowpass filter satisfying the specifications $f_p = 4$ kHz and $Q_p = 10$. Choose $C_1/C_2 = 5$.

Choosing a convenient value of 1 nF for C_2 and using (4.22), we have $R_2 = 3.979$ kΩ and $R_1 = 79.577$ kΩ. The value of C_1 should, of course, be 5 nF. ■

The passive ω_p and Q_p sensitivities can be obtained from (4.20) by inspection:

$$S_{R_1}^{\omega_p} = S_{R_2}^{\omega_p} = S_{C_1}^{\omega_p} = S_{C_2}^{\omega_p} = -0.5$$

and

$$S_{R_1}^{Q_p} = S_{C_1}^{Q_p} = -S_{R_2}^{Q_p} = -S_{C_2}^{Q_p} = 0.5$$

As is obvious from the above equations, the passive ω_p and Q_p sensitivities are low (as mentioned earlier).

If the unity gain amplifiers are realized using the op amp circuits in Fig. 4.2*b*, then, including the frequency dependence of the op amps, the unity gain amplifiers can be described by

$$K_1 = K_2 = \frac{1}{1 + s\tau} \tag{4.23}$$

where both op amps are assumed to have the same time constants. Since a given set of specifications for ω_p and Q_p are realized using the basic design equations (4.20), using (4.20) and (4.23) for K_1 and K_2 in (4.19), we find the denominator polynomial of the transfer function to be

$$D(s,\tau) = s^2 + \frac{\omega_p}{Q_p}s + \omega_p^2 + 2s\tau\left[s^2 + \omega_p s\left(Q_p + \frac{1}{Q_p}\right) + \omega_p^2\right]$$
$$+ s^2\tau^2\left[s^2 + \omega_p s\left(Q_p + \frac{1}{Q_p}\right) + \omega_p^2\right]$$

The active pole sensitivity of the pole $p = -\omega_p/(2Q_p) + j\omega_p$ can be found by applying (3.63) to the above polynomial. It is possible to obtain a more accurate result using (3.67) for calculating $\Delta p/p$, however, using (3.63), we obtain

$$S_\tau^p = -\omega_p Q_p - j\frac{\omega_p}{2} \tag{4.24}$$

Note that $|S_\tau^p| = \omega_p\sqrt{Q_p^2 + \frac{1}{4}}$ and is independent of the choice of the circuit parameters. Comparing this active pole sensitivity with that of a reasonably good design for a Sallen and Key filter, we can conclude that this circuit cannot be used for Q_p greater than 2 or 3. Though the passive sensitivities are low, this circuit can be used only for low-Q_p filters, and this limitation is imposed by the large active pole sensitivity and the large element spread.

4.3 *Sallen and Key Bandpass Filter*

We shall consider the most popular Sallen and Key bandpass filter, which uses a single noninverting finite gain amplifier. The circuit is shown in Fig.

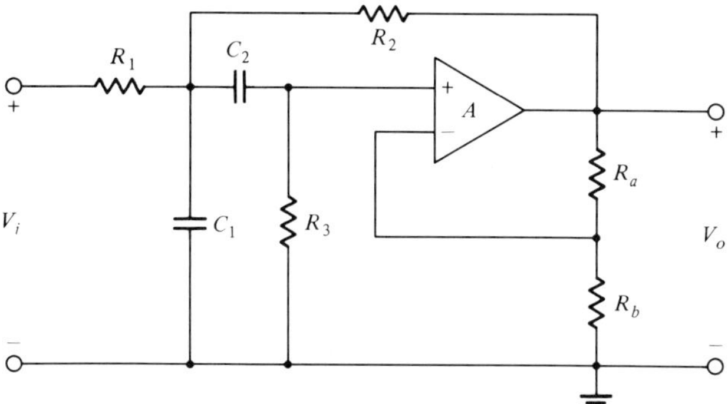

Figure 4.5 Sallen and Key bandpass filter.

4.5. When the op amp is ideal, the transfer function of the circuit is

$$\frac{V_0}{V_i} = \frac{(KG_1S_1)s}{s^2 + s[S_1(G_1 + G_2) + G_3(S_1 + S_2) - KG_2S_1] + G_3(G_1 + G_2)S_1S_2} \tag{4.25}$$

where $G_i = 1/R_i$, $S_i = 1/C_i$, and $K = 1 + R_a/R_b$. The basic design equations can be obtained by comparing this transfer function with the bandpass transfer function in Table 4.1:

$$H_0 \frac{\omega_p}{Q_p} = KG_1S_1 \tag{4.26a}$$

$$\omega_p^2 = G_3(G_1 + G_2)S_1S_2 \tag{4.26b}$$

$$\frac{\omega_p}{Q_p} = S_1(G_1 + G_2) + G_3(S_1 + S_2) - KG_2S_1 \tag{4.26c}$$

We shall develop three possible design procedures. In the first, we assume that $C_1 = C_2 = C$ and $\omega_p CR_3 = 1$. Then, for a convenient choice of the value of C, a unique solution for K, R_1, R_2, and R_3 can be obtained by solving (4.26):

$$K = 3 + \frac{H_0 - 1}{Q_p} \tag{4.27a}$$

$$R_1 = \frac{KQ_p}{H_0\omega_p C} \tag{4.27b}$$

$$R_2 = \frac{KQ_p}{(KQ_p - H_0)\omega_p C} \tag{4.27c}$$

and

$$R_3 = \frac{1}{\omega_p C} \tag{4.27d}$$

Example 4.8: Design a Sallen and Key bandpass filter to satisfy the specifications $f_p = 4$ kHz, $Q_p = 10$, and $H_0 = 7.5$.

Choosing 4.7 nF as a convenient value for C_1 and C_2, we find that $R_1 = 41.2$ kΩ, $R_2 = 10.655$ kΩ, and $R_3 = 8.4657$ kΩ. The possible values of R_a and R_b can be calculated with the required value of $K = 3.65$. One set of values may be $R_a = 53$ kΩ and $R_b = 20$ kΩ. ■

In the next design procedure, we fix the capacitance values but do not necessarily make them equal. That is, we set $C_2 = C$ and $C_1 = C/M$, where M may have any convenient value. We also fix K at a convenient value. Then, for a given set of specifications for H_0, ω_p, and Q_p, we can use the following set of formulas to find the conductance (hence the resistance) values:

$$G_1 = \frac{H_0}{KMQ_p}\omega_p C \tag{4.28a}$$

$$G_2 = \frac{Y}{2} + \sqrt{\left(\frac{Y}{2}\right)^2 + Z} - G_1 \tag{4.28b}$$

where

$$Y = \frac{(H_0 - 1)\omega_p C}{(K-1)MQ_p} \qquad \text{and} \qquad Z = \frac{(M+1)(\omega_p C)^2}{(K-1)M^2}$$

and

$$G_3 = \frac{(\omega_p C)^2}{M(G_1 + G_2)} \tag{4.28c}$$

Example 4.9: Given that $H_0 = 10$, $Q_p = 20$, and $f_p = 4$ kHz, find a realization of a Sallen and Key bandpass filter. Choose $C_1 = C_2 = 10$ nF and $K = 2$.

Note that $M = 1$. Substituting the different values in (4.28), we have $R_1 = 15.915$ kΩ, $R_2 = 2.8279$ kΩ, and $R_3 = 6.593$ kΩ. The possible choices for R_a and R_b are any two equivalued resistors: for example, 20 kΩ could be used for each. ■

Two free parameters were chosen arbitrarily in the previous two design procedures. The third possible design procedure is one in which the magnitude of the active pole sensitivity is minimized. That is, in this design procedure, the choices for the two free parameters, namely, M and K, are made to minimize the magnitude of the active pole sensitivity. The first step in developing this design procedure is to find the denominator polynomial with $A^{-1} = s\tau$. Since the nominal design satisfies the basic design equations (4.26), the denominator polynomial for a given ω_p and Q_p can be derived as

$$D(s, \tau) = s^2 + s\frac{\omega_p}{Q_p} + \omega_p^2 + Ks\tau\left(s^2 + as + \omega_p^2\right) \tag{4.29}$$

where

$$a = S_1(G_1 + G_2) + G_3(S_1 + S_2)$$

Applying (3.63) to the above denominator polynomial, the semilogarithmic sensitivity of the pole $p = -\omega_p/2Q_p + j\omega_p$ can be found to be

$$S_\tau^p = \frac{-K(a - \omega_p/Q_p)p}{2j\omega_p}$$

The nominal element values from which a is evaluated must satisfy (4.26) [in particular (4.26c)], and therefore

$$S_\tau^p = \frac{K^2}{2}\left(\frac{-j}{2Q_p} - 1\right)G_2 S_1 \tag{4.30}$$

Assume that G_1, G_2, and G_3 are chosen to satisfy a given set of specifications for H_0, Q_p, and ω_p and for a given set of values for M and K. Then, G_2 can be found from (4.28b). Therefore, substituting (4.28b) into (4.30), we have

$$|S_\tau^p| = F\omega_p \tag{4.31}$$

where

$$F = \frac{K^2}{4Q_p}\left[\frac{H_0 - 1}{K - 1} - \frac{2H_0}{K} + \sqrt{\left(\frac{H_0 - 1}{K - 1}\right)^2 + \frac{4Q_p^2(M + 1)}{K - 1}}\,\right] \tag{4.32}$$

For a given ω_p, the magnitude of the active pole sensitivity can be minimized by minimizing F with respect to K and M. For any value of K, the minimum of F lies on the line $M = 0$, which is physically impossible. This condition is similar to the one we faced in the case of a Sallen and Key lowpass filter. Therefore we have to choose a small value for M, and then we can try to minimize F with respect to K. This is a single-parameter minimization. Differentiating F with respect to K and equating it to zero, we can solve the resulting equation for K. Such an equation is

$$E\left[K(K - 2)(E - H_0 - 1) - 2H_0\right] + 2K^2(K - 1)(1 + M)Q_p^2 = 0 \tag{4.33}$$

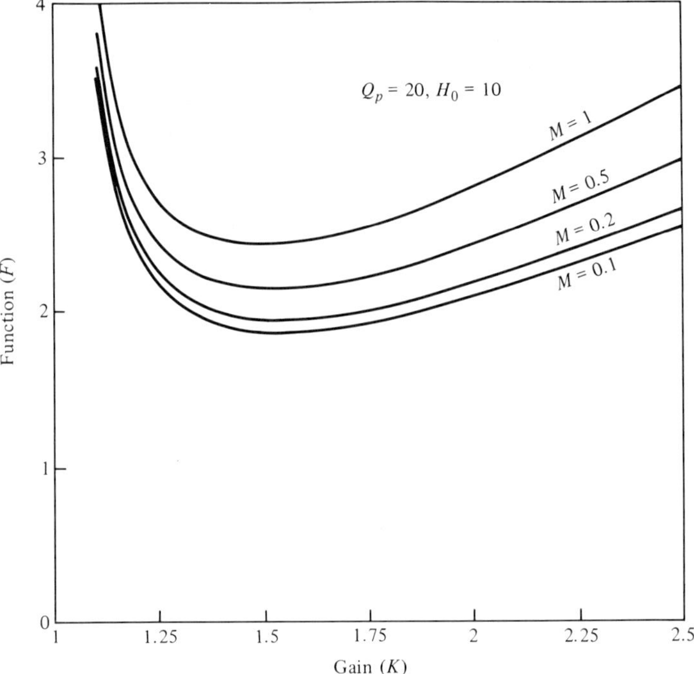

Figure 4.6 Plots of (4.32) as function of the gain K.

where

$$E = \sqrt{(H_0 - 1)^2 + 4Q_p^2(1 + M)(K - 1)}$$

The above equation is highly nonlinear, and it is difficult to find a mathematical solution for K. For a given set of values for H_0, Q_p, and M, it is possible to obtain a numerical solution for K. For example, if $H_0 = 10$, $Q_p = 20$, and $M = 0.5$, then $K = 1.50959$. When the value of H_0 alone is increased to 20, the required value of K is 1.68897. In Fig. 4.6 the function F, as given by (4.32), is plotted as a function of K for various values of M, with $H_0 = 10$ and $Q_p = 20$. One can observe from this figure that, for any capacitance ratio, F is shallow near $1.5 < K < 1.75$. This implies that, for any Q_p, H_0, and M, a reasonable choice of K can be made in this range with relatively no impunity of increasing the function. The following practical procedure can be used in designing this filter.

1. Choose a convenient capacitance ratio M $(1 > M > 0)$. In fact, $M = 0.2$ is quite sufficient. Lowering the value of M to less than 0.2 does not really decrease the function substantially. For practical reasons, it is not good to have a large spread in the capacitance values.

2. Choose a convenient value of K such that $1.5 < K < 1.75$.
3. Choose a convenient capacitance value for $C_2 = C$.
4. Use (4.28) to obtain the values of R_1, R_2, and R_3.

Example 4.10: Find a realization of a Sallen and Key bandpass filter that satisfies the specifications $H_0 = 10$, $Q_p = 20$, and $f_p = 4$ kHz. Choose $C_2 = 2.2$ nF, $K = 1.6$, and $M = 0.2$. Note that $1.5 < K < 1.75$.

Since $M = 0.2$, $C_1 = C_2/M = 11$ nF. Next, using (4.28), we have $R_1 = 11.575$ kΩ, $R_2 = 2.371$ kΩ, and $R_3 = 33.243$ kΩ. The possible values for R_a and R_b are 12 and 20 kΩ, respectively. The active pole sensitivity of this design is

$$S_\tau^p = 1.953\left(-1 - \frac{j}{2Q_p}\right)\omega_p$$

If the time constant of the op amp is equal to $0.5/\pi \times 10^{-6}$, which is a typical value for a μA 741 op amp, then

$$\frac{\Delta p}{p} = 1.953\left(-1 - \frac{j}{2Q_p}\right)(0.004) = -7.811 \times 10^{-3} - j\frac{7.811 \times 10^{-3}}{2Q_p}$$

Then,

$$\frac{\Delta\omega_p}{\omega_p} = -7.811 \times 10^{-3}$$

and

$$\frac{\Delta Q_p}{Q_p} \approx 7.811 \times 10^{-3}$$ ■

As calculated in Example 4.10, the effect of a nonzero value of τ is to shift the pole positions. When $\tau_n = \omega_p\tau$ is small, the shifts in the dominant poles may be small. However, when τ_n is sufficiently large, it may cause drastic changes in the transfer function magnitude in the passband. In fact, the ripple factor R is 0.312 in Example 4.10. Such changes in the pole positions can be compensated for by using the predistortion technique discussed in Chap. 3. It is possible to design a filter for a given set of values of M, H_0, ω_p, Q_p, and τ_n and to predistort the design using G_1, G_2, and G_3, at the same time achieving minimum active pole sensitivity using K. This has been implemented in the program SKBPN in App. 4B, in which the initial design is obtained with an assumption of $\omega_p = 1$ r/s and with a given set of values for τ_n, H_0, Q_p, and M. However, one can also obtain denormalized element values by inputting the values of f_p and C_2. (See App. 4B for details on this program.) With the specifications $H_0 = 10$, $Q_p = 20$, $M = 0.2$, $\omega_p = 1$ r/s, and $\tau_n = \omega_p \tau = 0.01$, the program predicts the approximate value of K for the magnitude of the active pole sensitivity to be a minimum. The corresponding nominal values with the assumption of $\tau_n = 0$ and $\Omega_p = 1$ are found with an arbitrary magnitude scaling of 1000 for impedances and are as follows:

Possible value of $K = 1.53$
$R_1 = 0.6120$ kΩ
$R_2 = 0.1204$ kΩ
$R_3 = 1.9880$ kΩ
Possible value of $R_a = 0.534$ kΩ
Possible value of $R_b = 1.000$ kΩ
$C_1 = 0.005$ F
$C_2 = 0.001$ F

Then predistortion is carried out, and at the same time the magnitude of the active pole sensitivity is minimized. The corresponding element values are as follows:

$\Omega_p = 1$
$K = 1.534$
$R_1 = 0.59011$ kΩ
$R_2 = 0.11834$ kΩ
$R_3 = 1.95130$ kΩ
Possible value of $R_a = 0.534$ kΩ
Possible value of $R_b = 1.000$ kΩ
$C_1 = 0.005$ F
$C_2 = 0.001$ F
Minimum magnitude of active pole sensitivity = 1.914E − 2

The above element values can also be denormalized by inputting the values of f_p and C_2. For example, with $f_p = 10$ kHz and $C_2 = 1$ nF, the following denormalized element values are obtained:

Desired $f_p = 10$ kHz
$R_1 = 9.3919$ kΩ
$R_2 = 1.8834$ kΩ
$R_3 = 31.056$ kΩ
Possible value of $R_a = 10.680$ kΩ
Possible value of $R_b = 20.000$ kΩ
$C_1 = 5$ nF
$C_2 = 1$ nF

We conclude this section with a discussion of the passive ω_p and Q_p sensitivities, which can be obtained with the following formulas:

$$S_{R_1}^{\omega_p} = -\frac{G_1}{2(G_1 + G_2)} \qquad S_{R_2}^{\omega_p} = -\frac{G_2}{2(G_1 + G_2)} \qquad S_{R_3}^{\omega_p} = -\frac{1}{2}$$

$$S_{C_1}^{\omega_p} = S_{C_2}^{\omega_p} = -\frac{1}{2} \qquad S_{R_a}^{\omega_p} = S_{R_b}^{\omega_p} = 0$$

$$S_{R_1}^{Q_p} = \frac{H_0}{K} - \frac{G_1}{2(G_1 + G_2)} \qquad S_{R_2}^{Q_p} = \frac{G_2 M(1 - K)Q_p}{\omega_p C} - \frac{G_2}{2(G_1 + G_2)}$$

$$S_{R_3}^{Q_p} = \frac{G_3(1 + M)Q_p}{\omega_p C} - \frac{1}{2} \qquad S_{C_1}^{Q_p} = -S_{C_2}^{Q_p} = \frac{1}{2} - \frac{G_3 Q_p}{\omega_p C}$$

$$S_{R_a}^{Q_p} = \frac{(K - 1)G_2 M Q_p}{\omega_p C} = -S_{R_b}^{Q_p}$$

If we substitute for G_1, G_2, and G_3 from (4.28), we can obtain the formulas in terms of H_0, M, Q_p, and K explicitly.

Example 4.11: Find the worst-case percentage changes that can be expected in ω_p and Q_p in the designs in Examples 4.9 and 4.10. Assume that the tolerance in the resistances and the capacitances is 0.5%.

The passive ω_p and Q_p sensitivities of the designs in Examples 4.9 and 4.10 have been calculated using the formulas given earlier and are listed in Table 4.2. The worst-case changes in ω_p and Q_p can be determined using

$$\left.\frac{\Delta y}{y}\right|_{WC} = \sum_{i=1}^{n} |S_{x_i}^{y}| \left|\frac{\Delta x_i}{x_i}\right|$$

Thus we obtain the worst-case percentage changes in ω_p and Q_p in the two designs as follows.

For Example 4.9:

$$\frac{\Delta\omega_p}{\omega_p} = 1\% \qquad \text{and} \qquad \frac{\Delta Q_p}{Q_p} = 68.25\%$$

Table 4.2 Passive ω_p and Q_p sensitivities of the designs in Examples 4.9 and 4.10

Design	Circuit element	R_1	R_2	R_3	R_a	R_b	C_1	C_2
Example 4.9	ω_p sensitivity	−0.0754	−0.4246	−0.5	0	0	−0.5	−0.5
	Q_p sensitivity	4.92	−28.565	23.64	28.14	−28.14	−11.54	11.54
Example 4.10	ω_p sensitivity	−0.085	−0.415	−0.5	0	0	−0.5	−0.5
	Q_p sensitivity	6.165	−18.722	12.557	18.31	−18.31	−10.381	10.381

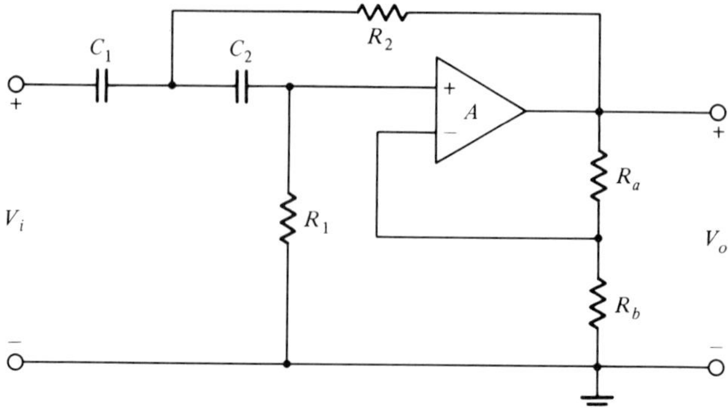

Figure 4.7 Sallen and Key highpass filter.

For Example 4.10:

$$\frac{\Delta\omega_p}{\omega_p} = 1\% \qquad \text{and} \qquad \frac{\Delta Q_p}{Q_p} = 47.41\%$$

The lower number for $\Delta Q_p/Q_p$ in the second case is mainly due to the capacitance spread we allowed in Example 4.10. It can thus be concluded that a capacitance spread of $M < 1$ not only helps reduce the active pole sensitivity but also reduces the passive Q_p sensitivities. ■

4.4 *Sallen and Key Highpass Filter*

A Sallen and Key highpass filter can be obtained by using *RC-CR* transformation on a lowpass filter. However, changing R_a and R_b to C_a and C_b has no advantage. Thus, retaining R_a and R_b as they are, the Sallen and Key highpass filter will be as shown in Fig. 4.7. The voltage transfer ratio of this circuit can be derived to be

$$\frac{V_0}{V_i} = \frac{Ks^2}{s^2 + s\left[G_1(S_1 + S_2) + G_2S_1(1 - K)\right] + G_1G_2S_1S_2} \tag{4.34}$$

where the op amp is assumed to be ideal; and $G_i = 1/R_i$, $S_i = 1/C_i$, and $K = 1 + R_a/R_b$.

Note the similarity between the transfer function of the lowpass filter and that of the highpass filter in terms of the denominator polynomials. Thus the design procedures for this network are similar to those for the lowpass network. The basic design equations are

$$\omega_p^2 = G_1G_2S_1S_2 \tag{4.35a}$$

and

$$G_1(S_1 + S_2) + S_1G_2(1 - K) = \frac{\omega_p}{Q_p} \tag{4.35b}$$

To develop a design procedure, we let

$$R_2 = \frac{R_1}{m^2} \quad \text{and} \quad C_2 = \frac{C_1}{n^2} \tag{4.36}$$

Then, for convenient choices of C_1 and n (hence C_2), R_1 and m are

$$R_1 = \frac{mn}{\omega_p C_1} \tag{4.37a}$$

and

$$m = \frac{-n}{2Q_p(K-1)} + \sqrt{\left[\frac{n}{2Q_p(K-1)}\right]^2 + \frac{n^2+1}{K-1}} \tag{4.37b}$$

Example 4.12: Design a Sallen and Key highpass filter to meet the specifications $f_p = 1$ kHz and $Q_p = 5$. Choose $K = 2$ and $n = 1$.

Choose a convenient value for C_1 such as 10 nF. Then, since $n = 1$, $C_2 = 10$ nF. Using (4.37*b*), we find that $m = 1.318$. Therefore $R_1 =$ 20.973 kΩ and $R_2 = 12.078$ kΩ. The possible values for R_a and R_b are any two equivalued resistors; for example, 20 kΩ could be used for each. ■

A highpass realization like the one shown in Fig. 4.7 is not really a highpass filter over the entire frequency range 0 to ∞. At the high-frequency level, the frequency-dependent characteristic of the op amp influences the characteristic of the filter and therefore, as $\omega \to \infty$, the gain drops to zero. It is only in the finite frequency range that the circuit behaves as a highpass filter.

Finding the design procedure for a highpass filter that minimizes the influence of the amplifier near ω_p, and calculation of the passive ω_p and Q_p sensitivities, are left as exercises for the reader.

4.5 *Filter Network for Realizing jω-Axis Zeros*

Elliptic filter functions possess $j\omega$-axis zeros and are in the form of lowpass and highpass notch (LPN and HPN, respectively) filter functions, as given in Table 4.1. Another useful filter function is the notch function itself, in

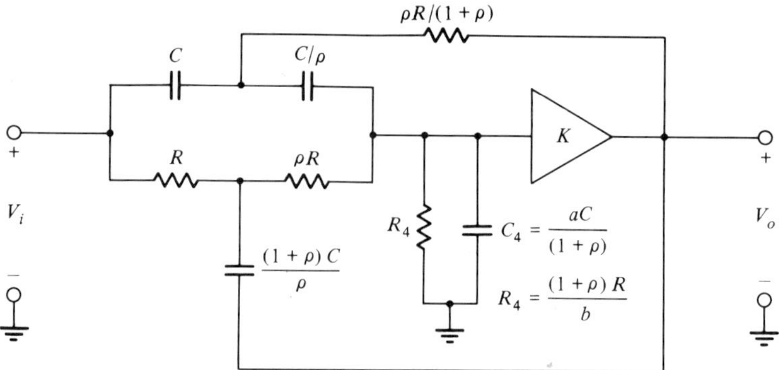

Figure 4.8 Sallen and Key filter for realizing lowpass and highpass notch filters.

which $\omega_p = \omega_z$. A Sallen and Key type network for realizing HPN and LPN functions is shown in Fig. 4.8. The VCVS of gain K can be realized using the circuit in Fig. 4.2a. The passive part of the network is a twin-T network combined with a loading network formed by R_4 and C_4. A more general form of the network in Fig. 4.8 is shown in Fig. 4.9. When the amplifier of gain K in the network in Fig. 4.8 is assumed to be ideal, it is only a particular case of the network shown in Fig. 4.9 in which $\beta_1 = \beta_2 = K$. The transfer function of the circuit, shown in Fig. 4.9, can be derived as

$$\frac{V_0}{V_i} = H_0 \frac{s^2 + \omega_z^2}{s^2 + (\omega_p/Q_p)s + \omega_p^2} \tag{4.38}$$

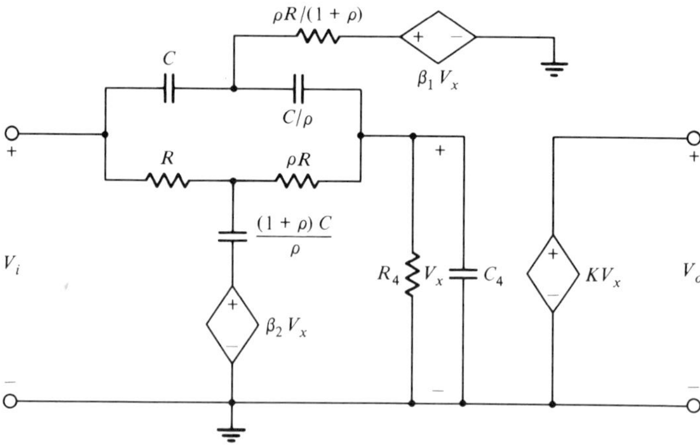

Figure 4.9 More general form of the network shown in Fig. 4.8.

where

$$H_0 = \frac{K}{1 + a} \tag{4.39a}$$

$$\omega_z = \frac{1}{RC} \tag{4.39b}$$

$$\omega_p = \omega_z \sqrt{\frac{1 + b}{1 + a}} \tag{4.39c}$$

$$Q_p = \frac{\alpha\sqrt{(1 + a)(1 + b)}}{2 - \beta_1 - \beta_2 + \alpha(a + b)} \tag{4.39d}$$

and

$$\alpha = \frac{\rho}{1 + \rho} \tag{4.39e}$$

There are several possible ways in which both LPN and HPN networks can be realized. We shall consider one possible design procedure for each case. In each design procedure, we set $\beta_1 = \beta_2 = K$, which means that we use the network in Fig. 4.8 for the design of both a LPN and a HPN.

Lowpass Notch Network

In a LPN filter function, $\omega_p < \omega_z$. Then we can set $b = 0$ in addition to $\beta_1 = \beta_2 = K$. This implies that the loading resistor R_4 can be removed from the network in Fig. 4.8 (as well as that in Fig. 4.9). All other elements will remain the same as those given in Fig. 4.8. Then, for convenient choices for values of C and ρ (hence α), the other element values can be obtained using the equations

$$R = \frac{1}{\omega_z C} \tag{4.40a}$$

$$a = \frac{\omega_z^2}{\omega_p^2} - 1 \tag{4.40b}$$

and

$$K = 1 + \frac{\alpha}{2}\left(a - \frac{\omega_z}{\omega_p Q_p}\right) \tag{4.40c}$$

A convenient choice is $\rho = 1$, since such a value will lead to equivalued capacitors and resistors.

Highpass Notch Network

The design of an HPN network can be achieved by setting $a = 0$ in (4.39) in addition to $\beta_1 = \beta_2 = K$. This is equivalent to removing the loading capacitor C_4 from the network in Fig. 4.8. Then, for convenient choices of C and ρ, the following equations give the element values:

$$R = \frac{1}{\omega_z C} \tag{4.41a}$$

$$b = \frac{\omega_p^2}{\omega_z^2} - 1 \tag{4.41b}$$

and

$$K = 1 + \frac{\alpha}{2}\left(b - \frac{\omega_p}{Q_p \omega_z}\right) \tag{4.41c}$$

Again $\rho = 1$ (hence $\alpha = 0.5$) is a convenient choice.

Example 4.13: Design an LPN network of the form shown in Fig. 4.8 that satisfies the specifications $f_p = 2$ kHz, $f_z = 2.5$ kHz, and $Q_p = 20$.

Choosing 10 nF and 1 as convenient values for C and ρ, respectively, we obtain

$$R = 6.3662 \text{ k}\Omega \qquad a = 0.5625 \qquad K = 1.125$$

Then, the network realization will be as shown in Fig. 4.10. The VCVS in the circuit can be replaced with the op amp circuit in Fig. 4.2*a*, and the possible choices for R_a and R_b are 2.5 and 20 kΩ, respectively. ■

A notch function is obtained if we choose $\omega_p = \omega_z$ in (4.38). One possible design procedure for this network function is to choose $b = a = 0$

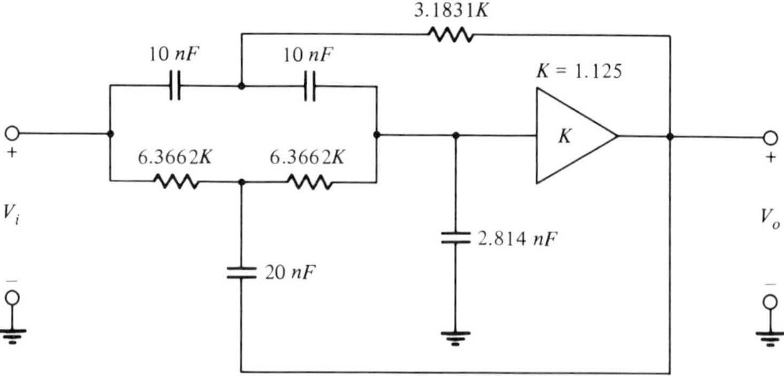

Figure 4.10 The LPN network designed to meet the specifications in Example 4.13.

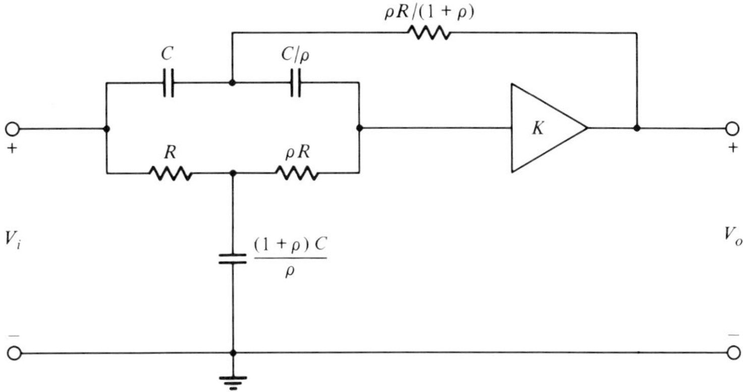

Figure 4.11 Sallen and Key type filter for realizing a notch filter function.

for the network in Fig. 4.9. In addition, either β_1 or β_2 can also be set to zero. Then, one possible choice is $\beta_1 = K$ and $\beta_2 = 0$, and another possible choice is $\beta_1 = 0$ and $\beta_2 = K$. With $\beta_1 = K$, $\beta_2 = 0$, and $a = b = 0$, the network in Fig. 4.9 is reduced to the one shown in Fig. 4.11. The required design equations for the network in Fig. 4.11 are

$$R = \frac{1}{\omega_z C} \tag{4.42a}$$

and

$$K = 2 - \frac{\alpha}{Q_p} \tag{4.42b}$$

Of course $\rho = 1$, hence $\alpha = 0.5$ is a convenient choice.

The design of the networks discussed in this section that realize complex conjugate zeros and poles is too complex. The reason for this is that the elements used in these networks require exactness and tighter tolerance. Actually this is a third-order filter circuit. In matching the passive elements there is a pole-zero cancellation so that the third-order network function is reduced to a second-order function. The exact cancellation of a pole and a zero requires exactness in element values, and this is one reason why unity is a good choice for ρ. For these reasons, when designing elliptic filter functions, one uses mostly networks with two or more amplifiers that avoid pole-zero cancellation. We consider such networks in later chapters.

4.6 *Higher-Order Filters*

Most of the techniques presented so far are for second-order realizations. As mentioned at the beginning of this chapter, higher-order filters are

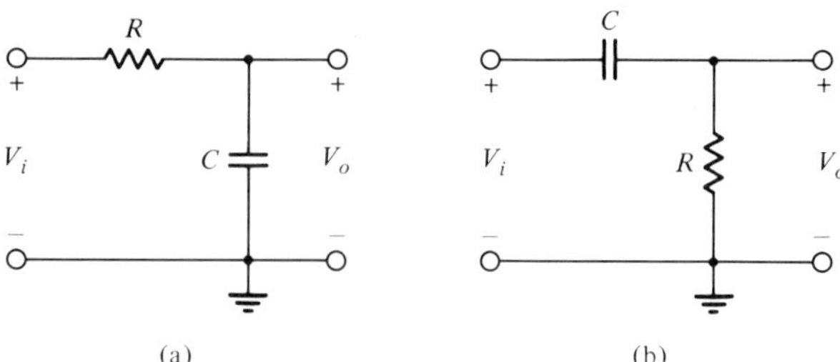

Figure 4.12 First-order filter sections. (*a*) Lowpass. (*b*) Highpass.

realized using second-order building blocks. One possible design procedure is to connect second- and first-order sections in cascade to realize the higher-order filter functions. In this case, if the given higher-order $H(s)$ is of even order (n is even), then exactly $n/2$ second-order networks will be required. However, if the order of $H(s)$ is odd (n is odd), then $H(s)$ will have a negative real pole in addition to the $(n - 1)/2$ complex conjugate pairs of poles. This means that $(n - 1)/2$ complex conjugate pairs of poles can be realized using $(n - 1)/2$ second-order networks. To realize a real pole, circuits of the type shown in Fig. 4.12 can be used. The network shown in Fig. 4.12*a* is a first-order lowpass section, and its voltage transfer function is given by

$$\frac{V_o}{V_i} = \frac{1}{1 + sRC} \tag{4.43}$$

The second network shown in Fig. 4.12*b* is a first-order highpass section, and its voltage transfer function is

$$\frac{V_o}{V_i} = \frac{sRC}{1 + sRC} \tag{4.44}$$

These sections can be added as a last section in the chain of networks, and the amplifier output of the last section will provide the necessary isolation. However, there is a disadvantage in this type of arrangement: The output of the filter will not be the amplifier output and thus the overall circuit will not have the necessary isolation. One way to avoid this problem is to combine this real pole with a second-order factor. This will result in a third-order denominator, and such a third-order transfer function can be realized as one whole unit with an amplifier output.

We shall consider the realization of a third-order lowpass filter function in this section. The third-order highpass realization is left as an exercise at the end of this chapter. We consider a third-order lowpass network with equivalued capacitors, as shown in Fig. 4.13.

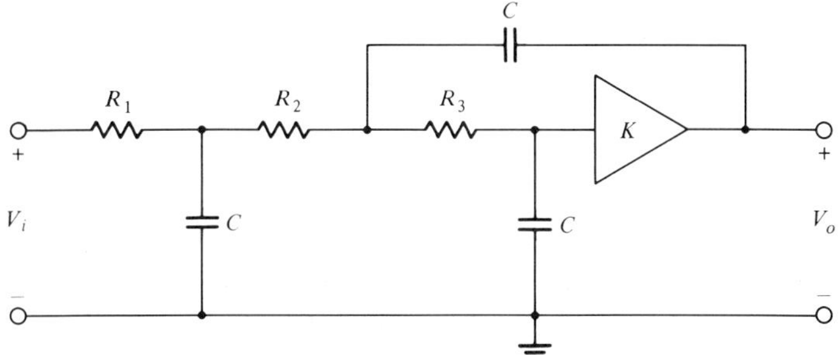

Figure 4.13 Third-order lowpass filter network.

The transfer function of this circuit can be obtained as

$$\frac{V_o}{V_i} = \frac{K}{\tau_1\tau_2\tau_3 s^3 + \left[2\tau_1\tau_3 + \tau_2\tau_3 + \tau_1\tau_2(2-K)\right]s^2 + \left[\tau_1 + \tau_3 + (\tau_1 + \tau_2)(2-K)\right]s + 1} \tag{4.45}$$

where $\tau_i = R_i C$, $i = 1, 2, 3$. Any third-order all-pole lowpass transfer function can always be written as

$$H(s) = \frac{H_0}{a_3 s^3 + a_2 s^2 + a_1 s + 1} \tag{4.46}$$

By comparing the numerators of (4.45) and (4.46), we find that $H_0 = K$. Usually the gain constant is arbitrary, hence K is available as a free parameter. For two reasons, $K = 2$ is a convenient choice: (1) the op amp circuit in Fig. 4.2*a* can be realized with equivalued resistors for R_a and R_b; and (2) such a choice will lead to simplified design equations. With $K = 2$, the design equations can be obtained by comparing the denominators of (4.45) and (4.46):

$$a_3 = \tau_1\tau_2\tau_3 \tag{4.4a}$$

$$a_2 = 2\tau_1\tau_3 + \tau_2\tau_3 \tag{4.47b}$$

and

$$a_1 = \tau_1 + \tau_3 \tag{4.47c}$$

We have three equations to be solved for the three variables τ_1, τ_2, and τ_3. We start with (4.47*c*) and see that

$$\tau_1 = a_1 - \tau_3 \tag{4.48}$$

Substituting (4.48) into (4.47*a*), we have

$$\tau_2 = \frac{a_3}{\tau_3(a_1 - \tau_3)} \tag{4.49}$$

Next, substituting τ_1 and τ_2 from (4.48) and (4.49) into (4.47*b*), we obtain an equation in terms of τ_3:

$$\tau_3^3 - 2a_1\tau_3^2 + \left(a_1^2 + 0.5a_2\right)\tau_3 - 0.5\left(a_1a_2 - a_3\right) = 0 \qquad (4.50)$$

It can be shown that there is a positive real solution for τ_3 such that $0 < \tau_3 < a_1$. Thus we solve (4.50) to find this real positive root of the third-order equation. Then, by back-substituting the value of τ_3 into (4.48) and (4.49), we can find the required values of τ_1 and τ_2. The next step is to choose a convenient value for C and then find the value of $R_i = \tau_i/C$, $i = 1, 2, 3$.

Example 4.14: The specifications for a lowpass filter are given as

$$A_p \le 0.5 \text{ db} \quad \text{and} \quad f_p = 1 \text{ kHz}$$
$$A_a \ge 35 \text{ dB} \quad \text{and} \quad f_a = 2 \text{ kHz}$$

These specifications can be met by the following transfer function obtained using Chebyshev approximation:

$$H(s) = \frac{H_0}{(s + \sigma_o)\prod_{i=1}^{2}\left(s^2 + A_is + B_i\right)}$$

where the coefficients are given in Table 4.3. It was decided to use a cascade combination of second- and third-order filter networks of the Sallen and Key type. Find a realization that meets the above specifications.

The given network function $H(s)$ can be split:

$$H(s) = H_1(s)H_2(s)$$

where

$$H_1(s) = \frac{H_{01}}{(s + \sigma_o)\left(s^2 + A_1s + B_1\right)} \qquad (4.51)$$

Table 4.3. Transfer function coefficients of a Chebyshev approximation that satisfies the specifications in Example 4.14[a]

i	A_i	B_i
1	1.40697E + 3	4.08911E + 7
2	3.68349E + 3	1.88220E + 7

[a] $H_0 = 1.75213\text{E} + 18$ is required to give a maximum passband gain of 0 dB; ripple factor $\varepsilon = 0.349311$; order of filter function, $N = 5$; $\sigma_o = 2.27652\text{E} + 3$.

and

$$H_2(s) = \frac{H_{02}}{s^2 + A_2 s + B_2} \tag{4.52}$$

There are two pairs of complex poles, and one can select either one to combine with the real pole and obtain the third-order function. In this example, it was decided to split the second-order factors such that, in (4.51),

$$\sigma_o = 2.27652 \times 10^3 \qquad A_1 = 3.68349 \times 10^3 \qquad B_1 = 18.8220 \times 10^6$$

Therefore, in (4.52), the values of A_2 and B_2 are

$$A_2 = 1.40697 \times 10^3 \qquad \text{and} \qquad B_2 = 40.8911 \times 10^6$$

When the approximation procedure is used, we usually obtain the third-order transfer function in the form of (4.51). Note that the design procedure for the network shown in Fig. 4.13 starts with a network function of the type given in (4.46). The first step is to convert an equation in the form of (4.51) to the form of (4.46) by dividing the numerator and denominator by $B_1\sigma_o$ and then to multiply out the factors to find a_1, a_2, and a_3. Then, we can follow the procedure described in this section. Finally, after finding the values of τ_1, τ_2, and τ_3, we can choose a convenient value for C and determine the resistance values. This entire procedure has been implemented in the program THIRD, listed in App. 4C. The inputs for the program are the values of σ_o, A_1, and B_1, and the outputs are the values of τ_1, τ_2, and τ_3. Then, by inputting the value of C, we can also obtain the values of R_1, R_2, and R_3. Using this program for our example, we find the following:

Inputs to the Program THIRD:
$\sigma_o = 2.27652\text{E} + 03$
$A_1 = 3.68349\text{E} + 03$
$B_1 = 1.88220\text{E} + 07$
Results:
$\tau_1 = 5.468074\text{E} - 04$
$\tau_2 = 4.841194\text{E} - 04$
$\tau_3 = 8.816088\text{E} - 05$
$C = 10$ nF
$R_1 = 5.468074\text{E} + 01$ kΩ
$R_2 = 4.841194\text{E} + 01$ kΩ
$R_3 = 8.816088\text{E} + 00$ kΩ

The design of the second-order section can be carried out using the lowpass filter circuit in Fig. 4.3. To conform with the previous design, if we want $C_1 = C_2$ and $K = 2$, the resistive element spread will be

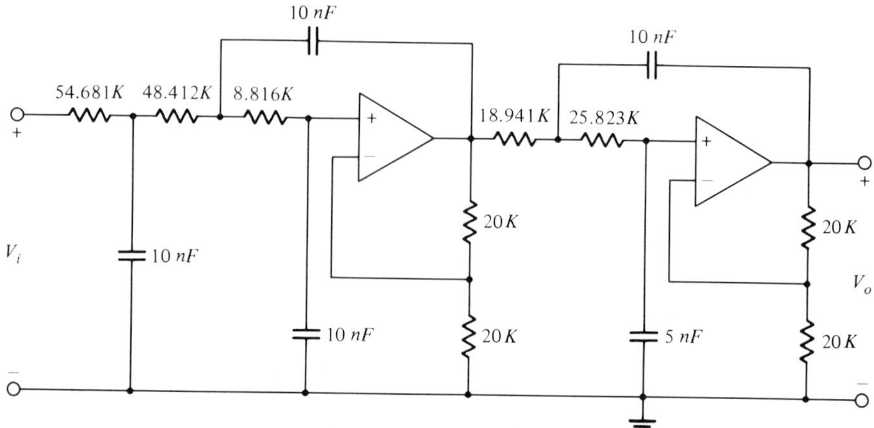

Figure 4.14 Network designed in the Example 4.14.

high (see Example 4.3a). Therefore we should allow some capacitor spread so that the resistor spread is not too large. Thus we choose $n^2 = C_2/C_1 = 2$ and $K = 2$. Using (4.8), we find that $m = 0.85645$. Then, using (4.7a) and choosing $C_1 = 5$ nF, we have $R_2 = 18.941$ kΩ and $R_1 = 25.823$ kΩ. Of course, $C_2 = 10$ nF. Thus the entire realization of the network to meet the specifications of this example will be as shown in Fig. 4.14. ■

4.7 *Filters with Zero Active Pole Sensitivity*

All types of filters using a single noninverting VCVS can be represented in the form of the general network shown in Fig. 4.15. Based on the assumption that the amplifier is ideal, the voltage transfer function of this filter can be derived as

$$\frac{V_o}{V_i} = \frac{KT_{21}}{1 - KT_{23}} \tag{4.53}$$

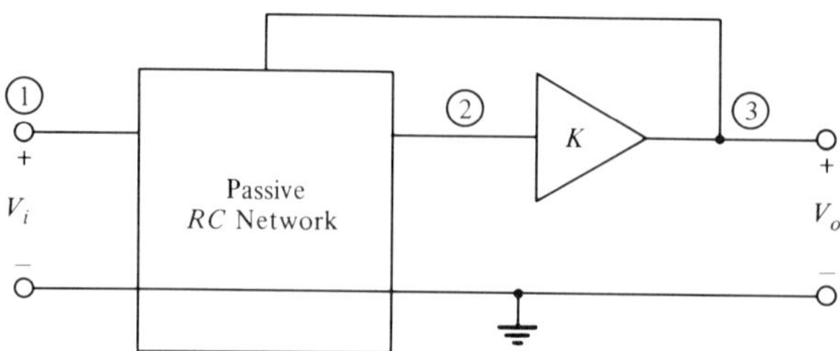

Figure 4.15 Most general form of an active filter using a single noninverting VCVS.

where

$$T_{21} = \left.\frac{V_2}{V_i}\right|_{V_3=0}$$

and

$$T_{23} = \left.\frac{V_2}{V_3}\right|_{V_i=0}$$

When the VCVS is realized using the op amp circuits in 4.2, and if we assume that the op amp is ideal except for frequency dependence, we can find the active pole sensitivity by replacing K with $K/(1 + Ks\tau)$. Assume that $T_{32} = n_{32}/d_{32}$ and $T_{12} = n_{12}/d_{12}$, where n_{ij} and d_{ji} are the numerator and denominator polynomials. Except in some special cases, $d_{32} = d_{12}$. Using these facts, we can write (4.53) is

$$\frac{V_0}{V_i} = \frac{Kn_{12}}{(d_{32} - Kn_{32}) + Ks\tau d_{32}} \tag{4.54}$$

Since we design any network with the assumption that $\tau = 0$, the denominator polynomial of the transfer function of the overall filter network as given by (4.54) with $\tau = 0$ must satisfy

$$D_0(s) = d_{32}(s) - Kn_{32}(s)$$
$$- K_1\left(s^2 + \frac{\omega_p}{Q_p}s + \omega_p^2\right) \tag{4.55}$$

where K_1 is some constant. Applying (3.63) to the denominator polynomial of (4.54), we have

$$S_\tau^p = \frac{-Kd_{32}(p)}{j2K_1\omega_p} \tag{4.56}$$

With $s = p$, $D_0(p)$ must be zero, and therefore

$$d_{32}(p) - Kn_{32}(p) = 0 \tag{4.57}$$

That is,

$$d_{32}(p) = Kn_{32}(p)$$

Thus

$$S_\tau^p = \frac{-K^2n_{32}(p)}{j2K_1\omega_p} \tag{4.58}$$

Looking at the configuration of every filter we have considered so far, we can easily see that $n_{32}(p)$ can never be zero because $n_{32}(s)$ will have all its

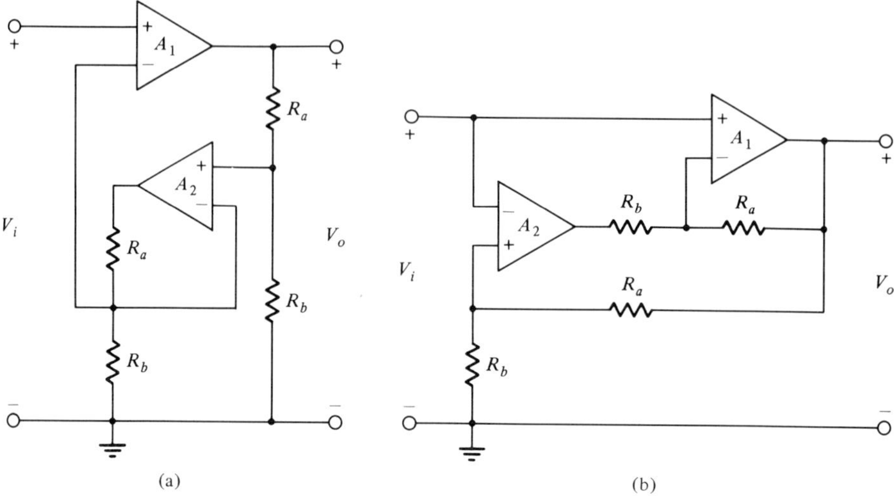

Figure 4.16 Actively compensated noninverting finite-gain amplifiers.

zeros on the negative real axis and p is complex conjugate. Thus the active pole sensitivity of all of these filters cannot be zero.

Next assume that we have an op amp circuit for realizing a VCVS with a gain in the form

$$K\frac{1 + as\tau}{1 + as\tau + bs^2\tau^2}$$

when the op amp time constants are included. Two such circuits are shown in Fig. 4.16, and for both, when $A_1, A_2 \to \infty$, the nominal gain K is given by

$$K = 1 + \frac{R_a}{R_b} \tag{4.59}$$

similar to the case of a single-amplifier circuit. However, in both circuits, when $A_1^{-1} = A_2^{-1} = s\tau$, the voltage gains of these circuits are in the form of

$$K\frac{1 + as\tau}{1 + as\tau + bs^2\tau^2}$$

The gain of the circuit in Fig. 4.16*b* reduces to $1/(1 + s\tau)$ when $K = 1$. This is the same gain as that of the voltage follower circuit in Fig. 4.2*b*. In fact, the circuit shown in Fig. 4.16*b* works better than the circuit in Fig. 4.16*a* only when $K > 2$, along with some associated stability problems caused by the higher-order poles of the op amps. Returning to the problem of active pole sensitivity of the filters, when we use the circuits in Fig. 4.16 in place of the VCVS in Fig. 4.15, the overall transfer function of the filter circuit in Fig. 4.15 can be found to be

$$\frac{V_0}{V_i} = \frac{K(1 + as\tau)n_{12}}{(d_{32} - Kn_{32})(1 + as\tau) + bs^2\tau^2 d_{32}} \tag{4.60}$$

Now, applying (3.63) to the denominator of (4.60) and using (4.57), we can show that

$$S_\tau^p = 0 \tag{4.61}$$

This means that, by replacing the conventional op amp circuits in Fig. 4.2*a* and *b* with the new actively compensated amplifier circuits in Fig. 4.16, the active pole sensitivity can be made zero. Remember that we have included only the first-order effects of τ in defining the active pole sensitivity. Furthermore, the actual value of τ is nonzero. The second- and higher-order effects will affect the performance of the circuits. The effectiveness of the use of actively compensated amplifiers in Sallen and Key filters is shown in the following example.

Example 4.15: Consider the design of a Sallen and Key bandpass filter using the conventional amplifier circuit in Fig. 4.2*a*. In Example 4.10, to meet the specifications $H_0 = 10$, $Q_p = 20$, and $f_p = 4$ kHz, we obtained a design in which the active pole sensitivity was a minimum suboptimally. We calculated the values of $\Delta\omega_p/\omega_p$ and $\Delta Q_p/Q_p$ assuming a nonzero value of $\tau = 0.5/\pi$ μs. Now assume that we are using the actively compensated amplifier circuit in Fig. 4.16*a* in place of the conventional amplifier circuit in Fig. 4.2*a*. Find the per-unit changes in ω_p and Q_p in the same network when the op amps have equal time constants of $\tau = 0.5/\pi$ μs. Also, find the ripple factor.

In order to calculate $\Delta p/p$, we must find the denominator polynomial of the transfer function of the filter circuit in Fig. 4.5 when the amplifier circuit in Fig. 4.16*a* is used in the circuit as a function of τ. Once we find the gain of the noninverting amplifier circuit in Fig. 4.16*a*, then K in (4.25) can be replaced with this gain. When $A_1^{-1} = A_2^{-1} = s\tau$, the gain of the noninverting amplifier circuit in Fig. 4.16*a* can be derived as

$$\frac{V_o}{V_i} = K\frac{1 + Ks\tau}{1 + Ks\tau + K^2s^2\tau^2}$$

where $K = 1 + R_a/R_b$.

Replacing K in (4.25) with the above gain, we can obtain the denominator polynomial of the filter circuit as

$$D(s, \tau) = \left(s^2 + \frac{\omega_p}{Q_p}s + \omega_p^2\right)(1 + Ks\tau) + K^2s^2\tau^2(s^2 + as + \omega_p^2)$$

where

$$a = S_1(G_1 + G_2) + G_3(S_1 + S_2)$$

Since $D_1(s) = K[s^2 + (\omega_p/Q_p)s + \omega_p^2]$, $D_1(p) = 0$. Therefore we cannot use (3.62) to find $\Delta p/p$. We have to use the more accurate formula (3.67) in this case and to use this formula we need

$$D_1'(p) = K\left(2p + \frac{\omega_p}{Q_p}\right)$$

and

$$D_2(p) = K^2\left(a - \frac{\omega_p}{Q_p}\right)p$$

where $p = -\omega_p/2Q_p + j\omega_p$. Substituting the above, we have the following per-unit change in p of this pole:

$$\frac{\Delta p}{p} = \frac{-K^2p^2\tau^2(a - \omega_p/Q_p)}{(2p + \omega_p/Q_p)(1 + Kp\tau)}$$

Substituting for a and p and using (4.26c), we obtain

$$\frac{\Delta p}{p} = \frac{-K^2\tau_n^2\left[(-1/2Q_p) + j\right]^2 KG_2S_1}{(j2\omega_p)(1 + jK\tau_n)} \tag{4.62}$$

where $\tau_n = \omega_p\tau$.

Now, substituting the different element values from Example 4.10 into (4.62), we have

$$\frac{\Delta p}{p} = (1.8924 - j43.386) \times 10^{-6}$$

Therefore, using (3.44) and (3.45), we obtain

$$\frac{\Delta\omega_p}{\omega_p} = 1.8924 \times 10^{-6} \qquad \text{and} \qquad \frac{\Delta Q_p}{Q_p} = 1.7354 \times 10^{-3}$$

The ripple factor is

$$R \approx 2Q_p\left|\frac{\Delta p}{p}\right| = 1.7371 \times 10^{-3}$$ ■

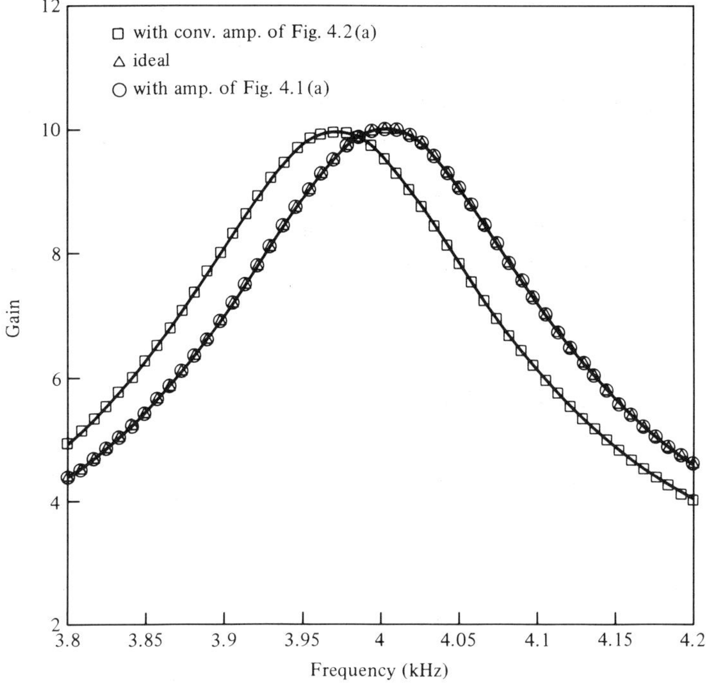

Figure 4.17 Magnitude responses of the circuit designed in Example 4.10.

We calculated the ripple factor in Example 4.10 and found it to be 0.312. Remember that this factor brings out the maximum change in the magnitude characteristic. Let us compare these numbers in their proper perspective. When we use a conventional amplifier in the filter circuit designed in Example 4.10, we expect a maximum change of 31.2% in the magnitude, whereas such a change is only 0.174% when we use an actively compensated amplifier in the same circuit. The active pole sensitivity is reduced by about 180 times, resulting in a great improvement in circuit performance. Similar improvements can be shown to be possible for other filters as well when such actively compensated amplifiers are used. To illustrate the full impact of these results, the magnitude responses of the filter designed in Example 4.10, when conventional and actively compensated amplifier circuits are used, are shown in Fig. 4.17. These characteristics were obtained using the program SPICE and the macromodel of the op amps with nonzero values of τ. The ideally expected characteristic is also given in the same figure for purposes of comparison. Note that, when the actively compensated amplifier is used, the filter characteristic is completely coincident with the ideal characteristic.

Before concluding this section, we consider the design of the Sallen and Key bandpass filter in Fig. 4.5 when an actively compensated amplifier is used. The advantage that we saw in Example 4.15 may not hold if ω_p is large. As the value of ω_p increases, τ_n also increases. Before we examine the impact of increasing the operating frequency of the actively compensated filter network, let us develop a design procedure for such a network in which the influence of the op amps can be minimized; that is, $|\Delta p/p|$ as given by (4.62), will be a minimum. The method of analysis is similar to that used for this filter in Sec. 4.3. Substituting G_2 from (4.28) in terms of $\omega_p C$ and using $C_1 = C_2/M = C/M$, we have

$$\frac{\Delta p}{p} = \frac{jK^3\tau_n^2(j - 1/2Q_p)^2}{4Q_p(1 + jK\tau_n)}$$

$$\times\left[\frac{H_0 - 1}{K - 1} - \frac{2H_0}{K} + \sqrt{\left(\frac{H_0 - 1}{K - 1}\right)^2 + \frac{4Q_p^2(M + 1)}{K - 1}}\,\right] \tag{4.63}$$

For a given set of specifications in terms of H_0, ω_p, and Q_p and for a given set of op amps, in order to minimize $|\Delta p/p|$, we have to minimize

$$F_2 = \frac{K^3}{4Q_p}\frac{1}{\left(1 + K^2\tau_n^2\right)^{1/2}}$$

$$\times\left[\frac{H_0 - 1}{K - 1} - \frac{2H_0}{K} + \sqrt{\left(\frac{H_0 - 1}{K - 1}\right)^2 + \frac{4Q_p^2(M + 1)}{K - 1}}\,\right] \tag{4.64}$$

This function F_2 is similar to the function F we gave earlier for this filter using a single noninverting op amp. The value of F_2 can be minimized with respect to the two parameters K and M. For any value of K, the minimum value of F_2 occurs with $M = 0$, which is not a practical solution. Therefore we can minimize F_2 only with respect to the single parameter K after choosing a convenient (but small) value for M. The plot of this function is shown in Fig. 4.18 for various capacitance ratios with $Q_p = 20$, $H_0 = 10$, and $\tau_n = 0.01$ to find a general pattern for these curves and to make a practical choice for K.

From the curves in Fig. 4.18, one can observe that it is better to choose a capacitor ratio with $M < 1$, and a reasonable choice may be 0.5 or 0.2. Choosing $M < 0.2$ does not really decrease the active sensitivity much. For any capacitor ratio, the choice of K should be close to 1.3 because near $K = 1.3$ the curves are flat. This gain is also convenient in terms of realization (values of 3 kΩ and 10 kΩ can be used for R_a and R_b, respectively).

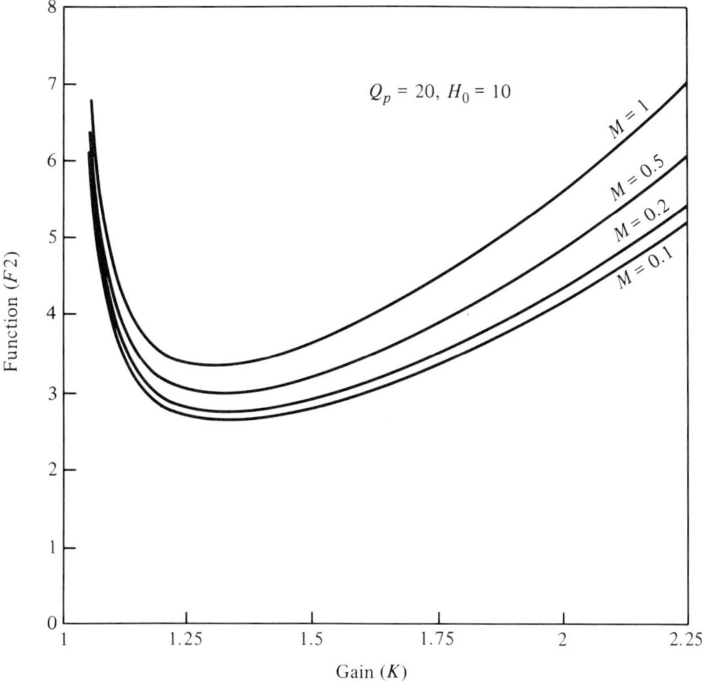

Figure 4.18 Plots of (4.64) as a function of the gain K.

Example 4.16: Design a Sallen and Key bandpass filter using the actively compensated noninverting amplifier in Fig. 4.16*a* to satisfy the specifications $f_p = 10$ kHz, $Q_p = 20$, and $H_0 = 5$. Use the design method for minimizing the influence of the op amp time constants. Find the per-unit changes in ω_p and Q_p and the ripple factor R. The op amp time constants are both equal to $0.5/\pi$ μs.

Choosing $M = 0.5$ and $K = 1.3$, and a convenient C_2 value of 1 nF, we find $C_1 = 2$ nF. The element values of R_1, R_2, and R_3 can be found using (4.28):

$$R_1 = 41.38\ \text{k}\Omega \qquad R_2 = 3.313\ \text{k}\Omega \qquad R_3 = 41.287\ \text{k}\Omega$$

Since we know the values of G_2 and S_1, it is easy to use (4.62) to find $\Delta p/p$, and one can verify the answer by using (4.63). Now, from (4.62) and with $\tau_n = 0.01$, we find that

$$\frac{\Delta p}{p} = (9.757 - j263.7) \times 10^{-6}$$

Therefore

$$\frac{\Delta\omega_p}{\omega_p} = 9.757 \times 10^{-6} \qquad \text{and} \qquad \frac{\Delta Q_p}{Q_p} = 10.55 \times 10^{-3}$$

Thus

$$R = 10.55 \times 10^{-3}$$

■

For a given H_0 and Q_p, it is possible to increase the value of ω_p simply by scaling the capacitance values while maintaining the same value of M. Therefore, to find the effect of increasing ω_p in this network using a particular design procedure, we write (4.63) as

$$\frac{\Delta p}{p} = \frac{j(j - 1/2Q_p)^2(1 + jK\tau_n)}{\left(1 + K^2\tau_n^2\right)^{1/2}} F_2\tau_n^2$$

and therefore

$$\Delta p \approx \left[1 + j\left(\frac{3}{2Q_p} - K\tau_n\right)\right]\omega_p F_2\tau_n^2 \tag{4.65}$$

when we assume that $K\tau_n$ and $3/(2Q_p)$ are small compared to unity.

The value of F_2 depends on τ_n (that is, ω_p) to a small extent. However, when we assume that $K\tau_n \ll 1$, then $K^2\tau_n^2 \ll 1$ and therefore F_2 is relatively independent of τ_n. Over the range of frequencies in which we are interested, this assumption is perfectly valid, and therefore we can ignore $K^2\tau_n^2$ in the denominator of (4.64) so that it will lead to a simplified analysis. Proceeding further, we can find the actual pole position p' as

$$p' = p + \Delta p = -\frac{\omega_p}{2Q_p}\left(1 - 2Q_pF_2\tau_n^2\right) + j\omega_p\left[1 + \left(\frac{3}{2Q_p} - K\tau_n\right)F_2\tau_n^2\right] \tag{4.66}$$

Examine the real part of p'. As the value of $\omega_p\tau = \tau_n$ increases, the real part becomes zero, and beyond a certain value of ω_p, the real part may become positive. If F_2 is assumed to be independent of ω_p, then the circuit will break into oscillation for a value of ω_p given by

$$\omega_p = \frac{1}{\tau}\sqrt{\frac{1}{2Q_pF_2}} \tag{4.67}$$

Consider Example 4.16, where we used a design in which F_2 is almost a minimum for any ω_p. The value of $F_2 \approx 2.638$. Then, using (4.67), we can show that the circuit will break into oscillation as we reach $f_p = 97.34$ kHz. Note that this frequency is less than $\frac{1}{10}$ of the gain bandwidth products of the op amps used in the circuit. One should note that, even before we reach this frequency, there will be a rapid increase in Q_p because of the decrease in the real part of p'. An increase in Q_p is called *Q enhancement* and this condition can be observed by calculating the actual values of ω_p and Q_p

denoted by ω_p' and Q_p':

$$\omega_p' \approx \omega_p\left[1 + \left(\frac{3}{2Q_p} - K\tau_n\right)F_2\tau_n^2\right] \tag{4.68a}$$

and

$$Q_p' \approx \frac{Q_p}{1 - 2Q_pF_2\tau_n^2} \tag{4.68b}$$

One can observe from (4.68b) that, as ω_p increases, Q_p' increases. Finally, as ω_p increases to the value given by (4.67), the circuit breaks into oscillation. In fact, the circuit may break into oscillation even before we reach this frequency because of slew rate limitations. Now, to determine the actual limitation, assume that the circuit is useful only up to a frequency where the ripple factor R is limited to X. Recall that this ripple factor represents the maximum change in the magnitude function. To find R, we use (4.63) in terms of F_2 and (3.42) and thus, for large Q_p (> 5),

$$R \approx 2Q_pF_2\tau_n^2$$

Then, the maximum value of ω_p for which $R = X$ can be found as

$$\omega_p = \frac{1}{\tau}\sqrt{\frac{X}{2Q_pF_2}} \tag{4.69}$$

For example, if $X - 0.1$ (representing a maximum change of 10% in the magnitude), for the design in Example 4.16, the value of f_p should be less than or equal to 30.61 kHz. It is only fair to ask, What is the maximum value of ω_p up to which the Sallen and Key bandpass filter in Example 4.10, using the single op amp noninverting amplifier in Fig. 4.2*a*, will have a ripple factor less than or equal to 0.1? To answer this question we first find $|\Delta p/p|$ using (4.31):

$$\left|\frac{\Delta p}{p}\right| = F\omega_p\tau = F\tau_n$$

where F is defined in (4.32). For $Q_p > 5$, we have the ripple factor

$$R \approx 2Q_pF\tau_n$$

Therefore the maximum value of ω_p for a given R is

$$\omega_p = \frac{R}{2Q_pF\tau} \tag{4.70}$$

For the filter circuit we designed in Example 4.10 with minimum active pole sensitivity, we have $F = 1.953$. With $R = 0.1$ and $Q_p = 20$, the maximum value of $f_p = 1.28$ kHz. The above calculations clearly show how the active compensation of conventional filters improves their performance in the low-frequency range and, for a given performance, how such active compensation extends the frequency of operation by an order of magnitude.

Let us make a critical observation about the effects of nonzero values of τ when we use actively compensated amplifiers in a Sallen and Key bandpass filter. The effect on ω_p is almost negligible, as observed from (4.68a) and from the calculations in Example 4.16. The substantial effect of the nonzero value of τ is only on Q_p. Then, we can safely assume that $\omega_p' \approx \omega_p$ for all practical purposes. This assumption holds true in particular when the choice of M and K is made such that F_2 is close to a minimum even when τ_n is as large as 0.05. This is because $\Delta\omega_p/\omega_p$ is found to be proportional to $1/Q_p$, whereas $\Delta Q_p/Q_p$ is found to be proportional to Q_p. We shall next present a very simple predistortion technique (not as complicated as that in Sec. 3.8) for this actively compensated filter with the assumption that $\omega_p' = \omega_p$ and τ_n is small. Assume that the specifications given are ω_{ps}, Q_{ps}, and H_{0s}. Also assume that we design the filter with certain values of ω_p, Q_p, and H_0 (unknown values for now) and with the assumption that $\tau = 0$. Then, if $\tau \neq 0$, $\omega_p = \omega_{ps}$, $Q_p = Q_{ps}$, and $H_0 = H_{0s}$, we know that $\omega_p' \neq \omega_{ps}$, $Q_p' \neq Q_{ps}$, and $H_0' \neq H_{0s}$. However, we can use (4.68) to calculate the actual pole frequency ω_p' and the pole Q factor Q_p' if the values of ω_p, Q_p, H_0, and τ are known and the actual dominant poles are given by

$$p_{1,2}' = \frac{-\omega_p'}{2Q_p'} \pm j\omega_p'$$

In turn, the denominator polynomial can be written as

$$D(s,\tau) = K_1\left[s^2 + \frac{\omega_p'}{Q_p'}s + (\omega_p')^2\right](1 + as + bs^2)$$

The second-order factor $1 + as + bs^2$ corresponds to the parasitic poles. Since $\omega_p' = \omega_p$ in this network and $K_1 = 1$, the entire transfer function is

$$H(s) = \frac{KG_1S_1(1 + Ks\tau)s}{(s^2 + \omega_p/Q_p's + \omega_p^2)(1 + as + bs^2)} \tag{4.71}$$

If we assume that the effects of the parasitic poles and the parasitic zero can be neglected, then the above transfer function can be approximated as

$$H(s) \approx \frac{KG_1S_1s}{s^2 + \omega_p/Q_p's + \omega_p^2} \tag{4.72}$$

In the above transfer function, observe that the numerator of (4.72) is same as that of the ideal transfer function. Thus the substantial effect of nonzero values of τ is only to cause a change in the pole Q factor from Q_p to Q'_p. Let us design a filter with a pole Q factor of Q_p and with the assumption that $\tau = 0$ [so that we will be able to use (4.28)] such that Q'_p meets the specifications rather than Q_p itself. In order to accomplish this, what should the value of Q_p be? Since Q'_p is known along with ω_p, K, τ, M, and H_0, we can substitute these values into (4.68*b*) to find the value of Q_p with which we should start the design. Since $|H(j\omega_p)| = H_0$, from (4.72), we find that

$$|H(j\omega_p)| = H_0 = \frac{KG_1S_1}{\omega_p}Q'_p$$

or

$$G_1 = \frac{H_0}{KMQ'_p}\omega_p C \tag{4.73a}$$

where $C = C_2$ and $C_1 = C/M$, as defined earlier in Sec. 4.3, and H_0, Q'_p, and ω_p are the specified values. If ω_p and Q_p are the design values, then we must satisfy (4.26*b*) and (4.26*c*), reproduced here as

$$\omega_p^2 = G_3(G_1 + G_2)S_1S_2 \tag{4.26b}$$

and

$$\frac{\omega_p}{Q_p} = (G_1 + G_2)S_1 + G_3(S_1 + S_2) - KG_2S_1 \tag{4.26c}$$

Substituting for G_3 from (4.26*b*) and for G_1 from (4.73*a*) in (4.26*c*), we solve a quadratic equation for $G_1 + G_2$ and obtain

$$G_2 = \frac{y}{2} + \sqrt{\left(\frac{y}{2}\right)^2 + Z} - G_1 \tag{4.73b}$$

where

$$y = \frac{(H_0/Q'_p - 1/Q_p)\omega_p C_2}{(K-1)M}$$

and

$$Z = \frac{(\omega_p C_2)^2(1+M)}{M^2(K-1)}$$

Once we find G_1 and G_2 using (4.73*a*) and (4.73*b*), respectively, then G_3 can be calculated as

$$G_3 = \frac{(\omega_p C_2)^2}{M(G_1 + G_2)} \tag{4.73c}$$

To find the value of G_2 from (4.73*b*), we must know the value of Q_p. This requires the solution of (4.68*b*), given the values of Q'_p (the specified value), ω_p, K, τ, M, and H_0. Substituting F_2 from (4.64) into (4.68*b*) and rearranging the terms, we have

$$Q'_p - \frac{Q'_p K^3 \tau_n^2}{2\sqrt{1 + K^2\tau_n^2}} \left[\frac{H_0 - 1}{K - 1} - \frac{2H_0}{K} + \sqrt{\left(\frac{H_0 - 1}{K - 1}\right)^2 + \frac{4Q_p^2(1 + M)}{K - 1}} \right] - Q_p = 0 \quad (4.74)$$

In (4.74) Q_p is the only unknown quantity that must be solved for. Since (4.74) is a nonlinear equation in Q_p, we use an iterative procedure to obtain the solution for Q_p from (4.74). The initial value with which one can start to find the solution of (4.74) is approximately given by

$$Q_p = Q'_p \left[\frac{1 - K^3\tau_n^2 \left(\dfrac{H_0 - 1}{K - 1} - \dfrac{2H_0}{K} \right)}{1 + Q'_p K^3 \tau_n^2 \left(\dfrac{1 + M}{K - 1} \right)^{1/2}} \right] \quad (4.75)$$

Note that (4.75) does *not* give the solution of (4.74) for Q_p. It only suggests a good starting value for finding this solution. We emphasize the fact that this simple technique works in this case because $\omega_p = \omega'_p$ and the numerator of (4.72) is same as that of the ideal transfer function.

Example 4.17: Design a Sallen and Key bandpass filter that uses the actively compensated noninverting amplifier in Fig. 4.16*a* to meet the specifications $f_p = 50$ kHz, $Q_p = 20$, and $H_0 = 5$. Use 741 type op amps with a nominal value of $\tau = 0.5/\pi$ μs and choose $M = 0.5$ and $K = 1.3$. Note that, except for f_p, these specifications are same as those given in Example 4.16.

If we design the filter with a nominal value of $\tau = 0$ and with $Q_p = 20$, then the value of Q'_p can be calculated using (4.68*b*) and will be 27.94. This means that there will be a 39.69% increase in Q_p. Therefore we decide to use the simple predistortion technique developed in this section. To do this, we assume that $Q'_p = 20$ and find the value of Q_p with which we want to carry out the nominal design with $\tau = 0$. First, we simplify (4.74) with the substitution of $K = 1.3$, $\tau_n = 0.05$, $M = 0.5$, $Q'_p = 20$, and $H_0 = 5$. Then we have

$$19.6908 - 54.809 \times 10^{-3} \sqrt{\left(\frac{4}{0.3}\right)^2 + 20Q_p^2} - Q_p = 0$$

With a possible starting value of $Q_p = 15.8145$, we solved the above equation iteratively and found that

$$Q_p = 15.7594$$

Next, if we obtain G_1, G_2, and G_3 using (4.73) with $Q_p = 15.7594$, hopefully, the circuit gain at $\omega = \omega_p$ will be $H_0 = 5$ and the circuit pole Q factor will be $Q'_p = 20$.

Choosing a capacitance value of $C_2 = 500$ pF (we need such a small value because of the high frequency of 50 kHz) and using (4.73*a*), we first find that $R_1 = 16.552$ kΩ. Then, using (4.73*b*) and (4.73*c*) with $Q_p = 15.7594$, we obtain the values of G_2 and G_3, hence the values $R_2 = 1.340$ kΩ and $R_3 = 16.352$ kΩ. The entire circuit realization is shown in Fig. 4.19. ■

This simple technique is possible here because $\omega'_p = \omega_p$ and the numerator of (4.72) is same as that of the ideal transfer function. Otherwise, the methods of Chap. 3 must be used. From the designer's point of view, there are many questions to be answered. First is Does this predistortion work? To answer this question, we provide the magnitude characteristics of the bandpass filters in Fig. 4.19, which were obtained using the program SPICE. In Fig. 4.20, curve *a* represents the ideal characteristic, and curve *b* represents the magnitude characteristic of the predistorted design, that is, of the network in Fig. 4.19. Curve *c* represents the characteristic if we had designed the filter without using the predistortion, as in Example, 4.16, but

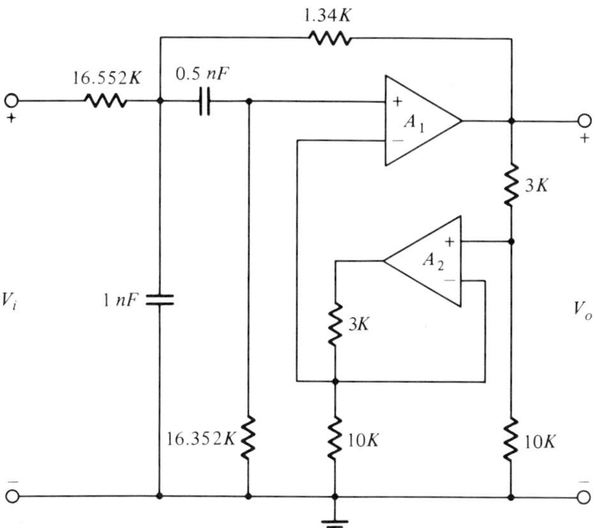

Figure 4.19 Circuit realized in Example 4.17.

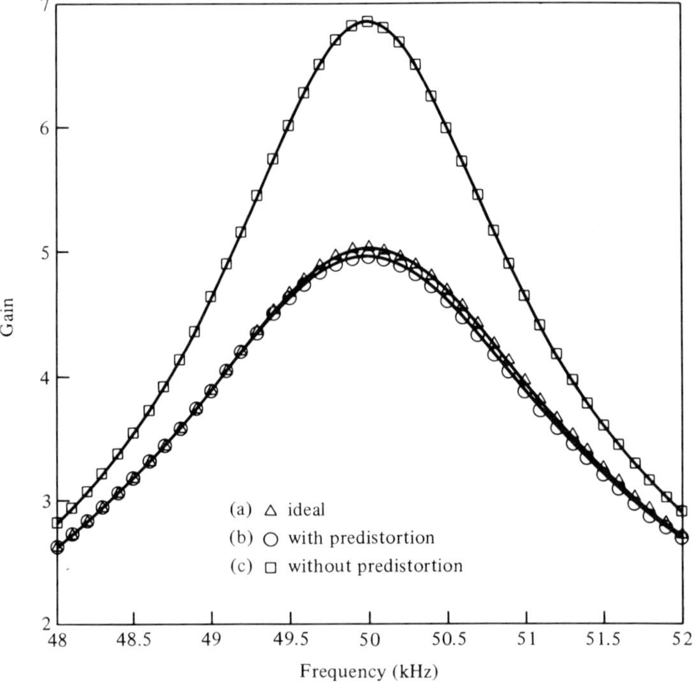

Figure 4.20 Magnitude responses of the bandpass filters designed in Example 4.17.

to meet the specifications in Example 4.17. Curves a and b are not distinguishable, and this proves the success of the predistortion technique. In fact, it also proves the validity of (3.67), using which (4.68) was derived even though the value of τ_n is large ($\tau_n = 0.05$). The second question is What is the maximum percentage of change in the magnitude function that can be expected for a specified change in the value of τ from its nominal value? To answer this question, we reduce the value of τ from $0.5/\pi$ to $0.375/\pi$ μs and obtain the magnitude characteristics. This is equivalent to a 33.33% increase in the gain bandwidth products of the op amps. The two magnitude characteristics, with $\tau = 0.5/\pi$ and $0.375/\pi$ μs, of the network in Fig. 4.19 are shown in Fig. 4.21. The maximum change occurs at $\omega = \omega_p$, and the magnitude decreases from 5 to 4.4852 at this frequency. This is equivalent to a maximum change of -10.3%. This brings out an important point under consideration. Once we predistort the design taking the nominal value of τ into account, any future changes in τ do not cause much change in the magnitude function. Note that in this example a 33.33% increase in the gain bandwidth product of the amplifiers causes a maximum change of only 10.3%. It is also possible to predict this change using the concept of ripple factor. The third question is, What are the passive ω_p and Q_p sensitivities of the bandpass filters that use the actively compensated

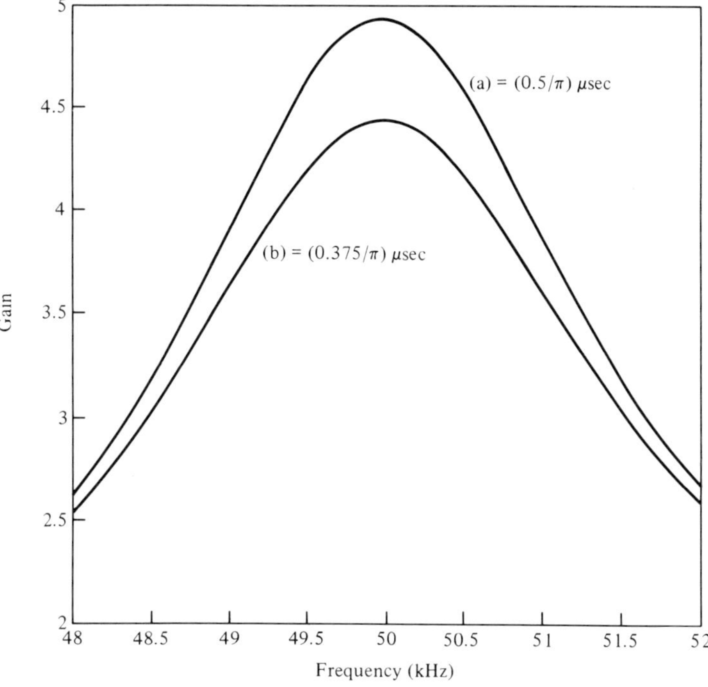

Figure 4.21 Magnitude responses of the bandpass filter in Fig. 4.19 with two different values of τ.

amplifier? Since passive ω_p and Q_p sensitivities can be obtained using the formulas developed in Sec. 4.3, we can calculate these sensitivities for the design in Example 4.17 with $K = 1.3$, $Q_p = 15.7594$, $M = 0.5$, and $\omega_p = 100\pi \times 10^3$ r/s and with the values of the different passive elements as calculated in Example 4.17:

$$S_{R_1}^{\omega_p} = -0.036 \qquad S_{R_2}^{\omega_p} = -0.464 \qquad S_{R_3}^{\omega_p} = -0.5$$

$$S_{C_1}^{\omega_p} = S_{C_2}^{\omega_p} = -0.5 \qquad S_{R_a}^{\omega_p} = S_{R_b}^{\omega_p} = 0$$

$$S_{R_1}^{Q_p} = 3.811 \qquad S_{R_2}^{Q_p} = -12.314 \qquad S_{R_3}^{Q_p} = -8.259$$

$$S_{C_1}^{Q_p} = -S_{C_2}^{Q_p} = -5.339$$

$$S_{R_a}^{Q_p} = -S_{R_b}^{Q_p} = 11.85$$

Compared to those for the design in Example 4.10 with the same required pole Q factor of 20, these passive sensitivities are generally lower. Thus the overall sensitivity performance of the Sallen and Key bandpass filters using an actively compensated filter will not only be much better in terms of the active sensitivity but also in terms of the passive sensitivities. Note also that an actively compensated filter, particularly with predistortion, can be used even at high frequencies without any difficulty.

4.8 *Conclusions*

In this chapter we have described the design of the popular Sallen and Key type active filters in which a noninverting VCVS is used as the active device. We have provided many alternative design procedures for different networks, indicating the advantages as well as the disadvantages of each. In addition, we have also discussed design procedures, for some proven second-order filters, by which the active sensitivity can be minimized. In general, such designs also possess lower passive sensitivities than those of designs obtained using other alternatives. Finally, in Sec. 4.7, some recently developed active compensation techniques were suggested to show how the influence of the op amps can be lowered by an order of magnitude. Such filters can be used even at high frequencies without much difficulty, particularly with predistortion. We used an in-depth analysis of a Sallen and Key bandpass filter to demonstrate the improvements caused by active compensation. However, such improvements can be shown to occur using other filter configurations as well. One final point is that there are noninverting amplifiers using three op amps that possess better properties than those of the two-op amp circuit in Fig. 4.16.

EXERCISES

4.1. Consider op amp realization of the noninverting amplifier in Fig. 1.47*a*. When the op amp is ideal, we know that $V_o/V_i = K = 1 + R_1/R_2$. We also know that, when $A^{-1} = s\tau$, then $V_o/V_i = K/(1 + Ks\tau)$. Another nonidealness of an op amp is its output impedance R_o. Including this R_o and the finite gain A, where $A^{-1} = s\tau$, we can represent this circuit with a VCVS and a finite output impedance Z_o as shown in Fig. E4.1. Find the values of K_1 and Z_o if $\tau = 0.5/\pi\ \mu\text{s}$,

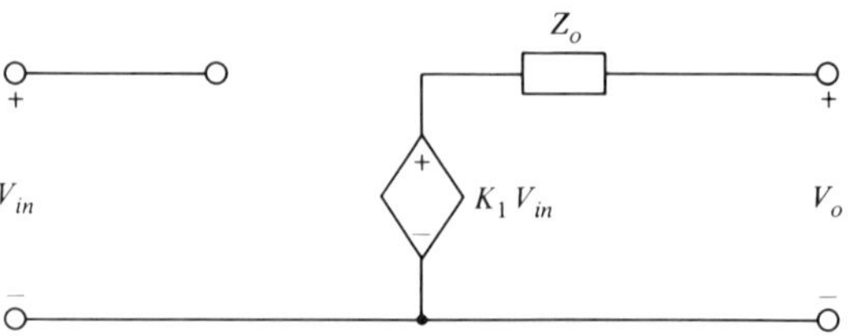

Figure E4.1 Possible equivalent circuit of a finite-gain amplifier considered in Exercise 4.1.

$R_o = 100\ \Omega$, and (a) $R_1 = R_2 = 1\ \text{k}\Omega$ and (b) $R_1 = R_2 = 56\ \text{k}\Omega$. Comment on the results obtained in parts (a) and (b).

4.2. Consider the circuit in Fig. 4.3.
(a) Using this circuit, design a Sallen and Key lowpass filter that meets the specifications $f_p = 10$ kHz and $Q_p = 10$. Choose $K = 2$ and $C_2/C_1 = 5$.
(b) Repeat Exercise 4.2*a* to achieve minimum active pole sensitivity. Choose $C_2/C_1 = 5$.
(c) Assume that the op amp has a nonzero value of $\tau = 0.5/\pi\ \mu\text{r/s}$. Find the per-unit changes in the realized values of the pole frequency and the pole Q factor. Apply (3.62) to the denominator polynomial to obtain the answers.
(d) Consider the design in part (b). Assume that the op amp τ is the same as that given in part (c). Use predistortion and obtain a new set of values for R_1 and R_2 so that the realized values of ω_p and Q_p will be the desired ones.

4.3. Consider the Sallen and Key bandpass filter circuit in Fig. 4.5.
(a) For a given set of values for $C_2 = C$, $C_1 = C/M$, and K, show that the element values of R_1, R_2, and R_3 can be obtained using (4.28).
(b) Assume that $C_1 = C_2 = 1$ nF and $K = 2$. Obtain the resistance values to meet the specifications $H_0 = 10$, $Q_p = 20$, and $f_p = 10$ kHz.
(c) Repeat part (b) with $C_1 = 10$ nF, $C_2 = 1$ nF, and $K = 1.6$.
(d) Obtain a design using the program SKBPN, assuming that $\tau = 0$.
(e) Compare the values of $|S_\tau^P|$ for the designs in parts (b), (c), and (d).

4.4. Consider the bandpass filter circuit in Fig. 4.5.
(a) For the design in Exercise 4.3*b* find the per-unit changes in ω_p and Q_p due to the nonzero value of $\tau = 0.5/\pi\ \mu\text{s}$. Also, find the per-unit changes in $|H(j\omega)|$ at the center frequency and at the 3-dB frequencies.
(b) Assume that the finite gain amplifier in the circuit is realized using the circuit in Fig. 4.16*a*. Both op amps have the same nonzero time constant $\tau = 0.5/\pi\ \mu\text{s}$. Repeat the Exercise 4.4*a*.
(c) Repeat part (b) using the finite gain amplifier circuit in Fig. 4.16*b*.

4.5. In the bandpass filter circuit in Fig. 4.5, assume that $G_1 = 1$, $G_2 = 3$, $G_3 = 0.5$, $C_1 = 1$, $C_2 = 2$, and $K = 1.55$.
(a) Obtain $S_x^{|H(j\omega)|}$ as a function of ω, where $x = G_1, G_2, G_3, C_1, C_2$, and K.
(b) The resistances have a tolerance of 1%, and the capacitances have a tolerance of 0.25%. Assume that K has been realized exactly.

Find the worst-case change $\Delta|H(j\omega)|/|H(j\omega)|$ to be expected at $\omega = 0.95$, 1, and 1.05 r/s.

(c) The temperature coefficient of the resistances is 0.01%/°C and that of the capacitances is −0.02%/°C. Find the expected variation in $|H(j\omega)|$ with an expected temperature increase of 25°C of C at $\omega = 0.95$ 1.05 r/s.

4.6. Use the Sallen and Key bandpass filter network in Fig. 4.5 to design a bandpass filter that meets the specifications $H_0 = 1$, $f_p = 2$ kHz, and $Q_p = 10$. Reverse the roles of the input and the ground terminals and derive the transfer function of the new network. Identify the type of the new transfer function.

4.7. Repeat Exercise 4.6 with $H_0 = 2$.

4.8. Consider the circuit in Fig. 4.7.

(a) Using this circuit, design a highpass filter that meets the specifications $f_p = 2$ kHz and $Q_p = 5$. Choose $R_1 = R_2$ and $C_1 = C_2$.

(b) Develop a design procedure for this circuit with minimum active pole sensitivity. Follow the method used for the Sallen and Key lowpass filter to design a highpass filter that meets the specifications in part (a).

(c) Compare the $|S_\tau^P|$ values of both designs.

4.9. Using the network in Fig. 4.8 design an HPN filter that meets the specifications $f_p = 2.5$ kHz, $Q_p = 10$, and $f_z = 2$ kHz.

4.10. Applying *RC-CR* transformation to the lowpass network in Fig. 4.13 results in a third-order highpass filter circuit. That is, by replacing resistances with capacitances and capacitances with resistances in the circuit in Fig. 4.13, one can obtain a highpass filter circuit. Develop a procedure for designing a highpass filter using such a circuit. The design procedure will be similar to that for the lowpass filter circuit in Fig. 4.13. A convenient choice for K is unity in the case of the highpass filter. Use this method to design a third-order highpass filter whose transfer function is

$$H(s) = \frac{H_0 s^3}{s^3 + a_2 s^2 + a_1 s + a_0}$$

where $a_2 = 5244$, $a_1 = 16.998 \times 10^6$, and $a_0 = 34.049 \times 10^9$.

4A

Program SKLPN

THIS APPENDIX PROVIDES DETAILS of the program SKLPN, written in GW BASIC, which can be run on an IBM PC or its compatibles. This program can be used to design the Sallen and Key lowpass filter circuit in Fig. 4.3, and the notation for the different elements can be found in that figure. The design procedure uses a predistortion technique* in which the influence of a nonzero τ value of the op amp used is compensated for and, at the same time, the magnitude of active pole sensitivity of the filter, defined as

$$S_\tau^p = \frac{\partial p}{\partial \tau}\frac{\tau}{p}$$

is minimized.

The procedure for using this program is simple: Just enter it and run it. The inputs for the program are Q_p, $n^2 = C_2/C_1$, and $\tau_n = \omega_p\tau$. These inputs are supplied through the keyboard when the appropriate question is displayed on the screen.

The design is carried out initially with the assumption that $\omega_p = 1$ r/s and $C_1 = 1$ F. However, one can obtain denormalized element values by inputting the values of ω_p and C_1 through the keyboard. If needed, the passive ω_p and Q_p sensitivities can also be found for the specified design using this program. The accuracy of the design can be verified by obtaining

*S. Natarajan, A Simple Predistortion Technique for Active Filters, *IEEE Trans. Circuits Syst.*, vol. CAS-31, no. 4, pp. 398–402, April 1984.

the normalized pole frequency and pole Q factor of the dominant poles of the predistorted design.

The program is followed by output from a sample session for the design of a filter circuit satisfying the specifications $f_p = 20$ kHz, $Q_p = 10$, and $n^2 = 2$, with $C_1 = 1$ nF, $\tau = 0.5/\pi$ μs, and $\tau_n = \omega_p\tau = 4\pi \times 10^4 \times (0.5/\pi) \times 10^{-6} = 0.02$.

```
10 '
20 '
30 '              PROGRAM LISTING OF 'SKLPN'
40 '              -------------------------
50 '
60 '
70 'Program to design Sallen and Key lowpass filter with min. active pole sens.
along with predistortion.
80 PRINT"THIS PROGRAM CAN BE USED TO DESIGN SALLEN AND KEY LOWPASS FILTER WITH M
INIMUM   ACTIVE POLE SENSITIVITY  ALONG WITH PREDISTORTION.":PRINT
90 DIM E$(7),X(4),E(7),D(4),Z(4)
100 E$(1)="R1":E$(2)="R2":E$(3)="C1":E$(4)="C2":E$(5)="Ra":E$(6)="Rb":E$(7)="K"
110 PRINT"INPUT THE SPECIFICATIONS FOR fp and Qp AS PROMPTED":PRINT
120 INPUT" fp in Hz";FP:INPUT" Q";Q
130 PI2=8!*ATN(1):WP=FP*PI2:RE=-1/(2*Q):IM=SQR(1-RE*RE)
140 PRINT"R1,R2 AND Rb ARE CALCULATED BY THE PROGRAM. INPUT THE OTHER REQUIRED P
ARAMETERS.":PRINT
150 INPUT"VALUE OF C1";E(3):C1=E(3)
160 INPUT"VALUE OF C2";E(4):C2=E(4)
170 INPUT"VALUE OF Ra";E(5):RA=E(5)
180 PRINT"INPUT THE NON-IDEAL PARAMETERS OF THE OP.AMP.":PRINT
190 INPUT"INPUT THE INVERSE OF THE GB PRODUCT OF THE OP.AMP.IN Hz^(-1).IN AN IDE
AL OP.AMP. IT IS ZERO.";T:T=T*FP
200 INPUT"INPUT THE NON-ZERO OUTPUT RESISTANCE OF THE OP.AMP. IDEALLY ITS VALUE
IS ZERO.";R0:AL=R0/RA:T1=R0*C2*WP
210 PRINT"STRAY CAPACITANCES FROM THE INVERTING AND NON-INVERTING INPUT TERMINAL
S TO GROUND ARE ZEROS IN IDEAL OP.AMPS.. HOWEVER THEIR VALUES MAY RANGE FROM 1 T
O 3 pF. IF YOU WANT TO INCLUDE THEM INPUT THEIR VALUES.":PRINT
220 INPUT"STRAY CAP.FROM NON-INVERTING TERMINAL TO GROUND";CS1
230 INPUT"STRAY CAP.FROM INVERTING TERMINAL TO GROUND";CS2:T2=RA*CS2*WP
240 LPRINT"DESIGN OF A SALLEN AND KEY LOWPASS FILTER":LPRINT STRING$(52,"_")
250 LPRINT:LPRINT"THE SPECIFICATIONS ARE":LPRINT:I=0
260 LPRINT"fp in Hz=";USING"##.#####^^^^";FP:LPRINT
270 LPRINT"Qp=";USING"##.#####^^^^";Q:LPRINT
280 LPRINT"C2/C1=";C2/C1:LPRINT
290 LPRINT"THE NON-IDEAL PARAMETERS OF THE OP.AMP. INCLUDED FOR PREDISTORTION"
300 LPRINT:LPRINT"RATIO fp/GB=";USING"##.#####^^^^";T:LPRINT
310 LPRINT"OUTPUT RESISTANCE=";USING"####.##";R0:LPRINT
320 LPRINT"STRAY CAP.FROM NON-INVERTING INPUT TERMINAL TO GROUND=";USING"##.####
#^^^^";CS1:LPRINT
330 LPRINT"STRAY CAP.FROM INVERTING INPUT TERMINAL TO GROUND=";USING"##.#####^^^
^";CS2:LPRINT STRING$(52,"_"):LPRINT
340 IF I=1 THEN C1=C1+CS1
350 N2=C2/C1:N=SQR(N2):Q1=N/(6*Q):M=Q1+SQR(Q1^2+(1+N2)/3):K=(M*M+N2+1-M*N/Q)/N2:
K=(INT(1000*K)+1)/1000
360 'Nominal Design equations
370 A3=T*T1*T2:J=1:J1=0:J2=1:J3=0
380 IF K<(1+1/N2) THEN K=1+1/N2+1/(Q*Q):X=-N*RE:M=X+SQR(X*X+(K-1)*N2-1)
390 R1=1/(M*N*WP*C1):R2=M*M*R1:RB=RA/(K-1):E(1)=R1:E(2)=R2:E(6)=RB:E(7)=K
400 IF I=0 THEN 410 ELSE 440
410 I=1:LPRINT"NOMINAL DESIGN PARAMETERS FOR MIN.ACTIVE POLE SENS.":LPRINT
420 FOR L=1 TO 7:LPRINT E$(L);"=";USING"##.#####^^^^";E(L):LPRINT:NEXT L
430 LPRINT STRING$(52,"_"):LPRINT
440 'Normalization of Element values
450 GN=WP*C1:C2=N2:R1=GN*R1:R2=GN*R2
460 A1=(K*(1+AL)-AL)*T:A2=((1+AL)*T2+K*T1)*T
470 ' Denominator coefficients
480 W0=R1*R2*C2:W1=R1+R2+C2*R1:W2=R1*C2
490 X(0)=1:X(1)=W1-K*W2+A1:X(2)=W0+W1*A1+A2-W2*A2:X(3)=A3+A2*W1+A1*W0-K*W2*T*T1
500 X(4)=A3*(W1-W2)+(1+AL)*W0*T*T2:ND=4:GOSUB 1010
510 X1=RD1:X2=RD0
520 X(4)=A3+(1+AL)*R2*C2*T*T2:X(3)=A2*(1+C2)+A1*R2*C2-K*C2*T1*T
530 X(2)=R2*C2+A1*(1+C2)-C2*T2:X(1)=1+C2-K*C2:X(0)=0:GOSUB 1010
540 A=RD1:B=RD0
550 X(4)=A3+(1+AL)*R1*C2*T*T2:X(3)=A2+A1*R1*C2:X(2)=R1*C2+A1:X(1)=1:X(0)=0
560 GOSUB 1010
```

```
570 C=RD1:D=RD0:DR2=(X2/B-X1/A)/(C/A-D/B):DR1=-(X1+C*DR2)/A:R1=R1+DR1:R2=R2+DR2
580 IF ABS(DR1/R1)>.0001 OR ABS(DR2/R2)>.0001 THEN 480
590 W0=R1*R2*C2:W1=R1+R2+C2*R1:W2=R1*C2
600 X(4)=((W1-W2)*T1+(1+AL)*W0)*T2:X(3)=(A3+A2*W1+A1*W0)/T-K*W2*T1
610 X(2)=(W1*A1+A2)/T:X(1)=A1/T:X(0)=0:GOSUB 1010
620 A=RD1:B=RD0
630 X(4)=X(4)*T:X(3)=X(3)*T:X(2)=X(2)*T+W0-W2*T2:X(1)=A1+(W1-K*W2):X(0)=1
640 IF J2=0 AND J3=0 THEN 650 ELSE 660
650 RX=R1:RY=R2:D(4)=X(4):D(3)=X(3):D(2)=X(2):D(1)=X(1):D(0)=X(0):J3=1:K4=K
660 X(0)=X(1):X(1)=2*X(2):X(2)=3*X(3):X(3)=4*X(4):ND=3:GOSUB 1010
670 C=RD1:D=RD0:S=SQR(((A*RE+B)^2+(A*IM)^2)/((C*RE+D)^2+(C*IM)^2))
680 IF J=0 THEN 700
690 J=0:K=K*1.0001:S1=S:GOTO 460
700 IF J1=2 THEN 770
710 K=K/1.0001:P1=(S-S1)*10000!/K:S7=S1:K1=K:P4=P1:P3=S7
720 IF J2=0 THEN 840 ELSE IF J1=1 THEN 820
730 IF ABS(P1)<.001 THEN 830 ELSE IF ABS(P1/K)>.1 THEN 750
740 V=1-P1/K:GOTO 760
750 IF P1>0 THEN V=.9 ELSE V=1.1
760 K=V*K:J1=2:J=1:GOTO 380
770 K=K/1.0001:P2=(S-S12)*10000!/K:S8=S1:K2=K:P4=P2:P3=S8
780 IF ABS(P2)<.001 THEN 830 ELSE IF ABS(P2)<ABS(P1) THEN 800
790 K=K1-P1*(K2-K1)/(P2-P1):K3=K1:P3=S7:GOTO 810
800 K=K2-P2*(K2-K1)/(P2-P1):K3=K2:P3=S8
810 IF ABS(K3/K-1)<.001 THEN 830 ELSE J1=1:J=1:GOTO 380
820 IF ABS(P3/S1-1)<.001 THEN 830 ELSE 760
830 K=(INT(1000*K)+1)/1000:J2=0:J=1:J1=0:GOTO 380
840 LPRINT"ELEMENT VALUES WITH MINIMUM ACTIVE POLE SENSITIVITY":LPRINT
850 LPRINT"AFTER PREDISTORTION IS APPLIED TO THE LOWPASS FILTER":LPRINT
860 'Denormalization of element values.
870 R1=RX/GN:R2=RY/GN:RB=RA/(K4-1):E(1)=R1:E(2)=R2:E(6)=RB:E(7)=K4
880 FOR L=1 TO 7:LPRINT E$(L);"=";USING"##.#####^^^^";E(L):LPRINT:NEXT L
890 LPRINT"MAGNITUDE OF MIN. ACTIVE POLE SENSITIVITY=";USING"##.#####^^^^";S7*T
900 PRINT"DO YOU WANT TO KNOW THE Wp AND Qp SENSITIVITIES WITH RESPECT TO PASSIV
E         ELEMENTS? Y/N";:INPUT A$
910 IF A$="N" OR A$="n" THEN 990
920 LPRINT STRING$(52,"_"):LPRINT"Wp AND Qp SENSITIVITIES ARE":LPRINT
930 N=SQR(E(4)/E(3)):M=SQR(E(2)/E(1)):P=M/N
940 E(1)=-.5:E(2)=-.5:E(3)=-.5:E(4)=-.5:E(5)=0:E(6)=0
950 FOR L=1 TO 6:LPRINT"Wp sens. w.r.t. ";E$(L);"=";E(L):LPRINT:NEXT L
960 E(1)=-.5+Q*P:E(2)=-E(1):E(3)=-.5+Q*(P+1/(M*N)):E(4)=-E(3)
970 E(5)=(K-1)*Q/P:E(6)=-E(5)
980 FOR L=1 TO 6:LPRINT"Qp sens. w.r.t. ";E$(L);"=";E(L):LPRINT:NEXT L
990 LPRINT STRING$(52,"_"):END
1000 'subroutine to divide a poly. with quadratic factor of s^2+s/Q+1
1010 Z(ND)=0:Z(ND-1)=0
1020 FOR L=ND-2 TO 0 STEP -1
1030 Z(L)=X(L+2)-Z(L+1)/Q-Z(L+2)
1040 NEXT L
1050 RD1=X(1)-Z(0)/Q-Z(1):RD0=X(0)-Z(0)
1060 RETURN
```

```
               A SAMPLE SESSION
DESIGN OF A SALLEN AND KEY LOWPASS FILTER
____________________________________________________

THE SPECIFICATIONS ARE

fp in Hz= 2.00000E+04

Qp= 1.00000E+01

C2/C1= 2

THE NON-IDEAL PARAMETERS OF THE OP.AMP. INCLUDED FOR PREDISTORTION

RATIO fp/GB= 2.00000E-02

OUTPUT RESISTANCE=   0.00

STRAY CAP.FROM NON-INVERTING INPUT TERMINAL TO GROUND= 0.00000E+00

STRAY CAP.FROM INVERTING INPUT TERMINAL TO GROUND= 0.00000E+00
____________________________________________________
NOMINAL DESIGN PARAMETERS FOR MIN.ACTIVE POLE SENS.

R1= 5.49591E+03
```

```
R2= 5.76117E+03
C1= 1.00000E-09
C2= 2.00000E-09
Ra= 1.00000E+04
Rb= 1.05042E+04
K= 1.95200E+00
```

```
ELEMENT VALUES WITH MINIMUM ACTIVE POLE SENSITIVITY
AFTER PREDISTORTION IS APPLIED TO THE LOWPASS FILTER
R1= 5.20222E+03
R2= 5.47791E+03
C1= 1.00000E-09
C2= 2.00000E-09
Ra= 1.00000E+04
Rb= 1.04932E+04
K= 1.95300E+00
MAGNITUDE OF MIN. ACTIVE POLE SENSITIVITY= 5.00484E-02
```

```
Wp AND Qp SENSITIVITIES ARE
Wp sens. w.r.t. R1=-.5
Wp sens. w.r.t. R2=-.5
Wp sens. w.r.t. C1=-.5
Wp sens. w.r.t. C2=-.5
Wp sens. w.r.t. Ra= 0
Wp sens. w.r.t. Rb= 0
Qp sens. w.r.t. R1= 6.756013
Qp sens. w.r.t. R2=-6.756013
Qp sens. w.r.t. C1= 13.64685
Qp sens. w.r.t. C2=-13.64685
Qp sens. w.r.t. Ra= 13.13393
Qp sens. w.r.t. Rb=-13.13393
```

4B

Program SKBPN

THIS APPENDIX PROVIDES DETAILS of the program SALPN, written in GW BASIC, which can be run on an IBM PC or its compatibles. This program can be used to design the Sallen and Key bandpass filter circuit in Fig. 4.5, and the notation for the different elements can be found in that figure. The design procedure uses a predistortion technique† in which the influence of the nonzero τ value of the op amp is compensated for and, at the same time, the magnitude of active pole sensitivity of the filter, defined as

$$S_\tau^p = \frac{\partial p}{\partial \tau}\frac{\tau}{p}$$

is minimized.

The procedure for using this program is simple: Just enter it and run it. The inputs for the program are H_0, Q_p, $M = C_2/C_1$, and $\tau_n = \omega_p \tau$. These inputs are supplied through the keyboard when the appropriate question is displayed on the screen.

The design is carried out initially with the assumption that $\omega_p = 1$ r/s and $C_2 = 1$ F. However, one can obtain denormalized element values by inputting the values of ω_p and C_2 through the keyboard. If needed, the passive ω_p and Q_p sensitivities can also be found for the specified design using this program. The accuracy of the design can be verified by obtaining

†S. Natarajan, A Simple Predistortion Technique for Active Filters, *IEEE Trans. Circuits Syst.*, vol. CAS-31, no. 4, pp. 398–402, April 1984.

the normalized pole frequency and pole Q factor of the dominant poles of the predistorted design.

The program is followed by the output from a sample session for the design of a filter circuit satisfying the specifications $f_p = 50$ kHz, $Q_p = 30$, $H_o = 10$, and $M = 0.2$, with $C_2 = 0.5$ nF, $\tau = 0.5/\pi\ \mu$s, and $\tau_n = \omega_p \tau = 10\pi \times 10^4 \times (0.5/\pi) \times 10^{-6} = 0.05$.

```
10 '
20 '
30 '                  PROGRAM LISTING OF 'SKBPN'
40 '                  -------------------------
50 '
60 '
70 'Program to design Sallen and Key bandpass filter with min. active pole sens.
 along with predistortion.
80 PRINT"THIS PROGRAM CAN BE USED TO DESIGN SALLEN AND KEY BANDPASS FILTER WITH
MINIMUM   ACTIVE POLE SENSITIVITY  ALONG WITH PREDISTORTION.":PRINT
90 DIM E$(8),X(4),E(8),D(4),Z(4)
100 E$(1)="R1":E$(2)="R2":E$(3)="R3":E$(4)="C1":E$(5)="C2":E$(6)="Ra":E$(7)="Rb"
:E$(8)="K"
110 PRINT"INPUT THE SPECIFICATIONS FOR fp,Qp and H0 AS PROMPTED":PRINT
120 INPUT" fp in Hz";FP:INPUT" Qp";Q:INPUT" H0=";H
130 PI2=8!*ATN(1):WP=FP*PI2:RE=-1/(2*Q):IM=SQR(1-RE*RE)
140 PRINT"R1,R2,R3 AND Rb ARE CALCULATED BY THE PROGRAM. INPUT THE OTHER REQUIRE
D PARAMETERS.":PRINT
150 INPUT"VALUE OF C1";E(4):C1=E(4)
160 INPUT"VALUE OF C2(C2<C1)";E(5):C2=E(5)
170 INPUT"VALUE OF Ra";E(6):RA=E(6)
180 PRINT"INPUT THE NON-IDEAL PARAMETERS OF THE OP.AMP.":PRINT
190 INPUT"INPUT THE INVERSE OF THE GB PRODUCT OF THE OP.AMP.IN Hz^(-1).IN AN IDE
AL OP.AMP. IT IS ZERO.";T:T=T*FP
200 INPUT"INPUT THE NON-ZERO OUTPUT RESISTANCE OF THE OP.AMP. IDEALLY ITS VALUE
IS ZERO.";R0:AL=R0/RA:AL1=1+AL
210 PRINT"STRAY CAPACITANCES FROM THE INVERTING AND NON-INVERTING INPUT TERMINAL
S TO GROUND ARE ZEROS IN IDEAL OP.AMPS.. HOWEVER THEIR VALUES MAY RANGE FROM 1 T
O 3 pF. IF YOU WANT TO INCLUDE THEM INPUT THEIR VALUES.":PRINT
220 INPUT"STRAY CAP.FROM NON-INVERTING TERMINAL TO GROUND";CS1
230 INPUT"STRAY CAP.FROM INVERTING TERMINAL TO GROUND";CS2:T2=RA*CS2*WP
240 LPRINT"DESIGN OF A SALLEN AND KEY BANDPASS FILTER":LPRINT STRING$(52,"_")
250 LPRINT:LPRINT"THE SPECIFICATIONS ARE":LPRINT:I=0
260 LPRINT"fp in Hz=";USING"##.#####^^^^";FP:LPRINT
270 LPRINT"Qp=";USING"##.#####^^^^";Q:LPRINT
280 LPRINT"H0=";USING"##.#####^^^^";H:LPRINT
290 LPRINT"C2/C1=";C2/C1:LPRINT
300 LPRINT"THE NON-IDEAL PARAMETERS OF THE OP.AMP. INCLUDED FOR PRDISTORTION"
310 LPRINT:LPRINT"RATIO fp/GB=";USING"##.#####^^^^";T:LPRINT
320 LPRINT"OUTPUT RESISTANCE=";USING"####.##";R0:LPRINT
330 LPRINT"STRAY CAP.FROM NON-INVERTING INPUT TERMINAL TO GROUND=";USING"##.####
#^^^^";CS1:LPRINT
340 LPRINT"STRAY CAP.FROM INVERTING INPUT TERMINAL TO GROUND=";USING"##.#####^^^
^";CS2:LPRINT STRING$(52,"_"):LPRINT
350 C2S=C2+CS1:C=C1+C2:CX=SQR(C*C2S-C2*C2):SC=WP*CX:R0=SC*R0
360 C2S=C2S/CX:C=C/CX:C2=C2/CX
370 K=1.6:Q2=Q*Q:M=C/(C-C2S)
380 A=SQR((H-1)^2+4*Q2*M*(K-1)):F1=A*(K*(K-2)*(A-H-1)-2*H)+2*K*K*(K-1)*M*Q2
390 F3=F2*(K*(K-2)*(2*A-H-1)-2*H)+2*A*(K-1)*(A-H-1)+4*K*(1.5*K-1)*M*Q2
400 K1=K-F1/F3
410 IF ABS(K1/K-1)<.001 THEN 420 ELSE K=K1:GOTO 380
420 K=(INT(1000*K)+1)/1000
430 J=1:J1=0:J2=1:J3=0:TT2=T*T2
440 'Nominal Design equations
450 Z1=(H-1)/(2*Q*(K*C2-C2S)):Z2=C/(K*C2-C2S):G=Z1+SQR(Z1*Z1+Z2)
460 G3=1/G:G1=H/(K*Q*C2):G2=G-G1
470 IF I=0 THEN 480 ELSE 530
480 LPRINT:E(1)=1/(SC*G1):E(2)=1/(SC*G2):E(3)=1/(SC*G3):E(7)=RA/(K-1):E(8)=K
490 I=1:LPRINT"NOMINAL DESIGN PARAMETERS FOR MIN.ACTIVE POLE SENS.":LPRINT
500 FOR L=1 TO 8:LPRINT E$(L);"=";USING"##.#####^^^^";E(L):LPRINT:NEXT L
510 LPRINT STRING$(52,"_"):LPRINT
520 ' Denominator coefficients and their derivatives w.r.t. G2 and G3.
530 BE=R0*G2:A2=AL1+BE:A1=K*A2-AL:G=G1+G2:D0=G*G3:D1=C*G3+C2S*G
540 X(0)=D0:X(1)=D1-K*C2*G2+(A1*D0-K*BE*G2*G3)*T
550 X(2)=1+(A1*D1-K*BE*C2S*G3)*T+(A2*D0-BE*G2*G3)*TT2-G2*C2*T2
560 X(3)=A1*T+(A2*D1-BE*C2S*G3)*TT2:X(4)=A2*TT2:ND=4:GOSUB 1110
```

```
570 X1=RD1:X2=RD0
580 X(0)=G3:X(1)=C2S-K*C2+(A1*G3+K*R0*D0-2*K*BE*G3)*T
590 X(2)=(A1*C2S+K*R0*(D1-C2S*G3))*T-C2*T2+(A2-2*BE)*G3*TT2
600 X(3)=K*R0*T+(A2*C2S+(D1-C2S*G3)*R0)*TT2:X(4)=R0*TT2:GOSUB 1110
610 A=RD1:B=RD0
620 X(0)=G:X(1)=C+(A1*G-K*BE*G2)*T:X(2)=(A1*C-K*BE*C2S)*T+(A2*G-BE*G2)*TT2
630 X(3)=(A2*C-BE*C2S)*TT2:X(4)=0:GOSUB 1110
640 Y=RD1:D=RD0:DG3=(X2/B-X1/A)/(Y/A-D/B):DG2=-(X1+Y*DG3)/A:G2=G2+DG2:G3=G3+DG3
650 IF ABS(DG2/G2)>.0001 OR ABS(DG3/G3)>.0001 THEN 530
660 BE=R0*G2:X1=K*Q/H:GX=G2/(X1*(C2/G3+BE*T)-1)
670 IF ABS(GX/G1-1)<.0001 THEN 680 ELSE G1=GX:GOTO 530
680 G1=GX:BE=R0*G2:A2=AL1+BE:A1=K*A2-AL:G=G1+G2:D0=G*G3:D1=C*G3+C2S*G
690 X(0)=0:X(1)=A1*D0-K*BE*G2*G3:X(2)=(A1*D1-K*BE*C2S*G3)+(A2*D0-G2*G3*BE)*T2
700 X(3)=A1+(A2*D1-BE*C2S*G3)*T2:X(4)=A2*T2:ND=4:GOSUB 1110
710 A=RD1:B=RD0
720 X(4)=X(4)*T:X(3)=X(3)*T:X(2)=X(2)*T+1-G2*C2*T2
730 X(1)=D1-K*C2*G2+X(1)*T:X(0)=D0
740 IF J2=0 AND J3=0 THEN 750 ELSE 770
750 D(4)=X(4):D(3)=X(3):D(2)=X(2):D(1)=X(1):D(0)=X(0):J3=1:K4=K
760 E(1)=1/(SC*G1):E(2)=1/(SC*G2):E(3)=1/(SC*G3):E(7)=RA/(K-1)
770 X(0)=X(1):X(1)=2*X(2):X(2)=3*X(3):X(3)=4*X(4):ND=3:GOSUB 1110
780 Y=RD1:R=RD0:S=SQR(((A*RE+B)^2+(A*IM)^2)/((Y*RE+R)^2+(Y*IM)^2))
790 IF J=0 THEN 810
800 J=0:K=K*1.0001:S1=S:GOTO 530
810 IF J1=2 THEN 880
820 K=K/1.0001:P1=(S-S1)*10000!/K:S7=S1:K1=K:P4=P1:P3=S7
830 IF J2=0 THEN 950 ELSE IF J1=1 THEN 930
840 IF ABS(P1)<.001 THEN 940 ELSE IF ABS(P1/K)>.1 THEN 860
850 V=1-P1/K:GOTO 870
860 IF P1>0 THEN V=.9 ELSE V=1.1
870 K=V*K:J1=2:J=1:GOTO 450
880 K=K/1.0001:P2=(S-S12)*10000!/K:S8=S1:K2=K:P4=P2:P3=S8
890 IF ABS(P2)<.001 THEN 940 ELSE IF ABS(P2)<ABS(P1) THEN 910
900 K=K1-P1*(K2-K1)/(P2-P1):K3=K1:P3=S7:GOTO 920
910 K=K2-P2*(K2-K1)/(P2-P1):K3=K2:P3=S8
920 IF ABS(K3/K-1)<.001 THEN 940 ELSE J1=1:J=1:GOTO 450
930 IF ABS(P3/S1-1)<.001 THEN 940 ELSE 870
940 K=(INT(1000*K)+1)/1000:J2=0:J=1:J1=0:GOTO 450
950 LPRINT"ELEMENT VALUES WITH MINIMUM ACTIVE POLE SENSITIVITY":LPRINT
960 LPRINT"AFTER PREDISTORTION IS APPLIED TO THE LOWPASS FILTER":LPRINT
970 E(7)=RA/(K4-1):E(8)=K4
980 FOR L=1 TO 8:LPRINT E$(L);"=";USING"##.#####^^^^";E(L):LPRINT:NEXT L
990 LPRINT"MAGNITUDE OF MIN. ACTIVE POLE SENSITIVITY=";USING"##.#####^^^^";S7*T
1000 PRINT"DO YOU WANT TO KNOW THE Wp AND Qp SENSITIVITIES WITH RESPECT TO PASSI
VE          ELEMENTS? Y/N";:INPUT A$
1010 IF A$="N" OR A$="n" THEN 1090
1020 LPRINT STRING$(52,"_"):LPRINT"Wp AND Qp SENSITIVITIES ARE":LPRINT
1030 G1=1/(E(1)*SC):G2=1/(SC*E(2)):G3=1/(SC*E(3)):M1=C/(C-C2):M=M1-1
1040 E(1)=-G1/(2*(G1+G2)):E(2)=-.5-E(1):E(3)=-.5:E(4)=-.5:E(5)=-.5:E(6)=0:E(7)=0

1050 FOR L=1 TO 7:LPRINT"Wp sens. w.r.t. ";E$(L);"=";E(L):LPRINT:NEXT L
1060 E(1)=H/K+E(1):E(2)=E(2)+G2*M*(1-K)*Q/C2:E(3)=-.5+Q*G3*M1/C2:E(4)=.5-G3*Q/C2

1070 E(5)=-E(4):E(6)=(K-1)*G2*M*Q/C2:E(7)=-E(6)
1080 FOR L=1 TO 7:LPRINT"Qp sens. w.r.t. ";E$(L);"=";E(L):LPRINT:NEXT L
1090 LPRINT STRING$(52,"_"):END
1100 'subroutine to divide a poly. with quadratic factor of s^2+s/Q+1
1110 Z(ND)=0:Z(ND-1)=0
1120 FOR L=ND-2 TO 0 STEP -1
1130 Z(L)=X(L+2)-Z(L+1)/Q-Z(L+2)
1140 NEXT L
1150 RD1=X(1)-Z(0)/Q-Z(1):RD0=X(0)-Z(0)
1160 RETURN
```

```
               A SAMPLE SESSION

DESIGN OF A SALLEN AND KEY BANDPASS FILTER
____________________________________________________

THE SPECIFICATIONS ARE

fp in Hz= 5.00000E+04

Qp= 3.00000E+01

H0= 1.00000E+01

C2/C1= .2
```

```
THE NON-IDEAL PARAMETERS OF THE OP.AMP. INCLUDED FOR PREDISTORTION
RATIO fp/GB= 5.00000E-02
OUTPUT RESISTANCE=   0.00
STRAY CAP.FROM NON-INVERTING INPUT TERMINAL TO GROUND= 0.00000E+00
STRAY CAP.FROM INVERTING INPUT TERMINAL TO GROUND= 0.00000E+00
```

```
NOMINAL DESIGN PARAMETERS FOR MIN.ACTIVE POLE SENS.
R1= 5.60735E+03
R2= 7.37473E+02
R3= 1.24367E+04
C1= 2.50000E-09
C2= 5.00000E-10
Ra= 1.00000E+04
Rb= 2.13675E+04
K= 1.46800E+00
```

```
ELEMENT VALUES WITH MINIMUM ACTIVE POLE SENSITIVITY
AFTER PREDISTORTION IS APPLIED TO THE LOWPASS FILTER
R1= 4.64559E+03
R2= 6.59038E+02
R3= 1.16274E+04
C1= 2.50000E-09
C2= 5.00000E-10
Ra= 1.00000E+04
Rb= 2.13220E+04
K= 1.46900E+00
MAGNITUDE OF MIN. ACTIVE POLE SENSITIVITY= 8.61059E-02
```

```
Wp AND Qp SENSITIVITIES ARE
Wp sens. w.r.t. R1=-6.211917E-02
Wp sens. w.r.t. R2=-.4378809
Wp sens. w.r.t. R3=-.5
Wp sens. w.r.t. C1=-.5
Wp sens. w.r.t. C2=-.5
Wp sens. w.r.t. Ra= 0
Wp sens. w.r.t. Rb= 0
Qp sens. w.r.t. R1= 6.745234
Qp sens. w.r.t. R2=-27.62065
Qp sens. w.r.t. R3= 19.21066
Qp sens. w.r.t. C1=-15.92555
Qp sens. w.r.t. C2= 15.92555
Qp sens. w.r.t. Ra= 27.18277
Qp sens. w.r.t. Rb=-27.18277
```

4C

Program THIRD

THIS APPENDIX DESCRIBES the program THIRD written in GW BASIC, which can be run in IBM PC or its compatibles. This program can be used to design the third-order lowpass filter network of Fig. 4.13. See Fig. 4.13 and the equation (4.51) for notations used.

Inputs Necessary

σ_o, A_1, B_1, and C.

Outputs

τ_1, τ_2, and τ_3 and the resistance values R_1, R_2, and R_3.

```
10 '
20 '
30 '              PROGRAM LISTING OF 'THIRD'
40 '
50 '
60 'THIS PROGRAM CAN BE USED TO DESIGN THE THIRD ORDER LOWPASS FILTER OF THE
70 'NETWORK SHOWN IN FIG.4.13.
80 LPRINT"DESIGN OF THIRD ORDER FILTER OF FIG.4.13."
90 LPRINT STRING$(52,"_"):LPRINT
100 PRINT CHR$(229);"0,A1,B1";:INPUT A0,A1,B1
110 LPRINT"   INPUTS TO THE PROGRAM ARE":LPRINT STRING$(52,"_")
120 LPRINT"       ";CHR$(229);"0 = ";USING "##.#####^^^^";A0
130 LPRINT"       ";"A1 = ";USING"##.#####^^^^";A1
140 LPRINT"       ";"B1 = ";USING"##.#####^^^^";B1
150 D2 = A0+A1:D1=B1+A1*A0:D0=A0*B1:A3=1/D0:A2=D2/D0:A1=D1/D0
160 B2=-2*A1:B1=A1*A1+.5*A2:B0=-.5*(A1*A2-A3):X=A1-A3/A2
170 F=B0+X*(B1+X*(B2+X)):F1=B1+X*(2*B2+3*X):X3=X-F/F1
180 IF ABS(X3/X-1)>.000001 THEN X = X3:GOTO 170
190 X1=A1-X3:X2=A3/(X3*X1):LPRINT:LPRINT"       THE RESULTS ARE":LPRINT
200 LPRINT"      ";CHR$(231);"1 = ";USING"##.######^^^^";X1
210 LPRINT"      ";CHR$(231);"2 = ";USING"##.######^^^^";X2
220 LPRINT"      ";CHR$(231);"3 = ";USING"##.######^^^^";X3
230 INPUT"DO YOU WANT TO FIND THE VALUES OF R1,R2,R3? Y/N";A$
240 IF A$="Y" OR A$="y" THEN 260 ELSE IF A$="N" OR A$="n" THEN 320 ELSE 250
250 PRINT" Y OR N EXPECTED":GOTO 230
260 INPUT"EQUIVALUED CAPACITANCES IS ASSUMED. INPUT VALUE OF C";C
270 C1 = C/1E-09:LPRINT:LPRINT"      C(nF)=";USING"##.####^^^^";C1
280 R1=X1/(1000*C):R2=X2/(1000*C):R3=X3/(1000*C)
290 LPRINT:LPRINT"     R1(K";CHR$(234);") = ";USING"##.#####^^^^";R1
300 LPRINT:LPRINT"     R2(K";CHR$(234);") = ";USING"##.#####^^^^";R2
310 LPRINT:LPRINT"     R3(K";CHR$(234);") = ";USING"##.#####^^^^";R3
320 LPRINT:LPRINT STRING$(52,"_"):END
```

5

Active Filters Using a Single Operational Amplifier

IN THE PRECEDING CHAPTER, we introduced active *RC* filters using a single VCVS of finite gain. The gain of such VCVSs was on the order of 1 to 3. Such a low-gain VCVS can be obtained by providing resistive feedback around the op amp, as shown in Fig. 4.2. Note that the op amp itself can be considered a VCVS whose ideal gain is ∞. In this chapter, we start with filters in which the op amp is used in an infinite gain mode. These types of filters, considered in Secs. 5.1 and 5.2, have the advantage that their passive ω_p and Q_p sensitivities are very low. However, they can be used only for realizing low-Q_p filters because of element spread and the active sensitivity. The most general configuration of filters in which the op amp is used in an infinite gain mode is shown in Fig. 5.1. In later sections we will consider filter configurations in which both positive and negative feedbacks are used. In such filter configurations, the passive sensitivities are higher, but the element spread is not a problem. The active pole sensitivities are also low, as in the Sallen and Key filter. For some of these filters, we will provide design techniques in which the influence of the op amp time constant on the pole can be minimized, and we will also give details of some of these filters in which the active pole sensitivity can be made zero.

The transfer function of the network shown in Fig. 5.1 can be found to be

$$H(s) = \frac{V_o}{V_i} = -\frac{T_{12}}{T_{32} + 1/A} \tag{5.1}$$

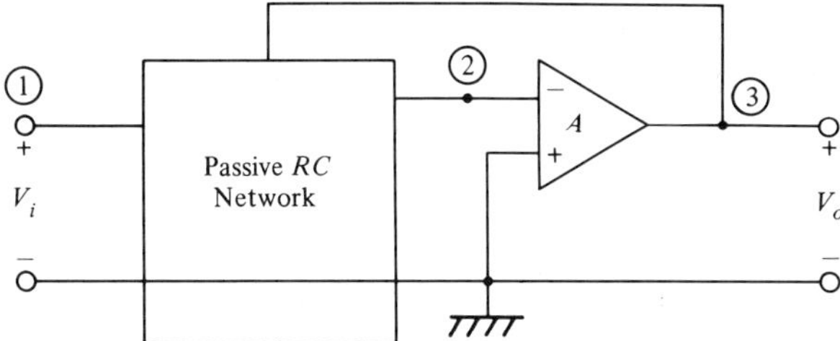

Figure 5.1 General network configuration using a single op amp in its infinite-gain mode.

where

$$T_{12} = \left.\frac{V_2}{V_i}\right|_{V_3=0} \quad \text{and} \quad T_{32} = \left.\frac{V_2}{V_3}\right|_{V_i=0}$$

Under ideal conditions for the op amp, where $A \to \infty$, $H(s)$ becomes

$$H(s) = \frac{V_o}{V_i} = \frac{-T_{12}}{T_{32}} \tag{5.2}$$

Note from (5.2) that, under ideal conditions where $A \to \infty$, the transfer function is completely controlled by the RC network.

5.1 *Second-Order Lowpass, Bandpass, and Highpass Filters*

The network required to realize second-order lowpass, bandpass, and highpass filters using the op amp in an infinite gain mode is shown in Fig. 5.2. In this network, the admittances Y_1 through Y_5 represent a single type of element, either capacitive or resistive. If we consider each of the individual feedbacks from the output of the op amp through Y_2 and Y_5 as

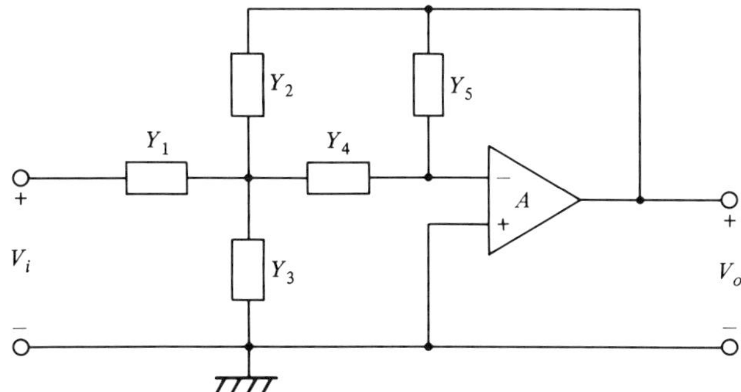

Figure 5.2 General form of multiple-feedback filters.

separate, we can think of this network as having two feedbacks. For this reason, it is also called a *multiple-feedback filter*. If the op amp is assumed to be ideal except for its finite gain of A, the transfer function of the network shown in Fig. 5.2 can be derived as

$$H(s) = \frac{V_o}{V_i} = \frac{-Y_1Y_4}{\begin{array}{l} Y_2Y_4 + Y_5(Y_1 + Y_2 + Y_3 + Y_4) \\ +A^{-1}[(Y_4 + Y_5)(Y_1 + Y_2 + Y_3 + Y_4) - Y_4^2] \end{array}} \tag{5.3}$$

Under ideal conditions where $A^{-1} \to 0$, the above transfer function can be reduced to

$$\frac{V_o}{V_i} = \frac{-Y_1Y_4}{Y_2Y_4 + Y_5(Y_1 + Y_2 + Y_3 + Y_4)} \tag{5.4}$$

Lowpass Filter

A multiple-feedback lowpass filter circuit is shown in Fig. 5.3. Assuming that the op amp is ideal, we can obtain the transfer function for this network. For this purpose we may use (5.4), since the network in Fig. 5.3 is only a particular case of the network in Fig. 5.2 in which $Y_1 = G_1$, $Y_2 = G_2$, $Y_3 = C_1s$, $Y_4 = G_3$, and $Y_5 = C_2s$. Thus, when $A^{-1} \to 0$, the transfer function for the circuit in Fig. 5.3 can be obtained as

$$\frac{V_o}{V_i} = \frac{-G_1G_3S_1S_2}{s^2 + s(G_1 + G_2 + G_3)S_1 + G_2G_3S_1S_2} \tag{5.5}$$

where $G_i = 1/R_i$ and $S_i = 1/C_i$.

Note that there is a negative sign in the numerator of the above transfer function. This means that the output signal will have a $-180°$ additional phase shift compared to the lowpass filter circuit we discussed in Chap. 4.

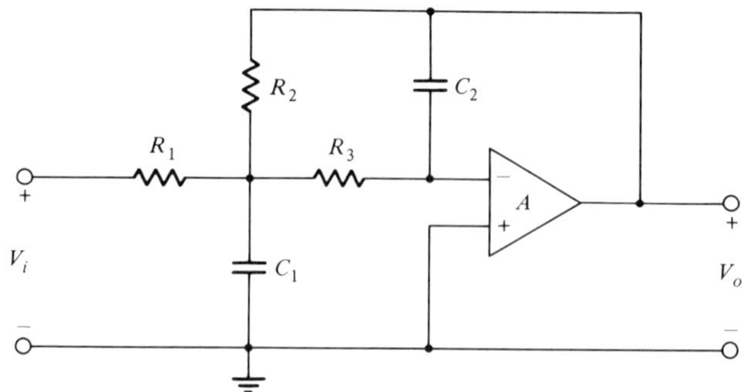

Figure 5.3 Single-amplifier lowpass filter.

This is true not only for this filter but also for every filter we will discuss in this chapter. It does not in any way hamper the filtering functions of these circuits. When we compare the transfer functions of these networks with those in Table 4.1, we do not have to worry about the negative sign and simply ignore it.

Therefore, without loss of generality, we simply ignore this sign and compare the transfer functions of the circuits with those in Table 4.1. Now, comparing the transfer function in (5.5) with the lowpass transfer function in Table 4.1, we obtain the following basic design equations:

$$G_1G_3S_1S_2 = H_o\omega_p^2 \tag{5.6a}$$

$$G_2G_3S_1S_2 = \omega_p^2 \tag{5.6b}$$

and

$$(G_1 + G_2 + G_3)S_1 = \frac{\omega_p}{Q_p} \tag{5.6c}$$

One possible (perhaps the only possible) solution is to assume values for the capacitances such that $C_2 = m^2C_1$ and C_1 is a convenient choice. Then one can solve (5.6) and obtain

$$G_2 = \frac{2m^2Q_p(\omega_pC_1)}{1 \pm \left(1 - 4m^2Q_p^2(1 + H_o)\right)^{1/2}} \tag{5.7a}$$

$$G_1 = H_oG_2 \tag{5.7b}$$

and

$$G_3 = \frac{\left(m\omega_pC_1\right)^2}{G_2} \tag{5.7c}$$

The choice of m^2 should be made such that

$$0 < m^2 \leq \frac{1}{4Q_p^2(1 + H_o)} \tag{5.8}$$

in order for G_2 to be real. Furthermore, we can have either a plus or a minus sign in the denominator of (5.7*a*) as long as G_2 is positive.

Example 5.1: Design a multiple-feedback lowpass filter that satisfies the specifications $f_p = 300$ Hz, $Q_p = 5$, and $H_o = 1$.

Since the choice of m^2 should satisfy (5.8), calculating the right side of (5.8) for this sample, we have $m^2 \leq 5 \times 10^{-3}$. We choose $m^2 = 5 \times 10^{-3}$ so that there is minimum capacitor spread. Choosing 0.1 μF for C_1, we have $C_2 = 500$ pF, $R_2 = R_1 = 106.1$ kΩ, and $R_3 = 53.05$ kΩ. ■

Note the capacitance spread even with a low Q_p of only 5. This example and inequality (5.8) clearly show that this filter is useful only for filter designs with a very low Q_p. The passive ω_p and Q_p sensitivities can easily be shown to be less than or equal to 0.5 with the use of (5.6*b*) and (5.6*c*). This is the major and perhaps the only advantage of this circuit. Next we find the active pole sensitivity. Substituting $A^{-1} = s\tau$ into (5.3) along with substitutions for the admittances, we obtain the following denominator polynomial:

$$D(s,\tau) = s^2 + s(G_1 + G_2 + G_3)S_1 + G_2G_3S_1S_2$$
$$+ s\tau\{s^2 + [G_3S_2 + (G_1 + G_2 + G_3)S_1]s + G_3(G_1 + G_2)S_1S_2\}$$

Applying (3.63) to the above denominator polynomial, we find that

$$S_\tau^p = \frac{-[G_3S_2p + G_1G_3S_1S_2]}{j2\omega_p}$$

Using (5.7), we have

$$S_\tau^p = \frac{j}{2}\left\{\left(\frac{-1}{2Q_p} + j\right)\frac{1}{2Q_pm^2}\left[1 \pm \sqrt{1 - 4m^2Q_p^2(H_o + 1)}\right] + H_o\right\}\omega_p$$

The minimum for the above expression can be obtained with a choice of $m \to \infty$. However, for physical realizability we have to satisfy (5.8) also. Therefore the minimum active pole sensitivity can be obtained with

$$m^2 = \frac{1}{4Q_p^2(1 + H_o)} \tag{5.9}$$

which is also the condition that minimizes the capacitance spread and is the one we used in Example 5.1. If we use this condition, then

$$S_\tau^p|_{\min} = -[Q_p(H_o + 1) + j0.5]\omega_p$$

and, because of the nonzero value of τ, we have

$$\frac{\Delta p}{p} = -[Q_p(H_o + 1) + j0.5](\omega_p\tau)$$

For medium- and high-Q_p filters, even the minimum active pole sensitivity of this circuit may be very high. So both the element spread and the active pole sensitivity of this circuit clearly indicate that we will be able to use it only for designing very low-Q_p filters.

Bandpass Filter

The circuit in Fig. 5.4 is a bandpass filter using the op amp in an infinite gain mode. When $A^{-1} \to 0$, the transfer function of this circuit can be obtained from (5.4) with the substitution of $Y_1 = G_1$, $Y_2 = C_2 s$, $Y_3 = G_3$, $Y_4 = C_1 s$, and $Y_5 = G_2$. Thus we have

$$\frac{V_o}{V_i} = \frac{-(G_1 S_2)s}{s^2 + G_2(S_1 + S_2)s + G_2(G_1 + G_3)S_1 S_2} \tag{5.10}$$

where $G_i = 1/R_i$ and $S_i = 1/C_i$. The basic design equations are

$$G_2(G_1 + G_3)S_1 S_2 = \omega_p^2 \tag{5.11a}$$

$$G_2(S_1 + S_2) = \frac{\omega_p}{Q_p} \tag{5.11b}$$

and

$$G_1 S_2 = H_o \frac{\omega_p}{Q_p} \tag{5.11c}$$

We shall develop the design procedure for a given set of capacitance values. Therefore we define $C_2 = n^2 C_1$. Then, from (5.11), we have

$$G_1 = n^2 \frac{H_o}{Q_p} \omega_p C_1 \tag{5.12a}$$

$$G_2 = \frac{n^2}{n^2 + 1} \frac{1}{Q_p} \omega_p C_1 \tag{5.12b}$$

$$G_3 = \left[Q_p(n^2 + 1) - n^2 \frac{H_o}{Q_p} \right] \omega_p C_1 \tag{5.12c}$$

In order that $G_3 \geq 0$, the following condition must be satisfied for the

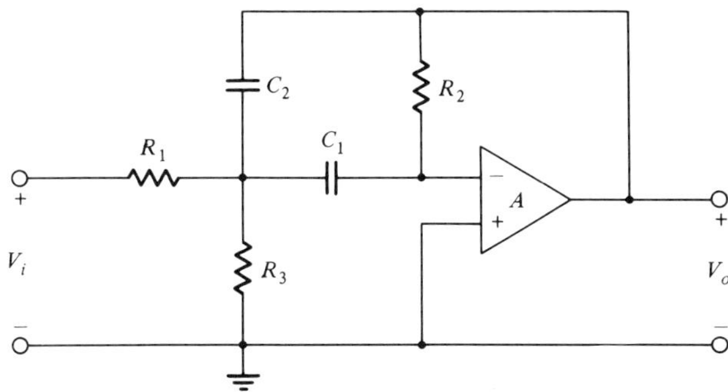

Figure 5.4 Single-amplifier bandpass filter.

choice of n^2:

$$n^2 \geq \frac{-Q_p^2}{Q_p^2 - H_o} \tag{5.13}$$

Under most conditions, it is not difficult to satisfy the above inequality.

Example 5.2: Design a bandpass filter of the type shown in Fig. 5.4 that satisfies the specifications $f_p = 3.4$ kHz, $Q_p = 10$, and $H_o = 5$. Choose $C_2/C_1 = 1$.

Selecting $C_1 = 10$ nF, we have $C_2 = 10$ nF. Then, using (5.12), we have $R_1 = 9.362$ kΩ, $R_2 = 93.62$ kΩ, and $R_3 = 0.2401$ kΩ. ■

In this network, the minimum resistive spread occurs when $n^2 \approx 1$ and is on the order of $4Q_p^2$. The magnitude of the passive ω_p and Q_p sensitivities can be shown to be less than or equal to 0.5, and the active pole sensitivity of this filter can be found to be

$$S_\tau^p = \frac{\omega_p Q_p (n^2 + 1)}{2n^2}\left(-1 - \frac{j}{2Q_p}\right) \tag{5.14}$$

Again we note that the active pole sensitivity is proportional to Q_p, as it was in the case of the lowpass filter discussed earlier. This fact and the element spread limit the use of this design to low-Q_p filters.

A highpass filter circuit, using the op amp in an infinite gain mode, is shown in Fig. 5.5. The analysis and synthesis of this network are similar to that for the previous two networks. The design problems associated with the highpass filter circuit are given as problems at the end of this chapter.

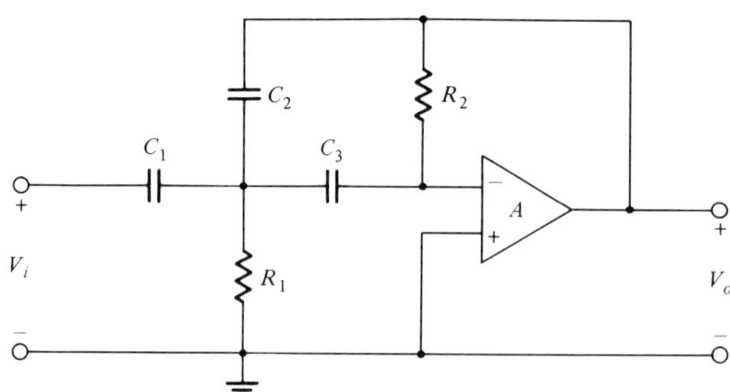

Figure 5.5 Single-amplifier highpass filter.

5.2 Notch Filters

In this section we consider a filter configuration using the op amp in an infinite gain mode, which can be used not only to realize notch filter functions but other filter functions also. In this sense, this configuration can be considered as a universal filter network. It can be used to realize a transfer function of the form

$$H(s) = -H_o \frac{s^2 + (\omega_z/Q_z)s + \omega_z^2}{s^2 + (\omega_p/Q_p)s + \omega_p^2} \tag{5.15}$$

Assuming that the op amp is ideal, the voltage transfer function of the circuit shown in Fig. 5.6 [1] can be derived as

$$\frac{V_o}{V_i} = -\frac{fs^2 + (g/T)s + (b/T^2)}{s^2 + (ad/T)s + [1 + a(b+2)]/T^2} \tag{5.16}$$

where

$$T = RC \tag{5.17a}$$

$$b + 2 = g + e \tag{5.17b}$$

and

$$f + 2 = d \tag{5.17c}$$

The versatility of this circuit can be seen by noting that the numerator coefficients of the transfer function are related to the circuit components in a one-to-one correspondence. Usually, in the case of medium- and high-Q_p filters, the value of a is very small compared to unity. Therefore the

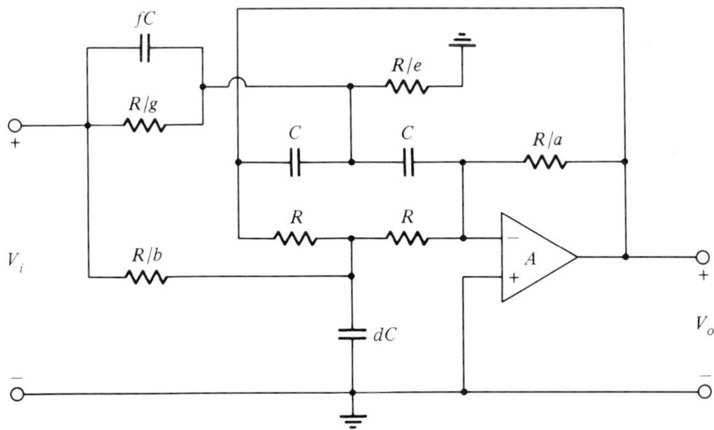

Figure 5.6 Single-amplifier universal biquad.

denominator coefficients can also be controlled almost independent of each other. Comparing (5.16) with (5.15), the basic design equations can be obtained:

$$\omega_p = \frac{\sqrt{1 + a(b + 2)}}{T} \tag{5.18a}$$

$$Q_p = \frac{\sqrt{1 + a(b + 2)}}{ad} \tag{5.18b}$$

$$\omega_z = \frac{\sqrt{b/f}}{T} \tag{5.18c}$$

$$Q_z = \frac{\sqrt{bf}}{g} \tag{5.18d}$$

and

$$H_o = f \tag{5.18e}$$

Using this network configuration, one can realize all types of filter functions. Table 5.1 shows how to select different parameters for realizing the different types.

Equations (5.18) can be solved for a given set of specifications and for a filter of a particular type after selecting some of the parameters as given in Table 5.1. Here we consider the design of notch filters only. In this case, $Q_z \to \infty$, and therefore we choose $g = 0$ as indicated by (5.18d). Further-

Table 5.1 Choice of parameters and type of transfer functions realized by the network in Fig. 5.6

Type	Choice of parameters	Transfer function
Lowpass	$f = g = 0$ $d = 2$ $b + 2 = e$	$\dfrac{V_o}{V_i} = -\dfrac{b/T^2}{s^2 + \dfrac{2a}{T}s + \dfrac{1 + a(2 + b)}{T^2}}$
Highpass	$g = b = 0$ $e = 2$ $d = f + 2$	$\dfrac{V_o}{V_i} = -\dfrac{fs^2}{s^2 + \dfrac{ad}{T}s + \dfrac{1 + 2a}{T^2}}$
Bandpass	$f = b = 0$ $d = 2$ $g + e = 2$	$\dfrac{V_o}{V_i} = -\dfrac{(g/T)s}{s^2 + \dfrac{2a}{T}s + \dfrac{1 + 2a}{T^2}}$
Notch	$g = 0$ $e = b + 2$ $d = f + 2$	$\dfrac{V_o}{V_i} = -\dfrac{fs^2 + b/T^2}{s^2 + \dfrac{ad}{T}s + \dfrac{1 + a(2 + b)}{T^2}}$

more, if the gain constant is not important, we can select a convenient value for f, say, unity. If we choose $f = 1$, then $d = 3$. With these values, we can solve (5.18) to obtain an equation for $\omega_p T$:

$$(\omega_p T)^2 = \frac{1 + 2\omega_p T/3Q_p}{1 - (\omega_z^2/\omega_p^2)(\omega_p T)/3Q_p} \tag{5.19}$$

Once we know the values of ω_p, ω_z, and Q_p, we can solve (5.19) to obtain the value of $\omega_p T$, hence the value of T. The remaining parameters can be found using the equations

$$a = \frac{\omega_p T}{3Q_p} \tag{5.20a}$$

$$b = \omega_z^2 T^2 \tag{5.20b}$$

and

$$e = b + 2 \tag{5.20c}$$

Example 5.3: Using the circuit in Fig. 5.6, design a lowpass notch (LPN) filter that satisfies the transfer function

$$H(s) = -\frac{s^2 + 4.4772 \times 10^9}{s^2 + 860.8s + 4.1223 \times 10^9}$$

From the above transfer function, we find that

$$\omega_p^2 = 4.1223 \times 10^9 \rightarrow \omega_p = 64.205 \times 10^3 \text{ r/s}$$

$$\frac{\omega_p}{Q_p} = 860.8 \rightarrow Q_p = 74.588$$

$$\omega_z^2 = 4.4772 \times 10^9 \rightarrow \omega_z = 66.912 \times 10^3 \text{ r/s}$$

Using the above values of ω_p, ω_z, and Q_p in (5.19), we solve this equation for $\omega_p T$ and obtain

$$\omega_p T = 1.006954$$

Then,

$$a = 4.500 \times 10^{-3}$$
$$b = 1.1013$$

and

$$e = 3.1013$$

Now,

$$RC = T = \frac{1.006954}{\omega_p} = 15.683 \times 10^{-6} \text{ s}$$

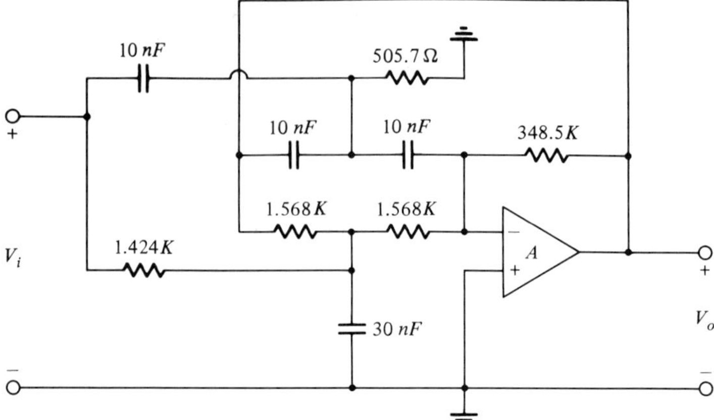

Figure 5.7 Network realizing the LPN function in Example 5.3.

Choosing a standard value of capacitance as $C = 10$ nF, we have

$$R = 1.5683 \text{ k}\Omega$$

Then, the network realizing the transfer function in Example 5.3 will be as shown in Fig. 5.7. ■

5.3 *Single-Amplifier Filter Networks with Both Negative and Positive Feedbacks*

In all the networks we have considered so far, we have been faced with two main problems, namely, a large element spread and high active pole sensitivity, and these two problems limit the usefulness of these circuits to only low-Q_p filters. In this section we will seek remedies for these problems. This is done by providing positive feedback in addition to the negative multiple-feedback already available in the network in Fig. 5.2. To realize all second-order filter functions, namely, lowpass, bandpass, and highpass, a network of the form shown in Fig. 5.8 is used.

If the op amp is ideal except for the finite gain, the transfer function of this circuit can be derived as

$$\frac{V_o}{V_i} = -\frac{Y_1 Y_4}{\begin{array}{r} Y_4(Y_2 + \beta Y_4) + (Y_1 + Y_2 + Y_3 + Y_4)[Y_5 - \beta(Y_4 + Y_5)] \\ + A^{-1}[(Y_4 + Y_5)(Y_1 + Y_2 + Y_3 + Y_4) - Y_4^2] \end{array}} \tag{5.21}$$

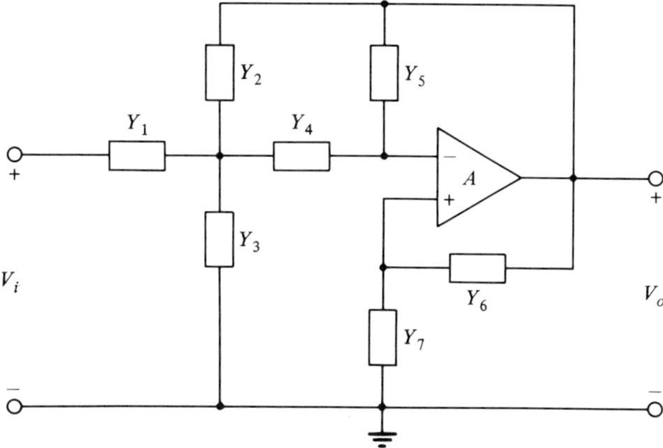

Figure 5.8 Single-amplifier network configuration with both positive and negative feedback.

where

$$\beta = \frac{Y_6}{Y_6 + Y_7} \tag{5.22}$$

Lowpass Filter

To realize a lowpass function, we choose $Y_1 = G_1$, $Y_2 = G_2$, $Y_3 = C_1 s$, $Y_4 = G_3$, $Y_5 = C_2 s$, $Y_6 = G_B$, and $Y_7 = G_D$ in the network in Fig. 5.8. The resulting network will be as shown in Fig. 5.9. When the op amp is ideal in the circuit in Fig. 5.9, the transfer function of the circuit can be obtained from (5.21) for this particular case by substituting the different admittances

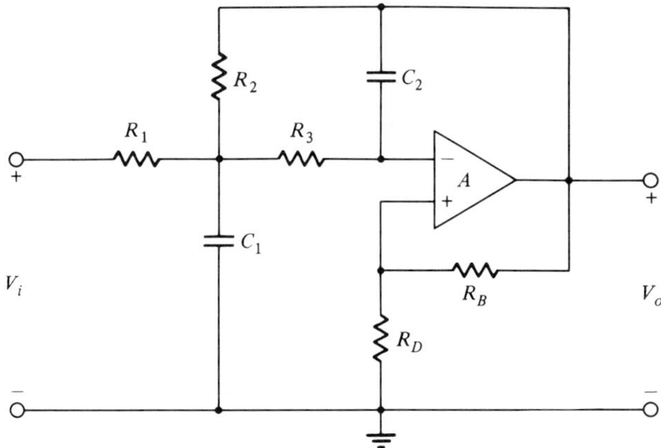

Figure 5.9 Single-amplifier lowpass filter with positive feedback.

and $A^{-1} = 0$. Thus we have

$$\frac{V_o}{V_i} = \frac{-G_1G_3S_1S_2/(1-\beta)}{s^2 + s\left[(G_1+G_2+G_3)S_1 - \dfrac{\beta}{1-\beta}G_3S_2\right] + G_3S_1S_2\left(G_2 - \dfrac{\beta}{1-\beta}G_1\right)} \tag{5.23}$$

where

$$\beta = \frac{G_B}{G_B + G_D}$$

From (5.23), a basic set of design equations can be formed:

$$G_3\left(G_2 - \frac{\beta}{1-\beta}G_1\right)S_1S_2 = \omega_p^2 \tag{5.24a}$$

$$(G_1+G_2)S_1 + G_3\left(S_1 - \frac{\beta}{1-\beta}S_2\right) = \frac{\omega_p}{Q_p} \tag{5.24b}$$

and

$$\frac{G_1}{(1-\beta)G_2 - \beta G_1} = H_o \tag{5.24c}$$

Assuming that our design must meet a given set of specifications for H_o, ω_p, Q_p, a given set of capacitance values, and β, we can solve (5.24) uniquely for G_1, G_2, and G_3. We first define $C_1 = C_2/M$. Then, solving for G_1, G_2, and G_3, we have

$$G_3 = \frac{\omega_pC_2}{2Q_p(M-\lambda)}\left[1 \pm \sqrt{1 - 4Q_p^2(H_o+1)(M-\lambda)}\right] \tag{5.25a}$$

$$G_1 = \frac{H_o(1-\beta)}{M}\,\frac{(\omega_pC_2)^2}{G_3} \tag{5.25b}$$

and

$$G_2 = \frac{1 + H_o\beta}{M}\,\frac{(\omega_pC_2)^2}{G_3} \tag{5.25c}$$

where

$$\lambda = \frac{\beta}{1-\beta} = \frac{G_B}{G_D}$$

In order for G_3 to be real, we have to satisfy the condition [derived from (5.25*a*])

$$0 < M \leq \lambda + \frac{1}{4Q_p^2(H_o + 1)} \tag{5.26}$$

This expression is similar to condition (5.8) for a capacitance ratio without positive feedback. However, (5.26) is much less restrictive compared to (5.8), and in this case it is possible to have equivalued capacitances by choosing β appropriately. Furthermore, we can select either a positive or a negative sign in (5.25*a*). However, choosing a negative sign may result in a smaller element spread and, selecting this sign in (5.25*a*), we have

$$G_3 = \frac{\omega_p C_2}{2Q_p(\lambda - M)}\left[-1 + \sqrt{1 + 4Q_p^2(H_o + 1)(\lambda - M)}\right] \tag{5.27}$$

Once we find the value of G_3 using either (5.27) or (5.25*a*), we can obtain the values of G_1 and G_2 using (5.25*b*) and (5.25*c*), respectively.

Example 5.4: Using the network shown in Fig. 5.9, design a lowpass filter that meets the specifications $f_p = 10$ kHz, $Q_p = 10$, and $H_o = 1$. Choose $\beta = 0.5$ and $M = 0.5$.

The choice of $\beta = 0.5$ will lead to equivalued resistors for R_B and R_D and, of course, $\lambda = 1$. Choosing $C_2 = 1$ nF and $C_1 = 2$ nF such that $M = 0.5$, we obtain $R_3 = 8.366$ kΩ. Then, using (5.25*b*) and (5.25*c*), we find that $R_1 = 30.279$ kΩ and $R_2 = 10.093$ kΩ. The choice for R_B and R_D can be any convenient set of equivalued resistors, for example, 10 kΩ could be used for each one. Note that with positive feedback the element spread is not more than 4. This is one important advantage of a design employing positive feedback compared to one based on the network in Fig. 5.3, which lacks positive feedback. ■

The passive ω_p and Q_p sensitivities of this lowpass filter can be derived as

$$S_{G_1}^{\omega_p} = \frac{-\lambda G_1}{2(G_2 - \lambda G_1)} \qquad S_{G_2}^{\omega_p} = \frac{1}{2} - S_{G_1}^{\omega_p} \qquad S_{G_3}^{\omega_p} = \frac{1}{2}$$

$$S_{C_1}^{\omega_p} = S_{C_2}^{\omega_p} = -\frac{1}{2} \qquad S_{G_D}^{\omega_p} = -S_{G_B}^{\omega_p} = \frac{0.5\lambda G_1}{G_2 - \lambda G_1}$$

$$S_{G_1}^{Q_p} = S_{G_1}^{\omega_p} - \frac{MG_1}{d_1 C_2} \qquad S_{G_2}^{Q_p} = S_{G_2}^{\omega_p} - \frac{MG_2}{d_1 C_2} \qquad S_{G_3}^{Q_p} = \frac{1}{2} + \frac{(\lambda - M)G_3}{d_1 C_2}$$

$$S_{C_1}^{Q_p} = -S_{C_2}^{Q_p} = \frac{M(G_1 + G_2 + G_3)}{d_1 C_2} - \frac{1}{2} \qquad S_{G_B}^{Q_p} = -S_{G_D}^{Q_p} = \frac{\lambda G_3}{d_1 C_2} - S_{G_D}^{\omega_p}$$

where

$$d_1 = (G_1 + G_2 + G_3)S_1 - \lambda G_3 S_2$$

Next, let us find the active pole sensitivity. The denominator polynomial of the transfer function can be obtained with $A^{-1} = s\tau$:

$$\begin{aligned} D(s,\tau) = s^2 &+ s\left[(G_1 + G_2 + G_3)S_1 - \frac{\beta}{1-\beta}G_3S_2\right] \\ &+ G_3S_1S_2\left(G_2 - \frac{\beta}{1-\beta}G_1\right) \\ &+ \frac{s\tau}{1-\beta}\{s^2 + s[(G_1 + G_2 + G_3)S_1 + G_3S_2] \\ &+ G_3(G_1 + G_2)S_1S_2\} \end{aligned} \tag{5.28}$$

Applying (3.63) to (5.28) for the pole $p = -\omega_p/2Q_p + j\omega_p$ and using the basic design equations (5.24), we obtain the semilogarithmic active pole sensitivity as

$$S_\tau^p = \frac{j}{2\omega_p}\frac{1}{(1-\beta)^2}\left[G_3S_2\left(-\frac{1}{2Q_p} + j\right)\omega_p + G_3G_1S_1S_2\right]$$

Next, using (5.27) and (5.25*b*) in the above equation, we have

$$S_\tau^p = \frac{j}{2(1-\beta)^2}\left[\left(-\frac{1}{2Q_p} + j\right)\frac{-1 + \sqrt{1 + 4Q_p^2(H_o + 1)(\lambda - M)}}{2Q_p(\lambda - M)} + H_o(1-\beta)\right]\omega_p \tag{5.29}$$

We shall develop another design procedure in which $|S_\tau^p|$ will be minimum. We first find $|S_\tau^p|$:

$$|S_\tau^p| \approx \frac{\omega_p}{2(1-\beta)^2}\left\{\frac{1}{4Q_p^2(\lambda - M)^2}\left[-1 + \sqrt{1 + 4Q_p^2(H_o + 1)(\lambda - M)}\right]^2 + H_o^2(1-\beta)^2\right\}^{1/2}$$

To minimize $|S_\tau^p|$, we can minimize the function

$$F = \frac{1}{2(1-\beta)^2}\left\{\frac{1}{4Q_p^2(\lambda - M)^2}\left[-1 + \sqrt{1 + 4Q_p^2(H_o + 1)(\lambda - M)}\right]^2 + H_o^2(1-\beta)^2\right\}^{1/2} \tag{5.30}$$

F becomes a minimum when $M \to 0$ for any choice of β, which again is physically impossible. Therefore, after choosing a reasonable but small value of M, we can minimize F with respect to β. Thus differentiating F with respect to β and equating it to zero, we obtain the following highly nonlinear equation:

$$16F + \frac{1}{(1-\beta)^3 F}\left\{-2H_o^2(1-\beta) + \frac{F_1}{(1-\beta)^2(\lambda - M)^2} \times \left[\frac{H_o + 1}{F_1 + 1} - \frac{F_1}{2Q_p^2(\lambda - M)}\right]\right\} = 0 \quad (5.31)$$

where $F_1 = -1 + \sqrt{1 + 4Q_p^2(H_o + 1)(\lambda - M)}$ and F is given by (5.30). For given values of M, H_o, and Q_p, one must resort to numerical techniques to solve (5.31) for β. For example, with $M = 0.2$, $H_o = 1$, $Q_p = 10$, the numerical solution for β is approximately 0.3. Consider the following example.

Example 5.5: Obtain a lowpass filter that meets the specifications in Example 5.4. Choose $M = 0.2$ and $\beta = 0.3$. Note that we are selecting a value of β to minimize the influence of τ on the pole. Obtain the ripple factors of this design and that in Example 5.4 due to a nonzero value of $\tau = 0.5/\pi$ μs. Compare these ripple factors.

With $\beta = 0.3$, we have $\lambda = \frac{3}{7}$. Choosing $C_2 = 1$ nF, we obtain $C_1 = 5$ nF and $R_3 = 5.793$ kΩ. Using this value of R_3 in (5.25*b*) and (5.25*c*), we find that $R_1 = 12.483$ kΩ and $R_2 = 6.727$ kΩ.

Because of the nonzero value of τ,

$$\frac{\Delta p}{p} = S_\tau^p \tau$$

and therefore

$$\left|\frac{\Delta p}{p}\right| = F\omega_p \tau$$

where F is given by (5.30). For this design, $\beta = 0.3$, $Q_p = 10$, $M = 0.2$, $H_o = 1$, and $\lambda = \frac{3}{7}$. Using these numbers in (5.30), we have

$$\left|\frac{\Delta p}{p}\right| = 28.93 \times 10^{-3} \qquad \text{and} \qquad R \approx 0.5786$$

For the design in Example 5.4, we have

$$\left|\frac{\Delta p}{p}\right| = 39.342 \times 10^{-3} \qquad \text{and} \qquad R = 0.7868$$

■

Note that, although we did not worry about $|\Delta p/p|$ for the design in Example 5.4, it is still close to what we obtained in Example 5.5. This is because F is almost flat near $\beta = 0.3$. The reduction in R in Example 5.5 is mainly due to the decrease in the capacitance ratio. The program SALPN in App. 5A can be used to design this lowpass filter with minimum active pole sensitivity. For given values of $\omega_p\tau$, Q_p, and H_o, predistortion is also used as part of this design procedure. Denormalized element values and passive ω_p and Q_p sensitivities can also be obtained for the designed network using the same program.

Example 5.6: Find the magnitudes of the passive pole sensitivities of the two networks designed in Examples 5.4 and 5.5. Assume that $|\Delta R/R| = 0.1\%$ and $|\Delta C/C| = 0.2\%$. Furthermore, assume that $|\Delta\tau/\tau_o| = 15\%$ near a nominal value of $\tau_o = 0.5/\pi$ μs. Find the worst-case per-unit change in $|\Delta p/p|$ in both cases and compare these values.

We know that

$$\left|\frac{\Delta p}{p}\right|_{WC} = \sum_{x_i} |S_{x_i}^p| \left|\frac{\Delta x_i}{x_i}\right|$$

$$= \sum_{R_i} |S_{R_i}^p| \left|\frac{\Delta R_i}{R_i}\right| + \sum_{C_i} |S_{C_i}^p| \left|\frac{\Delta C_i}{C_i}\right| + |S_\tau^p| \left|\frac{\Delta\tau}{\tau_o}\right|$$

where every sensitivity is a logarithmic sensitivity including S_τ^p.

We know how to find the semilogarithmic sensitivity of the pole with respect to τ, as we did in the previous example. Can we find the logarithmic sensitivity using the semilogarithmic sensitivity? To answer this question, consider

Logarithmic sensitivity of p with respect to τ

$$= \left.\frac{\partial p}{\partial \tau}\right|_{\tau=\tau_o} \frac{\tau_o}{p}$$

$$\approx \left.\frac{\partial p}{\partial \tau}\right|_{\tau\approx 0} \frac{\tau_o}{p}$$

$$= (\text{semilogarithmic sensitivity of } p \text{ with respect } \tau) \times \tau_o$$

Thus, for the networks designed in Examples 5.4 and 5.5, the magnitudes of the logarithmic active pole sensitivities are 39.324×10^{-3} and 28.93×10^{-3}, respectively.

Next we must find the pole sensitivities with respect to the passive elements. Consider

$$S_x^p \approx S_x^{\omega_p} - \frac{j}{2Q_p} S_x^{Q_p}$$

Table 5.2 The passive pole sensitivities for Example 5.6

Network design	x	G_1	G_2	G_3	G_B	G_D	C_1	C_2
Example 5.4	$\lvert S_x^p \rvert$	0.288	0.847	0.708	1.053	1.053	1.097	1.097
Example 5.5	$\lvert S_x^p \rvert$	0.202	0.681	0.604	0.647	0.647	0.792	0.792

and therefore

$$|S_x^p|^2 = |S_x^{\omega_p}|^2 + \frac{1}{4Q_p^2}|S_x^{Q_p}|^2$$

Thus, once we calculate $S_x^{\omega_p}$ and $S_x^{Q_p}$, we can obtain $|S_x^p|$. These values have been calculated and are given in Table 5.2 for these two networks.

Therefore the worst-case value of $|\Delta p/p|$ can be found to be

$$\left|\frac{\Delta p}{p}\right|_{WC} = 1.427\% \qquad \text{for Example 5.4}$$
$$= 1.032\% \qquad \text{for Example 5.5}$$

■

Bandpass Filter

A bandpass circuit with positive feedback is shown in Fig. 5.10. The transfer function of this circuit can be obtained from (5.21) by substituting $Y_1 = G_1$, $Y_2 = C_2 s$, $Y_3 = G_3$, $Y_4 = C_1 s$, $Y_5 = G_2$, $Y_6 = G_B$, and $Y_7 = G_D$. Thus we find the transfer function to be

$$\frac{V_o}{V_i} = \frac{-[G_1 S_2/(1-\beta)]s}{s^2 + [G_2(S_1 + S_2) - \lambda(G_1 + G_3)S_2]s + G_2(G_1 + G_3)S_1 S_2} \tag{5.32}$$

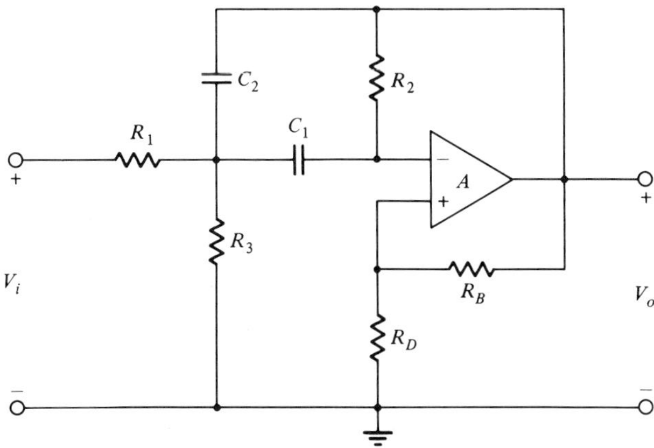

Figure 5.10 Single-amplifier bandpass filter with positive feedback.

when $A^{-1} \to 0$ and $\lambda = \beta/(1-\beta) = G_B/G_D$, $G_i = 1/R_i$, and $S_i = 1/C_i$. The basic design equations can be obtained from (5.32):

$$H_o \frac{\omega_p}{Q_p} = \frac{G_1 S_2}{1-\beta} \tag{5.33a}$$

$$\omega_p^2 = G_2(G_1 + G_3)S_1 S_2 \tag{5.33b}$$

and

$$\frac{\omega_p}{Q_p} = G_2(S_1 + S_2) - \frac{\beta}{1-\beta}(G_1 + G_3)S_2 \tag{5.33c}$$

In the first design procedure we develop for this network, it is assumed that $C_1 = C_2 = C$ and $G_2 = G_1 + G_3 = G$. Then we have a unique solution for the different elements:

$$G = G_2 = \omega_p C \tag{5.34a}$$

$$G_1 = \frac{H_o}{Q_p}(1-\beta)G \tag{5.34b}$$

and

$$G_3 = \left[1 - \frac{H_o}{Q_p}(1-\beta)\right]G \tag{5.34c}$$

where

$$\beta = \frac{2 - 1/Q_p}{3 - 1/Q_p} \tag{5.34d}$$

In order that $G_3 \geq 0$ in this design procedure, the following inequality must be satisfied:

$$H_o + 1 \leq 3Q_p \tag{5.35}$$

Example 5.7: Using the network configuration in (5.10), design a bandpass filter that meets the specifications $H_o = 10$, $Q_p = 20$, and $f_p = 4$ kHz.

We use the design procedure just developed above, and we choose $C_1 = C_2 = C = 10$ nF. Then, using (5.34*a*), we find that $R_2 = 3.979$ kΩ. Before we determine the other element values, we must find the value of β. Using (5.34*d*), we obtain $\beta = 0.66102$. Then we have $R_1 = 23.475$ kΩ and $R_3 = 4.791$ kΩ. Possible values for R_B and R_D are 10 and 19.5 kΩ, respectively. ■

Another possible, perhaps better, design procedure can be developed by selecting convenient values for β and the capacitor ratio C_2/C_1. For example, one may fix the value of β as 0.5 so that R_B and R_D can be realized with any set of equivalued resistors. To develop this design procedure, we first define two parameters m and n:

$$G_2 = m^2(G_1 + G_3) \tag{5.36}$$

and

$$C_2 = n^2 C_1 = n^2 C \tag{5.37}$$

Substituting the above in (5.33*a*) and (5.33*b*), we have

$$G_1 \doteq n^2(1 - \beta)\frac{H_o}{Q_p}\omega_p C \tag{5.38a}$$

and

$$G_2 = mn(\omega_p C) \tag{5.38b}$$

For a given pair of values for β and n^2, if we know the value of m, we will be able to find G_2 and G_3. We use (5.36), (5.37), and (5.38*b*) in (5.33*c*), and the result is a quadratic equation for m [see (5.44)]. Solving this equation, we find that

$$m = \frac{1}{2\rho Q_p} + \sqrt{\left(\frac{1}{2\rho Q_p}\right)^2 + \frac{\beta}{1-\beta}\frac{1}{\rho n}} \tag{5.39}$$

where $\rho = n + 1/n$.

Once we know the values of G_1 and G_2, then we can find G_3 with

$$G_3 = n\left[\frac{1}{m} - n\frac{(1-\beta)H_o}{Q_p}\right](\omega_p C) \tag{5.40}$$

Example 5.8: Using the network configuration in Fig. 5.10, design a bandpass filter that satisfies the specifications in Example 5.7. Choose $\beta = 0.5$ and $n^2 = 1$.

We find the required value of m using (5.39), and it is $m = 0.71972$. Choosing a convenient value of 10 nF for C_1 ($= C_2$ in this case), we find that $R_1 = 15.915$ kΩ, $R_2 = 5.5284$ kΩ, and $R_3 = 3.492$ kΩ. Note that with positive feedback the resistance ratio is considerably lower compared to the case without positive feedback. ■

In the previous two design procedures, two parameters were arbitrarily selected. For example, in the design in Example 5.8, we choose β and n as arbitrary parameters. Instead of selecting them arbitrarily, such as $\beta = 0.5$ and $n^2 = 1$, we can choose them so that the active pole sensitivity is minimized. To develop such a design procedure, we obtain the denominator polynomial from (5.21) with substitution for the different elements and $A^{-1} = s\tau$, where τ is the op amp time constant. Then, in this polynomial, we use (5.36), (5.37), and (5.38b) for G_2. The resultant polynomial $D(s, \tau)$ is of the form

$$D(s, \tau) = \left(s^2 + a\omega_p s + \omega_p^2\right) + \frac{s\tau}{1-\beta}\left(s^2 + b\omega_p s + \omega_p^2\right) \qquad (5.41)$$

where

$$a = \frac{m}{n}\left(n^2 + 1 - \frac{\beta}{1-\beta}\frac{1}{m^2}\right) \qquad (5.42)$$

and

$$b = \frac{m\left(n^2 + 1 + 1/m^2\right)}{n} \qquad (5.43)$$

Since the nominal design is carried out with the assumption that $\tau = 0$, we obtain

$$a = \frac{1}{Q_p} = \frac{m\{n^2 + 1 - \beta/[(1-\beta)m^2]\}}{n} \qquad (5.44)$$

The solution of (5.44) for m gives (5.39). Now, applying (3.63) to (5.41) and using (5.42), (5.43), and (5.44), we obtain the active pole sensitivity of the pole, $p = -\omega_p/2Q_p + j\omega_p$ as

$$S_\tau^p = \frac{-[1 + j/2Q_p]\omega_p}{2mn(1-\beta)^2}$$

Also,

$$\frac{\Delta p}{p} = \frac{-[1 + j/2Q_p](\omega_p \tau)}{2mn(1-\beta)^2} \qquad (5.45)$$

The magnitude of the active pole sensitivity is

$$|S_\tau^p| = \frac{\omega_p}{2mn(1-\beta)^2} = \frac{1}{F}\omega_p \qquad (5.46)$$

where

$$F = 2mn(1-\beta)^2$$

or

$$F = 2(1-\beta)^2\left[\left(\frac{\gamma}{2Q_p}\right) + \sqrt{\left(\frac{\gamma}{2Q_p}\right)^2 + \frac{\gamma\beta}{1-\beta}}\,\right] \tag{5.47}$$

where $\gamma = n^2/(n^2+1) < 1$. To obtain F in terms of β and n only, (5.39) has been used to eliminate m from the expression for F.

The magnitude of the active pole sensitivity can be minimized by selecting β and n^2 such that $1/F$ is minimized. Equivalently, F must be maximized with respect to the two variables β and n^2. For any value of β, F is a maximum when $\gamma = 1$ (by inspection), which is again physically not possible, as in all other previous filters. The expression $\gamma = 1$ implies that $C_2/C_1 \to \infty$. Again we are left with a single-parameter minimization of $1/F$ with respect to β. Since $F \neq 0$, the condition that minimizes $1/F$ is $\partial F/\partial\beta = 0$. Using this fact, we obtain the following equation which can be solved to find the value of β:

$$\frac{\gamma}{2} - 2(1-\beta)\left(\frac{\gamma}{2Q_p}\right)^2 - 2\gamma\beta - \frac{(1-\beta)\gamma}{Q_p}\left[\left(\frac{\gamma}{2Q_p}\right)^2 + \frac{\gamma\beta}{1-\beta}\right]^{1/2} = 0 \tag{5.48}$$

When Q_p is large (> 5), an approximate solution for β is 0.25. However, to find an exact solution for β, we must adopt a numerical method to solve (5.48) for a given Q_p and n^2. The initial value of β for such a numerical solution can be 0.25. Here, F is a nonlinear function of β, and $\partial F/\partial\beta$ is also a nonlinear function. It is neither necessary nor required to find an exact value of β for which $1/F$ will be minimized. To obtain an idea of the approximate range for the value of β, plots of $1/F$ are shown in Fig. 5.11 with $Q_p = 20$ and with various values of n^2. These curves are almost all the same, even with different values of Q_p, since Q_p (> 5) has a negligible effect on $1/F$. For any capacitance ratio, the fact that these curves are shallow in the range $0.2 \le \beta \le 0.3$ allows us to select any convenient value for β in this range with almost a negligible increase in the value of $1/F$. Perhaps the best choice for β is 0.25, so that the ratio of R_B to R_D is 3 : 1, which is a convenient ratio. Furthermore, any minor variation in the value of β near $\beta = 0.25$ will not cause substantial increase in $1/F$, and hence in $|S_\tau^p|$.

We next address the problem of the capacitance ratio n^2. It can be seen from Fig. 5.11 that, as n^2 increases, $1/F$ decreases, especially when $0.2 \le \beta \le 0.3$. However, such an improvement moderates as n^2 increases by more than 2. The plot of $20\log(1/F)$ as a function of $\log n^2$, with $\beta = 0.25$ and $Q_p = 20$, is shown in Fig. 5.12. As can be seen from this plot, the curve falls rapidly for low values of C_2/C_1, and for high values of C_2/C_1 there is no

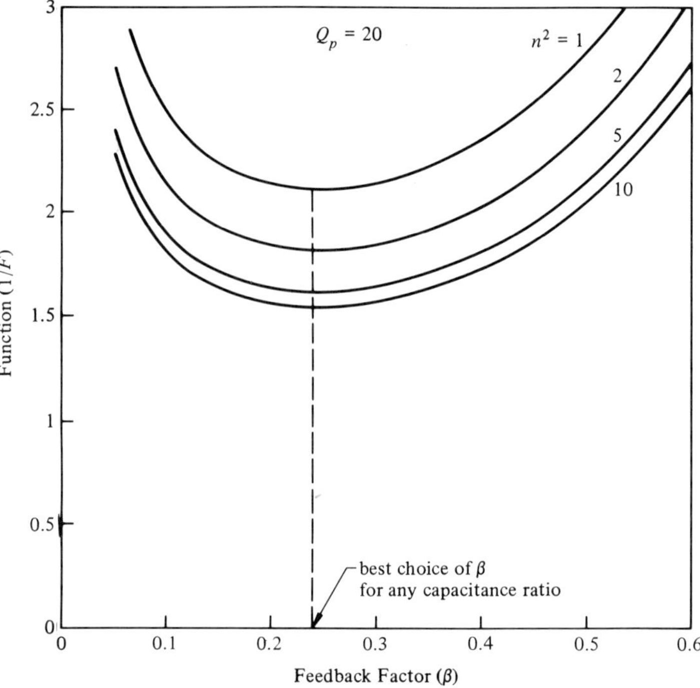

Figure 5.11 Plots of the function $1/F$ as a function of β.

substantial decrease in $1/F$. For practical reasons, it is not good to allow a large capacitance spread. However, a compromise value can be obtained by drawing two asymptotes to this curve as shown in Fig. 5.12. Thus a possible range for n^2 may be $2 \leq n^2 \leq 10$, since there is no substantial decrease in $1/F$ for $n^2 \geq 10$. In conclusion, a suboptimal solution to the problem of minimizing the active pole sensitivity in a single-amplifier bandpass filter can be obtained by choosing β and n^2 such that, for any Q_p value,

$$0.2 \leq \beta \leq 0.3 \qquad \text{and} \qquad 2 \leq n^2 \leq 10$$

Example 5.9: Design a single-amplifier bandpass (SAB) filter that meets the specifications in Example 5.7 such that it has minimum active pole sensitivity. Choose $C_1 = 2$ nF and $C_2 = 10$ nF. Assume that the filter networks designed in this example and Example 5.8 use a 741 type op amp whose typical nonzero value of $\tau = 0.5/\pi$ μs. Find the per-unit changes in ω_p and Q_p and also the ripple factors and then compare these values.

Since $C_2 = 10$ nF and $C_1 = 2$ nF, we have $n^2 = 5$. Hence, $\gamma = \frac{5}{6}$. Using this value of γ and $Q_p = 20$ and starting with $\beta = 0.25$, we solve (5.48) numerically to obtain $\beta = 0.23985$. Since this is an inconvenient

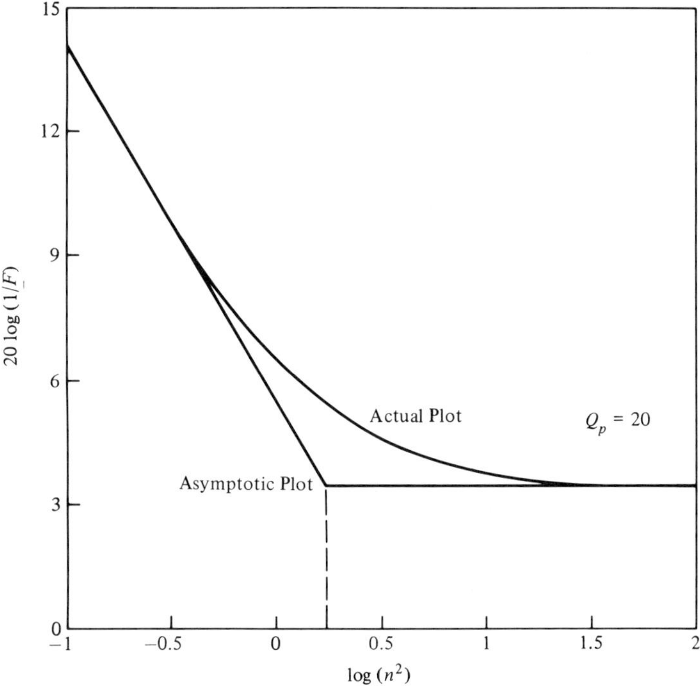

Figure 5.12 Plot of $20\log(1/F)$ as a function of $\log(n^2)$ with $\beta = 0.25$.

value for β, we choose $\beta = 0.24$. With $\beta = 0.24$, $Q_p = 20$, and $n - \sqrt{5}$, we find the value of m using (5.39), and it is $m = 0.23892$. Then, using (5.38) and (5.40), we find that $R_1 = 10.471$ kΩ, $R_2 = 37.238$ kΩ, and $R_3 = 2.6672$ kΩ. The possible choices for R_B and R_D are 19 and 6 kΩ, respectively.

Using (5.45), we find $\Delta p/p$ for the network designed in this example:

$$\frac{\Delta p}{p} = -\left(1 + \frac{j}{2Q_p}\right)(6.4813 \times 10^{-3})$$

Therefore

$$\frac{\Delta\omega_p}{\omega_p} = -6.4813 \times 10^{-3} \qquad \text{and} \qquad \frac{\Delta Q_p}{Q_p} \approx 6.4813 \times 10^{-3}$$

Also, the ripple factor is

$$R \approx 2Q_p\left|\frac{\Delta p}{p}\right| = 0.25925$$

Next, we obtain the above quantities for the network designed in Example 5.8:

$$\frac{\Delta p}{p} = -\left(1 + \frac{j}{2Q_p}\right)(11.115 \times 10^{-3})$$

and, from the above equation,

$$\frac{\Delta\omega_p}{\omega_p} = -11.115 \times 10^{-3} \qquad \text{and} \qquad \frac{\Delta Q_p}{Q_p} \approx 11.115 \times 10^{-3}$$

The ripple factor for this design, because of the nonzero value of τ, is then

$$R = 0.446$$

The effect of the nonzero value of τ on the network designed in Example 5.9 will be only 58.3% of its effect on the network designed in Example 5.8. Note that both networks satisfy the same specifications, so the network designed in Example 5.9 is clearly better. ■

Next we provide formulas for the passive ω_p and Q_p sensitivities:

$$S_{G_1}^{\omega_p} = \frac{1}{2}\frac{G_1}{G_1 + G_3} \qquad S_{G_3}^{\omega_p} = \frac{1}{2} - S_{G_1}^{\omega_p} \qquad S_{G_2}^{\omega_p} = \frac{1}{2}$$

and

$$S_{C_1}^{\omega_p} = S_{C_2}^{\omega_p} = -\tfrac{1}{2}$$

$$S_{G_1}^{Q_p} = S_{G_1}^{\omega_p} + \frac{\lambda G_1}{n^2(d_1 C)} \qquad S_{G_3}^{Q_p} = S_{G_3}^{\omega_p} + \frac{\lambda G_3}{n^2(d_1 C)}$$

$$S_{G_2}^{Q_p} = \frac{1}{2} - \frac{(1 + 1/n^2)G_2}{d_1 C}$$

$$S_{C_1}^{Q_p} = -S_{C_2}^{Q_p} = \frac{G_2}{d_1 C} - \frac{1}{2}$$

and

$$S_{G_B}^{Q_p} = -S_{G_D}^{Q_p} = \frac{\lambda}{n^2}\frac{G_1 + G_3}{d_1 C}$$

where

$$d_1 = G_2(S_1 + S_2) - \lambda(G_1 + G_3)S_2$$

If a nominal design is used, $d_1 = \omega_p/Q_p$. Furthermore, one can use (5.36) through (5.38) to find all the sensitivities in terms of m, n, Q_p, and β:

$$S_{G_1}^{\omega_p} = \frac{1}{2}\frac{H_o}{Q_p}mn(1-\beta) \qquad S_{G_2}^{\omega_p} = \frac{1}{2} \qquad S_{G_3}^{\omega_p} = \frac{1}{2} - S_{G_1}^{\omega_p}$$

and

$$S_{C_1}^{\omega_p} = S_{C_2}^{\omega_p} = -\frac{1}{2}$$

$$S_{G_1}^{Q_p} = H_o\left[\beta + \frac{n(1-\beta)}{2Q_p}\right] \qquad S_{G_2}^{Q_p} = \frac{1}{2} - \rho m Q_p$$

$$S_{G_3}^{Q_p} = -\left(S_{G_1}^{Q_p} + S_{G_2}^{Q_p}\right) \qquad S_{C_1}^{Q_p} = \frac{mnQ_p - 1}{2} = -S_{C_2}^{Q_p}$$

and

$$S_{G_B}^{Q_p} = \frac{\lambda Q_p}{mn} = -S_{G_D}^{Q_p}$$

Assume that we use the predistortion method in Sec. (3.8) to compensate for the nonzero nominal value of τ. Then, the change in the poles will be due to future changes in resistances and capacitances, and to the op amp time constant τ. Such changes are random in nature. However, in integrated circuits, there is a definite correlation between future changes in the same types of elements (such as resistances and capacitances). In the previous design technique for the bandpass filter, we minimized only the influence of the op amp time constant. However, in order to minimize the overall circuit variability, it is necessary to minimize the variance of $|\Delta p/p|$ due to the changes in all the elements, which includes the changes due to resistances, capacitances, and τ. That is, we have to minimize

$$\sigma^2 = E\left(\left|\frac{\Delta p}{p}\right|^2\right) \tag{5.49}$$

We know that

$$\frac{\Delta p}{p} = \sum_{R_i} S_{R_i}^p \frac{\Delta R_i}{R_i} + \sum_{C_i} S_{C_i}^p \frac{\Delta C_i}{C_i} + S_\tau^p \frac{\Delta\tau}{\tau_o}$$

where all the sensitivities are logarithmic sensitivities. If we assume that we have zero cross-correlation, then

$$\sigma^2 = E\left(\left|\frac{\Delta p}{p}\right|^2\right)$$

$$= \sum_{R_i} |S_{R_i}^p|^2 \sigma_R^2 + \sum |S_{C_i}^p|^2 \sigma_C^2 + |S_\tau^p|^2 \sigma_\tau^2 \qquad (5.50)$$

where

$$\sigma_R^2 = E\left(\left|\frac{\Delta R_i}{R_i}\right|^2\right) \qquad \sigma_C^2 = E\left(\left|\frac{\Delta C_i}{C_i}\right|^2\right) \qquad \sigma_\tau^2 = E\left(\left|\frac{\Delta \tau}{\tau_o}\right|^2\right)$$

We will be able to calculate all the sensitivities for given values of ω_p, Q_p, and H_o in terms of n and β. Then, (5.50) can be minimized with respect to β and n. This minimization procedure has been documented by Fleischer [2] and therefore we shall use it. For a given β, he found that the minimum of σ^2 lies on the line $n^2 \to \infty$, which is not physically possible. However, for a chosen $n^2 = 1$, he has provided a simple method of design consisting of the following steps:

1. Calculate

$$Q = \left(\frac{8\sigma_R^2 + \sigma_C^2}{8\sigma_\tau^2 \omega_p^2 \tau_o^2}\right)^{1/4} \qquad (5.51)$$

2. Calculate

$$\beta = \frac{Q_p - Q}{2Q_p(Q)^2 + Q_p - Q} \qquad (5.52)$$

3. Use this value of β and $n^2 = 1$ to design the circuit in Fig. 5.10.

Example 5.10: Using the method of Fleisher, design an SAB bandpass filter that meets the specifications $H_o = 10$, $Q_p = 20$, and $f_p = 4$ kHz. The nonzero value of $\tau = 0.5/\pi$ μs. Assume that $\sigma_\tau = 0.15$, $\sigma_R = 10^{-3}$, and $\sigma_C = 2 \times 10^{-3}$.

Using $\omega_p \tau = 0.004$ and the values of σ_R, σ_C, and σ_τ in (5.51), we have $Q = 1.4287$. Then, from (5.52) we find that $\beta = 0.1853$. Using this value of β and $n^2 = 1$ (that is, $n = 1$) in (5.39), we find that $m = 0.34996$. From (5.38) and (5.40), we obtain $R_1 = 9.768$ kΩ, $R_2 = 11.369$ kΩ, and $R_3 = 1.624$ kΩ. ■

Example 5.11: Calculate the variance for $|\Delta p/p|$ in the networks designed in Examples 5.9 and 5.10. Assume that the values of σ_R, σ_C, and σ_τ are the same as those given in Example 5.10. The value of τ is $0.5/\pi$ μs in each case.

To answer this question, we must calculate the passive ω_p and Q_p sensitivities from which we can obtain $|S_x^p|$. The ω_p and Q_p sensitivities using the formulas developed earlier are given in Table 5.3 for these two networks. The pole sensitivities have also been calculated using (3.41) and are given in Table 5.3.

To obtain a meaningful interpretation of these numbers, we calculate the different parts of the variance as follows.

For Example 5.9:

$$\sum_{R_i} |S_{R_i}^p|^2 \sigma_R^2 = 1.859 \times 10^{-6}$$

$$\sum_{C_i} |S_{C_i}^p|^2 \sigma_C^2 = 4.112 \times 10^{-6}$$

$$|S_\tau^p|^2 \sigma_\tau^2 = 0.945 \times 10^{-6}$$

$$\sum_{R_i} |S_{R_i}^p|^2 \sigma_R^2 + \sum_{C_i} |S_{C_i}^p|^2 \sigma_C^2 = 5.971 \times 10^{-6}$$

and

$$\sigma^2 = 6.916 \times 10^{-6}$$

For Example 5.10:

$$\sum_{R_i} |S_{R_i}^p|^2 \sigma_R^2 - 1.875 \times 10^{-6}$$

$$\sum_{C_i} |S_{C_i}^p|^2 \sigma_C^2 = 4.048 \times 10^{-6}$$

$$|S_\tau^p|^2 \sigma_\tau^2 = 1.668 \times 10^{-6}$$

$$\sum_{R_i} |S_{R_i}^p|^2 \sigma_R^2 + \sum_{C_i} |S_{C_i}^p|^2 \sigma_C^2 = 5.923 \times 10^{-6}$$

and

$$\sigma^2 = 7.591 \times 10^{-6}$$

■

Table 5.3 Pole sensitivity calculations for Example 5.11

Network design	x	G_1	G_2	G_3	G_B	G_D	C_1	C_2	τ
Example 5.9	$S_x^{\omega_p}$	0.425	0.5	0.075	0	0	−0.5	−0.5	
	$S_x^{Q_p}$	2.825	−12.322	9.497	11.822	−11.822	4.842	−4.842	
	$\lvert S_x^p \rvert$	0.431	0.587	0.249	0.296	0.296	0.514	0.514	6.48×10^{-3}
Example 5.10	$S_x^{\omega_p}$	0.204	0.5	0.296	0	0	−0.5	−0.5	
	$S_x^{Q_p}$	2.057	−13.498	11.441	12.998	−12.998	3.00	−3.00	
	$\lvert S_x^p \rvert$	0.210	0.603	0.412	0.325	0.325	0.506	0.506	8.61×10^{-3}

Note that we designed both networks to meet the same set of specifications. Therefore we can draw some conclusions from these numbers:

1. The contributions due to the passive components are almost identical in both networks and are *almost irreducible* (irrespective of the choice of β, m, and n). This is an interesting point.
2. The lower overall variability in the network designed in Example 5.9 is mainly due to the decreased influence of the op amp time constant on the pole frequency. This is achieved mainly by allowing some capacitor spread (see Figs. 5.11 and 5.12). As a matter of interest, it has also been shown [3] that, by allowing such capacitor spread, the noise performance of the circuit in this network can also be improved.
3. The contributions due to the passive components are *almost irreducible*, but the contribution due to the op amp time constant can be minimized. This is one important reason why we concentrated mostly on minimizing $|S_\tau^p|$. Furthermore, assume that we increase the operating frequency of the networks by increasing the value of ω_p with the same value of Q_p, n, and β in each network. This can be achieved by simply rescaling the passive components. The contributions due to the passive components will remain the same. That is, when we rescale the networks to obtain a higher ω_p value, $\Sigma_{R,C}|S_{R,C}^p|^2\sigma_{R,C}^2$ will remain the same as calculated before. However, increasing the operating frequency will increase the contribution due to $|S_\tau^p|$ for a given op amp. For example, if we rescale the components to obtain an f_p value of 25 kHz in both networks without changing the values of β, n, and m in each network, then in Example 5.9,

$$|S_\tau^p|^2\sigma_\tau^2 = 39.914 \times 10^{-6}$$

and in Example 5.10,

$$|S_\tau^p|^2\sigma_\tau^2 = 65.156 \times 10^{-6}$$

These values are far higher than the total contributions of resistances and capacitances combined. Therefore it makes sense to try to minimize the influence of the op amp time constant, especially at higher operating frequencies rather than trying to minimize the total variability. Minimizing the total variability requires many more calculations than minimizing the influence of the op amp time constant alone, and at the same time we do not really achieve much by minimizing the total variability.

As a final matter of interest in regard to this network, consider the following example related to a design using predistortion to compensate for the nonzero value of τ.

Example 5.12: Design an SAB bandpass filter that meets the specifications $f_p = 20$ kHz, $Q_p = 20$, and $H_o = 10$. Choose $C_1 = 1$ nF, $C_2 = 5$ nF, and the value of β that minimizes the active pole sensitivity. Use the predistortion technique to compensate for the nonzero value $\tau = 0.5/\pi$ μs.

To use the predistortion technique we must obtain a design with the assumption that $\tau = 0$. The capacitor ratio is $n^2 = 5$, and $Q_p = 20$. For these values of n^2 and Q_p, we can choose $\beta = 0.24$ (see Example 5.9). Further, the nominal element values can also be found simply by scaling the element values of the resistors and capacitors obtained in Example 5.9. Thus, using the appropriate frequency and impedance magnitude scalings, the required resistors of the nominal design are $R_1 = 4.188$ kΩ, $R_2 = 14.895$ kΩ, and $R_3 = 1.067$ kΩ. To verify these results, (5.33) can be used as a check.

To carry out the predistortion, we must obtain the denominator polynomial in terms of the passive elements:

$$D(s,\tau) = s^2 + s\left[G_2(S_1 + S_2) - \lambda G S_2\right] + G_2 G S_1 S_2 + \frac{s\tau}{1-\beta}\left\{s^2 + s\left[G_2(S_1 + S_2) + G S_2\right] + G_2 G S_1 S_2\right\}$$

where $G = G_1 + G_3$ and $\lambda = \beta/(1 - \beta)$.

Observe that G_1 and G_3 always appear together as $G_1 + G_3$, and we shall treat $G_1 + G_3 = G$ as a single variable. Assuming that we want to use only resistances for predistortion, we have G_2 and G, which are the two variables needed. We can use (3.70) to achieve the predistortion. In using this equation, we substitute the nominal value $\tau = 0.5/\pi$ μs instead of using $\tau = 0$. Further, we calculate

$$\left.\frac{\partial D}{\partial G_2}\right|_{s=p} = G S_1 S_2\left[\frac{p(S_1 + S_2)}{G S_1 S_2} + 1\right]\left(1 + \frac{p\tau}{1-\beta}\right)$$

$$\left.\frac{\partial D}{\partial G}\right|_{s=p} = G_2 S_1 S_2\left[\frac{p\left(-\lambda + \dfrac{p\tau}{1-\beta}\right)}{G_2 S_1} + 1 + \frac{p\tau}{1-\beta}\right]$$

and

$$\left.\frac{\partial D}{\partial \tau}\right|_{s=p} = \frac{G S_2 p^2}{(1-\beta)^2}$$

Using the nominal values of resistors and capacitors other than those of G and G_2 in the above equations, and using $p = \omega_p[-1/2Q_p + j]$ and $\tau = 0.5/\pi$ μs, where the values of ω_p and Q_p are known from the specifications, we have

$$\left.\frac{\partial D}{\partial G_2}\right|_{s=p} = GS_1S_2(0.96647 + j0.66636)$$

$$\left.\frac{\partial D}{\partial G}\right|_{s=p} = G_2S_1S_2(0.96492 - j0.56705)$$

and

$$\left.\frac{\partial D}{\partial \tau}\right|_{s=p} = \frac{GS_2\omega_p}{(1-\beta)^2}(-19.98 \times 10^{-3} - j999.7 \times 10^{-6})$$

Now, using the above equations in (3.70), equating the real and imaginary parts of the resulting equation, and substituting for only the required values, we obtain two equations:

$$0.96647\frac{\Delta G_2}{G_2} + 0.96492\frac{\Delta G}{G} = 66.98 \times 10^{-3}$$

and

$$0.66636\frac{\Delta G_2}{G_2} - 0.56705\frac{\Delta G}{G} = 4.862 \times 10^{-3}$$

Solving these equations, we have

$$\frac{\Delta G}{G} = 33.53 \times 10^{-3}$$

and

$$\frac{\Delta G_2}{G_2} = 35.83 \times 10^{-3}$$

Therefore

$$G_2' = 69.54 \times 10^{-6}$$

and

$$G' = 1.216 \times 10^{-3}$$

Now we have to find G_1' and G_3' such that $G_1' + G_3' = G'$ and

$$\frac{G_1'}{1-\beta}\frac{1}{G_2'G'}\frac{1}{S_1} = \frac{H_o}{\omega_p Q_p}$$

The last equation fixes the required H_o. Thus

$$G_1' = 255.6 \times 10^{-6}$$

and

$$G_3' = 959.9 \times 10^{-6}$$

Thus to compensate for the nonzero values of τ, the resistance values should be $R_1 = 3.912$ kΩ, $R_2 = 14.38$ kΩ, and $R_3 = 1.042$ kΩ. To verify whether, by using these resistance values, the denominator polynomial has the required pole positions, the denominator polynomial and its zeros have been found:

$$D(s) = 209.4 \times 10^{-9} s^3 + 1.0684 s^2 + 10.228 \times 10^3 s + 16.906 \times 10^9$$

Whenever we have such large and such small numbers, calculation of the roots is not accurate. However, we can substitute $s = \omega_p p$ and normalize the denominator polynomial. Thus, carrying out this operation, we have

$$\begin{aligned} D(s = \omega_p p) &= \omega_p^2 (26.316 \times 10^{-3} p^3 + 1.068 p^2 \\ &\qquad + 81.338 \times 10^{-3} p + 1.0705) \\ &= 1.0666 \omega_p^2 (1 + ap)(p^2 + 51.533 \times 10^{-3} p + 1.0032) \end{aligned}$$

The actual pole frequency ω_p' will be

$$\omega_p' = 1.0016 \omega_p$$

and

$$Q_p' = 19.44$$

In order to obtain accurate positioning of the poles, we must use the second iterative method until the dominant roots are the required poles. ■

This circuit can be designed using the program SABPN listed in App. 5B. The necessary inputs are H_o, Q_p, the capacitance ratio n^2, and the value of $\tau_n = \omega_p \tau$. It provides a nominal design as well as a predistorted design with minimum active pole sensitivity. The design assumes a normalized value of $\omega_p = 1$ r/s, however, we can also obtain denormalized element values by using the actual values of ω_p and C_1. We can calculate the passive ω_p and Q_p sensitivities of the design as well and can verify whether the dominant poles have the required ω_p and Q_p values for the normalized design.

Highpass and Lowpass Notch Filters

A general biquad that can realize most filter functions except the lowpass filter function is shown in Fig. 5.13 [4]. The bandpass filter configuration is

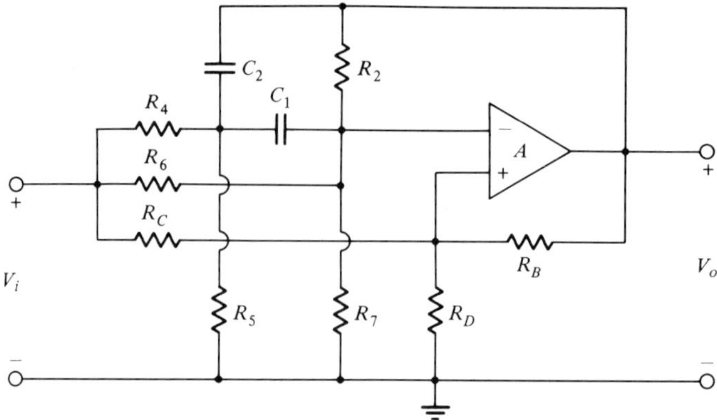

Figure 5.13 Friend's SAB circuit.

a particular case of this configuration, and the highpass filter function can also be realized using this configuration. If it is assumed that the op amp gain is finite, the transfer function of this circuit can be derived as

$$H(s) = \frac{n_2 s^2 + n_1 s + n_0}{D_0(s) + (1+\lambda)A^{-1}D_1(s)} \tag{5.53}$$

where the polynomials $D_0(s)$ and $D_1(s)$ are of the form

$$D_0(s) = s^2 + d_{01}s + d_{00}$$

and

$$D_1(s) = s^2 + d_{11}s + d_{10}$$

The above coefficients are as follows in terms of the element values.

$$d_{01} = (G_2 - \lambda G_3)(S_1 + S_2) - \lambda G_1 S_2 \tag{5.54a}$$

$$d_{00} = G_1(G_2 - \lambda G_3)S_1 S_2 \tag{5.54b}$$

$$d_{11} = G_1 S_2 + (G_2 + G_3)(S_1 + S_2) \tag{5.55a}$$

$$d_{10} = G_1(G_2 + G_3)S_1 S_2 \tag{5.55b}$$

$$n_2 = \sigma \tag{5.56a}$$

$$n_1 = \sigma[G_1 S_2 + (S_1 + S_2)(G_2 + G_3)] - (1+\lambda)[G_4 S_2 + G_6(S_1 + S_2)] \tag{5.56b}$$

and

$$n_0 = G_1[\sigma(G_2 + G_3) - (1+\lambda)G_6]S_1 S_2 \tag{5.56c}$$

where

$$G_1 = G_4 + G_5 \tag{5.57}$$

$$G_3 = G_6 + G_7 \tag{5.58}$$

$$G_A = G_C + G_D \tag{5.59}$$

$$\lambda = \frac{G_B}{G_A} \tag{5.60}$$

and

$$\sigma = \frac{G_C}{G_A} \tag{5.61}$$

Observe from (5.53) that, when $A^{-1} \to 0$, we will be able to realize a transfer function of the form

$$H(s) = H_o \frac{s^2 + (\omega_z/Q_z)s + \omega_z^2}{s^2 + (\omega_p/Q_p)s + \omega_p^2} \tag{5.62}$$

Therefore, assuming an ideal op amp and comparing (5.53) and (5.62), we have the following basic set of design equations:

$$H_o = n_2$$

$$\frac{\omega_z}{Q_z} = \frac{n_1}{n_2}$$

$$\omega_z^2 = \frac{n_0}{n_2}$$

$$\frac{\omega_p}{Q_p} = d_{01}$$

and

$$\omega_p^2 = d_{00}$$

where the different coefficients are given by (5.54) through (5.61) in terms of the element values. Given the values of H_o, ω_z, Q_z, ω_p, and Q_p, we have more elements to choose from than number of equations available and therefore we fix some of the quantities and then solve for the element values in terms of the quantities defined through (5.62). A convenient design procedure is to choose the capacitance values of C_1 and C_2 and the conductance values of G_B and G_A. Given these values, we can solve for other element values:

$$G_1 = \frac{\omega_p C_2}{2\lambda Q_p}\left[-1 + \sqrt{1 + 4Q_p^2(1 + M)\lambda}\right] \tag{5.63}$$

$$G_4 = \frac{H_o}{(1 + \lambda)}\left[G_1 + \frac{(1 + M)\omega_z^2 C_2^2}{G_1} - \frac{\omega_z C_2}{Q_z}\right] \tag{5.64a}$$

For notch filters, where $Q_z = \infty$ and $n_1 = 0$,

$$G_4 = \frac{H_o}{1+\lambda}\left[G_1 + \frac{(1+M)\omega_z^2 C_2^2}{G_1}\right] \tag{5.64b}$$

$$G_3 = \frac{MC_2^2(\omega_z^2 - \omega_p^2)}{(1+\lambda)(1-\delta/H_o)G_1} \tag{5.65}$$

$$G_2 = \frac{M\omega_p^2 C_2^2}{G_1} + \lambda G_3 \tag{5.66}$$

$$G_5 = G_1 - G_4 \tag{5.67}$$

$$G_C = H_o G_A \tag{5.68}$$

and

$$G_D = G_A - G_C \tag{5.69}$$

where

$$M = \frac{C_1}{C_2} \tag{5.70}$$

$$\delta = \frac{G_6}{G_6 + G_7} \tag{5.71}$$

δ is set to 1 for highpass notch (HPN) filters and to 0 for LPN filters. If $\delta = 0$, then $G_6 = 0$, and we do not need resistor R_6 for LPN filters. If $\delta = 1$, then $G_7 = 0$, and we do not need resistor R_7 for HPN filters. The value of G_3 as found from (5.65) will be either equal to G_7 for LPN filters (as $G_6 = 0$) or equal to G_6 for HPN filters (as $G_7 = 0$). This means that, for LPN filter design,

$$G_6 = 0 \tag{5.72a}$$

and

$$G_3 = G_7 = \frac{MC_2^2(\omega_z^2 - \omega_p^2)}{(1+\lambda)G_1} \tag{5.72b}$$

and, for HPN filter design,

$$G_7 = 0 \tag{5.73a}$$

and

$$G_3 = G_6 = \frac{MC_2^2(\omega_p^2 - \omega_z^2)}{(1+\lambda)(1/H_o - 1)G_1} \tag{5.73b}$$

Note that for HPN filter design, we should have

$$H_o < 1 \tag{5.74}$$

and, for any notch filter,

$$H_o \leq \frac{1 + \lambda}{1 + (1 + M)(\omega_z C_2/G_1)^2} \tag{5.75}$$

In fact, G_1 can be eliminated from (5.75) using (5.63), and the condition can be obtained in terms of the specifications and the inputs λ and M. However, it is easy to evaluate G_1 using (5.63) because it does not involve the value of H_o, and then we can verify whether (5.75) is valid. Furthermore, H_o is not really critical and, if necessary, we can lower its value to satisfy the inequalities (5.74) and (5.75) and then proceed with the design.

Example 5.13: Design a highpass notch filter that meets the specifications defined by the following transfer function:

$$H(s) = H_o \frac{s^2 + 36.5337 \times 10^6}{s^2 + 431.484s + 42.4983 \times 10^6}$$

From the above transfer function, we find that

$$\omega_p = 6.5191 \times 10^3 \text{ r/s}$$

$$Q_p = 15.1085$$

and

$$\omega_z = 6.0443 \times 10^3 \text{ r/s}$$

Since $\omega_p > \omega_z$, it is an HPN function.

Assume that $C_1 = C_2 = 5$ nF, $G_A = 10^{-4}$, and $G_B = 10^{-4}/3$. We also have $\lambda = \frac{1}{3}$ and $M = 1$. Then,

$$G_1 = 76.6059 \times 10^{-6}$$

Therefore

$$R_1 = 13.054 \text{ k}\Omega$$

Since there is no requirement that we match the value of H_o, we can choose any value satisfying the two inequalities (5.74) and (5.75). First, we calculate the right side of (5.75), and if it is less than 1, we can choose H_o with an equality sign in (5.75). Then, $G_4 = G_1$, which implies that $G_5 = 0$, and thus we save one resistor (R_5). We find that

$$\frac{1 + \lambda}{1 + (1 + M)(\omega_z C_2/G_1)^2} = 1.01682 > 1$$

Therefore we must make an arbitrary choice for H_o (< 1) and thus

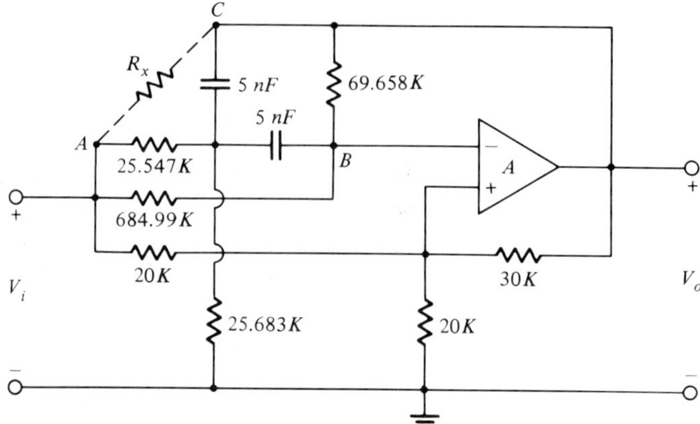

Figure 5.14 Circuit realization designed in Example 5.13.

assume that $H_o = 0.5$. Then,

$$G_4 = 37.6692 \times 10^{-6} \qquad \text{and} \qquad R_4 = 26.547 \text{ k}\Omega$$

For HPN filter design, we set $G_7 = 0$, and from (5.73*b*), we have

$$G_3 = G_6 = 1.45988 \times 10^{-6} \qquad \text{and} \qquad R_6 = 684.987 \text{ k}\Omega$$

Using (5.66), we obtain

$$G_2 = 14.3558 \times 10^{-6} \qquad \text{and} \qquad R_2 = 69.6584 \text{ k}\Omega$$

From (5.67) through (5.69), we have

$$G_5 = 38.9367 \times 10^{-6} \qquad \text{and} \qquad R_5 = 25.683 \text{ k}\Omega$$

$$G_C = 0.5G_A = 0.5 \times 10^{-4} \qquad \text{and} \qquad R_C = 20 \text{ k}\Omega$$

$$G_D = 0.5G_A = 0.5 \times 10^{-4} \qquad \text{and} \qquad R_D = 20 \text{ k}\Omega$$

Of course, $R_B = 30$ kΩ. The circuit realization of the HPN filter designed in this example will be as shown in Fig. 5.14. ■

The resistor spread is 34 to 1, which is acceptable. However, a 684.99-kΩ resistor may be too large. Such large resistances can be avoided by using the following procedure.

Place an arbitrary resistor R_x, as shown in Fig. 5.14, that does not affect the transfer function because it is between two zero impedance sources (the op amp output and the input). Note that R_x, 684.99 kΩ, and 69.658 kΩ form a delta network connected across *AB*, *BC*, and *CA*, which can be converted to an equivalent star network without changing the transfer

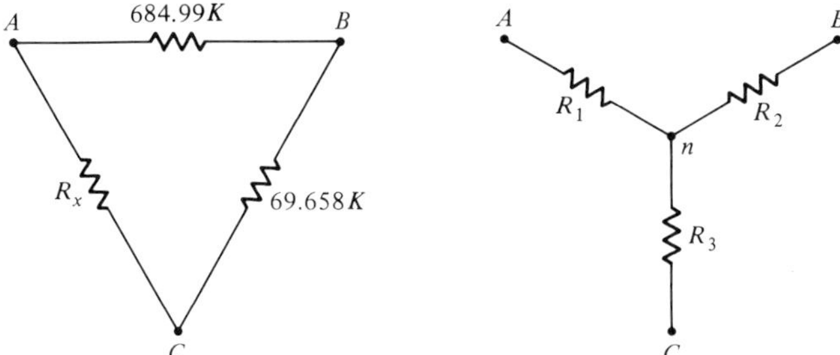

Figure 5.15 $\Delta - Y$ transformation for the network in Fig. 5.14.

function. This procedure creates an additional node and requires three resistors as opposed to two resistors as in Fig. 5.14.

Consider the delta network shown in Fig. 5.15 and its equivalent star network. We must choose R_x such that, when we transform the delta network into its equivalent, the resistances R_1, R_2, and R_3 do not have values that are too large or too small. Furthermore, R_x can be selected such that one of these resistors, R_1, R_2, or R_3, results in a standard value. Therefore, assuming that R_x is a few tens of kiloohms, then either R_1 or R_2 will be a maximum depending on whether R_x is greater or less than 69.658 kΩ. Furthermore, if we want either one to be a standard resistor, for example R_1, then we can choose R_x such that

$$\frac{R_x(684.99)}{R_x + 69.658 + 684.99} = R_1 = R_o$$

where R_o is the standard resistor value. For example, if we choose $R_1 = 56$ kΩ ($= R_o$), then R_x must be equal to 67.188 kΩ. With this value of R_x, we find that $R_2 = 58.059$ kΩ and $R_3 = 5.6948$ kΩ. If this transformed network is used, then the network in Fig. 5.14 will be like the one shown in Fig. 5.16. In the circuit in this figure, not only are the resistance values moderate but the resistance spread also has been reduced to an order of 10 to 1, an interesting fact.

Another design procedure may be developed in which the choices of λ and M are such that the active pole sensitivity will be a minimum. Observe that the zeros are not affected by the nonzero value of τ. First, we develop an expression for S_τ^p:

$$S_\tau^p = \frac{-(1+\lambda)[(d_{11} - d_{01})p + (d_{10} - d_{00})]}{j2\omega_p}$$

$$= \frac{j}{2\omega_p}(1+\lambda)^2\{[G_1S_2 + G_3(S_1 + S_2)]\,p + G_1G_3S_1S_2\} \quad (5.76)$$

The next step is to substitute for G_1 and G_3 from (5.63) and (5.65),

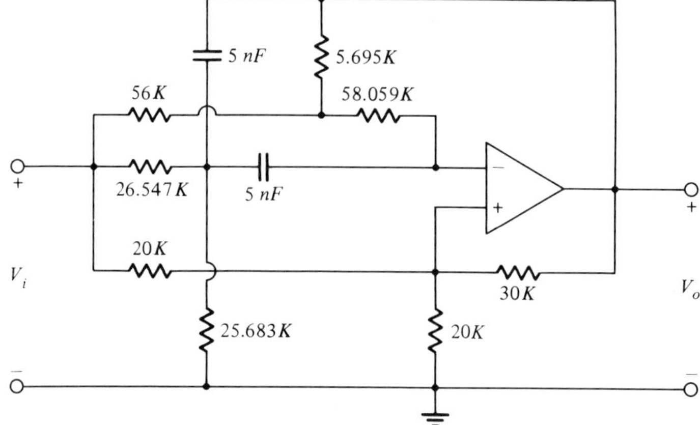

Figure 5.16 Equivalent but better circuit with reduced resistor spread for Example 5.13.

respectively. Then, using the fact that $C_1 = MC_2$, it is possible to find S_τ^p as a function of M, λ, δ, H_o, ω_p, and ω_z only. Then, for a given type of notch filter and for a given set of specifications, we can obtain $|S_\tau^p|$ as a function of M and λ. Such a function is highly nonlinear, and we will not attempt to minimize it here.

However, one can show that $|S_\tau^p|$ will be a minimum as $M \to 0$, which again is not possible. So one can choose a small value of M (but not too small for practical reasons). For any value of M where $|(\omega_z^2 - \omega_p^2)|$ is close to unity, the minimum of $|S_\tau^p|$ is obtained when λ is approximately equal to $\frac{1}{3}$. The actual value of λ for which $|S_\tau^p|$ is a minimum depends on the specifications.

Example 5.14: Calculate the active pole sensitivity of the highpass notch filter designed in Example 5.13. Also, find $\Delta\omega_p/\omega_p$ and $\Delta Q_p/Q_p$ if the nonzero value of $\tau = 0.5/\pi$ μs. Using (5.76), we have

$$S_\tau^p = \frac{j\left(\frac{4}{3}\right)^2\omega_p}{2}\left[\left(-\frac{1}{2Q_p} + j\right)(2.3502 + 0.106) + 0.1053\right]$$

$$= \frac{j}{2}\left(\frac{4}{3}\right)^2\omega_p\left(23.981 \times 10^{-3} + j2.456\right)$$

$$S_\tau^p = \left(-2.1831 + j21.317 \times 10^{-3}\right)(\omega_p)$$

Therefore

$$\frac{\Delta p}{p} = S_\tau^p \tau = \left(-2.1831 + j21.317 \times 10^{-3}\right)(\omega_p \tau)$$

$$= \left(-2.1831 + j21.317 \times 10^{-3}\right)(0.0010375)$$

Thus

$$\frac{\Delta\omega_p}{\omega_p} = -2.265 \times 10^{-3}$$

and

$$\frac{\Delta Q_p}{Q_p} = -0.66831 \times 10^{-3}$$ ■

5.4 *Actively Compensated Single-Amplifier Biquads*

As the value of f_p of single-amplifier filters increases, there may be a considerable shift in the dominant poles realized by these circuits [5]. Such shifts may be minimized by using the techniques described in earlier sections, and they may also be compensated for using the predistortion technique. However, the main problem of active sensitivity will still remain in the network. For higher values of f_p beyond a few kilohertz, the effect of the variation in the op amp time constant on the performance of the circuit may be the sole factor determining the usefulness of the network (see Example 5.11 and the comments following it). To extend the operating frequency range of these circuits, an active compensation technique can be used to improve their high-frequency performance. The general scheme for all the single-amplifier filters discussed in this chapter is shown in Fig. 5.17. When it is assumed that the op amp is ideal except for the finite frequency-dependent gain of A, the transfer function of this circuit can be derived as

$$H(s) = \frac{\sigma - T_{31}}{T_{32} - \beta + s\tau} \tag{5.77}$$

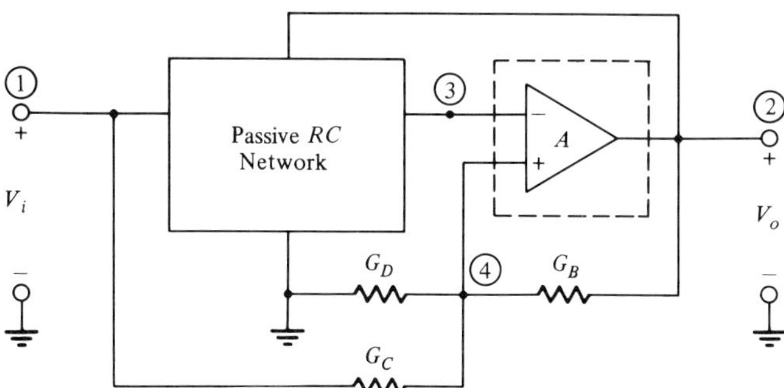

Figure 5.17 General network configuration for all single-amplifier circuits discussed in this chapter.

where

$$\sigma = \frac{G_C}{G_B + G_D + G_C} \qquad \beta = \frac{G_B}{G_B + G_D + G_C}$$

$$T_{32} = \left.\frac{V_3}{V_2}\right|_{V_1=0} \qquad T_{31} = \left.\frac{V_3}{V_1}\right|_{V_2=0}$$

and

$$A^{-1} = s\tau$$

The ideal transfer function can be obtained by substituting $\tau = 0$ into (5.77), and therefore

$$H_I(s) = \frac{\sigma - T_{31}}{T_{32} - \beta} \tag{5.78}$$

Using (5.77), we can easily show that neither the pole nor the transfer function sensitivity with respect to τ is equal to zero. Now, assume that we replace the single op amp in the circuit in Fig. 5.17 with another circuit consisting of two additional resistors and one additional op amp, as shown in Fig. 5.18. The network in Fig. 5.18 is an actively compensated version of the circuit in Fig. 5.17. Again assuming that the op amps used in the circuit in Fig. 5.18 are ideal except for the finite frequency-dependent gains of the op amps, we can obtain the transfer function of the circuit in Fig. 5.18 as

$$H(s) = \frac{N(s)}{D(s)} = \frac{\alpha(\sigma - T_{31}) - s\tau_1 T_{31}}{\alpha(T_{32} - \beta) + s\tau_1[T_{32} - (1 - \alpha) + s\tau_2]} \tag{5.79}$$

where $\alpha = Y_1/(Y_1 + Y_2)$, the other parameters σ, T_{31}, β, and T_{32} are as defined in (5.77), $A_1^{-1} = s\tau_1$, and $A_2^{-1} = s\tau_2$. When $\tau_1 = \tau_2 = 0$, the transfer function of (5.79) reduces to the one given by (5.78). Thus, if the op amps are assumed to be completely ideal, the networks in Figs. 5.17 and

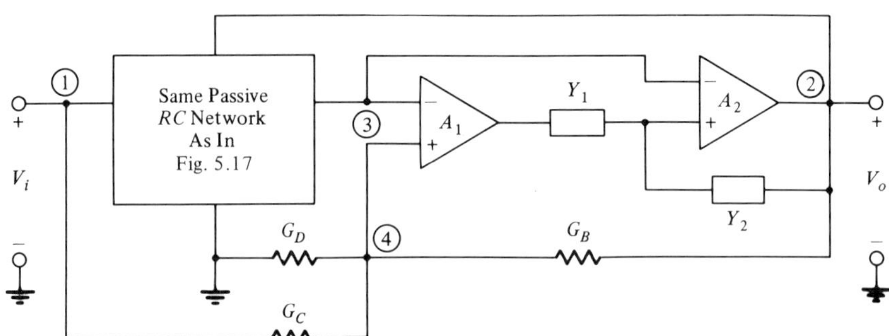

Figure 5.18 Active compensation scheme for single-amplifier filters.

5.18 will realize the same transfer function. In the network in Fig. 5.18, α is a free parameter and can be chosen to minimize and, if possible, to eliminate the sensitivity of $H(s)$ with respect to τ_1 and τ_2. In finding the appropriate value of α, we first determine the *semilogarithmic sensitivities of* $H(s)$ with respect to both τ_1 and τ_2:

$$S_{\tau_1}^{H(s)} = -s\left[\frac{T_{31}}{\sigma - T_{31}} + \frac{T_{32} - (1 - \alpha)}{T_{32} - \beta}\right] \tag{5.80}$$

and

$$S_{\tau_2}^{H(s)} = 0 \tag{5.81}$$

Note that the semilogarithmic sensitivity of $H(s)$ with respect to τ_2 is zero, which will be further discussed later. For now, in order for the semilogarithmic sensitivity of $H(s)$, with respect to τ_1, to go to zero, we must satisfy the condition

$$\frac{T_{31}}{\sigma - T_{31}} + \frac{T_{32} - (1 - \alpha)}{T_{32} - \beta} = 0$$

Using the above condition, we find the required value of α as

$$\alpha = 1 - \beta - \frac{\sigma}{H_I(s)} \tag{5.82}$$

or, equivalently, with $\alpha = Y_1/(Y_1 + Y_2)$, we obtain the condition in terms of admittances as

$$\frac{Y_2}{Y_1} = \frac{\beta + \sigma/H_I(s)}{1 - \beta - \sigma/H_I(s)} \tag{5.83}$$

where $H_I(s)$ is defined in (5.78).

For the lowpass and bandpass filters in Figs. 5.9 and 5.10, respectively, $\sigma = 0$, since $G_C = 0$. Therefore, in these networks, the semilogarithmic sensitivity of the entire transfer function with respect to both τ_1 and τ_2 is zero when

$$\frac{Y_2}{Y_1} = \frac{\beta}{1 - \beta} = \frac{G_B}{G_D} \tag{5.84}$$

This is easy to implement with $Y_2 = G_B$ and $Y_1 = G_D$. For the notch filters in Fig. 5.13, the ratio Y_2/Y_1 must be frequency-dependent, as indicated by (5.83). However, by choosing Y_1 and Y_2 as indicated in (5.84), we can show that the semilogarithmic active pole sensitivities with respect to both τ_1 and τ_2 will be equal to zero, and that the zero of transmission of the notch filters will remain sensitive to the nonzero values of τ_1. This means that, even in the case of notch filters, the transfer function magnitude near ω_p, but not

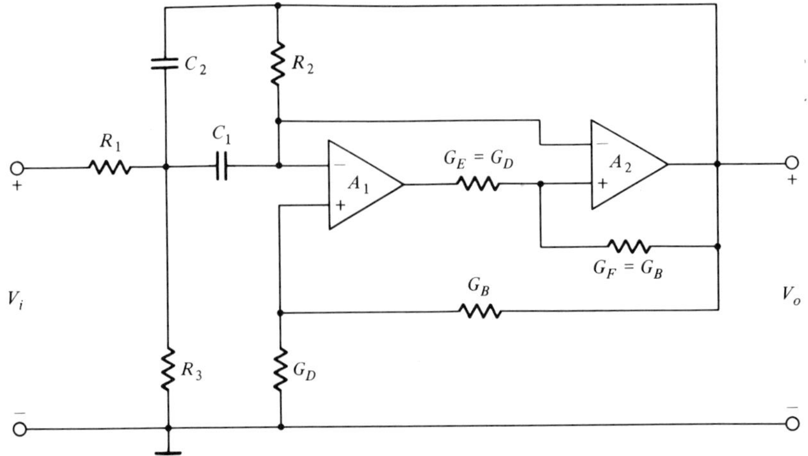

Figure 5.19 Actively compensated single-amplifier bandpass filter.

near ω_z, will be insensitive to τ_1 and τ_2 and that this can be considered acceptable.

Next we shall consider the design of a bandpass filter when active compensation is used. Such a circuit can be obtained by modifying the circuit in Fig. 5.10, and the actively compensated bandpass filter circuit will be as shown in Fig. 5.19. Note that, in this circuit, for the compensation to be successful we must choose R_E and R_F equal to R_D and R_B, respectively, as suggested by (5.84).

If the op amps are considered ideal except for their finite gains, the transfer function of this circuit can be derived as

$$H(s) = \frac{-G_1 S_2 s(1 + s\tau_1/\alpha)/\alpha}{D_0(s) + (s\tau_1/\alpha)D_1(s) + (s^2\tau_1\tau_2/\alpha^2)D_2(s)} \tag{5.85}$$

where

$$A_1^{-1} = s\tau_1 \qquad A_2^{-1} = s\tau_2$$

and

$$D_0(s) = D_1(s) = s^2 + s[G_2(S_1 + S_2) - \lambda S_2(G_1 + G_3)] + G_2(G_1 + G_3)S_1S_2$$

$$D_2(s) = s^2 + s[G_2(S_1 + S_2) + S_2(G_1 + G_3)] + G_2(G_1 + G_3)S_1S_2$$

$$\lambda = \frac{\beta}{1-\beta} = \frac{G_B}{G_D} = \frac{G_F}{G_E}$$

and

$$\alpha = \frac{G_E}{G_E + G_F} = \frac{G_D}{G_D + G_B} = 1 - \beta$$

Note that the above transfer function is reduced to the one given in (5.32) when $\tau_1 = \tau_2 = 0$. This means that the nominal design of the circuit in Fig. 5.19 can be carried out using the same design procedures developed for the bandpass circuit shown in Fig. 5.10. For example, after selecting convenient values for β and $n^2 = C_2/C_1$, the design equations (5.36) through (5.40) can be used to obtain the values of G_1, G_2, and G_3.

Since $D_1(s) = D_0(s)$ in (5.85), we can rewrite the transfer function given by (5.85) as

$$H(s) = \frac{-G_1 S_2 s/\alpha}{D_0(s) + \dfrac{s^2\tau_1\tau_2/\alpha^2}{1 + s\tau_1/\alpha} D_2(s)} \tag{5.86}$$

From (5.86), it can easily be seen that the entire transfer function is not affected by the first-order effects of τ_1 or τ_2 but only by the product of τ_1 and τ_2, that is, the second-order and higher-order effects of τ. Such second-order effects are reflected in the shifts in pole position. The second-order effects can be found from (3.67), and to use this equation without loss of generality let us assume that $\tau_1 = \tau_2 = \tau$. Then, from (5.85), the denominator polynomial will be of the form

$$D(s, \tau) = D_0(s) + \frac{s\tau}{\alpha} D_1(s) + \frac{s^2\tau^2}{\alpha^2} D_2(s) \tag{5.87}$$

where $D_0(s)$, $D_1(s)$, and $D_2(s)$ can be expressed in terms of β, n, and the nominal values of ω_p and Q_p as

$$D_0(s) = D_1(s) = s^2 + \frac{\omega_p}{Q_p} s + \omega_p^2$$

$$D_2(s) = s^2 + b\omega_p s + \omega_p^2$$

where b is defined by (5.43). Note that m and n in (5.43) are defined by (5.39) and (5.37), respectively. Now, applying (3.67) to (5.87), we can obtain the per-unit change in pole $p \approx -\omega_p/2Q_p + j\omega_p$ as

$$\frac{\Delta p}{p} = \frac{j\left(\dfrac{1}{2Q_p} + j\right)^2 \left(b - \dfrac{1}{Q_p}\right)(\omega_p\tau)^2}{2(1-\beta)^2\left(1 - \dfrac{\omega_p\tau}{2Q_p} + j\omega_p\tau\right)}$$

$$\approx \frac{\left[\left(\dfrac{1}{Q_p} - \omega_p\tau\right) - j\right]\left(\omega_p^2\tau^2\right)}{2mn(1-\beta)^3} \tag{5.88}$$

where terms on the order of $1/(4Q_p^2)$, $\omega_p\tau/(2Q_p)$, and $(\omega_p\tau)^2$ have been ignored compared to unity. Further, we have also used (5.43) and (5.44). From (5.88) it can be seen that, to minimize $\Delta p/p$, we must minimize the function $1/F$, where

$$F = 2mn(1 - \beta)^3$$

or

$$F = 2(1 - \beta)^3\left[\frac{\gamma}{2Q_p} + \sqrt{\left(\frac{\gamma}{2Q_p}\right)^2 + \frac{\gamma\beta}{1 - \beta}}\right] \tag{5.89}$$

where (5.39) has been used and $\gamma = n^2/(n^2 + 1) < 1$. The function F is similar to (but not the same as) the one given in (5.47) for the same bandpass filter without active compensation. The maximum of F (and therefore the minimum of $1/F$) occurs, for any value of β, when $\gamma = 1$, which is a physically impossible condition. Therefore, after choosing a large capacitance ratio such as $n^2 = 5$, we can maximize F with respect to β. Equating $\partial F/\partial\beta$ to zero, we solve the following equation to find the value of β:

$$-6(1 - \beta)\left\{\frac{\gamma}{2Q_p}\left[\left(\frac{\gamma}{2Q_p}\right)^2 + \frac{\gamma\beta}{1 - \beta}\right]^{1/2} + \left(\frac{\gamma}{2Q_p}\right)^2 + \frac{\gamma\beta}{1 - \beta}\right\} + \gamma = 0 \tag{5.90}$$

This is a nonlinear equation, and only a numerical solution for β can be obtained. For medium- and high-Q_p filters, we can ignore $(\gamma/2Q_p)^2$ compared to $\gamma\beta/(1 - \beta)$, and then we can obtain an algebraic solution for β in terms of λ:

$$\begin{aligned}\lambda &\approx \left(\sqrt{0.2} - \frac{0.3}{Q_p}\gamma^{1/2}\right)^2 \\ &\approx \tfrac{1}{5} \qquad \text{for high-}Q_p \text{ filters}\end{aligned} \tag{5.91}$$

In fact, it is not necessary to obtain an accurate value of β or λ, since the function $1/F$, for a given value of n^2 (hence γ), is almost flat when β ranges from 0.1 to 0.2. Therefore, without any substantial increase in the function $1/F$, we can choose any convenient value of β such that $0.1 \le \beta \le 0.2$ in this case. So the following simple design steps can be used to minimize the influence of the op amp time constants in the design of actively compensated bandpass filters.

1. Choose a moderate capacitance spread of n^2 such that $2 \le n^2 \le 5$.
2. Choose β in the range $0.1 \le \beta \le 0.2$. The lower value of β will also mean lower passive sensitivities.

The same design steps can also be used in the case of notch filters, because such a procedure leads to minimization of the dominant effects of the op amp on filter performance.

Example 5.15: Using the filter circuit in Fig. 5.19, design a bandpass filter that meets the specifications $f_p = 20$ kHz, $Q_p = 50$, and $H_o = 10$. Choose $\beta = 0.1$ and $n^2 = 2$. Calculate $\Delta p/p$, $\Delta\omega_p/\omega_p$, and $\Delta Q_p/Q_p$ due to the nonzero values of τ when $\tau_1 = \tau_2 = 0.5/\pi$ μs for this design. Also, find the magnitudes of the pole sensitivities with respect to the passive elements.

Since $n^2 = 2$, we have $\rho = 2.121$. Using (5.39), we find that $m = 0.1972$. Choosing $C_2 = 2$ nF and $C_1 = 1$ nF and using (5.38) and (5.40), we obtain $R_1 = 22.105$ kΩ, $R_2 = 28.531$ kΩ, and $R_3 = 1.1684$ kΩ. The values of R_B and R_D are 18 and 2 kΩ, respectively, and the same values can be used for R_F and R_E, respectively. From (5.88), we can find $\Delta p/p$, and since $\omega_p\tau = 1/Q_p$ in this particular example,

$$\frac{\Delta p}{p} = -j983.6 \times 10^{-6}$$

Therefore

$$\frac{\Delta\omega_p}{\omega_p} = 0$$

and

$$\frac{\Delta Q_p}{Q_p} = 98.36 \times 10^{-3}$$

The passive ω_p and Q_p sensitivities can be calculated with the formulas developed in Sec. 5.3 and are listed in Table 5.4. Then, using (3.41) we can find $|S_x^p|$, and these values are also given in Table 5.4. ■

The passive ω_p sensitivities are almost the same as those found in the networks in Examples 5.9 and 5.10. The passive Q_p sensitivities have increased but are not proportional to Q_p. Part of the reason is the reduced value of β allowed by the active compensation. Finally, comparing the magnitudes of the pole sensitivities, they are reduced in this design com-

Table 5.4 Calculation of the passive sensitivities for the design in Example 5.15

x	G_1	G_2	G_3	G_B	G_D	C_1	C_2
$S_x^{\omega_p}$	0.1273	0.5	0.3727	0	0	−0.5	−0.5
$S_x^{Q_p}$	1.127	−20.42	19.29	19.92	−19.92	6.473	−6.473
$\lvert S_x^p \rvert$	0.1278	0.5401	0.4197	0.1992	0.1992	0.5042	0.5042

pared to those in the designs in Examples 5.9 and 5.10, but not substantially. The reason for this is that the magnitudes of the pole sensitivities are dominated by the ω_p sensitivities, which are irreducible. If the passive sensitivities are irreducible, then the only way to improve the performance of the filters is to minimize the dominant sensitivities due to the op amp time constants. In this way, active compensation helps us to extend the operating frequency and the operating Q_p of these filters beyond those of their conventional counterparts using a single op amp. Finally, a comment should be made about the level of active pole sensitivity after active compensation has been used. In general, the logarithmic active pole sensitivity is

$$S_\tau^p \approx \text{semilogarithmic pole sensitivity} \times \tau = \frac{\Delta p}{p}$$

Therefore, in this case, $\Delta p/p$ as given by (5.88) also provides an approximate value of the logarithmic pole sensitivity with respect to τ. That is,

$$S_\tau^p \approx \frac{\left[\left(\dfrac{1}{Q_p} - \omega_p \tau\right) - j\right]\left(\omega_p^2 \tau^2\right)}{2mn(1-\beta)^3} \tag{5.92}$$

The above equation also implies that, for the design in Example 5.15 with $f_p = 20$ kHz (5 times the f_p value in Examples 5.9 and 5.10) and $Q_p = 50$ (2.5 times the Q_p value in Examples 5.9 and 5.10), the active pole sensitivity is 6.6 times lower than that of the design in Example 5.9. In fact, if we calculate the total variance of the pole frequency, it will be dominated by the irreducible passive pole sensitivities. Thus the active compensation technique is able to minimize (and almost eliminate) the dominant op amp influence even in the case of a high Q_p and at a moderately high operating frequency.

Before we conclude our discussion of this topic, we examine the effect of increasing the value of ω_p for a given set of op amps. From (5.88), we find that

$$\frac{\Delta \omega_p}{\omega_p} = \frac{(1/Q_p - \omega_p \tau)(\omega_p \tau)^2}{F} \tag{5.93a}$$

and

$$\frac{\Delta Q_p}{Q_p} = \frac{2Q_p}{F}(\omega_p \tau)^2 \tag{5.93b}$$

where F is defined in (5.89).

The per-unit change in ω_p will be very small, even for high values of ω_p, because it is proportional to $1/Q_p$. However, since $\Delta Q_p/Q_p$ is proportional to Q_p, it can reach large values for high values of $\omega_p \tau$. The changes predicted by (5.93) will occur only when the changes are small ($< 10\%$). However, a more accurate result can be obtained using the fact that the actual pole position p' is equal to $p + \Delta p$, where Δp is calculated using (5.88). Assume that $\omega_p \tau = \tau_n$ and $(\omega_p \tau)^2/F = x$. Then we can show that

$$p' = -\frac{\omega_p}{2Q_p}\left[1 - x\left(2Q_p - \frac{1}{Q_p} + \tau_n\right)\right] + j\omega_p\left[1 + \left(\frac{1.5}{Q_p} - \tau_n\right)x\right]$$

$$\approx \frac{-\omega_p}{2Q_p}(1 - 2Q_p x) + j\omega_p\left[1 + \left(\frac{1.5}{Q_p} - \tau_n\right)x\right]$$

because, usually, $2Q_p \gg \tau_n - 1/Q_p$.

By definition, $p' = -\omega_p'/2Q_p' + j\omega_p'$, where ω_p' and Q_p' are, respectively, the actual (realized) values of the pole frequency and the pole Q factor, respectively. Therefore we have

$$\frac{Q_p'}{Q_p} \cong \frac{1}{1 - 2Q_p x} \tag{5.94a}$$

and

$$\frac{\omega_p'}{\omega_p} \cong 1 + \left(\frac{3}{2Q_p} - \tau_n\right)x \tag{5.94b}$$

From (5.94a), we find that Q_p' will reach ∞ (that is, the circuit will oscillate), when

$$2Q_p x = 1$$

The above condition is equivalent to

$$\frac{\tau_n^2}{(1-\beta)^3\left\{\dfrac{\gamma}{2Q_p^2} + \left[\left(\dfrac{\gamma}{2Q_p^2}\right)^2 + \dfrac{\gamma\beta}{1-\beta}\dfrac{1}{Q_p^2}\right]^{1/2}\right\}} = 1$$

In most cases $(\gamma/2Q_p)^2 \ll \gamma\beta/(1-\beta)$. Therefore the approximate value of τ_n for which the circuit will oscillate is given by

$$\tau_{no} \approx \frac{1}{(Q_p)^{1/2}}(1-\beta)^{3/2}(\gamma\beta)^{1/4} \tag{5.95}$$

For example, when $\beta = 0.1$, $\gamma = \frac{2}{3}$ and $Q_p = 50$, as in Example 5.15, $\tau_{no} = 0.063$. This means that, when we use an op amp for which $\tau = 0.5/\pi$

μs and when we realize the filter circuit designed in Example 5.15 with $f_p = 63$ kHz, the circuit will oscillate. For the same example, when $\tau_n = 0.063$, then $\omega_p' \approx 0.99967\omega_p$. Thus we find that $\omega_p' \approx \omega_p$. Therefore, as in the design of the circuit in Fig. 5.19, even with minimized influence of the op amp time constant, when the operating frequency is increased to a large value, the circuit may suffer from severe Q enhancement. However, such Q enhancement due to the nonzero value of τ can be avoided by using a simple predistortion technique as illustrated next.

Since $\omega_p' \approx \omega_p$ in this case, we can find a design Q_p such that Q_p' is the required value. Assume that Q_p is the value of the pole Q factor with which the circuit is designed and Q_p' is the pole Q factor desired. Then we can use (5.94a) to determine the value of Q_p with which the circuit should be designed. If we use this Q_p, the resultant actual circuit pole Q factor will be close to the required value as given by Q_p'. Thus we can achieve the end result in a simple way. Such a procedure requires solution of the following equation for Q_p and for a given set of values for Q_p', τ_n, β, and γ:

$$Q_p\left\{1 + \frac{Q_p'\tau_n^2}{(1-\beta)^3}\,\frac{1}{\dfrac{\gamma}{2Q_p} + \left[\left(\dfrac{\gamma}{2Q_p}\right)^2 + \dfrac{\gamma\beta}{1-\beta}\right]^{1/2}}\right\} - Q_p' = 0 \quad (5.96)$$

Example 5.16: Using the filter circuit in Fig. 5.19, design a bandpass filter that meets the specifications $f_p = 50$ kHz, $Q_p = 50$, and $H_o = 10$. Choose $\beta = 0.1$ and $n^2 = 2$. Use the simple predistortion technique to compensate for the nonzero values $\tau_1 = \tau_2 = 0.5/\pi$ μs. Find the magnitude of the active pole sensitivity of this design.

First we set $Q_p' = 50$ and determine the value of Q_p using (5.96). To do this, we find that $\tau_n = 0.05$ and $\gamma = n^2/(n^2+1) = \frac{2}{3}$. Therefore, for this example, (5.96) is

$$Q_p\left\{1 + \frac{0.17147}{\dfrac{1}{3Q_p} + \left[\left(\dfrac{1}{3Q_p}\right)^2 + \dfrac{2}{27}\right]^{1/2}}\right\} - 50 = 0$$

The solution of this equation requires an iterative procedure, and the initial value of Q_p can be assumed to be $50/[(1 + 0.17147)(\sqrt{27/2}\,)] = 30.675$. Starting with this value and solving the above equation iteratively, we find that $Q_p = 31.139$. We choose C_1 as 500 pF and C_2 as 1 nF. Then, using $Q_p = 31.129$ in (5.38a) we find that $R_1 = 11.013$ kΩ, and using (5.39) we find that $m = 0.20017$. Therefore $R_2 = 22.489$ kΩ and $R_3 = 981.4\Omega$.

The denominator polynomial of the circuit designed above is

$$D(s, \tau) = \left(s^2 + a\omega_p s + \omega_p^2\right)\left(1 + \frac{s\tau}{1-\beta}\right) + \frac{s^2\tau^2}{(1-\beta)^2}\left(s^2 + b\omega_p s + \omega_p^2\right)$$

where $a = 1/31.139$, $b = 3.9572$, $\omega_p = 314.16 \times 10^3$ r/s, $\tau = 0.5/\pi$ μs, and $\beta = 0.1$.

For convenience, let us substitute $s = \omega_p z$, where z is another complex frequency variable. Then, the equivalent denominator polynomial is

$$\hat{D}(z, \tau_n) = (z^2 + az + 1)\left(1 + \frac{z\tau_n}{1-\beta}\right) + \frac{z^2\tau_n^2}{(1-\beta)^2}(z^2 + bz + 1)$$

$$\triangleq D_0(z) + \frac{z\tau_n}{1-\beta} D_1(z) + \frac{z^2\tau_n^2}{(1-\beta)^2} D_2(z)$$

where $\tau_n = 0.05$. Note that

$$D_0(z)|_{z=-0.01+j} = \frac{j}{31.139} \neq 0$$

In order to find $\Delta p/p$ for this design using (3.67), we must determine the pole position of $z = p$ such that $D_o(p) = 0$:

$$p \approx -\frac{a}{2} + j$$

Since $D_1(z) = D_0(z)$, we have $D_1(p) = 0$. Thus we obtain

$$\begin{aligned}\frac{\Delta p}{p} &= \frac{-p\tau_n^2 D_2(p)}{2j\omega_p(1 + p\tau_n)} \\ &\approx \frac{[(a - \tau_n) - j]\tau_n^2}{2mn(1-\beta)^3} \\ &= (-17.885 \times 10^{-3} - j)(6.0572 \times 10^{-3})\end{aligned}$$

Thus we find that the magnitude of the logarithmic sensitivity is

$$|S_\tau^p| = \left|\frac{\Delta p}{p}\right| \approx 6.0572 \times 10^{-3}$$

Since we have brought the pole positions, including the effect of the nonzero values of τ, to the required positions, what we are interested in now is the future variations in τ. In such cases, the future variation is given by

$$\left|\frac{\Delta p}{p}\right| = |S_{\tau}^{p}|\left|\frac{\Delta \tau}{\tau}\right|$$
$$= (6.0572 \times 10^{-3})\left|\frac{\Delta \tau}{\tau}\right| \tag{5.97}$$

■

Now consider the bandpass filter designed in Example 5.12 using a single-amplifier filter circuit. In this example, the specifications were $f_p = 20$ kHz, $Q_p = 20$, and $H_o = 10$. We chose $n^2 = 5$ (greater than the value used in Example 5.16) and $\beta = 0.24$ to minimize the active pole sensitivity. Let us calculate the active pole sensitivity for this example to evaluate the performance of this design with future changes in τ. The denominator polynomial of this circuit as a function of τ is

$$D(s,\tau) = s^2 + a\omega_p' s + (\omega_p')^2 + \frac{s\tau}{1-\beta}\left[s^2 + b\omega_p' s + (\omega_p')^2\right]$$
$$= D_0(s) + s\tau D_1(s)$$

where $\omega_p' = 130.01$ kr/s, $a = 51.476 \times 10^{-3}$, and $b = 2.5114$. To use (3.62) or (3.63), we must find the pole p such that $D_o(p) = 0$:

$$p = -\frac{a\omega_p'}{2} + j\omega_p'$$

Then, using (3.62), we find that

$$\left|\frac{\Delta p}{p}\right| = 33.487 \times 10^{-3}$$

The magnitude of the logarithmic pole sensitivity is

$$|S_{\tau}^{p}| = 33.487 \times 10^{-3}$$

and therefore the future variation in $|\Delta p/p|$ due to the future variation in τ will be

$$\left|\frac{\Delta p}{p}\right| = |S_{\tau}^{p}|\left|\frac{\Delta \tau}{\tau}\right| = (33.487 \times 10^{-3})\left|\frac{\Delta \tau}{\tau}\right| \tag{5.98}$$

The future variation in $|\Delta p/p|$ due to the future variation in τ will be 5.5 times lower in the actively compensated circuit compared to the same variation in the network designed in Example 5.12 even though both ω_p and Q_p for the actively compensated circuit are 2.5 times the same values in the network designed in Example 5.12. This means that the technique of active compensation of active filters extends the frequency of operation as well as the operating Q_p. Thus actively compensated filters can be used with higher values of ω_p and Q_p safely without much of a problem due to active sensitivity, particularly after predistortion is employed in their design.

5.5 *Conclusions*

We have considered the realization of biquadratic filter functions using single op amps. First we looked at filter networks having only negative feedback, however, such networks are useful for realizing only low-Q_p filters. This drawback can be avoided by adding positive feedback, and networks with both positive and negative feedback work better in terms of both active pole sensitivity and element spread even in medium-Q_p cases. Detailed design techniques for most useful networks have been provided wherein the dominant and reducible active pole sensitivity can be minimized. In Sec. 5.4, we explored the use of a simple active compensation scheme for compensating for the nonzero τ value for the op amp. This scheme does not require matched op amps and eliminates the first-order effects of the τ from the active pole sensitivity. Since the active pole sensitivity is proportional to $(\omega_p\tau)^2$, this compensation extends both the operating frequency and the Q_p value, and therefore such networks have greater potential than their corresponding counterparts that use a single op amp.

REFERENCES

1. Hamilton, T., and Sedra, A. S.: Single amplifier biquad active filter, Proc. Intl. Symposium on Circuits and Systems, pp. 355–359, April 1972.
2. Fleisher, P. E.: Sensitivity minimization in SAB circuits, *IEEE Trans. Circuits Syst.*, vol. CAS-23, pp. 45–55, January 1976.
3. Bacheler, H. J., and Guggenbul, W.: Noise and Sensitivity Optimization of a SAB, *IEEE Trans. Circuits Syst.*, vol. CAS-26, pp. 30–36, January 1979.
4. Friend, J. J.: A Single op amp Biquadratic Filter Section, *IEEE ISCT Digest*, Technical papers, p. 189, 1970.
5. Natarajan, S.: Active Sensitivity Minimization in SABs with Active Compensation and Optimization, *IEEE Trans. Circuits Syst.*, vol. 29, no. 4, pp. 239–245, April 1982.

EXERCISES

5.1. Using the single-amplifier lowpass filter circuit in Fig. 5.3, realize the specifications $f_p = 300$ Hz, $Q_p = 0.707$, and $H_o = 1$. Choose a convenient value for C_1 and choose C_2 to minimize the active pole sensitivity.

5.2. Using the circuit in Fig. 5.4, design a bandpass filter that meets the specifications $f_p = 1$ kHz, $Q_p = 5$, and $H_o = 2$. Choose $C_2 = C_1$. Also, find the active pole sensitivity of the design.

5.3. Consider the network shown in Fig. E5.3, which is an infinite gain realization. Find the transfer function V_2/V_1 and from this transfer function find the values of f_p, Q_p, and H_o.

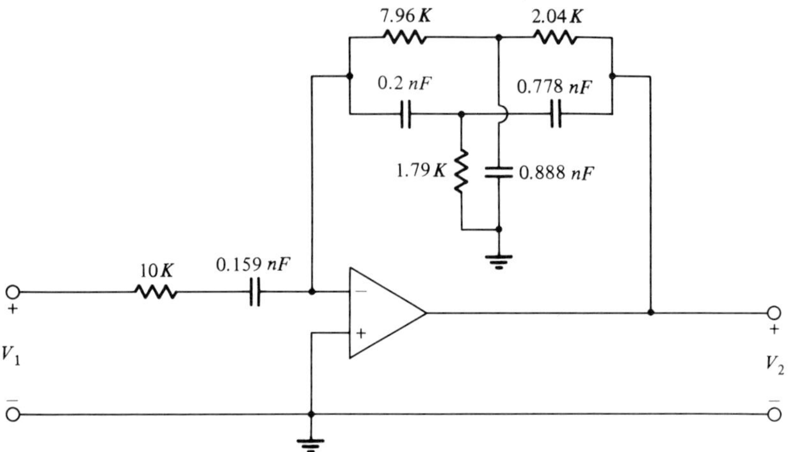

Figure E5.3 Network considered in Exercise 5.3.

5.4. Using the network in Fig. 5.6, design a bandpass filter that meets the specifications in Exercise 5.2.

5.5. Consider the circuit in Fig. 5.9.

(a) Using this circuit, design a lowpass filter that meets the specifications $f_p = 4$ kHz, $Q_p = 10$, and $H_o = 2$. Choose $\beta = 0.5$ and M to satisfy the inequality (5.26).

(b) Repeat the design choosing β to minimize the active pole sensitivity.

(c) Find the active pole sensitivities of both designs and compare them.

(d) Consider the design in part (b). Assume that the op amp has a nonzero value of $\tau = 0.5/\pi$ μs. Use the parameters R_2 and R_3 and predistort the design so that the actual values of f_p and Q_p are the required ones.

5.6. Prove the design equations (5.38) and (5.40) for the bandpass filter circuit in Fig. 5.10.

5.7. It is desired to design a bandpass filter using the circuit in Fig. 5.10. The design specifications are $f_p = 20$ kHz, $Q_p = 10$, and $H_o = 20$.

(a) Obtain a design in which the active pole sensitivity is minimized.

(b) Obtain the passive and active pole sensitivities for the design.

(c) If $\sigma_R = 0.001$, $\sigma_C = 0.002$, and $\sigma_\tau = 0.025$, find the total variance. Assume that $\tau = 0.5/\pi$ μs.

(d) Assume that we want to use predistortion to compensate for the nonzero value of τ in part (c) so that the realized values of ω_p, Q_p, and H_o are as desired. Use G_1, G_2, and G_3 for this purpose and obtain a predistorted design. You may verify your design using the program SABPN.

5.8. Consider the bandpass filter in Fig. 5.10 again.
(a) If $\beta = 0.5$, $C_1 = C_2 = 10$ nF, $R_1 = 6.366$ kΩ, $R_2 = 2.235$ kΩ, and $R_3 = 1.379$ kΩ, find $S_x^{|H(j\omega)|}$ where $x = \beta$, C_1, C_2, R_1, R_2, and R_3.
(b) Assume that $\Delta G_i/G_i = -0.01$, $\Delta C_i/C_i = -0.005$ for all i, and $\Delta\beta/\beta = 0.001$. Find the worst-case change in $|H(j\omega)|$ at the 3-dB frequencies.

5.9. Design an LPN filter using the network in Fig. 5.13 to realize the following transfer function:

$$H(s) = H_o \frac{s^2 + 42.4983 \times 10^6}{s^2 + 431.484s + 36.5337 \times 10^6}$$

Choose $\lambda = \frac{1}{3}$, $C_1 = C_2$, and a value of H_o that satisfies (5.75) with the equality constraint.

5.10. Consider the design of the bandpass filter circuit in Fig. 5.19. Assume that $\beta = G_B/(G_B + G_D)$ and that the two amplifiers have the same nonzero value of $\tau = 0.5/\pi$ μs. Also assume that the parameter values in this circuit are the same as those given in Prob. 5.8a.
(a) Find the active pole sensitivity.
(b) Use predistortion to compensate for the nonzero values of τ in this circuit and determine the new values of R_1, R_2, and R_3.

5.11. Consider the circuit in Fig. 5.19.
(a) Using this circuit, design a bandpass filter that meets the specifications $f_p = 30$ kHz, $Q_p = 20$, and $H_o = 10$. Choose $\beta = 0.1$ and $n^2 = 1$. Calculate $\Delta f_p/f_p$ and $\Delta Q_p/Q_p$ due to the nonzero value $\tau = 0.5/\pi$ μs.
(b) Repeat Exercise 5.10b for this circuit.

5A

Program SALPN

THIS APPENDIX PROVIDES THE DETAILS of the program SALPN, written in GW BASIC, which can be run on an IBM PC or its compatible minicomputer. This program can be used to design the SAB lowpass filter circuit in Fig. 5.9, and the notation for the different elements can be found in that figure. The design procedure uses a predistortion technique* in which the influence of a nonzero τ value for the op amp used is compensated for, and at the same time the magnitude of active pole sensitivity of the filter

$$S_{\tau}^{p} = \frac{\partial p}{\partial \tau}\frac{\tau}{p}$$

is minimized.

The procedure for using this program is simple. Just enter the program and run it. The inputs for the program are H_o, Q_p, $M = C_2/C_1$, and $\tau_n = \omega_p \tau$. These inputs are given through the keyboard when the appropriate question is displayed on the screen.

The design is carried out initially with the assumption that $\omega_p = 1$ r/s and $C_2 = 1$ F. However, one can obtain denormalized element values by inputting the values of ω_p and C_2 through the keyboard. With the use of this program, the passive ω_p and Q_p sensitivities can also be found for the specified design and the design accuracy can be verified by obtaining the

*S. Natarajan, A Simple Predistortion Technique for Active Filters, *IEEE Trans. Circuits Syst.*, vol. CAS-31, no. 4, pp. 398–402, April 1984.

normalized pole frequency and pole Q factor of the dominant poles of the predistorted design.

Following the program is the output for a sample session for the design of a filter circuit satisfying the specifications $f_p = 10$ kHz, $Q_p = 10$, $H_o = 1$, and $M = 0.5$, with $C_2 = 1$ nF, $\tau = 0.5/\pi$ μs, and $\tau_n = \omega_p\tau = 2\pi \times 10^4 \times (0.5/\pi) \times 10^{-6} = 0.01$.

```
10 '
20 '
30 '              PROGRAM LISTING OF 'SALPN'
40 '
50 '
60 'Program to design Single Amplifier Lowpass filter with min. active pole sens
. along with predistortion.
70 PRINT"THIS PROGRAM CAN BE USED TO DESIGN SINGLE AMPLIFIER LOWPASS FILTER WITH
 MINIMUM ACTIVE POLE SENSITIVITY  ALONG WITH PREDISTORTION.":PRINT
80 DIM E$(8),X(5),E(9),D(5),Z(5)
90 E$(1)="R1":E$(2)="R2":E$(3)="R3":E$(4)="C1":E$(5)="C2":E$(6)="RB":E$(7)="RD":
E$(8)=CHR$(225)
100 PRINT"INPUT THE SPECIFICATIONS FOR fp,Qp and H0 AS PROMPTED":PRINT
110 INPUT" fp in Hz";FP:INPUT" Qp";Q:INPUT" H0=";H
120 PI2=8!*ATN(1):WP=FP*PI2:RE=-1/(2*Q):IM=SQR(1-RE*RE)
130 PRINT"R1,R2,R3 AND RD ARE CALCULATED BY THE PROGRAM. INPUT THE OTHER REQUIRE
D PARAMETERS.":PRINT
140 INPUT"VALUE OF C1";E(4):C1=E(4)
150 INPUT"VALUE OF C2(C2<C1)";E(5):C2=E(5):M=C2/C1
160 INPUT"VALUE OF RB";E(6):RB=E(6)
170 PRINT"INPUT THE NON-IDEAL PARAMETERS OF THE OP.AMP.":PRINT
180 INPUT"INPUT THE INVERSE OF THE GB PRODUCT OF THE OP.AMP.IN Hz^(-1).IN AN IDE
AL OP.AMP. IT IS ZERO.";T:T=T*FP
190 INPUT"INPUT THE NON-ZERO OUTPUT RESISTANCE OF THE OP.AMP. IDEALLY ITS VALUE
IS ZERO.";R0:AL=R0/RB:AL1=1+AL:T3=R0*WP*C2
200 PRINT"STRAY CAPACITANCES FROM THE INVERTING AND NON-INVERTING INPUT TERMINAL
S TO GROUND ARE ZEROS IN IDEAL OP.AMPS.. HOWEVER THEIR VALUES MAY RANGE FROM 1 T
O 3 pF. IF YOU WANT TO INCLUDE THEM INPUT THEIR VALUES.":PRINT
210 INPUT"STRAY CAP.FROM NON-INVERTING TERMINAL TO GROUND";CS1:T2=RB*CS1*WP
220 INPUT"STRAY CAP.FROM INVERTING TERMINAL TO GROUND";CS2:TT2=T2*T
230 LPRINT"DESIGN OF A SINGLE AMPLIFIER LOWPASS FILTER":LPRINT STRING$(52,"_")
240 LPRINT:LPRINT"THE SPECIFICATIONS ARE":LPRINT:I=0
250 LPRINT"fp in Hz=";USING"##.#####^^^^";FP:LPRINT
260 LPRINT"Qp=";USING"##.#####^^^^";Q:LPRINT
270 LPRINT"H0=";USING"##.#####^^^^";H:LPRINT
280 LPRINT"C2/C1=";M:LPRINT
290 LPRINT"THE NON-IDEAL PARAMETERS OF THE OP.AMP. INCLUDED FOR PREDISTORTION"
300 LPRINT:LPRINT"RATIO fp/GB=";USING"##.#####^^^^";T:LPRINT
310 LPRINT"OUTPUT RESISTANCE=";USING"####.##";R0:LPRINT
320 LPRINT"STRAY CAP.FROM NON-INVERTING INPUT TERMINAL TO GROUND=";USING"##.####
#^^^^";CS1:LPRINT
330 LPRINT"STRAY CAP.FROM INVERTING INPUT TERMINAL TO GROUND=";USING"##.#####^^^
^";CS2:LPRINT STRING$(52,"_"):LPRINT:C3=C2+CS2:E(9)=C3
340 B=.5:H1=H+1:H2=H*H:Q2=4*Q*Q:Q3=H1*Q2:J=0
350 B1=1-B:B2=B1*B1:L=B/B1-M:F1=SQR(1+Q3*L)-1:F=SQR(F1*F1/(Q2*L*L)+H2*B2)/(2*B2)
360 F2=16*F+(-2*H2*B1+F1*(H1/(F1+1)-F1/(Q2*L/2))/(B2*L*L))/(B2*B1*F)
370 IF F2=0 THEN 410 ELSE IF J=1 THEN 390
380 B3=B:B=1.0001*B:J=1:F3=F2:GOTO 350
390 B4=B3*(F2-1.0001*F3)/(F2-F3):B=B4
400 IF ABS(B/B4-1)<.001 THEN 410 ELSE J=0:GOTO 350
410 B=(INT(1000*B)+1)/1000:E(8)=B
420 J=1:J1=0:J2=1:J3=0
430 'Nominal Design equations
440 F1=M-1/Q3:BX=F1/(F1+1):IF B>=BX THEN 450 ELSE B=BX
450 F=1-B:L=B/F:G=L-M:G3=(SQR(1+Q3*G)-1)/(2*Q*G):G1=H*F/(M*G3):G2=(1+H*B)/(M*G3)

460 IF I=0 THEN 470 ELSE 520
470 SC1=1/(WP*E(5)):E(1)=SC1/G1:E(2)=SC1/G2:E(3)=SC1/G3:E(7)=RB/(1/B-1)
480 I=1:LPRINT"NOMINAL DESIGN PARAMETERS FOR MIN.ACTIVE POLE SENS.":LPRINT
490 FOR L=1 TO 8:LPRINT E$(L);"=";USING"##.#####^^^^";E(L):LPRINT:NEXT L
500 LPRINT STRING$(52,"_"):LPRINT
510 ' Denominator coefficients and their derivatives w.r.t. G2 and G3.
520 K=1/B:CX=SQR(E(4)*(K*E(5)-E(9))):CY=E(5)/CX:SC=WP*CX:G1=G1*CY:G2=CY*G2
530 G3=G3*CY:R0=R0*SC:C1=E(4)/CX:C3=E(9)/CX:C2=E(5)/CX:CS2=C3-C2
540 BE=G2*R0:G=G1+G2+G3:DV=G2*G3:DX=C3*G+C1*G3:DY=G3*(G1+G2):DW=AL1+BE:R=K*DW-AL
550 X(0)=K*DV-DY:X(1)=K*C2*G-DX+T*(DY*R-K*BE*DV)
560 X(2)=1+T*(DX*R+K*(DY*T3-BE*G2*C3))+TT2*(DW*DY-BE*DV)+C2*G*T2
```

```
570 X(3)=T*(R*C1*C3+K*DX*T3)+TT2*(DW*DX-BE*G2*C3+DY*T3)
580 X(4)=K*T*T3*C1*CS2+TT2*(DX*T3+DW*C1*C3):X(5)=C1*CS2*TT2*T3:ND=5:GOSUB 1150
590 X1=RD1:X2=RD0
600 X(0)=(K-1)*G3:X(1)=(K*C2-C3)+T*G3*(R-2*BE*K):X(2)=C3*T*R+K*T*(G3*T3-2*BE*C3)

610 X(3)=K*CS2*T*T3+(DW*C3+T3*G3-2*BE*C3)*TT2:X(4)=TT2*(CS2*T3+BE*C1*C3/G2)
620 X(5)=0:GOSUB 1150
630 A=RD1:B=RD0
640 G=G-G3:X(0)=K*G2-G:X(1)=K*C2-C1-C3+T*(G*R-K*BE*G2):X(4)=TT2*T3*(C1+CS2)
650 X(2)=((C1+C3)*R+K*G*T3)*T+TT2*(DW*G-BE*G2)+C2*T2
660 X(3)=K*T*(C1+CS2)*T3+TT2*(DW*(C1+C2)-BE*C2+G*T3):X(5)=0:GOSUB 1150
670 Y=RD1:D=RD0:DG3=(X2/B-X1/A)/(Y/A-D/B):DG2=-(X1+Y*DG3)/A:G2=G2+DG2:G3=G3+DG3
680 IF ABS(DG2/G2)>.0001 OR ABS(DG3/G3)>.0001 THEN 540
690 GX=(K-1)*G2/(1+K/H):IF ABS(GX/G1-1)>.0001 THEN G1=GX:GOTO 540
700 G1=GX:BE=R0*G2:G=G1+G2+G3:DV=G2*G3:DX=C3*G+C1*G3:DY=G3*(G1+G2)
710 DW=AL1+BE:DZ=K*DW-AL:X(0)=0:X(1)=DZ*DY-K*BE*DV
720 X(2)=DX*DZ+K*(DY*T3-BE*G2*C3)+T2*(DW*DY-BE*DV)
730 X(3)=DZ*C1*C3+K*DX*T3+T2*(DW*DX-BE*G2*C3+DY*T3)
740 X(4)=K*T3*C1*CS2+(DX*T3+DW*C1*C3)*T2:X(5)=C1*CS2*T2*T3:GOSUB 1150
750 A=RD1:B=RD0
760 X(5)=X(5)*T:X(4)=X(4)*T:X(3)=X(3)*T:X(2)=X(2)*T+1+G*C2*T2
770 X(1)=X(1)*T+K*C2*N-DX:X(0)=K*G2*G3-DY
780 IF J2=0 AND J3=0 THEN 790 ELSE 810
790 D(4)=X(4):D(3)=X(3):D(2)=X(2):D(1)=X(1):D(0)=X(0):J3=1:K4=K
800 SCX=1/SC:E(1)=SCX/G1:E(2)=SCX/G2:E(3)=SCX/G3:E(7)=RB/(K-1)
810 X(0)=X(1):X(1)=2*X(2):X(2)=3*X(3):X(3)=4*X(4):X(4)=5*X(5):ND=4:GOSUB 1150
820 Y=RD1:R=RD0:S=SQR(((A*RE+B)^2+(A*IM)^2)/((Y*RE+R)^2+(Y*IM)^2))
830 IF J=0 THEN 850
840 J=0:K=K*1.0001:S1=S:GOTO 540
850 IF J1=2 THEN 920
860 K=K/1.0001:P1=(S-S1)*10000!/K:S7=S1:K1=K:P4=P1:P3=S7
870 IF J2=0 THEN 990 ELSE IF J1=1 THEN 970
880 IF ABS(P1)<.001 THEN 980 ELSE IF ABS(P1/K)>.1 THEN 900
890 V=1-P1/K:GOTO 910
900 IF P1>0 THEN V=.9 ELSE V=1.1
910 K=V*K:J1=2:J=1:B=1/K:GOTO 440
920 K=K/1.0001:P2=(S-S12)*10000!/K:S8=S1:K2=K:P4=P2:P3=S8
930 IF ABS(P2)<.001 THEN 980 ELSE IF ABS(P2)<ABS(P1) THEN 950
940 K=K1-P1*(K2-K1)/(P2-P1):K3=K1:P3=S7:GOTO 960
950 K=K2-P2*(K2-K1)/(P2-P1):K3=K2:P3=S8
960 IF ABS(K3/K-1)<.001 THEN 980 ELSE J1=1:J=1:B=1/K:GOTO 440
970 IF ABS(P3/S1-1)<.001 THEN 980 ELSE 910
980 K=(INT(1000*K)+1)/1000:J2=0:J=1:J1=0:B=1/K:GOTO 440
990 LPRINT"ELEMENT VALUES WITH MINIMUM ACTIVE POLE SENSITIVITY":LPRINT
1000 LPRINT"AFTER PREDISTORTION IS APPLIED TO THE LOWPASS FILTER":LPRINT
1010 E(7)=RB/(K4-1):E(8)=1/K4
1020 FOR L=1 TO 8:LPRINT E$(L);"=";USING"##.#####^^^^";E(L):LPRINT:NEXT L
1030 LPRINT"MAGNITUDE OF MIN. ACTIVE POLE SENSITIVITY=";USING"##.#####^^^^";S7*T

1040 PRINT"DO YOU WANT TO KNOW THE Wp AND Qp SENSITIVITIES WITH RESPECT TO PASSI
VE          ELEMENTS? Y/N";:INPUT A$
1050 IF A$="N" OR A$="n" THEN 1130
1060 LPRINT STRING$(52,"_"):LPRINT"Wp AND Qp SENSITIVITIES ARE":LPRINT
1070 G1=1/E(1):G2=1/E(2):G3=1/E(3):G=G1+G2+G3:L=1/(K4-1):E(1)=L*G1/(2*(G2-L*G1))
1080 E(2)=-E(1)-.5:E(3)=-.5:E(4)=-.5:E(5)=-.5:E(6)=2*E(1):E(7)=-E(6)
1090 FOR I=1 TO 7:LPRINT"Wp sens. w.r.t. ";E$(I);"=";E(I):LPRINT:NEXT I
1100 D1=G*M-L*G3:E(1)=M*G1/D1+E(1):E(2)=E(2)+M*G2/D1:E(3)=-.5-(L-M)*G3/D1
1110 E(4)=M*G/D1-.5:E(5)=-E(4):E(6)=-L*G3/D1+E(6):E(7)=-E(6)
1120 FOR I=1 TO 7:LPRINT"Qp sens. w.r.t. ";E$(I);"=";E(I):LPRINT:NEXT I
1130 LPRINT STRING$(52,"_"):END
1140 'subroutine to divide a poly. with quadratic factor of s^2+s/Q+1
1150 Z(ND)=0:Z(ND-1)=0
1160 FOR L=ND-2 TO 0 STEP -1
1170 Z(L)=X(L+2)-Z(L+1)/Q-Z(L+2)
1180 NEXT L
1190 RD1=X(1)-Z(0)/Q-Z(1):RD0=X(0)-Z(0)
1200 RETURN
```

A SAMPLE SESSION

```
DESIGN OF A SINGLE AMPLIFIER LOWPASS FILTER
____________________________________________________

THE SPECIFICATIONS ARE

fp in Hz= 1.00000E+04

Qp= 1.00000E+01

H0= 1.00000E+00

C2/C1= .5
```

```
THE NON-IDEAL PARAMETERS OF THE OP.AMP. INCLUDED FOR PREDISTORTION

RATIO fp/GB= 1.00000E-02

OUTPUT RESISTANCE=  75.00

STRAY CAP.FROM NON-INVERTING INPUT TERMINAL TO GROUND= 0.00000E+00

STRAY CAP.FROM INVERTING INPUT TERMINAL TO GROUND= 0.00000E+00
```

```
NOMINAL DESIGN PARAMETERS FOR MIN.ACTIVE POLE SENS.

R1= 3.11109E+04

R2= 1.07611E+04

R3= 7.92017E+03

C1= 2.00000E-09

C2= 1.00000E-09

RB= 1.00000E+04

RD= 9.45525E+03

β= 4.86000E-01
```

```
ELEMENT VALUES WITH MINIMUM ACTIVE POLE SENSITIVITY

AFTER PREDISTORTION IS APPLIED TO THE LOWPASS FILTER

R1= 3.02213E+04

R2= 1.04559E+04

R3= 7.54830E+03

C1= 2.00000E-09

C2= 1.00000E-09

RB= 1.00000E+04
RD= 9.45180E+03

β= 4.85909E-01

MAGNITUDE OF MIN. ACTIVE POLE SENSITIVITY= 1.77455E-02
```

```
Wp AND Qp SENSITIVITIES ARE

Wp sens. w.r.t. R1= .2429543

Wp sens. w.r.t. R2=-.7429543

Wp sens. w.r.t. R3=-.5

Wp sens. w.r.t. C1=-.5

Wp sens. w.r.t. C2=-.5

Wp sens. w.r.t. RB= .4859086

Wp sens. w.r.t. RD=-.4859086

Qp sens. w.r.t. R1= 3.314177

Qp sens. w.r.t. R2= 8.133984

Qp sens. w.r.t. R3=-11.44816

Qp sens. w.r.t. C1= 23.74451

Qp sens. w.r.t. C2=-23.74451

Qp sens. w.r.t. RB=-22.7586

Qp sens. w.r.t. RD= 22.7586
```

5B

Program SABPN

THIS APPENDIX PROVIDES THE DETAILS of program SABPN, written in GW BASIC, which can be run on an IBM PC or its compatible minicomputer. This program can be used to design the SAB bandpass filter circuit in Fig. 5.10, and the notation for the different elements can be found in that figure. The design procedure uses a predistortion technique[†] in which the influence of a nonzero τ value for the op amp is compensated for, and at the same time the magnitude of active pole sensitivity of the filter

$$S_\tau^p = \frac{\partial p}{\partial \tau}\frac{\tau}{p}$$

is minimized.

The procedure for using this program is simple. Just enter the program and run it. The inputs for the program are H_o, Q_p, $n^2 = C_2/C_1$, and $\tau_n = \omega_p \tau$. These inputs are given through the keyboard when the appropriate question is displayed on the screen.

The design is carried out initially with the assumption that $\omega_p = 1$ r/s and $C_1 = 1$ F. However, one can obtain the denormalized element values by inputting the values of ω_p and C_1 through the keyboard. With the use of this program, the passive ω_p and Q_p sensitivities can also be found for the specified design and the design accuracy verified by obtaining the normal-

[†]S. Natarajan, A Simple Predistortion Technique for Active Filters, *IEEE Trans. Circuits Syst.*, vol. CAS-31, no. 4, pp. 398–402, April 1984.

ized pole frequency and pole Q factor of the dominant poles of the predistorted design.

Following the program is the output for a sample session for the design of a filter circuit satisfying the specifications $f_p = 20$ kHz, $Q_p = 20$, $H_o = 10$, and $n^2 = 5$, with $C_1 = 1$ nF, $\tau = 0.5/\pi$ μs, and $\tau_n = \omega_p\tau = 4\pi \times 10^4 \times (0.5/\pi) \times 10^{-6} = 0.02$.

```
10 '
20 '
30 '              PROGRAM LISTING OF 'SABPN'
40 '
50 '
60 'Program to design Single Amplifier bandpass filter with min. active pole sen
s. along with predistortion.
70 PRINT"THIS PROGRAM CAN BE USED TO DESIGN SINGLE AMPLIFIER BANDPASS FILTER WIT
H MINIMUM ACTIVE POLE SENSITIVITY  ALONG WITH PREDISTORTION.":PRINT
80 DIM E$(8),X(5),E(9),D(5),Z(5)
90 E$(1)="R1":E$(2)="R2":E$(3)="R3":E$(4)="C1":E$(5)="C2":E$(6)="RB":E$(7)="RD":
E$(8)=CHR$(225)
100 PRINT"INPUT THE SPECIFICATIONS FOR fp,Qp and H0 AS PROMPTED":PRINT
110 INPUT" fp in Hz";FP:INPUT" Qp";Q:INPUT" H0=";H
120 PI2=8*ATN(1):WP=FP*PI2:RE=-1/(2*Q):IM=SQR(1-RE*RE)
130 PRINT"R1,R2,R3 AND RD ARE CALCULATED BY THE PROGRAM. INPUT THE OTHER REQUIRE
D PARAMETERS.":PRINT
140 INPUT"VALUE OF C1";E(4):C1=E(4)
150 INPUT"VALUE OF C2(C2>C1)";E(5):C2=E(5):M=C2/C1
160 INPUT"VALUE OF RB";E(6):RB=E(6)
170 PRINT"INPUT THE NON-IDEAL PARAMETERS OF THE OP.AMP.":PRINT
180 INPUT"INPUT THE INVERSE OF THE GB PRODUCT OF THE OP.AMP.IN Hz^(-1).IN AN IDE
AL OP.AMP. IT IS ZERO.";T:T=T*FP
190 INPUT"INPUT THE NON-ZERO OUTPUT RESISTANCE OF THE OP.AMP. IDEALLY ITS VALUE
IS ZERO.";R0:AL=R0/RB
200 PRINT"STRAY CAPACITANCES FROM THE INVERTING AND NON-INVERTING INPUT TERMINAL
S TO GROUND ARE ZEROS IN IDEAL OP.AMPS.. HOWEVER THEIR VALUES MAY RANGE FROM 1 T
O 3 pF. IF YOU WANT TO INCLUDE THEM INPUT THEIR VALUES.":PRINT
210 INPUT"STRAY CAP.FROM NON-INVERTING TERMINAL TO GROUND";CS1:T2=RB*CS1*WP
220 INPUT"STRAY CAP.FROM INVERTING TERMINAL TO GROUND";CS2:T3=WP*R0*CS2:TT2=T*T2

230 LPRINT"DESIGN OF A SINGLE AMPLIFIER BANDPASS FILTER":LPRINT STRING$(52,"_")
240 LPRINT:LPRINT"THE SPECIFICATIONS ARE":LPRINT
250 LPRINT"fp in Hz=";USING"##.#####^^^^";FP:LPRINT
260 LPRINT"Qp=";USING"##.#####^^^^";Q:LPRINT
270 LPRINT"H0=";USING"##.#####^^^^";H:LPRINT
280 LPRINT"C2/C1=";M:LPRINT
290 LPRINT"THE NON-IDEAL PARAMETERS OF THE OP.AMP. INCLUDED FOR PREDISTORTION"
300 LPRINT:LPRINT"RATIO fp/GB=";USING"##.#####^^^^";T:LPRINT
310 LPRINT"OUTPUT RESISTANCE=";USING"####.##";R0:LPRINT
320 LPRINT"STRAY CAP.FROM NON-INVERTING INPUT TERMINAL TO GROUND=";USING"##.####
#^^^^";CS1:LPRINT
330 LPRINT"STRAY CAP.FROM INVERTING INPUT TERMINAL TO GROUND=";USING"##.#####^^^
^";CS2:LPRINT STRING$(52,"_"):LPRINT
340 B=.25:L=M/(M+1):L1=(L/(2*Q))^2:L2=L/2
350 B1=1-B:L3=SQR(L1+L*B/B1):F1=L*(.5-2*B)-B1*(2*L1+L*L3/Q)
360 F2=2*L1-L*(2-L3/Q+L/(2*Q*B1*L3))
370 B2=B-F1/F2:IF ABS(B2/B-1)<.0001 THEN 380 ELSE B=B2:GOTO 350
380 B=(INT(1000*B2)+1)/1000:K=1/B:E(8)=B
390 'Nominal Design equations
400 N=SQR(M):ROE=N+1/N:A1=1/(2*ROE*Q):A2=B/((1-B)*ROE*N):EM=A1+SQR(A1*A1+A2)
410 SC=WP*C1:G1=M*(1-B)*H*SC/Q:G2=EM*N*SC:G3=G2/(EM*EM)-G1
420 E(1)=1/G1:E(2)=1/G2:E(3)=1/G3:E(7)=RB/(K-1)
430 LPRINT"NOMINAL DESIGN PARAMETERS FOR MIN.ACTIVE POLE SENS.":LPRINT
440 FOR L=1 TO 8:LPRINT E$(L);"=";USING"##.#####^^^^";E(L):LPRINT:NEXT L
450 LPRINT STRING$(52,"_"):LPRINT
460 'Normalization of element values.
470 CX=SQR((K-1)*C1*C2):SC=WP*CX:G1=G1/SC:G2=G2/SC:G3=G3/SC:R0=R0*SC
480 C=C1+C2:C3=C1+CS2:C42=C1*C2+CS2*(C1+C2):C1=C1/CX:C2=C2/CX
490 C3=C3/CX:C=C/CX:C42=C42/(CX*CX):A1=K-AL+K*AL
500 ' Denominator coefficients and their derivatives w.r.t. G2 and G3.
510 BE=R0*G2:G=G1+G3:D0=G*G2:D1=C*G2:D2=C3*G:D3=C2*G:GA=R0*G
520 X(0)=D0*(K-1):X(1)=(K-1)*D1-D2+A1*T*D0
530 X(2)=(K*C1*C2-C42)+T*(D1+D2)*A1+K*T*BE*(D3+D2)+D0*TT2*(1+AL)
540 X(3)=T*C42*A1+K*T*(BE*(C1*C2+C42)+C2*C3*GA)+TT2*((1+AL)*(D1+D2)+BE*(D3+D2))
550 X(4)=TT2*(C42*(1+AL+BE)+C2*(C1*BE+C3*GA))+K*T*T3*C1*C2
```

```
560 X(5)=TT2*T3*C1*C2:ND=5:GOSUB 930
570 X1=RD1:X2=RD0
580 X(0)=(K-1)*G:X(1)=(K-1)*C+G*T*A1:X(2)=C*T*A1:X(3)=0:X(4)=0
590 X(5)=0:GOSUB 930
600 A=RD1:B=RD0
610 X(0)=G2*(K-1):X(1)=-C3+G2*T*A1:X(2)=C3*T*A1+C3*K*T*BE:X(3)=0:X(4)=0
620 X(5)=0:GOSUB 930
630 Y=RD1:D=RD0:DG3=(X2/B-X1/A)/(Y/A-D/B):DG2=-(X1+Y*DG3)/A:G2=G2+DG2:G3=G3+DG3
640 IF ABS(DG2/G2)>.0001 OR ABS(DG3/G3)>.0001 THEN 510
650 XA=K*Q*C1/(H*(K-1)*G2)-1:GX=G3/XA:IF ABS(GX/G1-1)>.0001 THEN G1=GX:GOTO 510
660 G1=GX:BE=R0*G2:G=G1+G3:D0=G*G2:D1=C*G2:D2=C3*G:D3=C2*G:GA=R0*G
670 X(0)=0:X(1)=D0*A1:X(2)=(D1+D2)*A1+K*BE*(D3+D2)+T2*D0*(1+AL)
680 X(3)=C42*A1+K*BE*(C1*C2+C42)+K*GA*C2*C3+T2*((D1+D2)*(1+AL)+BE*(D2+D3))
690 X(4)=T2*(C42*(1+AL+BE)+C1*C2*BE+C2*C3*GA)+K*T3*C1*C2:X(5)=T2*T3*C1*C2
700 GOSUB 930
710 A=RD1:B=RD0
720 X(5)=X(5)*T:X(4)=X(4)*T:X(3)=X(3)*T:X(2)=X(2)*T+(K*C1*C2-C42)
730 X(1)=X(1)*T+(K-1)*D1-D2:X(0)=(K-1)*D0
740 D(4)=X(4):D(3)=X(3):D(2)=X(2):D(1)=X(1):D(0)=X(0)
750 SCX=1/SC:E(1)=SCX/G1:E(2)=SCX/G2:E(3)=SCX/G3
760 X(0)=X(1):X(1)=2*X(2):X(2)=3*X(3):X(3)=4*X(4):X(4)=5*X(5):ND=4:GOSUB 930
770 Y=RD1:R=RD0:S=SQR(((A*RE+B)^2+(A*IM)^2)/((Y*RE+R)^2+(Y*IM)^2))
780 LPRINT"ELEMENT VALUES WITH MINIMUM ACTIVE POLE SENSITIVITY":LPRINT
790 LPRINT"AFTER PREDISTORTION IS APPLIED TO THE BANDPASS FILTER":LPRINT
800 FOR L=1 TO 8:LPRINT E$(L);"=";USING"##.#####^^^^";E(L):LPRINT:NEXT L
810 LPRINT"MAGNITUDE OF MIN. ACTIVE POLE SENSITIVITY=";USING"##.#####^^^^";S*T
820 PRINT"DO YOU WANT TO KNOW THE Wp AND Qp SENSITIVITIES WITH RESPECT TO PASSIV
E        ELEMENTS? Y/N";:INPUT A$
830 IF A$="N" OR A$="n" THEN 910
840 LPRINT STRING$(52,"_"):LPRINT"Wp AND Qp SENSITIVITIES ARE":LPRINT
850 G1=1/E(1):G2=1/E(2):G3=1/E(3):G=G1+G3:L=1/(K-1):E(1)=-G1/(2*G)
860 E(2)=-.5:E(3)=-.5-E(1):E(4)=-.5:E(5)=-.5:E(6)=0:E(7)=0
870 FOR I=1 TO 7:LPRINT"Wp sens. w.r.t. ";E$(I);"=";E(I):LPRINT:NEXT I
880 D1=G2*(1+M)-L*G:E(1)=-L*G1/D1+E(1):E(2)=E(2)+(1+M)*G2/D1:E(3)=-.5-L*G3/D1
890 E(4)=M*G2/D1-.5:E(5)=-E(4):E(6)=-L*G/D1:E(7)=-E(6)
900 FOR I=1 TO 7:LPRINT"Qp sens. w.r.t. ";E$(I);"=";E(I):LPRINT:NEXT I
910 LPRINT STRING$(52,"_"):END
920 'subroutine to divide a poly. with quadratic factor of s^2+s/Q+1
930 Z(ND)=0:Z(ND-1)=0
940 FOR L=ND-2 TO 0 STEP -1
950 Z(L)=X(L+2)-Z(L+1)/Q-Z(L+2)
960 NEXT L
970 RD1=X(1)-Z(0)/Q-Z(1):RD0=X(0)-Z(0)
980 RETURN
```

A SAMPLE SESSION

DESIGN OF A SINGLE AMPLIFIER BANDPASS FILTER

THE SPECIFICATIONS ARE

fp in Hz= 2.00000E+04

Qp= 2.00000E+01

H0= 1.00000E+01

C2/C1= 5

THE NON-IDEAL PARAMETERS OF THE OP.AMP. INCLUDED FOR PREDISTORTION

RATIO fp/GB= 2.00000E-02

OUTPUT RESISTANCE= 100.00

STRAY CAP.FROM NON-INVERTING INPUT TERMINAL TO GROUND= 2.00000E-12

STRAY CAP.FROM INVERTING INPUT TERMINAL TO GROUND= 2.00000E-12

NOMINAL DESIGN PARAMETERS FOR MIN.ACTIVE POLE SENS.

R1= 4.18829E+03

R2= 1.48953E+04

```
R3= 1.06687E+03
C1= 1.00000E-09
C2= 5.00000E-09
RB= 1.00000E+04
RD= 3.15789E+03
β= 2.40000E-01
```

```
ELEMENT VALUES WITH MINIMUM ACTIVE POLE SENSITIVITY
AFTER PREDISTORTION IS APPLIED TO THE BANDPASS FILTER
R1= 3.91899E+03
R2= 1.43542E+04
R3= 1.04594E+03
C1= 1.00000E-09
C2= 5.00000E-09
RB= 1.00000E+04
RD= 3.15789E+03
β= 2.40000E-01
MAGNITUDE OF MIN. ACTIVE POLE SENSITIVITY= 3.25548E-02
```

```
Wp AND Qp SENSITIVITIES ARE
Wp sens. w.r.t. R1=-.1053328
Wp sens. w.r.t. R2=-.5
Wp sens. w.r.t. R3=-.3946672
Wp sens. w.r.t. C1=-.5
Wp sens. w.r.t. C2=-.5
Wp sens. w.r.t. RB= 0
Wp sens. w.r.t. RD= 0
Qp sens. w.r.t. R1=-2.375421
Qp sens. w.r.t. R2= 11.27579
Qp sens. w.r.t. R3=-9.005702
Qp sens. w.r.t. C1= 9.313158
Qp sens. w.r.t. C2=-9.313158
Qp sens. w.r.t. RB=-10.77579
Qp sens. w.r.t. RD= 10.77579
```

6

Multiple-Amplifier Filters

OPERATIONAL AMPLIFIERS ARE no longer expensive, because technology has advanced to the point where thousands of transistors can be packed onto a small chip. Thus the argument that networks using a single op amp are relatively cheap is no longer valid. Single-amplifier networks have many disadvantages, such as large element spread and complex design equations. Complex design equations also mean that the tuning of the filters can be difficult. Furthermore, one might require a tighter tolerance in element values, and such single-amplifier circuits are useful only in realizing low-and medium-Q_p filters. Multiple-amplifier circuits do not have these disadvantages. Also, as pointed out earlier, there is little advantage in minimizing active devices. To realize high-Q_p filters, one has to use multiple-amplifier networks for the reasons mentioned above. The first category of multiple-amplifier filters we shall consider uses generalized impedance converters (GIC) that require two op amps. The second category requires integrators and summing amplifiers as the basic components. These filters of the second category generally require three or more number of op amps.

6.1 *Generalized Impedance Converter Circuits*

Frequency-selective networks require resonant circuits for their operation. In passive networks both inductors and capacitors are needed to obtain such resonance characteristics. Since the use of inductors is ruled out in

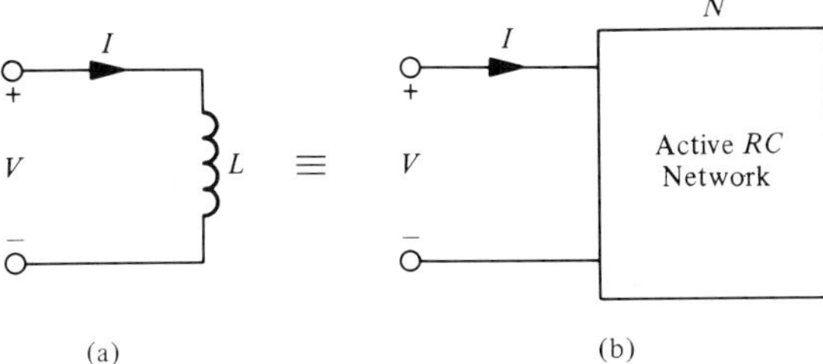

Figure 6.1 (*a*) Inductor. (*b*) Its active *RC* simulator.

these circuits, one method of realizing such resonance characteristics is to use *synthetic* or *simulated inductors*. A circuit that simulates the inductive impedance across one of its ports should employ only op amps, resistances, and capacitances, as illustrated in Fig. 6.1. With reference to these circuits, the driving point impedance across the input port of the inductor is

$$Z_{11}(s) = \frac{V}{I} = sL \tag{6.1}$$

If the active *RC* circuit in Fig. 6.1*b* simulates the volt-ampere relationship in (6.1), then this circuit can be used in place of the inductor. The network *N* should have at least one energy-storing element, and this element should be a capacitor. Also, this network should not contain more than one capacitor, because it is desired that the network be canonical. This capacitor can be extracted as a separate component, and the circuit in Fig. 6.1*b* can be treated as a two-port network terminated with a capacitor as shown in Fig. 6.2. The remaining network will be an active *R* network. In order for the driving point impedance of this network to be inductive, the admittance matrix description of the active *R* network should be in the following form:

$$Y = \begin{bmatrix} 0 & \pm G_2 \\ \mp G_1 & 0 \end{bmatrix} \tag{6.2}$$

where G_1 and G_2 are positive conductances. Any active two-port network that has the admittance matrix (6.2) is called a *gyrator*. If the two-port

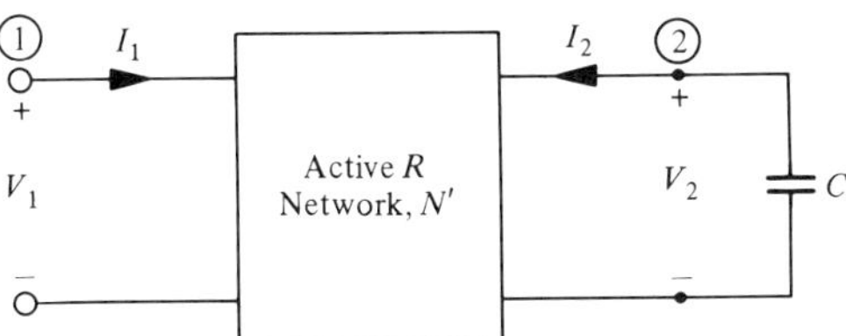

Figure 6.2 Inductance simulation using an active *R* network.

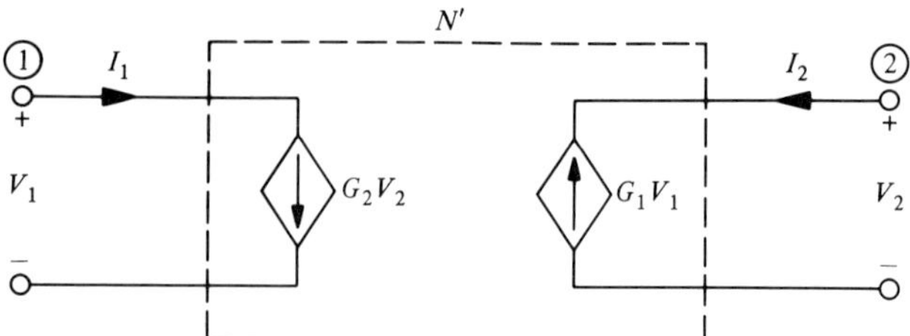

Figure 6.3 Controlled-source representation of a gyrator.

network has the above Y-matrix description, then the inductance value of the simulated inductor will be

$$L = \frac{C}{G_1 G_2} \tag{6.3}$$

The controlled source representation of (6.2) with $+G_2$ and $-G_1$ is shown in Fig. 6.3.

A two-amplifier circuit [1] for simulating the inductance is shown in Fig. 6.4. This circuit has emerged as the best possible circuit for simulating inductors after various topologies were analyzed by many researchers. This circuit uses general impedances Z_1 through Z_5. With this circuit, in addition to simulating inductance, one can obtain other types of impedance conversions by choosing appropriate values of Z_1 through Z_5. When the op

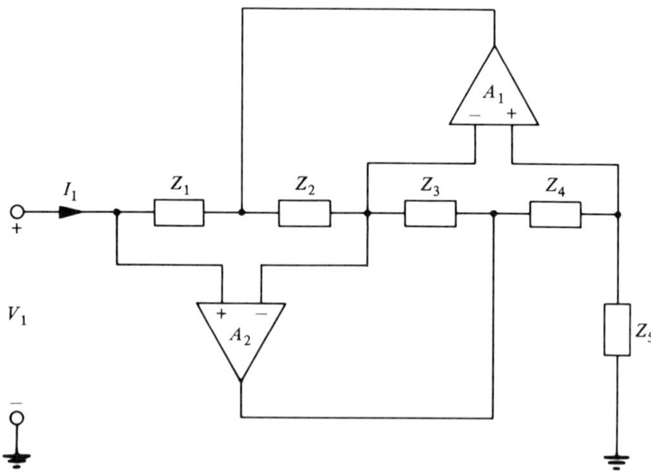

Figure 6.4 Two-amplifier circuit that can be used for inductance simulation.

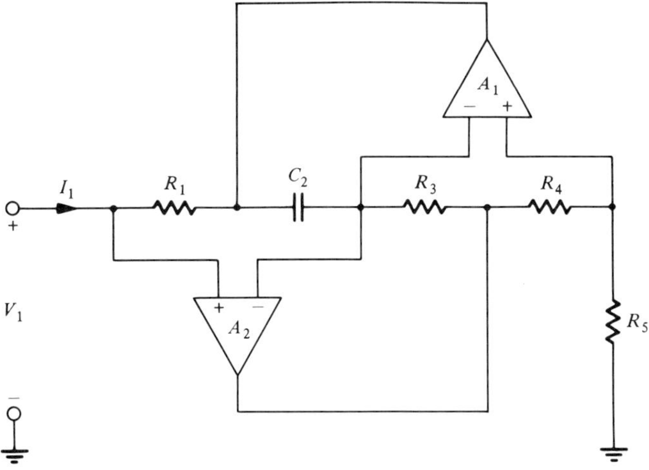

Figure 6.5 Inductance simulation circuit—type A.

amps are assumed to be ideal, the driving point impedance is

$$Z_{11}(s) = \frac{V_1}{I_1} = \frac{Z_1 Z_3 Z_5}{Z_2 Z_4} \tag{6.4}$$

In order for this impedance to have the form of (6.1), there are two possible choices for the impedances Z_1 through Z_5.

1. The choice of $Z_1 = R_1$, $Z_2 = 1/C_2 s$, $Z_3 = R_3$, $Z_4 = R_4$, and $Z_5 = R_5$ leads to the circuit in Fig. 6.5 with

$$Z_{11}(s) = C_2 \frac{R_1 R_3 R_5}{R_4} s \tag{6.5}$$

   The circuit in Fig. 6.5 is often referred to as type-A inductance simulation.
2. Similarly, the choice of $Z_1 = R_1$, $Z_2 = R_2$, $Z_3 = R_3$, $Z_4 = 1/C_4 s$, and $Z_5 = R_5$ leads to the circuit in Fig. 6.6 with

$$Z_{11}(s) = C_4 \frac{R_1 R_3 R_5}{R_2} s \tag{6.6}$$

   This circuit in Fig. 6.6 is called type-B inductance simulation.

Consider the port across which the capacitor is connected. Assume that the capacitor is removed and that this pair of terminals forms another port. Thus this network, after removal of the capacitor, becomes a two-port network, and this two-port network will form a gyrator as in Fig. 6.2. However, when the same circuits are considered as grounded two-port

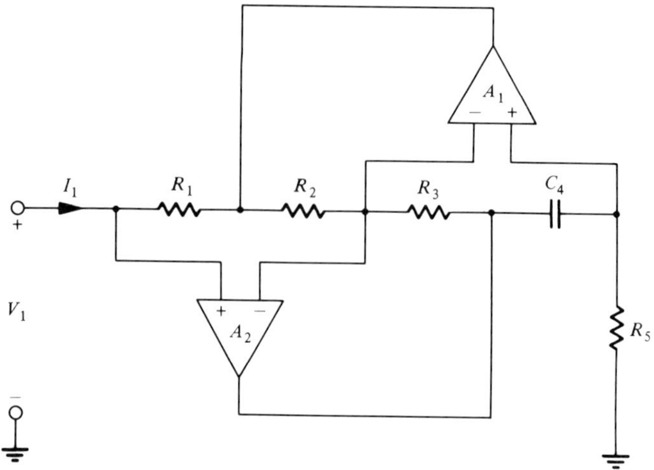

Figure 6.6 Inductance simulation circuit—type B.

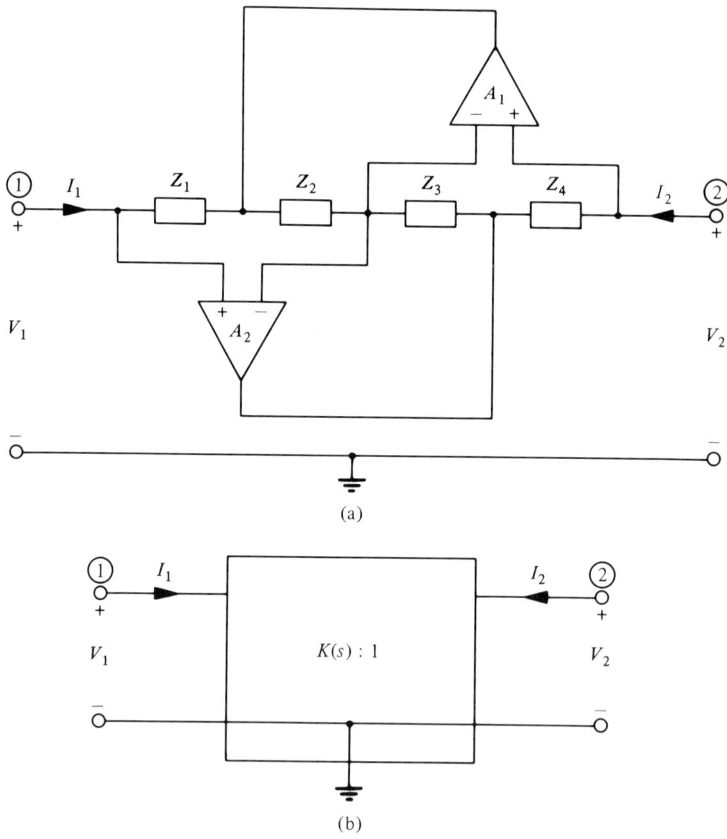

Figure 6.7 (*a*) Antoniou's generalized impedance converter. (*b*) A two-port representation.

networks, port 2 being the port across which Z_5 is connected, then these circuits can be thought of as *impedance converters*. In Fig. 6.4, the input impedance is

$$Z_{11}(s) = K(s)Z_5 \tag{6.7}$$

where the conversion factor is

$$K(s) = \frac{Z_1 Z_3}{Z_2 Z_4} \tag{6.8}$$

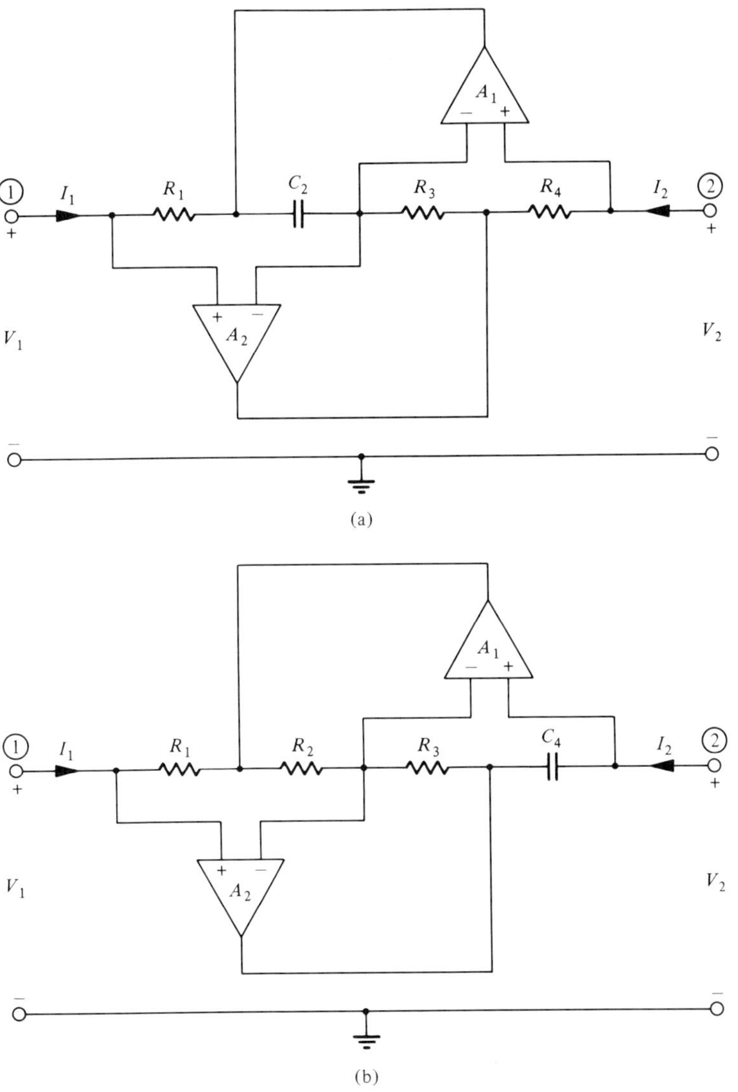

Figure 6.8 Two possible implementations of GIC with $K(s) = Ks$.

As one can see from (6.8), the conversion factor $K(s)$ can be any desired function. The circuit in Fig. 6.4, with the terminating impedance Z_5 removed, is called Antoniou's *generalized impedance converter* (GIC). The circuit corresponding to this GIC is shown in Fig. 6.7 along with its black box representation. This type of GIC circuit is commercially available in the form of an integrated circuit (AF 120 C). If port 2 of this circuit is terminated with an impedance of Z_{L2}, then the input impedance at port 1 will be $Z_{11}(s) = K(s)Z_{L2}$. If port 1 is terminated with Z_{L1}, then the input impedance at port 2 will be $Z_{22}(s) = Z_{L1}/K(s)$. The inductance simulation circuits in Figs. 6.5 and 6.6 have $K(s) = Ks$. Two GIC circuits with a conversion factor of $K(s) = Ks$ can be extracted from Figs. 6.5 and 6.6 and are shown in Fig. 6.8.

Now consider the GIC in Fig. 6.7a, where $Z_1 = 1/C_1s$, $Z_2 = R_2$, $Z_3 = R_3$, and $Z_4 = R_4$. Assume that we terminate port 2 of this circuit with a capacitance of value C, as shown in Fig. 6.9a. This circuit is called

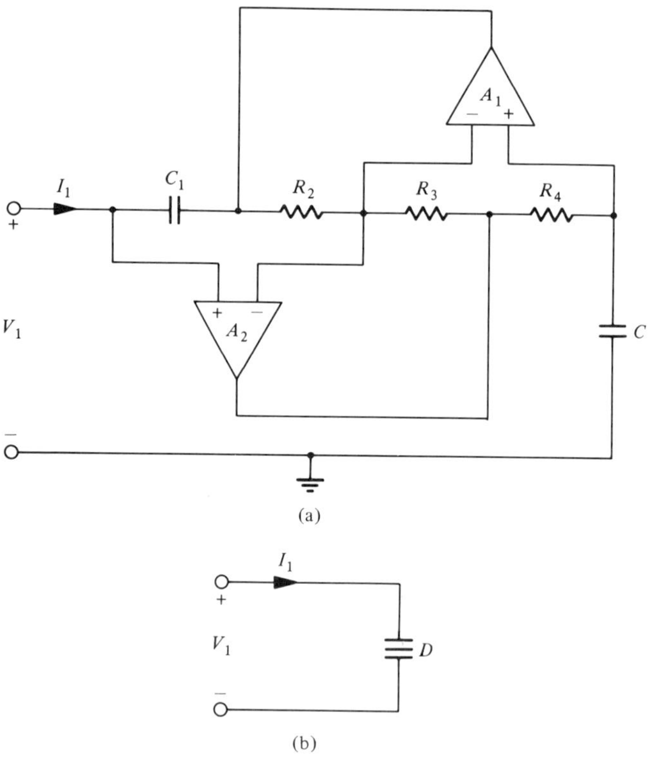

Figure 6.9 (a) Frequency-dependent negative resistor (FDNR) realization. (b) Its circuit symbol.

type-B FDNR circuit. The input impedance of this circuit is

$$Z_{11}(s) = \frac{Z_1 Z_3 Z_5}{Z_2 Z_4} = \frac{R_3}{CC_1 R_2 R_4 s^2} \tag{6.9}$$

and, for sinusoidal excitations,

$$Z_{11}(j\omega) = \frac{-1}{D\omega^2} \tag{6.10a}$$

where

$$D = \frac{CC_1 R_2 R_4}{R_3} \tag{6.10b}$$

Now we find that $Z_{11}(j\omega)$ is a frequency-dependent negative resistance (FDNR). This FDNR also plays an important role in the design of active filters, and the circuit symbol for an FDNR element is shown in Fig. 6.9*b*. There is another possible circuit realization for an FDNR (type A) element using the GIC.

6.2 *Sensitivity Properties of Simulated Inductance and Frequency-Dependent Negative Resistance Elements*

An important category of active *RC* filters uses these simulated elements as basic components, as we shall see in this and the following chapters. Biquads can also be derived using GICs. Since the sensitivity properties of such filters depend critically on the sensitivity properties of the simulated elements, it is worthwhile to study the sensitivities of these elements. The magnitude of the sensitivity of the simulated elements, with respect to any passive element in the circuit, is unity. Thus we need to consider only the active sensitivities of these elements. We can obtain both simulated inductance and FDNR elements in the circuit in Fig. 6.4 by choosing different impedances in different ways. Therefore we consider this circuit in our general sensitivity analysis. Including the finite gains of the op amps, we can find the input impedance as

$$Z_{11}(s) = \frac{Z_1 Z_3 Z_5}{Z_2 Z_4} \frac{N(s)}{D(s)} \tag{6.11}$$

where

$$N(s) = 1 + \left(1 + \frac{Z_4}{Z_5}\right)\left(A_2^{-1} + \frac{A_1^{-1} Z_2}{Z_3}\right) + A_1^{-1} A_2^{-1}\left(1 + \frac{Z_4}{Z_5}\right)\left(1 + \frac{Z_2}{Z_3}\right)$$

and

$$D(s) = 1 + \left(1 + \frac{Z_5}{Z_4}\right)\left(A_1^{-1} + \frac{A_2^{-1}Z_3}{Z_2}\right) + A_1^{-1}A_2^{-1}\left(1 + \frac{Z_5}{Z_4}\right)\left(1 + \frac{Z_3}{Z_2}\right)$$

For simplicity of analysis, let us assume that the nonzero time constants of the two op amps are equal. Substituting $A_1 = A_2 = 1/(s\tau)$,

$$Z_{11}(s) = \frac{Z_1 Z_3 Z_5}{Z_2 Z_4} \frac{1 + s\tau(1 + Z_4/Z_5)(1 + Z_2/Z_3)(1 + s\tau)}{1 + s\tau(1 + Z_5/Z_4)(1 + Z_3/Z_2)(1 + s\tau)} \qquad (6.12)$$

The semilogarithmic active sensitivity of $Z_{11}(s)$ is

$$S_\tau^{Z_{11}(s)} = \frac{\partial Z_{11}(s)}{\partial \tau} \frac{1}{Z_{11}(s)}\Bigg|_{\tau=0}$$

$$= s\left[\left(1 + \frac{Z_4}{Z_5}\right)\left(1 + \frac{Z_2}{Z_3}\right) - \left(1 + \frac{Z_5}{Z_4}\right)\left(1 + \frac{Z_3}{Z_2}\right)\right] \qquad (6.13)$$

Type A: Inductance Simulation

There are two possible inductance simulation circuits, as shown in Figs. 6.5 and 6.6. In the type-A circuit in Fig. 6.5, we have

$$Z_2 = \frac{1}{j\omega C_2} \qquad Z_3 = R_3$$

$$Z_4 = R_4 \qquad \text{and} \qquad Z_5 = R_5$$

Using the above in (6.13), the active sensitivity of this circuit is

$$S_\tau^{Z_{11}(j\omega)} = j\omega\left[\left(a - \frac{1}{a}\right) + \frac{1 + a + (\omega C_2 R_3)^2(1 + 1/a)}{j\omega C_2 R_3}\right]$$

where $a = R_4/R_5$.

The magnitude of this sensitivity is a minimum when $a = \omega C_2 R_3 = 1$. Since $\omega C_2 R_3$ cannot equal unity at all frequencies, one can satisfy this relationship only at some critical frequency ω_c. Thus the design steps for realizing an inductor with an inductance value of L_o and with minimum active sensitivity are

1. Choose any two equivalued resistors for R_4 and R_5.
2. Choose a convenient value for C_2 and choose $R_3 = 1/(\omega_c C_2)$.
3. Choose $R_1 = \omega_c L_o$.

This completes the design steps for the circuit in Fig. 6.5.

Type B: Inductance Simulation

The conditions for realizing the type-B inductance simulation circuit in Fig. 6.6 with minimum active sensitivity are $R_2 = R_3$, $R_5 = 1/\omega_c C_4$, and $R_1 = \omega_c L_o$ for a specified value of L_o at a critical frequency ω_c. These are the design formulas as well.

It is possible for the op amp time constants to differ. In such cases, the condition for minimum active sensitivity in a type-A circuit is $R_4\tau_2 = R_5\tau_1$. When the values of τ_1 and τ_2 are equal, this reduces to the condition $R_4 = R_5$. In a type-B circuit, even when $\tau_1 \neq \tau_2$, the condition $R_2 = R_3$ is sufficient for minimum active sensitivity. Next, we consider the FDNR realization shown in Fig. 6.9.

Type B: FDNR Circuit

The active sensitivity of the FDNR circuit in Fig. 6.9 can be obtained by substituting the element values into (6.13):

$$S_\tau^{Z_{11}(j\omega)} = j\omega\left\{b - \frac{1}{b} + j\omega CR_4\left[1 + b + \frac{1 + 1/b}{(\omega CR_4)^2}\right]\right\}$$

where $b = R_2/R_3$.

The above sensitivity can be minimized by choosing $R_2 = R_3$ and $\omega CR_4 = 1$. At a critical frequency ω_c, one can select $R_4 = 1/(\omega_c C)$. One more design step involves choosing $C_1 = \omega_c D$ for a given value of D.

In the realization of reactive or resistive circuits, the effects of the nonidealness of the op amps on the Q factor of these elements should be taken into account. Consider the inductance simulation. We should be able to realize an ideal inductive impedance of $Z_{11}(j\omega) = j\omega L_o$, where L_o is the ideally required inductance value. However, the actual impedance, including the nonzero time constant of the op amps, is generally

$$Z_{11a}(j\omega) = j\omega L_a + R_a \tag{6.14}$$

where R_a is ideally zero and the ideal value L_a should be equal to L_o. We can write

$$Z_{11a}(j\omega) = Z_{11}(j\omega) + \Delta Z(j\omega) \tag{6.15}$$

and the deviation in the impedance $\Delta Z(j\omega)$ for small changes can be found as

$$\begin{aligned}\Delta Z(j\omega) &= \left.\frac{\partial Z_{11}(j\omega)}{\partial \tau}\right|_{\tau=0} \tau \\ &= \tau Z_{11}(j\omega) S_\tau^{Z_{11}(j\omega)} \end{aligned} \tag{6.16}$$

Substituting (6.16) into (6.15) we have, to a first-order approximation,

$$Z_{11a}(j\omega) = Z_{11}(j\omega)\left[1 + \tau S_{\tau}^{Z_{11}(j\omega)}\right] \tag{6.17}$$

The above equation is valid for every circuit we have discussed. For the circuit in Fig. 6.5, we have

$$Z_{11a}(j\omega) = j\omega L_o\left\{1 + \frac{\tau\left[1 + a + (\omega C_2 R_3)^2(1 + 1/a)\right]}{C_2 R_3}\right\}$$
$$+\omega^2 L_o \tau\left(\frac{1}{a} - a\right) \tag{6.18}$$

Therefore the per-unit change in the inductance value and the Q factor of the simulated inductor are

$$\frac{\Delta L}{L_o} = \frac{\tau\left[1 + a + (\omega C_2 R_3)^2(1 + 1/a)\right]}{C_2 R_3} \tag{6.19a}$$

and

$$Q_a = \frac{a}{\omega\tau(1 - a^2)} \tag{6.19b}$$

With the nominal value of $a = 1$, the Q factor becomes infinite and the per-unit change in the inductance value is

$$\frac{\Delta L}{L_o} = 2\omega_c\tau\left[1 + \left(\frac{\omega}{\omega_c}\right)^2\right]$$

In the type-B inductance simulation circuit in Fig. 6.6, the condition $R_2 = R_3$ leads to an infinite Q factor and, even if this factor is not infinity, one can tune R_2 or R_3 to obtain this value. Again, with the nominal design values $R_2 = R_3$ and $\omega_c C_4 R_5 = 1$, the per-unit change in the inductance value can be shown to be the same as above:

$$\frac{\Delta L}{L_o} = 2\omega_c\tau\left[1 + \left(\frac{\omega}{\omega_c}\right)^2\right]$$

In an FDNR circuit, the Q factor is defined as the ratio of the real part to the imaginary part. Using this definition and the active sensitivity of the circuit in Fig. 6.9, for the nominal design of this circuit given earlier, we can show the per-unit change in the D value and the Q factor of this element to be

$$\frac{\Delta D}{D_o} = 2\omega_c\tau\left[1 + \left(\frac{\omega}{\omega_c}\right)^2\right]$$

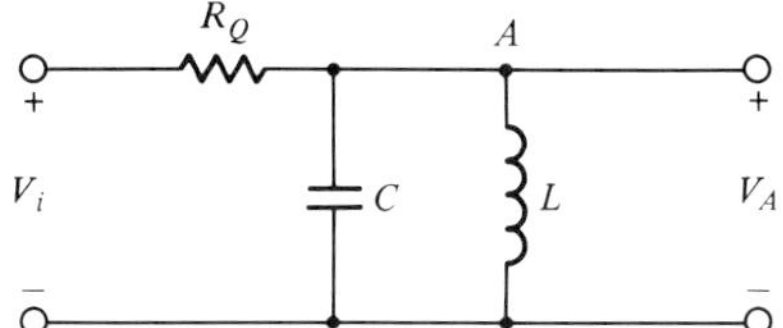

Figure 6.10 Second-order bandpass filter.

and

$$Q_a = \infty$$

Even if the Q factor is not infinity, in any of the circuits discussed above, one can tune a resistor to obtain a large Q value. However, in the inductance values of all the above elements, the frequency-dependent deviation cannot be made zero unless $\tau = 0$.

6.3 *Biquads Derived from Generalized Impedance Converter Circuits*

Simulated inductance and FDNR elements can be used for the realization of both second- and higher-order filters. The realization of higher-order filters using these elements will be discussed in the next chapter. For now, we consider the realization of second-order filters using these elements. We can employ synthetic inductors and FDNR elements in realizing resonant circuits along with other capacitances and resistances. For example, consider the *RLC* circuit in Fig. 6.10. This is a second-order bandpass filter. The transfer function of this circuit can be derived as

$$\frac{V_o}{V_i} = \frac{\left(\omega_p/Q_p\right)s}{s^2 + \left(\omega_p/Q_p\right)s + \omega_p^2} \tag{6.20}$$

where $\omega_p = 1/(LC)^{1/2}$ and $R_Q = Q_p(L/C)^{1/2}$.

The grounded inductor in the circuit in Fig. 6.10 may be replaced with any one of the synthetic inductors in Fig. 6.5 or 6.6. If this is done, the output will be from a high-impedance node, A. However, this problem can be overcome with the use of type-A inductance simulation circuit. In the type-A inductance simulation circuit, the output of op amp A_2 is just $(1 + R_4/R_5)V_1$ with ideal op amps. To see this clearly, assume that we replace the inductor in the circuit in Fig. 6.10 with a type-A simulated inductor. The resulting circuit is shown in Fig. 6.11. Here R_c is an external resistor, the purpose of which is to eliminate the possible Q enhancement in this circuit, and this will be explained later in this section. For now, assume

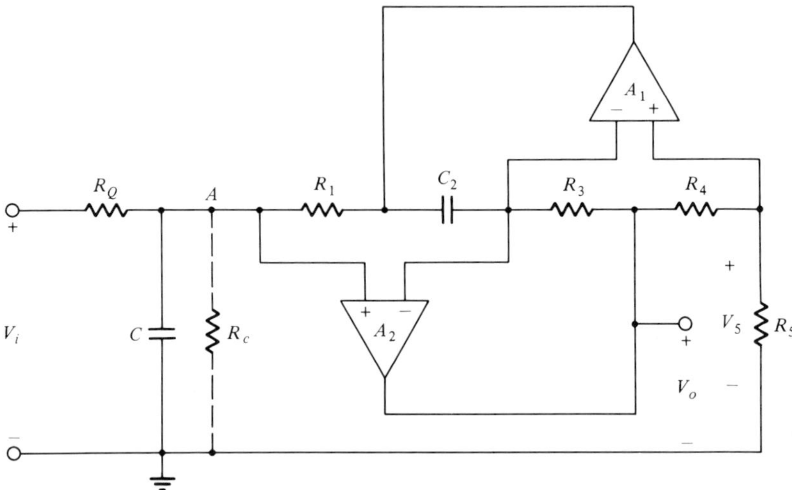

Figure 6.11 Second-order bandpass filter derived from a GIC.

that $G_c = 0$. According to Fig. 6.11, the output V_A is at a high-impedance node. However, with ideal op amps $V_5 = V_A = V_o R_5/(R_4 + R_5)$. Therefore $V_o = (1 + R_4/R_5)V_A$. Thus the filter characteristic of the circuit in Fig. 6.11, when the output is V_o, has the same ω_p and Q_p values that it had in Fig. 6.10 with a gain constant of $H = 1 + R_4/R_5$.

The voltage transfer ratio V_o/V_i for the circuit in Fig. 6.11, when the op amps are assumed to be ideal, is

$$\frac{V_o}{V_i} = \frac{H(\omega_p/Q_p)s}{s^2 + (\omega_p/Q_p)s + \omega_p^2} \tag{6.21}$$

where

$$H = 1 + \frac{R_4}{R_5} \tag{6.22a}$$

$$\frac{\omega_p}{Q_p} = \frac{1}{R_Q C} \tag{6.22b}$$

and

$$\omega_p^2 = \frac{R_4}{CC_2R_1R_3R_5} \tag{6.22c}$$

Using the degrees of freedom available, we can choose $C_2 = C$ and $R_1 = R_3$.

Then, the following choices should be made:

$$\frac{R_4}{R_5} = H - 1 \tag{6.23a}$$

$$R_1 = R_3 = \frac{(H-1)^{1/2}}{\omega_p C} \tag{6.23b}$$

$$R_Q = \frac{Q_p R_1}{(H-1)^{1/2}} \tag{6.23c}$$

The magnitudes of the passive ω_p sensitivities are equal to 0.5. The passive Q_p sensitivities are also small and are in the range 0.5 to 1. In fact, these sensitivities are at their theoretical minimum. Only the active sensitivities are of main concern. Since the active sensitivity of this circuit depends entirely on the active sensitivity of the simulated inductance, the active pole sensitivity can be minimized by choosing

$$R_4 = R_5 \qquad \text{implying that } H = 2 \tag{6.24a}$$

$$R_1 = R_3 = \frac{1}{\omega_p C} \tag{6.24b}$$

and

$$R_Q = Q_p R_1 \tag{6.24c}$$

If it is assumed that we have selected the passive components according to (6.23) and that $\tau_1 = \tau_2 = \tau$ for simplicity, the denominator polynomial of the transfer function of this circuit, including the effect of R_c, can be derived as

$$D(s,\tau) = s^2 + \left(\frac{1}{Q_p} + b\right)\omega_p s + \omega_p^2 + Hs\tau\left(a_3 s^3 + a_2 s^2 + a_1 s + a_0\right) \tag{6.25}$$

where

$$a_3 = \tau$$

$$a_2 = 1 + \omega_p \tau \left[\frac{1}{Q_p} + b + \frac{2}{(H-1)^{1/2}}\right]$$

$$a_1 = \omega_p \left\{\left[\frac{1}{Q_p} + b + \frac{2}{(H-1)^{1/2}}\right] + \omega_p \tau \left[\frac{1}{Q_p} + b + \frac{1}{(H-1)^{1/2}}\right]\right\}$$

$$a_0 = \frac{\omega_p^2\left[1/Q_p + b + 1/(H-1)^{1/2}\right]}{(H-1)^{1/2}}$$

and

$$b = \frac{1}{\omega_p C R_c}$$

Applying (3.62) to the above denominator polynomial, we can show that

$$\frac{\Delta p}{p} = \frac{H}{2}\left(-\left[b + \frac{2}{(H-1)^{1/2}} \right] + j\left\{ b\left[\frac{1}{(H-1)^{1/2}} - \frac{1}{2Q_p} \right] + \frac{2-H}{H-1} \right\} \right) \omega_p \tau \quad (6.26)$$

In the conventional realization $G_c = 0$, and therefore $b = 0$. Then (6.26) reduces to

$$\frac{\Delta p}{p} = \left[\frac{-H}{(H-1)^{1/2}} + \frac{j(H/2)(2-H)}{H-1} \right] \omega_p \tau \quad (6.27)$$

When the nonzero value of τ is included, $p' = p + \Delta p$ is then given by

$$p' = \frac{-\omega_p}{2Q_p}\left[1 - \frac{H(H-2)Q_p \omega_p \tau}{H-1} \right] + j\omega_p \left[1 - \frac{H\omega_p \tau}{(H-1)^{1/2}} \right]$$

The approximate expressions for the actual pole frequency and the pole Q factor may be obtained from the above equation and are

$$\omega_p' = \omega_p \left[1 - \frac{H\omega_p \tau}{(H-1)^{1/2}} \right] \quad (6.28a)$$

and

$$\frac{Q_p'}{Q_p} = \frac{1}{1 - H(H-2)Q_p \omega_p \tau/(H-1)} \quad (6.28b)$$

Example 6.1: Design a GIC-derived bandpass filter to meet the specifications $f_p = 10$ kHz, $Q_p = 10$, and $H = 8$. Choose $C = C_2 = 10$ nF. Assume that both op amps have the same nonzero τ value and that it is equal to $0.5/\pi$ μs. Find the actual pole frequency f_p' and the pole Q factor Q_p'.

Choosing $R_5 = 1$ kΩ and using (6.23a), we have $R_4 = 7$ kΩ. From (6.23b), we find that $R_1 = R_3 = 4.211$ kΩ. Finally, the value of R_Q can be obtained using (6.23c), and it is 15.915 kΩ.

Next, the actual pole frequency f_p' and the actual pole Q factor Q_p' can be found using (6.28):

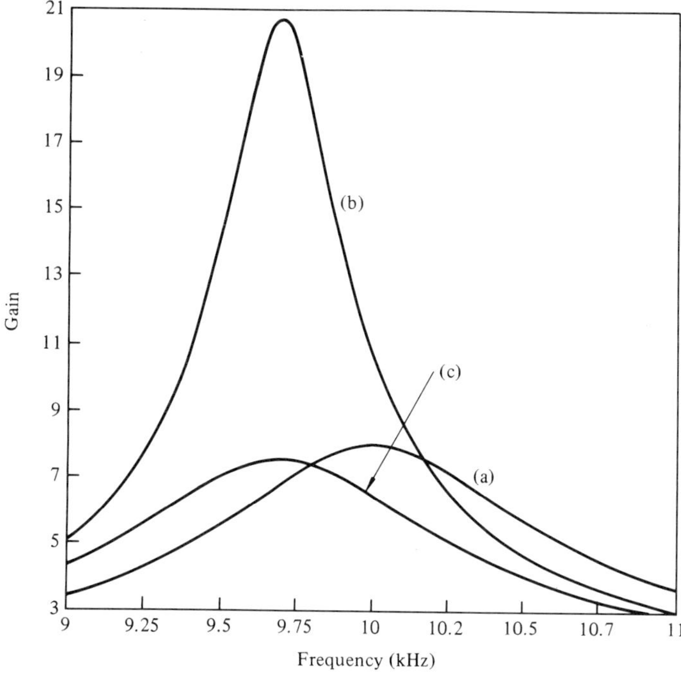

Figure 6.12 Magnitude responses of the GIC-based bandpass filters designed in Example 6.1. (*a*) Ideal response as well as the response of the network with compensation and predistortion. (*b*) With no compensation or predistortion. (*c*) With compensation but without predistortion.

$$f_p' = 9.698 \text{ kHz}$$

and

$$Q_p' = 31.818$$ ■

Whenever there is a large increase in the Q factor (in this example, it is more than three times greater than the required pole Q factor), the condition is called *Q enhancement*. The magnitude responses of the filter designed in this example are shown in Fig. 6.12. These characteristics were obtained using computer simulation. Curve a is the expected ideal characteristic if the op amps are ideal. Curve b indicates the response of the filter if the effects of the nonzero time constants of the op amps are included in finding the response. By comparing these two curves, one can clearly see the effect of the Q enhancement.

By referring to (6.28*b*), one can easily see that, when $H = 2$, there is no Q enhancement and the active sensitivity is a minimum. When $H = 2$, the biquad is called the *optimum biquad* because of the low active and passive sensitivities. However, when $H > 2$, the performance of the bandpass filter deteriorates considerably, and the major effect is Q enhancement. To avoid Q enhancement in the filter, it has been suggested [2] that a compensating resistor R_c be connected as shown in Fig. 6.11. In order to find a value of R_c that compensates for the effect of Q enhancement, we must satisfy the equality:

$$\left.\frac{\partial p}{\partial \tau}\right|_{\tau=b=0} \tau + \left.\frac{\partial p}{\partial b}\right|_{\tau=b=0} b = 0$$

or, equivalently,

$$\left.\frac{\partial D(s, \tau)}{\partial \tau}\right|_{\tau=b=0} \tau + \left.\frac{\partial D(s, \tau)}{\partial b}\right|_{\tau=b=0} b = 0$$

Applying the above formulas to the denominator polynomial of (6.25) and equating the real part of the equation to zero, we obtain an equation for b. Solving this equation, we find that

$$b = \frac{H(H-2)\omega_p\tau/(H-1)}{1 + 2H\omega_p\tau\left[1/(H-1)^{1/2} - 1/(2Q_p)\right]} \tag{6.29}$$

With the above choice of b, Q enhancement can be avoided. However, the change in the pole frequency cannot be avoided and, in fact, the per-unit

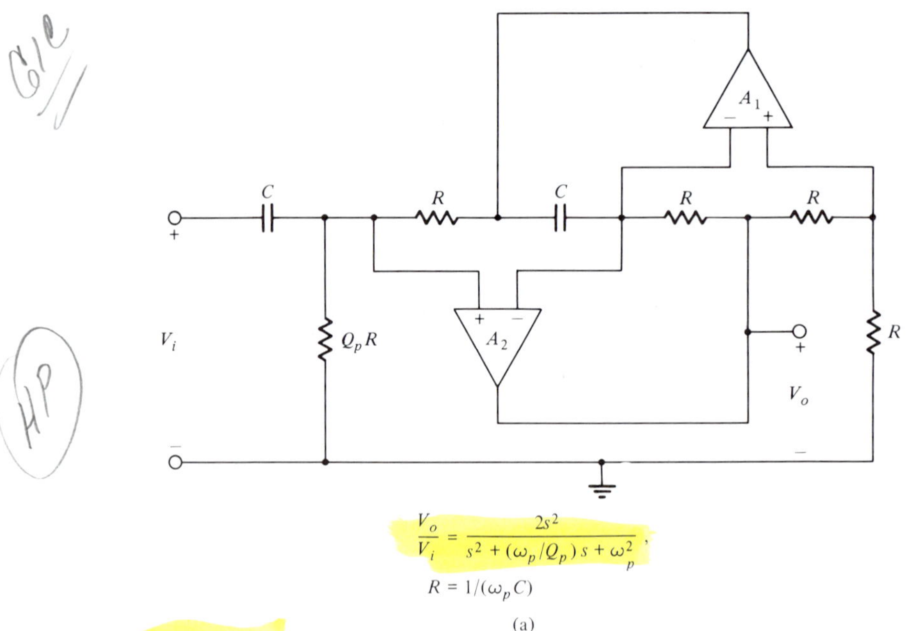

$$\frac{V_o}{V_i} = \frac{2s^2}{s^2 + (\omega_p/Q_p)s + \omega_p^2},$$

$R = 1/(\omega_p C)$

(a)

Figure 6.13 Optimum second-order filter sections derived from GIC. (a) Highpass. (b) Lowpass. (c) Lowpass notch. (d) Highpass notch. (e) All-pass.

change in the pole frequency will be

$$\frac{\Delta f_p}{f_p} = -\frac{H}{2}\frac{2}{(H-1)^{1/2}+b}\omega_p\tau \tag{6.30}$$

For example, in the network designed in Example 6.1, if we choose $b = 0.652$ (i.e., $R_c = 24.428\ \text{k}\Omega$), then Q enhancement can be avoided. The network designed in Example 6.1 was simulated on a computer with the above value of R_c, and the magnitude response is shown in Fig. 6.12 as curve *c*. One can clearly see that there is no Q enhancement in this characteristic. However, there is a frequency shift. Again, the predistortion methods of Chap. 3 may be used to design this circuit for a given set of specifications. This has been implemented in the program GICBPN in App. 6A. This program can be used to design a modified GIC-derived bandpass filter, which includes R_c, with minimum active pole sensitivity where predistortion is used simultaneously. For example, the values of the circuit components in Fig. 6.11 to meet the specifications in Example 6.1 are $R_1 = 4.2321\ \text{k}\Omega$, $R_3 = 4.2108\ \text{k}\Omega$, $R_4 = 75.2751\ \text{k}\Omega$, $R_5 = 10\ \text{k}\Omega$, $R_Q = 15.9155\ \text{k}\Omega$, $R_c = 21.1612\ \text{k}\Omega$, and $C = C_2 = 10$ nF. Finally, it should be mentioned that addition of the resistor R_c will help in avoiding Q enhancement but not in reducing active pole sensitivity.

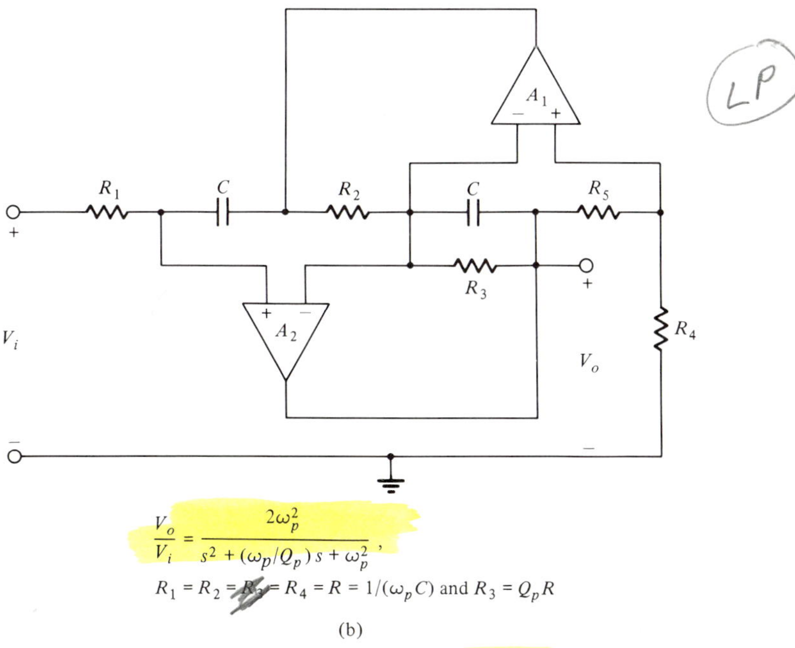

$$\frac{V_o}{V_i} = \frac{2\omega_p^2}{s^2 + (\omega_p/Q_p)s + \omega_p^2},$$

$R_1 = R_2 = R_5 = R_4 = R = 1/(\omega_p C)$ and $R_3 = Q_p R$

(b)

Figure 6.13 Continued.

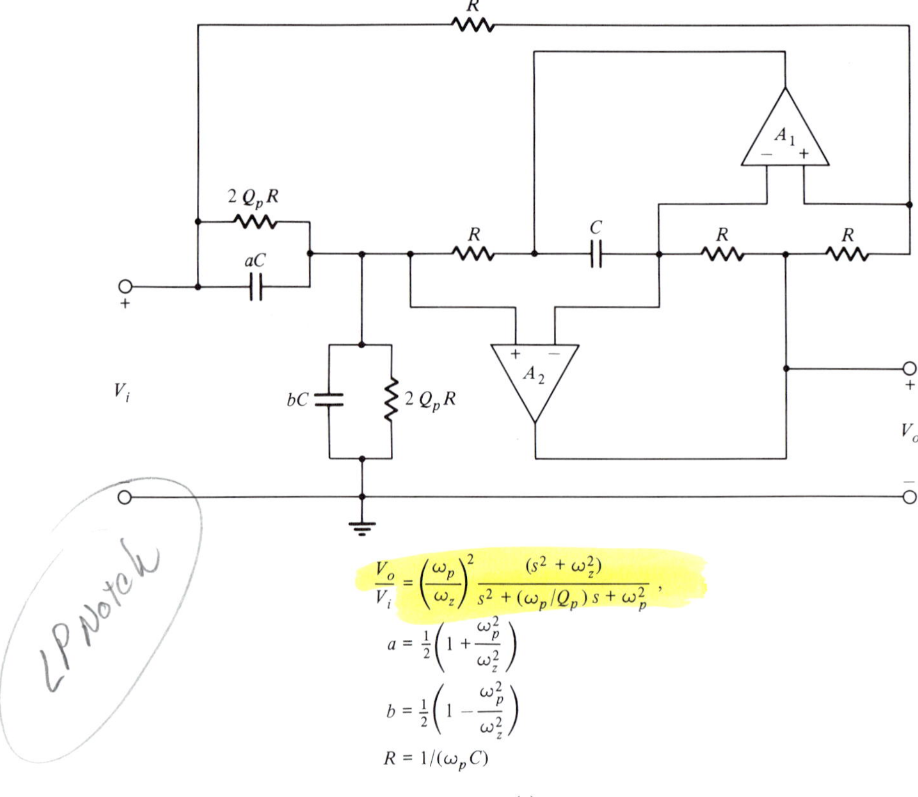

(c)

Figure 6.13 Continued.

So far we have dealt with a GIC-derived bandpass filter. There are other GIC-derived filters with low passive and active sensitivities. Figure 6.13 gives the other types of filters [3], and all of them have low sensitivities. The design equations and their corresponding transfer functions are also given in Fig. 6.13.

6.4 *Integrators and Their Properties*

The next category of filters considered belong to the group known as *analog computer simulation* techniques. The basic components required for this type of simulation technique are integrators and summing amplifiers. The most general form of a summing amplifier circuit is shown in Fig. 6.14.

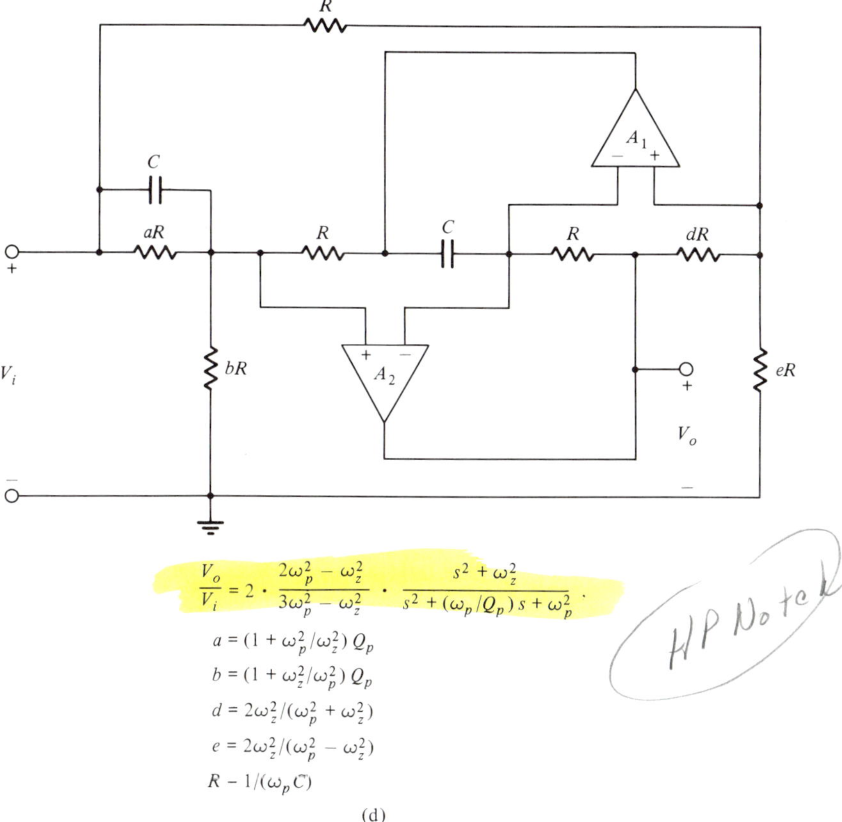

(d)

Figure 6.13 Continued.

When the op amp is assumed to be ideal, the output voltage can be shown to be

$$V_o = \left[1 + \frac{\sum^{k} G_j}{G}\right] \frac{\sum^{n} G_i' V_i'}{\sum^{n} G_i'} - \frac{\sum^{k} G_j V_j}{G} \tag{6.31}$$

As a particular application of this circuit, consider the case where $k = 2$ and $n = 2$ in Fig. 6.14. The op amp circuit corresponding to this case is shown in Fig. 6.15. For this circuit (6.31) reduces to

$$V_o = \frac{G + G_1 + G_2}{G(G_1' + G_2')} (G_1' V_1' + G_2' V_2') - \frac{G_1 V_1 + G_2 V_2}{G} \tag{6.32}$$

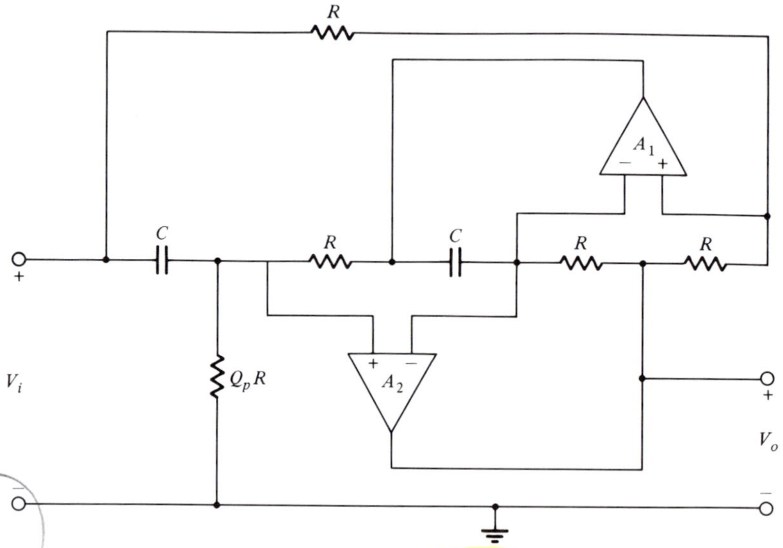

$$\frac{V_o}{V_i} = \frac{s^2 - (\omega_p/Q_p)s + \omega_p^2}{s^2 + (\omega_p/Q_p)s + \omega_p^2}$$

$$R = 1/(\omega_p C)$$

(e)

Figure 6.13 Continued.

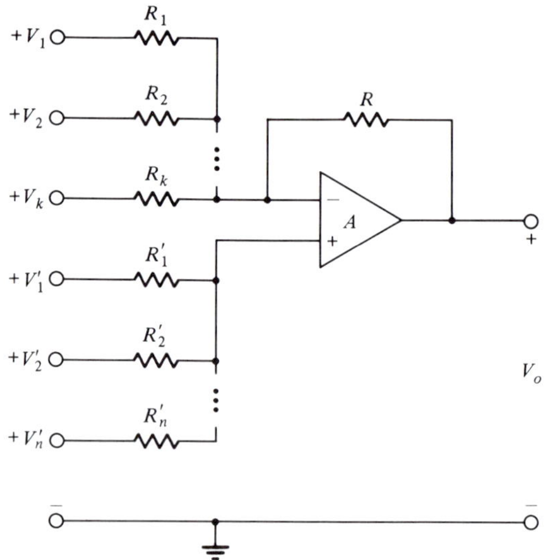

Figure 6.14 Operational amplifier summing amplifier.

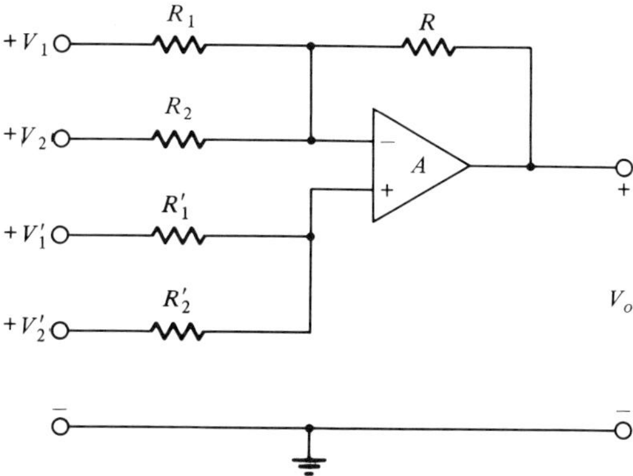

Figure 6.15 Operational amplifier summing amplifier with four inputs.

Integrators are of two types, namely, inverting and noninverting. An inverting integrator (also called a Miller integrator) is shown in Fig. 6.16. When the op amp is assumed to be ideal, this circuit has the transfer function

$$T_I(s) = \frac{V_o}{V_i} = \frac{-1}{RCs} = \frac{-\omega_c}{s} \tag{6.33}$$

where $\omega_c = 1/RC$ is the 0-dB crossover frequency of the integrator and it is an important frequency for integrators. For sinusoidal excitations,

$$T_I(j\omega) = |T|\angle\phi = \frac{\omega_c}{\omega}\angle 90° \tag{6.34}$$

From (6.34), one can see that an ideal inverting integrator should have as its magnitude and phase functions

$$|T_I(j\omega)| = \frac{\omega_c}{\omega} \qquad \text{and} \qquad \arg T_I(j\omega) = 90° \tag{6.35}$$

However, as in any other op amp circuit, the frequency-dependent gain of

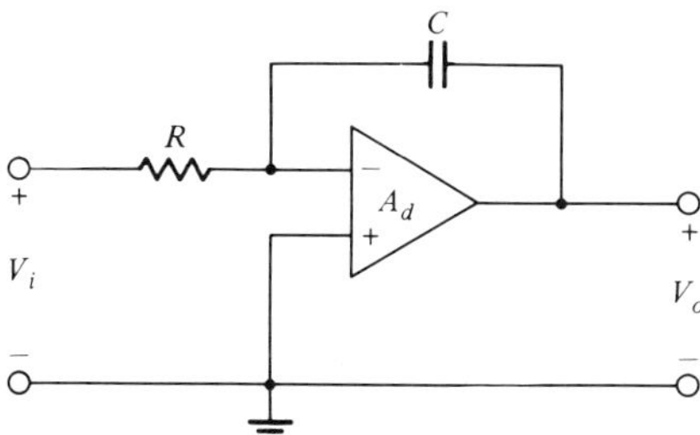

Figure 6.16 Inverting (Miller) integrator.

the op amp affects the performance of this component also. The nonzero time constant of the op amp affects both the magnitude and phase of the transfer function. Since the sensitivities of the filters using the integrator as a component are closely related to the sensitivity of the integrator used, it is worthwhile to study the effects of the nonzero time constant of the op amps in integrators. When the op amp gain is assumed to be $1/(s\tau)$ in terms of its time constant τ, the transfer function of the circuit in Fig. 6.16 is

$$T(s) = \frac{V_o}{V_i} = -\frac{\omega_c/s}{(1 + \omega_c\tau) + s\tau} \tag{6.36}$$

where $\omega_c = 1/RC$.

The semilogarithmic active sensitivity of $T(s)$ can be shown to be

$$S_\tau^{T(s)} = -(\omega_c + s) \tag{6.37}$$

and therefore, for sinusoidal excitations,

$$S_\tau^{T(j\omega)} = -(\omega_c + j\omega) \tag{6.38}$$

Using (6.38), we can find the first-order changes in the magnitude and phase functions due to the nonzero value of τ:

$$\frac{\Delta|T|}{|T|} = -\omega_c\tau \tag{6.39a}$$

and

$$\Delta\phi = -\omega\tau \tag{6.39b}$$

At the critical frequency, $\omega = \omega_c$, the deviations in both magnitude and phase are equal, and one is as bad as the other. In addition, the deviation in the phase is a function of the frequency.

One method of minimizing the errors caused by the nonzero τ value is to connect a compensating capacitor C_c across the resistor in the circuit in Fig. 6.16. This method of compensation is known as *passive compensation* and is shown in Fig. 6.17. The transfer function of this circuit can be derived as

$$T(s) = \frac{-\omega_c}{s}\,\frac{1 + RC_c s}{(1 + \omega_c\tau) + s\tau(1 + C_c/C)} \tag{6.40}$$

If we choose $C_c = \tau/R$, then (6.40) reduces to

$$T(s) = \frac{-\omega_c}{s}\,\frac{1}{1 + \omega_c\tau} \tag{6.41}$$

From (6.41), for sinusoidal excitations, we find that

$$\frac{\Delta|T|}{|T|} = -\omega_c\tau \tag{6.42a}$$

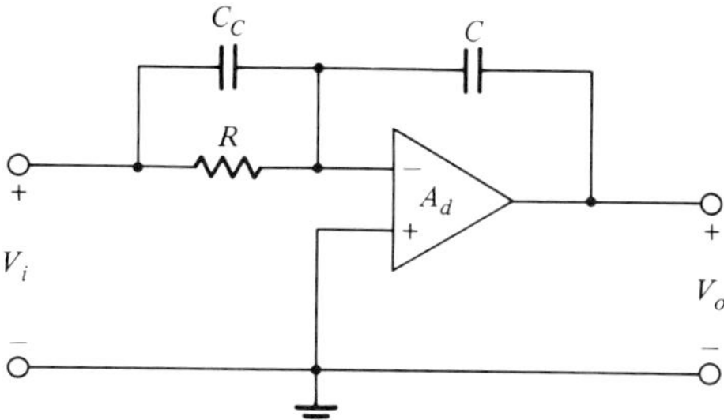

Figure 6.17 Passive compensation of a Miller integrator.

and

$$\Delta\phi = 0 \tag{6.42b}$$

However, $\partial\phi/\partial\tau$ is not equal to zero. As τ varies with temperature and power supply variations, the compensating capacitor C_c in this circuit can eliminate the initial change in the phase but cannot compensate for the subsequent changes in the value of τ. In fact, (6.37) and (6.38) can be used for this circuit also, but minor changes are needed in (6.39).

Now consider the circuits shown in Fig. 6.18. They are also inverting integrators and are actively compensated. The transfer function of the circuit in Fig. 6.18a [4] is

$$T(s) = \frac{V_o}{V_i} = \frac{-\omega_c}{s}\frac{1 + s\tau_2}{(1 + \omega_c\tau_1) + s\tau_1(1 + \omega_c\tau_2) + s^2\tau_1\tau_2} \tag{6.43}$$

Since the time constants of the two op amps can be expected to have the same values and can also be expected to track with each other (particularly in dual and quad op amps), we can assume that $\tau_1 = \tau_2 = \tau$ and $\Delta\tau_1/\tau_1 = \Delta\tau_2/\tau_2 = \Delta\tau/\tau$. Hence, substituting $\tau_1 = \tau_2 = \tau$ in (6.43), we have

$$T(s) = \frac{V_o}{V_i} = \frac{-\omega_c}{s}\frac{1 + s\tau}{(1 + \omega_c\tau)(1 + s\tau) + (s\tau)^2} \tag{6.44}$$

The semilogarithmic active sensitivity of $T(s)$ is

$$S_\tau^{T(s)} = -\omega_c \tag{6.45}$$

and

$$S_\tau^{T(j\omega)} = -\omega_c \tag{6.46}$$

By comparing (6.46) with (6.38), we can conclude that the imaginary part of the active sensitivity has been eliminated by using the active compensation

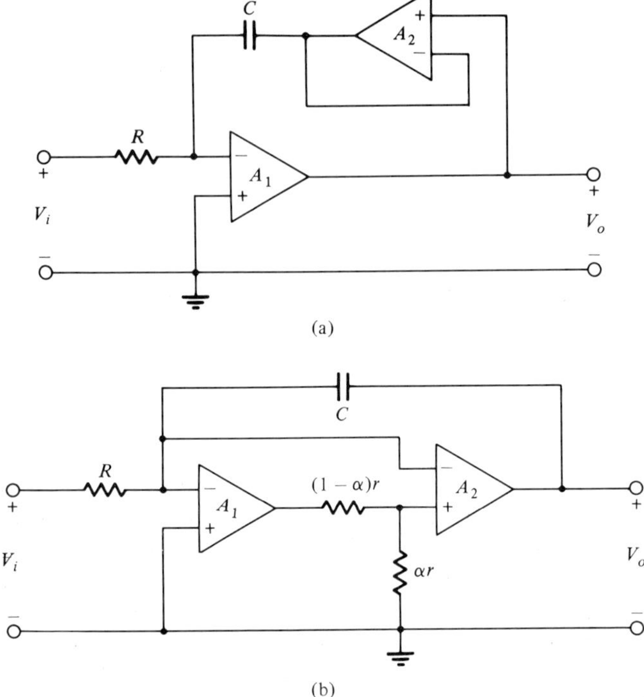

Figure 6.18 Actively compensated Miller integrators.

technique. Also using (6.46), we have

$$\frac{\Delta|T|}{|T|} = -\omega_c \tau \tag{6.47a}$$

and

$$\Delta\phi = 0 \tag{6.47b}$$

The circuit in Fig. 6.18*b* [5] has the transfer function

$$T(s) = \frac{V_o}{V_i} = \frac{-\omega_c}{s} \frac{1 + \alpha s \tau_1}{1 + \alpha s \tau_1(1 + \omega_c \tau_2) + \alpha s^2 \tau_1 \tau_2} \tag{6.48}$$

Since the nominal values of τ_1 and τ_2 are zeros, from the above equations we obtain the semilogarithmic sensitivities as

$$S_{\tau_1}^{T(s)} = \frac{\partial T(s)}{\partial \tau_1} \frac{1}{T(s)}\bigg|_{\tau_1 = \tau_2 = 0} = 0 \tag{6.49a}$$

and

$$S_{\tau_2}^{T(s)} = \frac{\partial T(s)}{\partial \tau_2} \frac{1}{T(s)}\bigg|_{\tau_1 = \tau_2 = 0} = 0 \tag{6.49b}$$

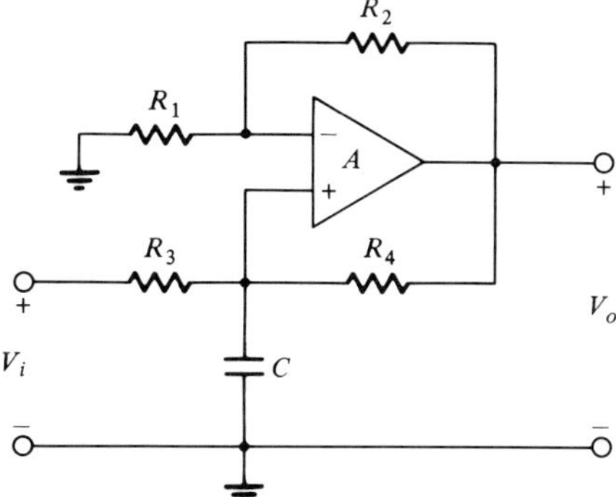

Figure 6.19 Deboo noninverting integrator.

From (6.49), several observations can be made about the circuit in Fig. 6.18*b*:

1. The sensitivities of $T(s)$ with respect to both τ_1 and τ_2 are zeros. We did not find such zero sensitivities for any of the previously discussed circuits. Even with nonzero time constants, these sensitivities will be very small.
2. We did not have to assume that $\tau_1 = \tau_2$, and the time constants of the two op amps need not track with each other for obtaining such very small sensitivities. This is because very small sensitivities are obtained with respect to both τ_1 and τ_2.
3. The choice of α is arbitrary. Of course, $\alpha = 1$ is a convenient choice because it eliminates the need for the two resistors $(1 - \alpha)R$ and αR and the output of A_1 can be directly connected to the noninverting input of A_2.

Thus the circuit in Fig. 6.18*b* is the best of all the inverting integrators we have discussed so far. Next, we shall consider the realization of noninverting integrators.

The noninverting integrator circuit shown in Fig. 6.19 was first described by Deboo [6] and uses only one op amp. The transfer function of this circuit can be derived as

$$T(s) = \frac{V_o}{V_i} = \frac{1 + a}{CR_3 s + (1 - ab) + s\tau(1 + b + CR_3 s)} \tag{6.50}$$

where $a = R_2/R_1$, $b = R_3/R_4$, and $A^{-1} = s\tau$. With nominal values of $\tau = 0$ and $ab = 1$, the transfer function reduces to

$$T(s) = \frac{1 + a}{CR_3 s} \tag{6.51}$$

This circuit is attractive because it uses a single op amp, but it has certain drawbacks. First, the circuit will become unstable if $ab > 1$. Also, if $ab < 1$, it will realize only a negative real pole (even assuming that $\tau = 0$) and not a pole at the origin as required by an ideal integrator. All these mean that it will require tighter tolerance of passive elements because the circuit operation is highly sensitive to passive elements. Therefore the practical usefulness of this circuit is very limited and we shall not consider it further.

The conventional method of realizing a noninverting integrator is shown in Fig. 6.20a. When the op amps are ideal, all the circuits in Fig. 6.20

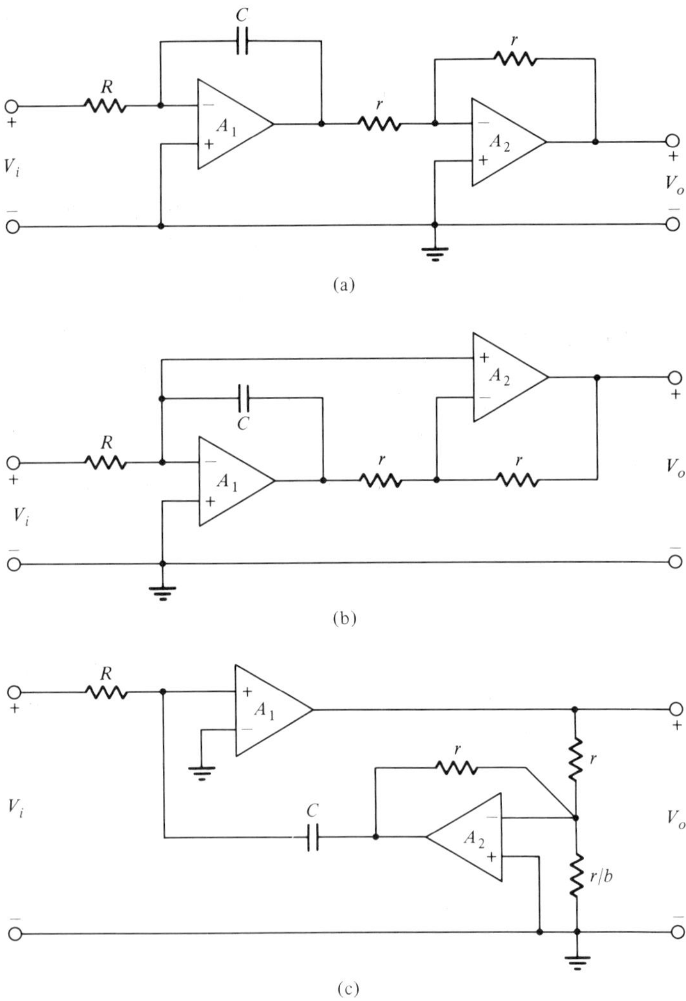

Figure 6.20 Other noninverting integrators.

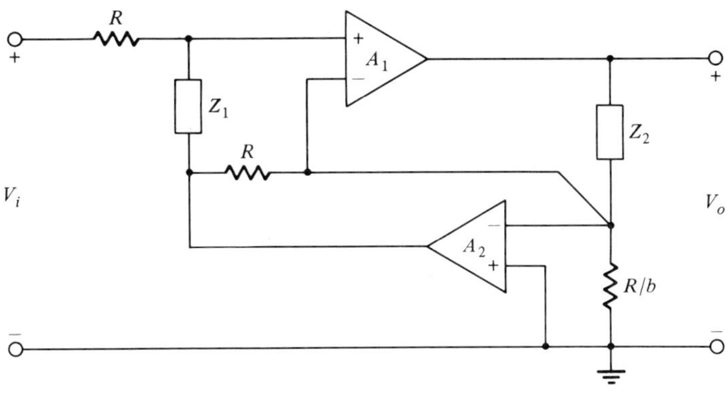

(a) $Z_1 = R$
$Z_2 = (1/Cs)$
$\omega_c = (1/RC)$

(b) $Z_1 = (1/Cs)$
$Z_2 = R$
$\omega_c = (1/RC)$

(d)

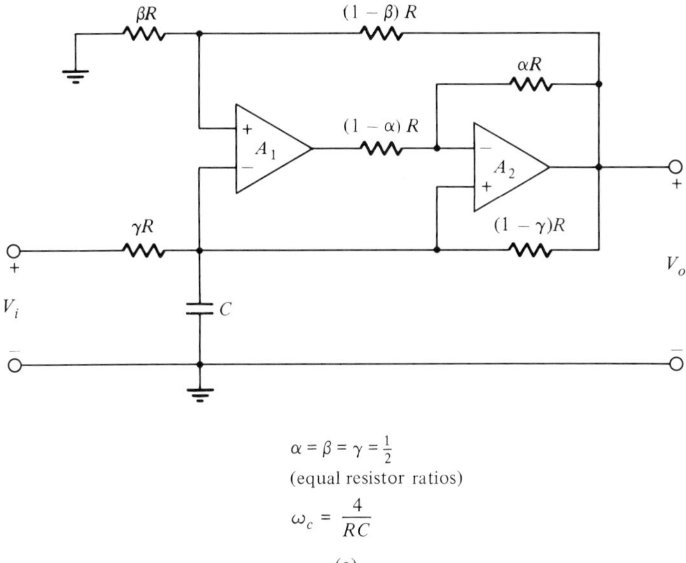

$\alpha = \beta = \gamma = \frac{1}{2}$
(equal resistor ratios)

$$\omega_c = \frac{4}{RC}$$

(e)

Figure 6.20 Continued.

Table 6.1 Comparison of the active sensitivities of the noninverting integrators

Circuit	$S_{\tau_1}^{T(j\omega)}$	$S_{\tau_2}^{T(j\omega)}$	$\Delta\|T\|/\|T\|$	$\Delta\|T\|/\|T\|$, $\tau_1 = \tau_2 = \tau$	$\Delta\phi$	$\Delta\phi$, $\tau_1 = \tau_2 = \tau$
6.20*a*	$-\omega_c - j\omega$	$-2j\omega$	$-\omega_c\tau_1$	$-\omega_c\tau$	$-j\omega(\tau_1 + 2\tau_2)$	$-3j\omega\tau$
6.20*b*	$-\omega_c + j\omega$	$-2j\omega$	$-\omega_c\tau_1$	$-\omega_c\tau$	$\omega(\tau_1 - 2\tau_2)$	$-\omega\tau$
6.20*c*	$-\omega_c - j\omega$	$j(2 + b)\omega$	$-\omega_c\tau_1$	$-\omega_c\tau$	$-\omega\tau_1 + (2 + b)\omega\tau_2$	$\omega(1 + b)\tau$ $\omega\tau$, if $b = 0$
6.20*d* (i)[a]	$-2\omega_c$	$(-\omega^2/\omega_c) + j\omega(b - 1)$	$-2\omega_c\tau_1 - \omega_c\tau_2$	$-3\omega_c\tau$	$\omega(b - 1)\tau_2$	$\omega(b - 1)\tau$ 0, if $b = 1$
6.20*d* (ii)	$-\omega_c - j\omega$	$-\omega_c + j(1 + b)\omega$	$-\omega_c(\tau_1 + \tau_2)$	$-2\omega_c\tau$	$\omega\{(1 + b)\tau_2 - \tau_1\}$	$b\omega\tau$ 0, if $b = 0$
6.20*e*	0	0	0	0	0	0

[a] In this case $\Delta|T|/|T|$ is frequency-dependent, but the value in the table was evaluated at the critical frequency $\omega = \omega_c$.

realize the transfer function

$$T(s) = \frac{V_o}{V_i} = \frac{\omega_c}{s} \tag{6.52}$$

where the corresponding ω_c values are also given in Fig. 6.20. All the networks, except the one in Fig. 6.20*e*, are absolutely stable. The network in Fig. 6.20*e* is conditionally stable, similar to Deboo's integrator circuit. The passive sensitivities of the transfer function of all the circuits are low and have about the same values. However, the active sensitivities of each one are quite different from those of others. Table 6.1 lists the active sensitivities and the effects the nonzero time constants of the op amps have on the magnitude and phase responses of their transfer functions when τ_1 and τ_2 are assumed to be small.

6.5 *Integrator Filters*

Second-order integrator filters are derived from block diagram representations of second-order transfer functions. Consider the transfer function of the form

$$T(s) = \frac{V_o}{V_i} = \frac{P(s)}{s^2 + (\omega_p/Q_p)s + \omega_p^2} \tag{6.53}$$

Rearranging (6.53), we have

$$V_o = -\frac{\omega_p/Q_p}{s}V_o - \frac{\omega_p^2}{s^2}V_o + \frac{P(s)}{s^2}V_i \tag{6.54}$$

Consider the particular case where $P(s) = H_o s^2$. Then, (6.54) can be represented by the block diagram shown in Fig. 6.21. This block diagram has two negative feedback loops, one controlling Q_p and the other controlling ω_p. The filters realizing this block diagram are called *two-integrator loop filters*. All the parameters of the filters can be controlled independently, and this is one of the major advantages of these types of filters compared to the

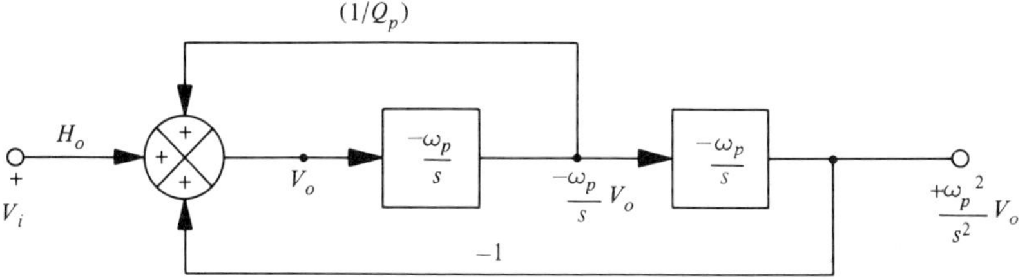

Figure 6.21 Block diagram representation of (6.54).

single-amplifier filters discussed in the previous chapters. Another advantage is their ability to provide different filtering functions simultaneously at different outputs in the same circuit.

Kerwin-Huelsman-Newcomb Biquad

The second-order filter circuit shown in Fig. 6.22 is called the *KHN biquad* [7] after its inventors, W. J. Kerwin, L. P. Huelsman, and R. W. Newcomb. This filter realization is also referred to as a *state variable* filter because this circuit is derived from state variable representation of the second-order transfer function shown in Fig. 6.21. Assuming that the op amps in the circuit are ideal, we obtain

$$V_{BP} = -\frac{V_{HP}}{R_1 C_1 s} \tag{6.55a}$$

$$V_{LP} = \frac{-V_{BP}}{R_2 C_2 s} = \frac{V_{HP}}{R_1 R_2 C_1 C_2 s^2} \tag{6.55b}$$

Using the above equations in the circuit analysis, we find the transfer functions

$$\frac{V_{HP}}{V_i} = \frac{H_1 s^2}{D(s)} \tag{6.56}$$

$$\frac{V_{BP}}{V_i} = \frac{-H_1 s/(R_1 C_1)}{D(s)} \tag{6.57}$$

and

$$\frac{V_{LP}}{V_i} = \frac{H_1/(R_1 R_2 C_1 C_2)}{D(s)} \tag{6.58}$$

where

$$H_1 = \frac{1 + R_6/R_5}{1 + R_3/R_4} \tag{6.59}$$

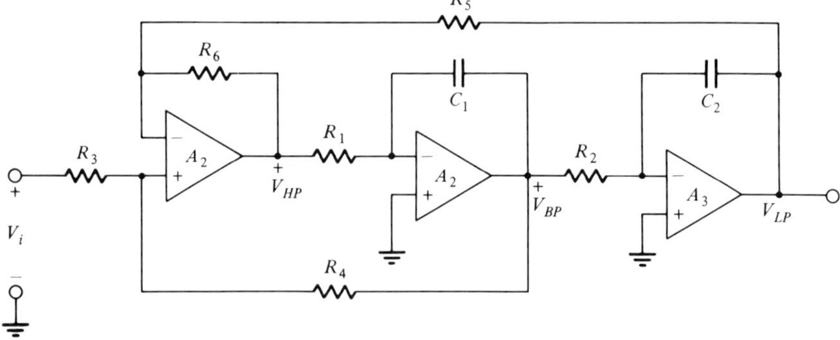

Figure 6.22 The KHN biquad.

and

$$D(s) = s^2 + \frac{s[(1 + R_6/R_5)/(1 + R_3/R_4)]}{(R_1C_1)} + \frac{R_6/R_5}{R_1R_2C_1C_2} \quad (6.60)$$

Equation (6.59) is one of the design equations, and two more design equations for ω_p and Q_p can be obtained from the denominator polynomial:

$$\omega_p = \left[\frac{R_6/R_5}{R_1R_2C_1C_2}\right]^{1/2} \quad (6.61a)$$

and

$$\frac{1}{Q_p} = \frac{1 + R_6/\{R_5}{1 + R_4/\{R_3} \frac{(R_5R_2C_2)^{1/2}\}}{(R_6R_1C_1)^{1/2}\}} \quad (6.61b)$$

The passive ω_p and Q_p sensitivities are

$$S^{\omega_p}_{R_1, R_2, R_5, C_1, C_2} = -0.5 = -S^{\omega_p}_{R_6} \quad (6.62a)$$

$$S^{Q_p}_{R_1, C_1} = -S^{Q_p}_{R_2, C_2} = 0.5 \quad (6.62b)$$

$$S^{Q_p}_{R_4} = -S^{Q_p}_{R_3} = \frac{R_4}{R_4 + R_3} < 1 \quad (6.62c)$$

and

$$S^{Q_p}_{R_5} = -S^{Q_p}_{R_6} = \frac{-Q_p}{2} \frac{R_5 - R_6}{1 + R_4/R_3} \frac{(R_2C_2)^{1/2}}{(R_5R_6R_1C_1)^{1/2}} \quad (6.62d)$$

The Q_p sensitivity with respect to R_5 and R_6 may be set to zero by choosing $R_5 = R_6$. Also, the Q_p sensitivities with respect to R_4 and R_3 are less than unity, and all other sensitivities are at their theoretical minimum of 0.5. This is one of the reasons for the popularity of KHN biquads. The simplest design procedure can be developed by choosing $R_5 = R_6$, $R_1 = R_2 = R$, and $C_1 = C_2 = C$. We also select a convenient value for C. Then,

$$R_1 = R_2 = R = \frac{1}{\omega_p C} \quad (6.63a)$$

and

$$R_4 = (2Q_p - 1)R_3 \quad (6.63b)$$

The gain constant is then restricted for lowpass and highpass filters, by

$$H_o = 2 - \frac{1}{Q_p} \quad (6.64a)$$

and for bandpass filters by

$$H_o = 1 - 2Q_p \tag{6.64b}$$

The values of ω_p and Q_p can be independently controlled by adjusting R_1 and/or R_2 (for ω_p) and R_4 (for Q_p).

Example 6.2: Using the state variable realization, design a bandpass filter that meets the specifications $f_p = 3$ kHz and $Q_p = 20$. Choose $C_1 = C_2 = C = 10$ nF.

Using (6.63), we find that $R_1 = R_2 = 5.305$ kΩ. Choose $R_3 = R_5 = R_6 = 1$ kΩ. Then $R_4 = 39$ kΩ. ■

In the KHN filter, there is a free parameter (R_6/R_5) which was arbitrarily set to unity earlier. However, we can use this parameter to minimize the magnitude of the active pole sensitivity. The derivation of the active pole sensitivity of this network is left as an exercise, however, one can show that the magnitude of the active pole sensitivity is minimized by having $R_6/R_5 = 0.732$. The program KHNB, which appears in App. 6B, can be used to design this filter circuit with minimum magnitude of the active pole sensitivity and with predistortion. The gain constant requirement is not met in this design.

The Tow-Thomas Biquad

Another implementation of the block diagram in Fig. 6.21 is shown in Fig. 6.23 [8–10]. As can be seen from the circuit in Fig. 6.23, the op amp outputs provide only bandpass and lowpass filter functions, and a highpass filter function is not available as an op amp output. This is not a major

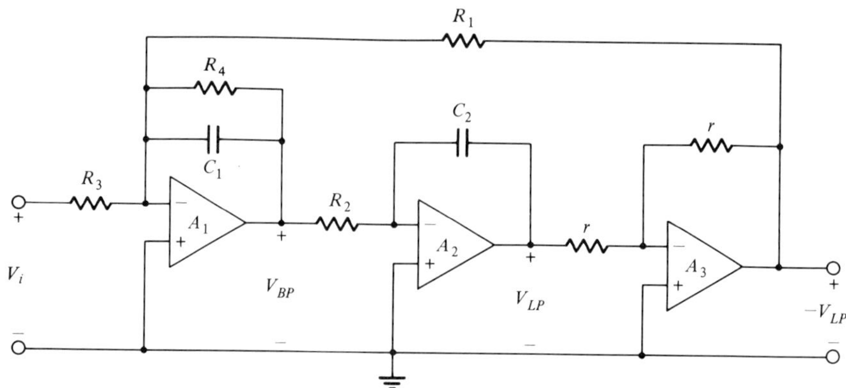

Figure 6.23 Tow-Thomas biquad.

disadvantage. As we will soon see, a modification of this circuit is able to provide a highpass output. The advantage of this circuit is that, in addition to the fact that ω_p and Q_p can be controlled independently as in the KHN biquad, Q_p and $|H_o|$ can be controlled using individual resistors rather than a combination of resistors as in the KHN biquad. Furthermore, all the noninverting input terminals of the op amps are grounded. When the op amps are ideal, the transfer functions of the network in Fig. 6.23 are

$$\frac{V_{BP}}{V_i} = \frac{-s/(R_3C_1)}{s^2 + s/(R_4C_1) + 1/(R_1R_2C_1C_2)} \tag{6.65a}$$

and

$$\frac{V_{LP}}{V_i} = \frac{1/(R_2R_3C_1C_2)}{s^2 + s/(R_4C_1) + 1/(R_1R_2C_1C_2)} \tag{6.65b}$$

Note that the transfer functions given by (6.65) are simpler compared to those in (6.56) through (6.60) for KHN biquads. The ω_p and Q_p sensitivities with respect to the passive elements have magnitudes equal to either 0.5 or 1. A design procedure can be developed by choosing $R_1 = R_2 = R$ and $C_1 = C_2 = C$. This choice also ensures that the maximum outputs of the op amps over the entire frequency range will be equal and that no op amp will be overdriven. This problem involves dynamic range and is addressed in the next chapter with respect to higher-order filter realizations. Proceeding further with the above choice, we have the following design equations:

$$R = \frac{1}{\omega_p C} \tag{6.66a}$$

$$R_4 = Q_p R \tag{6.66b}$$

and

$$R_3 = \frac{R}{H_o} \qquad \text{for lowpass} \tag{6.66c}$$

or

$$R_3 = \frac{Q_p}{H_o}R \qquad \text{for bandpass} \tag{6.66d}$$

We shall next find the active sensitivity of the pole $p = \omega_p(-1/2Q_p + j1)$ for this circuit. We assume that all the op amps have identical time constants equal to τ and that $A_1 = A_2 = A_3 = 1/(s\tau)$. Since we are interested in only the active pole sensitivity, we also assume that the passive components have been chosen to satisfy the given set of specifications using (6.66). Then, the denominator polynomial of the transfer function can be shown to be

$$D(s, \tau) = s^2 + \frac{\omega_p}{Q_p}s + \omega_p^2 + s\tau\left[\frac{\omega_p^2}{Q_p} + s\omega_p\frac{(2Q_p + H_o + 4)}{Q_p} + 4s^2\right] + \cdots$$

Applying (3.62) to the above denominator polynomial, we have

$$S_\tau^p = -\omega_p\left(1 + \frac{H_o}{2Q_p} + j2\right) \tag{6.67}$$

In Sallen and Key and single-amplifier filters, the imaginary part of the active pole sensitivity was almost negligible, and in the approximations we made we were able to ignore this part. However, in multiple-amplifier filters, we find that the imaginary part of the active pole sensitivity is comparable to the real part and we cannot ignore it. In order to see the effects of this sensitivity and the nonzero value of τ, let us calculate the shift in the dominant pole positions. Using these sensitivities we shall find the approximate changes in the ω_p and Q_p values. We know that

$$\Delta p = \frac{\partial p}{\partial \tau}\Delta\tau = \frac{\partial p}{\partial \tau}\tau = \tau p S_\tau^p \tag{6.68}$$

Substituting (6.67) into (6.68) with $p = \omega_p(-1/2Q_p + j1)$, we have

$$\Delta p = \omega_p^2\tau\left(2 - \frac{jH_o}{2Q_p}\right) \tag{6.69}$$

where terms on the order of $1/Q_p$ compared to unity have been ignored. The new pole position, $p + \Delta p$, to a first-order approximation, is

$$\begin{aligned} p' &= \frac{-\omega_p(1 - 4Q_p\omega_p\tau)}{2Q_p} + j\omega_p\left[1 - \omega_p\tau\left(1 + \frac{H_o}{2Q_p}\right)\right] \\ &\triangleq \frac{-\omega_p'}{2Q_p'} + j\omega_p' \end{aligned}$$

where ω_p' and Q_p' are the effective values of the pole frequency and the pole Q factor, respectively. From the above equations, we know, to a first-order approximation, that

$$\omega_p' = \omega_p\left[1 - \omega_p\tau\left(1 + \frac{H_o}{2Q_p}\right)\right] \tag{6.70}$$

and

$$Q_p' \simeq \frac{Q_p}{1 - 4Q_p\omega_p\tau} \tag{6.71}$$

In order to understand the meaning of (6.70) and (6.71), consider the following example.

Example 6.3: Design a Tow-Thomas (TTH) biquad to meet the specifications of a bandpass filter having $H_o = Q_p = 20$ with $f_p = 5$ kHz. Assume that the op amps are of the internally compensated 741 type whose nominal *GB* product values are 1 MHz. Calculate the actual values of f_p and Q_p.

Choosing 10 nF as a convenient value for C_1 and C_2, we require that $R_1 = R_2 = R_3 = 3.183$ kΩ and $R_4 = 63.66$ kΩ. We select any convenient value, such as 10 kΩ, for r. Using (6.70), we find that $f_p' = 4.963$ kHz. The actual value of Q_p will be Q_p' and is 33.33 (at a frequency range of only 5 kHz). This Q enhancement can be traced back to the imaginary part of the active pole sensitivity and finally back to the phase sensitivities of the transfer functions of the integrators and the inverter due to the nonzero time constants of the op amps in this network. ■

Example 6.3 illustrates that, in terms of active pole sensitivity, these biquads are the worst of all the types of filters we have considered so far. Furthermore, from (6.71), note that, when $\omega_p Q_p > 1/(4\tau) = B/4$, Q_p' becomes negative. This means that, when $\omega_p Q_p \geq B/4$, the filter circuit is unstable, which is an inherent problem associated with these types of filters. A simple method of precorrecting this situation is to design the filter with ω_p and Q_p such that ω_p' and Q_p' result in the desired values. This is just predistortion. For this filter, predistortion can be achieved by solving (6.70) and (6.71) for ω_p and Q_p in terms of ω_p' and Q_p' using a system of nonlinear equations. Solving for Q_p from (6.71), we have

$$Q_p = \frac{Q_p'}{1 + 4Q_p'\omega_p\tau} \tag{6.72}$$

Substituting this value of Q_p into (6.70), we obtain a third-order equation for ω_p:

$$\left[2H_o(\omega_p\tau)^2 + \left(1 + \frac{H_o}{2Q_p'}\right)(\omega_p\tau) - 1\right]\omega_p + \omega_p' = 0 \tag{6.73a}$$

or

$$\omega_p = \frac{\omega_p'}{1 - 2H_o(\omega_p\tau)^2 - (1 + H_o/2Q_p')(\omega_p\tau)} \tag{6.73b}$$

We can solve the above third-order equation (6.73*a*) or (6.73*b*) iteratively starting with $\omega_p = \omega_p'$ and using the desired pole Q factor and gain constant values for Q_p' and H_o, respectively. Equation (6.73*b*) is suitable for iteration. Once the design value of ω_p is found, we can substitute this value into (6.72) to find the design value of Q_p in one step. For example,

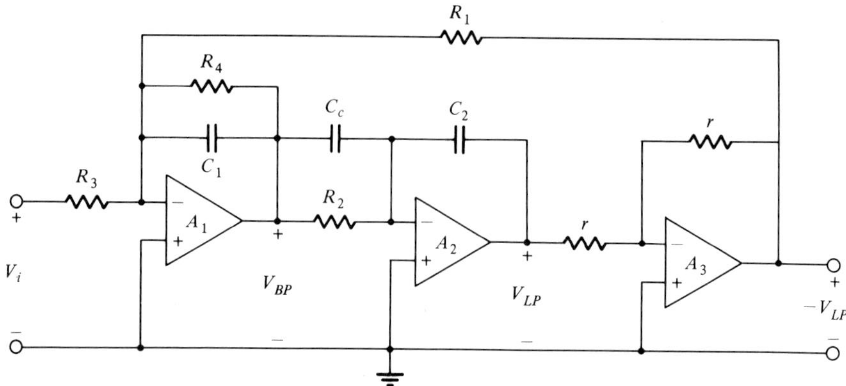

Figure 6.24 Passively compensated Tow-Thomas biquad with $C_c = 4\tau/R_2$.

consider the problem in Example 6.3. The desired or actual values of Q_p, f_p, and H_o should be 20 kHz, 5 kHz, and 20, respectively. Let us find the corresponding design values of Q_p and f_p. Substituting the desired pole Q factor values for Q_p' and H_o in (6.73*b*) and solving this equation iteratively, we find the design value of f_p to be 5.043 kHz. One usually obtains this value after two or three iterations. Using this value of f_p in (6.72), we find the design value of Q_p to be 14.25 in one step. Thus, if the filter circuit is designed with $Q_p = 14.25$ and $f_p = 5.043$ kHz, the filter is likely to have realized Q_p and f_p values close to the desired values, namely, $Q_p' = 20$ and $f_p' = 5$ kHz.

Another method of avoiding Q enhancement is to connect a compensating capacitor C_c across either R_1 or R_2, similar to the passive compensation technique discussed in the previous section for the inverting integrator. Assume that we connect the capacitance C_c across R_2 and the resulting circuit is shown in Fig. 6.24. Before connecting this capacitor, we can find the excess phase $\Delta\phi$, as explained in the previous section for each integrator and the inverting amplifier. For the first and second integrators, we have

$$\Delta\phi_1 = -\omega\tau_1 \quad \text{and} \quad \Delta\phi_2 = -\omega\tau_2 \tag{6.74}$$

For the inverting unity gain amplifier, it can be shown that the excess phase is

$$\Delta\phi_3 = -2\omega\tau_3 \tag{6.75}$$

Therefore the net phase lag introduced in the loop is

$$\begin{aligned}\Delta\phi &= \Delta\phi_1 + \Delta\phi_2 + \Delta\phi_3 \\ &= -\omega(\tau_1 + \tau_2 + 2\tau_3)\end{aligned} \tag{6.76}$$

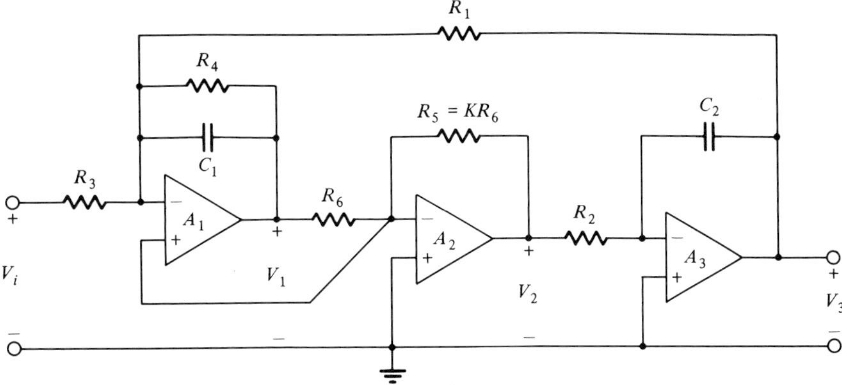

Figure 6.25 Modified Tow-Thomas biquad.

If the op amp time constants are identical and equal to τ, then

$$\Delta\phi = -4\omega\tau \tag{6.77}$$

This phase lag, caused by the nonzero time constants of the op amps in the loop, is responsible for the Q enhancement. Connecting the compensating capacitor C_c introduces an additional phase lead in the loop, and this phase lead is approximately equal to $\omega C_c R_2$ (for small values of $\omega C_c R_2$). If the value of $C_c = 4\tau/R_2$, then the net excessive phase introduced in the loop will be zero, hence the Q enhancement can be avoided.

Consider another circuit with a simple modification in TTH biquad as shown in Fig. 6.25. There is no increase in the number of components as compared to the circuit in Fig. 6.24. By choosing the value of K in the circuit in Fig. 6.25 appropriately, the magnitude of active pole sensitivity can be reduced considerably compared to what it is in the circuits in Figs. 6.23 and 6.24. In this circuit, V_1 is an inverting bandpass output, V_2 is a noninverting bandpass output, and V_3 is an inverting lowpass output. Just like the TTH biquad, this circuit provides multiple outputs. When the op amps are ideal and $K = 1$, the design equations of (6.66) can still be used without modification. However, we want to choose K to minimize the active pole sensitivity. We first find the denominator polynomial including the nonzero values of τ of the op amps used in this circuit. Assuming that $A_1 = A_2 = A_3 = 1/(s\tau)$, the transfer functions of the circuit can be obtained as

$$\frac{V_1}{V_i} = \frac{-asW(1 + W\tau + s\tau)[1 + (K + 1)s\tau]}{D(s, \tau)} \tag{6.78}$$

$$\frac{V_2}{V_i} = \frac{KasW(1 + W\tau + s\tau)}{D(s, \tau)} \tag{6.79}$$

and

$$\frac{V_3}{V_i} = \frac{-KaW^2}{D(s,\tau)} \tag{6.80}$$

where

$$D(s,\tau) = s^2 + \frac{W}{Q}s + KW^2 + s\tau\left\{3s^2 + sW\left[\frac{K+2}{Q} + 1 - (K-1)b\right] + \frac{W^2}{Q}\right\} + \cdots \tag{6.81}$$

$$W = \frac{1}{R_1C_1} = \frac{1}{R_2C_2}$$

$$Q = \frac{R_4}{R_1}$$

$$K = \frac{R_5}{R_6}$$

$$a = \frac{R_1}{R_3}$$

and

$$b = 1 + a + \frac{1}{Q}$$

When we assume ideal op amps, the design equations are

$$W = \frac{\omega_p}{K^{1/2}} \tag{6.82a}$$

$$Q = \frac{Q_p}{K^{1/2}} \tag{6.82b}$$

The choice for a, which depends on the output desired, is

$$a = \begin{cases} \dfrac{H_oK^{1/2}}{Q_p} & \text{for an inverting bandpass} \\ \dfrac{H_o}{Q_pK^{1/2}} & \text{for a noninverting bandpass} \\ H_o & \text{for an inverting lowpass} \end{cases} \tag{6.82c}$$

where K is an arbitrary parameter. We can also choose this parameter K to minimize the active pole sensitivity. First, we obtain the semilogarithmic active pole sensitivity by applying (3.62) to the denominator polynomial in (6.81) and then find the approximate expression to be

$$S_{\tau}^{p} = -\frac{1}{2}\left\{W\left[\frac{K+2}{Q} + 1 - (K-1)b\right] - \frac{3\omega_p}{Q_p}\right\} - j1.5\omega_p \quad (6.83)$$

Note that there are terms with both positive and negative signs in the real part of (6.83). Therefore this real part can be made zero with an appropriate choice of K. Since b is a complicated function of K that depends on the output desired, it is difficult to suggest in advance a particular choice for K. However, once the output is specified, K can be selected. In particular, if Q_p, $H_o \gg 1$, $K = 2$ will make the real part of the active pole sensitivity zero. In any case, henceforth we assume that this can be done, and then the active pole sensitivity reduces to

$$S_{\tau}^{p} = -j1.5\omega_p \quad (6.84)$$

Equation (6.84) means that the frequency shift due to the nonzero value of τ can be completely eliminated and there will be only Q enhancement in the modified TTH circuit if the choice for K is made appropriately. From (6.84), it is easy to show that

$$\omega_p' = \omega_p \quad (6.85a)$$

and

$$Q_p' = \frac{Q_p}{1 - 3Q_p\omega_p\tau} \quad (6.85b)$$

Obviously, a simple predistortion technique may be used to design this circuit. The design value for Q_p can be obtained from Q_p':

$$Q_p = \frac{Q_p'}{1 + 3Q_p'\omega_p\tau} \quad (6.86)$$

The above equation can be used for predistortion. The procedure for minimizing the active pole sensitivity and the predistortion procedure for designing this modified TTH biquad have been implemented in the program TTHN listed in App. 6C. For large ω_p and/or Q_p, there may be a large Q enhancement and the poles may even migrate into the right half of the s plane. Therefore, the numerical predistortion may not even converge, and this is one of the difficulties involved in this and other similar circuits. In the program TTHN, the nominal design uses (6.86), making it useful for any set of values for ω_p and Q_p.

Both techniques, predistortion and passive compensation, take care of the initial deviations in Q_p and perhaps in ω_p. However, they fail to eliminate the active pole sensitivity, and it may be still large in three-amplifier circuits. Next, we consider other circuits where the active pole sensitivity can be minimized to a large extent. By referring to the TTH biquad shown in Fig. 6.23, it is easy to see that the inverting integrator formed by A_2 and the inverting amplifier formed by A_3 together just serve as a noninverting integrator. By replacing this combination with a noninverting integrator whose transfer function sensitivity is a minimum, we can minimize the overall sensitivity. Similarly, the first inverting integrator can also be replaced with an actively compensated inverting integrator circuit. In fact, there are many possibilities, but we will discuss only a few tested circuits, which are shown in Fig. 6.26. When the op amps are ideal, each one realizes the given set of H_o, ω_p, and Q_p exactly. However, the active pole sensitivities are entirely different for each one. In the analysis of these circuits, let us assume that $\tau_1 \neq \tau_2 \neq \tau_3 \neq \tau_4$. Let us first consider the Akerberg and Mossberg circuit [11] in Fig. 6.26*a*. Ignoring the terms on the order of $1/Q_p$ compared to unity and second-order terms of $\tau_i\tau_j$ compared to τ_i or τ_j, we find the denominator polynomial of the transfer function of this circuit to be

$$D(s,\tau_1,\tau_2,\tau_3,\tau_4) = \omega_p^2 + \frac{\omega_p}{Q_p}s(1+\omega_p Q_p \tau_3)$$

$$+s^2\left[1+(K+1)\omega_p\tau_1+\omega_p\tau_2\right]$$

$$+s^3(\tau_1+\tau_2)+s^4(\tau_1+b\tau_3)\tau_2+b\tau_1\tau_2\tau_3 s^5 \quad (6.87)$$

where $b = 2 + r/R_X$.

The pole sensitivities $p = \omega_p(-1/2Q_p + j)$ with respect to τ_1, τ_2, and τ_3 can be found using (6.87). When all the time constants deviate from their nominal values of zero to τ_1, τ_2, and τ_3, let us find the total change in p, which is Δp. This deviation, to a first-order approximation, is

$$\Delta p = \frac{\partial p}{\partial \tau_1}\tau_1 + \frac{\partial p}{\partial \tau_2}\tau_2 + \frac{\partial p}{\partial \tau_3}\tau_3 \quad (6.88)$$

Now

$$\frac{\partial p}{\partial \tau_i} = \left.\frac{-\partial D(s,\tau_1,\tau_2,\tau_3,\tau_4)/\partial \tau_i}{\partial D(s,\tau_1,\tau_2,\tau_3,\tau_4)/\partial s}\right|_{s=p;\ \text{all}\ \tau_i=0} \quad (6.89)$$

Applying (6.89) to (6.87), we have

$$\frac{\partial p}{\partial \tau_1} = \omega_p^2\frac{1-j(K+1)}{2} \quad (6.90a)$$

$$\frac{\partial p}{\partial \tau_2} = \omega_p^2\frac{1-j}{2} \quad (6.90b)$$

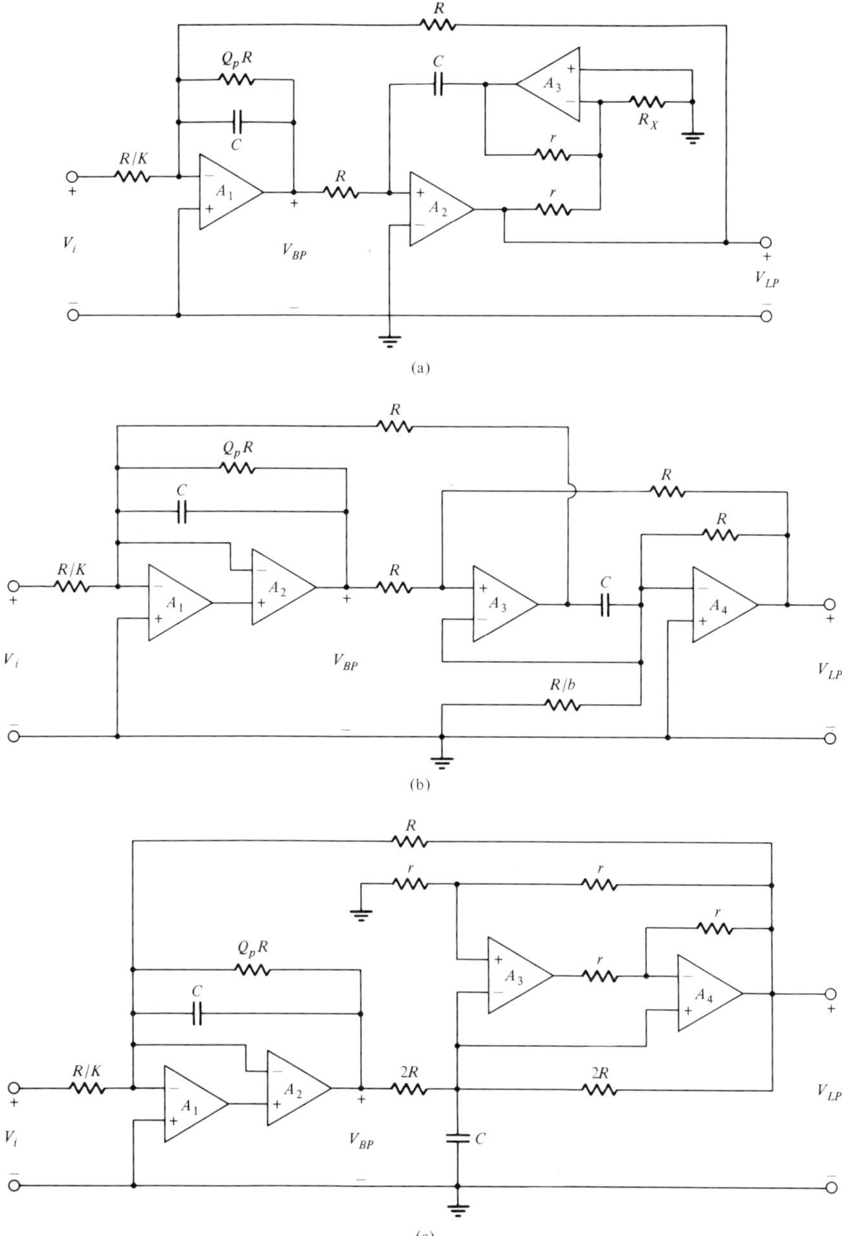

Figure 6.26 Biquads using double-integrator loops, where $K = H_0$ for lowpass, $K = H_0/Q_p$ for bandpass, and $RC = 1/\omega_p$.

and

$$\frac{\partial p}{\partial \tau_3} = -\frac{b\omega_p^2}{2} \tag{6.90c}$$

Using (6.90) in (6.88), we obtain

$$\Delta p = \{(\tau_1 + \tau_2 - b\tau_3) - j[(K+1)\tau_1 + \tau_2]\}\left(\frac{\omega_p^2}{2}\right) \tag{6.91}$$

It is the real part of (6.91) that causes the Q enhancement. To see this, let us find the new position of the dominant pole p':

$$\begin{aligned} p' &= p + \Delta p \\ &= \frac{-\omega_p\left[1 - \omega_p Q_p(\tau_1\tau_2 - b\tau_3)\right]}{2Q_p} + j\omega_p\left\{1 - \frac{\omega_p}{2}[(K+1)\tau_1 + \tau_2]\right\} \end{aligned}$$

From the above equation we obtain the approximate expressions for the effective pole frequency and pole Q factor:

$$\omega_p' = \omega_p\left[1 - \frac{\omega_p}{2}(K+1)\tau_1 + \tau_2\right] \tag{6.92a}$$

and

$$Q_p' \simeq \frac{Q_p}{1 - \omega_p Q_p(\tau_1 + \tau_2 - b\tau_3)} \tag{6.92b}$$

Note that there are terms in the denominator of (6.92b) with positive and negative signs. Thus Q enhancement can be avoided if we choose $b = (\tau_1 + \tau_2)/\tau_3$. For matched op amps, when $\tau_1 = \tau_2 = \tau_3$, this condition becomes $b = 2$, implying that $r/R_X = 0$. This can be achieved by setting the value of R_X to infinity and removing this resistor from the circuit. Note that Q enhancement can be avoided only if $(\tau_1 + \tau_2)/\tau_3 \geq 2$. Even after compensating for Q enhancement, the active sensitivity is not completely eliminated from the circuit because, after we have chosen $b = (\tau_1 + \tau_2)/\tau_3$, we are left with

$$\Delta p = -j[(K+1)\tau_1 + \tau_2]\frac{\omega_p^2}{2} \tag{6.93}$$

indicating only partial success in eliminating the active sensitivity in the circuit. However, note that in this circuit the major problem of Q enhancement has been avoided along with reduction in the active sensitivity, unlike the situation in the predistortion or passive compensation technique. One of the disadvantages of this filter circuit is that A_2 and A_3 require stricter compensation. This means that the slope of the magnitude decibel plot must be 6 dB/octave at the 0-dB crossover frequency of the amplifiers. In practical tests, high-frequency oscillations have been observed at the outputs of A_2 and A_3. The reason for this is that, at high frequencies, the capacitor connected between the output of A_3 and the input of A_2 acts as a short circuit and one amplifier feeds the other, thus giving rise to high-frequency oscillations. These high-frequency oscillations are limited by the slew rate of the op amps and do not appear at the bandpass output. Another problem has also been observed in this network. Self-oscillations at a frequency close to ω_p may arise at the instant of switching on the power supply. These self-oscillations can be prevented by connecting a pair of silicon diodes back-to-back across the input of A_2.

Next, considering the circuit in Fig. 6.26*b* [12], the denominator polynomial of any one of the transfer functions can be derived as

$$D(s, \tau_1, \tau_2, \tau_3, \tau_4) \simeq \omega_p^2 + \left[\frac{1}{Q_p} + \omega_p\tau_1 + (1+b)\omega_p\tau_4\right]\omega_p s$$
$$+ (1 + 2\omega_p\tau_3 + \omega_p\tau_4)s^2 + (2\tau_4 + \tau_1)s^3$$
$$+ \tau_1(\tau_2 + 2\tau_4)s^4 + 2\tau_1\tau_2\tau_4 s^5 \qquad (6.94)$$

The total change in $p = \omega_p(-1/2Q_p + j)$, due to the nonzero values of τ, can be determined by first finding $\partial p/\partial\tau_i$, $i = 1, 2, 3, 4$. Applying (6.89) to (6.94) and ignoring the terms on the order of $1/Q_p$ compared to unity, we obtain the derivatives

$$\frac{\partial p}{\partial \tau_1} = \frac{\partial p}{\partial \tau_2} = 0 \qquad (6.95a)$$

$$\frac{\partial p}{\partial \tau_3} = -j\omega_p^2 \qquad (6.95b)$$

and

$$\frac{\partial p}{\partial \tau_4} = [(1-b) - j]\frac{\omega_p^2}{2} \qquad (6.95c)$$

Using (6.95), we find that

$$\Delta p = -j\omega_p^2\left(\tau_3 + \frac{\tau_4}{2}\right) + \frac{(1-b)\omega_p^2\tau_4}{2} \qquad (6.96)$$

The Q enhancement is caused by the real part of the right side of (6.96). This can be made zero by choosing $b = 1$, and this cancelation is independent of the values of the op amp time constants. Once the real part of Δp is made zero, the active pole sensitivity is reduced to the extent of changing only the pole frequency ω_p. Thus the effective pole frequency ω_p' is

$$\omega_p' = \omega_p\left[1 - \omega_p\left(\tau_3 + \frac{\tau_4}{2}\right)\right] \tag{6.97}$$

With reference to Fig. 6.26*c*, the circuit [5] uses noninverting and inverting integrators whose transfer functions have zero sensitivities with respect to the op amp time constants. Because of this, the active pole sensitivities with respect to all the time constants τ_1, τ_2, τ_3, and τ_4 can be shown to be zero without requiring matched op amps. This means that, to a first-order approximation, the actual pole Q factor and pole frequency have their designed values. That is,

$$Q_p' = Q_p \qquad \text{and} \qquad \omega_p' = \omega_p \tag{6.98}$$

However, note that a noninverting integrator has an inherent stability problem. This problem depends on the matching of equivalued resistors, and therefore it is not a serious one in the operation of the circuit once it is tuned. The circuits in Fig. 6.26*b* and *c* (filters I and II, respectively, for convenience of later reference) do not require matched op amps to avoid Q enhancement. In filter I the active pole sensitivity has been only partially eliminated, whereas in filter II the active pole sensitivity has been eliminated completely. The two filter circuits were built and tested with the nominal values $Q_p = 100$, $f_p = 10$ kHz, and $H_o = 1$. The bandpass output characteristics were experimentally obtained with LM 307 op amps and are shown in Fig. 6.27. The op amps had *GB* products in the range of only 500 to 800 kHz, though their nominal values are supposed to be 1 MHz. The characteristics in Fig. 6.27*a* and *b* are those of filters II and I, respectively, when the supply voltages to the op amps were maintained at ± 15 V. In filter I the measured values of Q_p and ω_p were only 94.5 and 9.83 kHz, respectively, while in filter II the same values were 106.2 and 9.98 kHz. As can be easily seen, the decrease in the value of Q_p in filter I and the increase in the value of Q_p in filter II are caused by the second-order effects of the time constants of the op amps. The experimental characteristic of filter II is close to the ideally desired one, whereas the characteristic of filter I has a frequency shift close to two bandwidths. The experimental characteristics were obtained, in both filters, after changing the supply voltages to the op amps to ± 6 V. The characteristics in Fig. 6.27*c* and *d* refer to the those of filters II and I, respectively, the supply voltage being ± 6 V. The decrease in the supply voltage to the op amps causes a decrease in the *GB* products of the op amps (i.e., an increase in the time constant of the op amps). The change in the power supply voltage brings out the effects of the active pole sensitivities. Thus the change in the magnitude characteristic is an indication of the active sensitivity. Therefore the maximum changes in

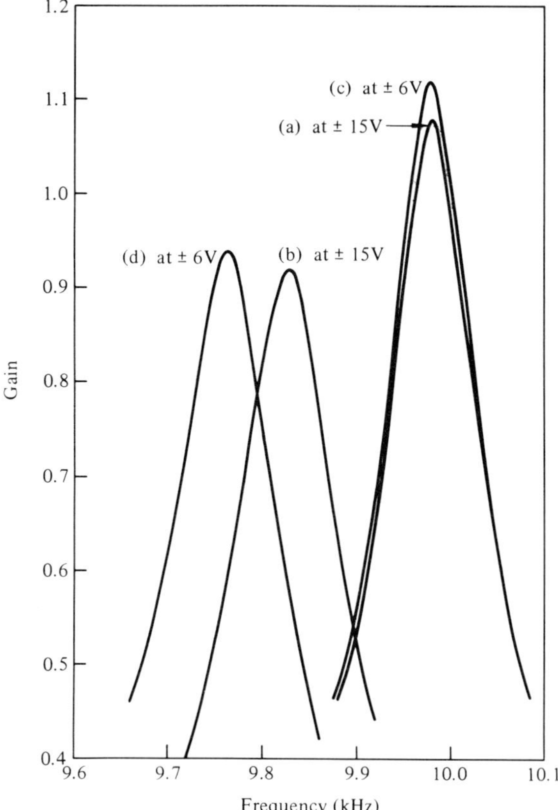

Figure 6.27 Magnitude characteristics of double-integrator loop filters. (*a* and *c*) For filter II. (*b* and *d*) For filter I.

the magnitude were measured in the two filter circuits and found to be 3.7% in filter II and 41.5% in filter I. Clearly, filter II is superior to filter I in terms of active sensitivity.

6.6 *Realization of Finite Transmission Zeros Using Double-Integrator Loops*

A KHN biquad realizes LP, BP, and HP functions, while other biquads realize LP and BP functions only. To obtain finite zeros of transmission, either a summation method or a feedforward technique can be used. The *summation method* is used with a KHN biquad and is illustrated in Fig. 6.28. When the op amps in the circuit in Fig. 6.28 are ideal, the voltage

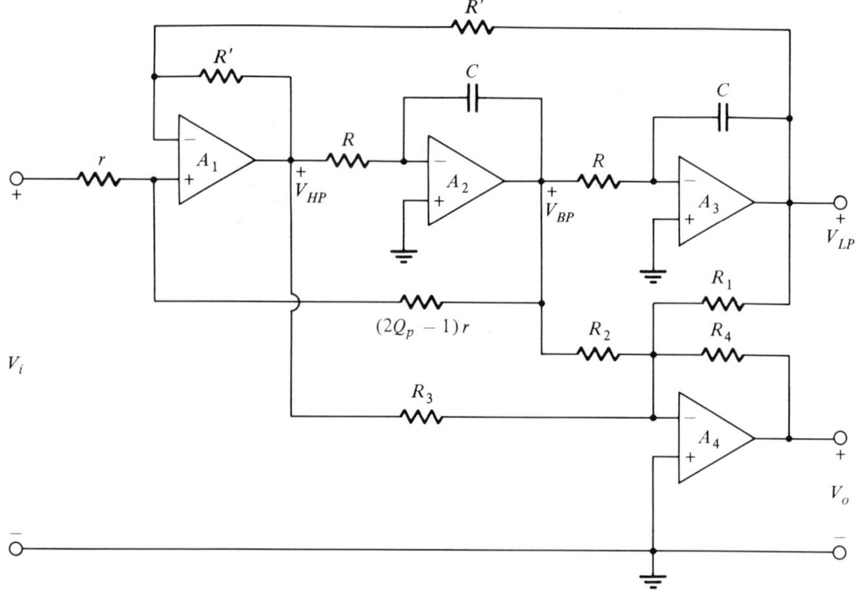

Figure 6.28 Realization of arbitrary transmission zeros using the KHN biquad.

transfer function is

$$\frac{V_o}{V_i} = \frac{-R_4}{R_3} \frac{2Q_p - 1}{Q_p} \frac{s^2 - (R_3/R_2)\omega_p s + (R_3/R_1)\omega_p^2}{s^2 + (\omega_p/Q_p)s + \omega_p^2} \tag{6.99}$$

where $\omega_p = 1/(RC)$.

Lowpass and highpass notch filters can be realized using the network in Fig. 6.28 with

$$R_2 = \infty \tag{6.100a}$$

$$R_3 = \frac{R_4(2 - 1/Q_p)}{H_o} \tag{6.100b}$$

and

$$R_1 = R_3 \frac{\omega_p^2}{\omega_z^2} \tag{6.100c}$$

for any arbitrary R_4.

An all-pass filter of the form

$$\frac{V_o}{V_i} = -H_o \frac{s^2 - (\omega_p/Q_p)s + \omega_p^2}{s^2 + (\omega_p/Q_p)s + \omega_p^2} \tag{6.101}$$

can also be realized by choosing the components appropriately in the network in Fig. 6.28.

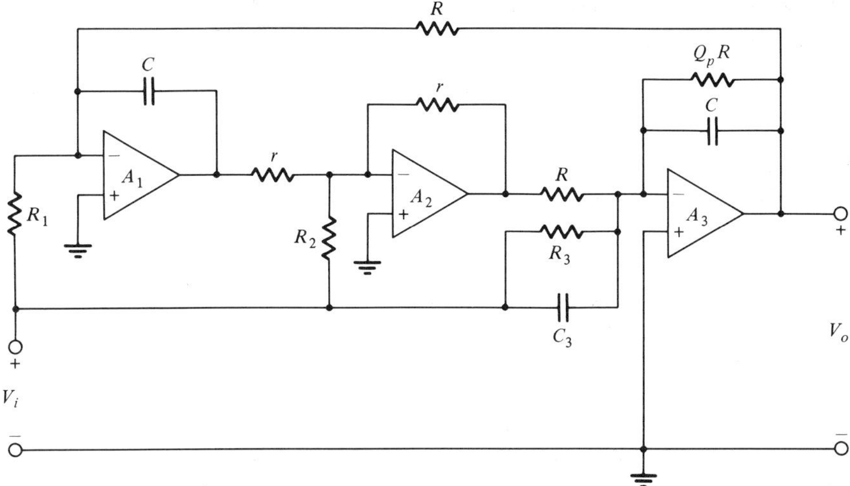

Figure 6.29 Realization of arbitrary transmission zeros using the Tow-Thomas biquad.

A TTH biquad can also be used to realize arbitrary transmission zeros. Here the technique employed is called the *feedforward method*. A TTH biquad for realizing arbitrary zeros of transmission is shown in Fig. 6.29. Note that the feedforward technique does not require a fourth op amp, however, it requires an additional capacitor C_3. When the op amps are ideal, the transfer function of this circuit is

$$\frac{V_o}{V_i} = \frac{-C_3}{C}\frac{s^2 + s[1 - rR_3/(RR_2)]/(R_3C_3) + (RC)\omega_p^2/(R_1C_3)}{s^2 + s(\omega_p/Q_p) + \omega_p^2} \tag{6.102}$$

where $\omega_p = 1/(RC)$.

If we do not have to satisfy the requirement for H_o, we can choose the value of C_3 to be equal to C in some cases. Then, the design of various types of filters can be carried out as indicated in Table 6.2.

We can also employ the feedforward technique in the circuits in Fig. 6.26 to realize lowpass and highpass notch functions and obtain an all-pass transfer function using the circuit in Fig. 6.26*a*. As an example, we describe the technique of realizing notch filter functions using the network in Fig. 6.26*b*. This network has been modified to realize notch functions, as shown in Fig. 6.30. When the op amps are assumed to be ideal, the transfer function of this circuit is

$$\frac{V_o}{V_i} = \frac{-C_1}{C}\frac{s^2 + \omega_z^2}{s^2 + s(\omega_p/Q_p) + \omega_p^2} \tag{6.103}$$

Table 6.2 Element constraints and design equations for the circuit in Fig. 6.29 (TTH biquad)

Type of transfer function desired	Element constraints	Choice of R_1, R_2, and R_3
Lowpass	$R_2 = R_3 = \infty$, $C_3 = 0$	$R_1/R = 1/H_o$
Bandpass	R_3 or $R_2 = \infty$, $R_1 = \infty$, $C_3 = 0$	(a) $R_2 = \infty$, $R_3/r = H_o/Q_p$ (b) $R_3 = \infty$, $R_2/R = H_o/Q_p$ (inverting)
Highpass	$R_1 = R_2 = R_3 = \infty$	$H_o = 1$ (inverting)
Notch	$R_2 = R_3 = \infty$	$R_1 = R(\omega_p^2/\omega_z^2)$, $H_o = 1$ (inverting)
All-pass	$R_3 = \infty$	$R_1 = R$, $R_2 = Q_p r$

where $\omega_p = 1/(RC)$ and $\omega_z^2 = \omega_p/(R_2C_1)$. For the ideal filter design, $R_2C_1 = \omega_p/\omega_z^2$ and $RC = 1/\omega_p$. However, in order to determine the value of R/b, the compensating resistor, one must find the transfer function V_o/V_i, including the effects of the nonzero time constants of the op amps. Retaining only the dominant effects of the time constants, the transfer function of this circuit can be shown to be

$$\frac{V_o}{V_i} = \frac{-C_1}{C}\,\frac{(2+d)\tau_4 s^3 + s^2\left[1 + (2+d)\omega_p\tau_3 + \left(\omega_z^2/\omega_p\right)\tau_4\right] + s(1+b)\omega_z^2\tau_4 + \omega_z^2}{(2+d)\tau_4 s^3 + s^2[1 + (2+d)\omega_p\tau_3 + \omega_p\tau_4(1 + 1/Q_p)] + s(\omega_p/Q_p)[1 + (2+d)\omega_p\tau_3 + (1+b)\omega_p Q_p\tau_4] + \omega_p^2} \tag{6.104}$$

where $d = R/R_2$.

It should be noted that the sensitivities of the entire transfer function are zero with respect to both τ_1 and τ_2. However, the nonzero time constants τ_3 and τ_4, as can be easily seen from (6.104), affect not only the poles but also the zeros of the transfer function.

The active pole and zero sensitivities can be found by applying (6.89) to both the denominator and numerator polynomials. Thus we obtain, from the denominator polynomial,

$$\frac{\partial p}{\partial \tau_3} = -j\left(1 + \frac{d}{2}\right)\omega_p^2 \tag{6.105a}$$

and

$$\frac{\partial p}{\partial \tau_4} = \frac{\omega_p^2}{2}[(1 + d - b) - j] \tag{6.105b}$$

In order to minimize the active pole sensitivity in this network, we must

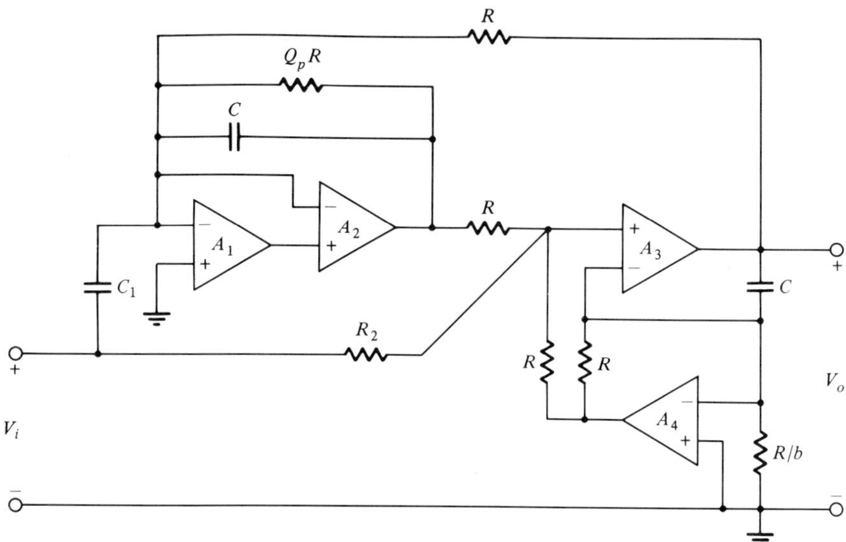

Figure 6.30 Circuit for realizing notch filter functions.

choose $b = 1 + d$. This differs from the compensation we obtained for the bandpass network of the same type earlier. There we had $d = 0$, $R_2 = \infty$, and $b = 1$. Now, if $b = 1 + d$, then

$$\frac{\partial p}{\partial \tau_4} = -\frac{j\omega_p^2}{2} \tag{6.106}$$

Using the above, we can find the total change Δp, due to the nonzero time constants of τ_3 and τ_4, to be

$$\Delta p = -j\omega_p\left[\left(1 + \frac{d}{2}\right)\omega_p\tau_3 + \frac{\omega_p\tau_4}{2}\right] \tag{6.107}$$

Using (6.107), we can show that

$$Q_p' = Q_p \tag{6.108a}$$

and

$$\omega_p' = \omega_p\left\{1 - \left[\left(1 + \frac{d}{2}\right)\omega_p\tau_3 + \frac{\omega_p\tau_4}{2}\right]\right\} \tag{6.108b}$$

Ideally, the zeros of the transfer function should be $s = z = \pm j\omega_z$ and $1/Q_z = 0$. Now, because of the nonzero time constants, the zeros will also be out of place, and therefore we shall find the changes in ω_z and $1/Q_z$. For this we apply (6.89) to the numerator polynomial of the transfer function in

(6.104) and obtain the derivatives

$$\frac{\partial z}{\partial \tau_3} = -j\left(1 + \frac{d}{2}\right)\omega_z\omega_p \tag{6.109a}$$

and

$$\frac{\partial z}{\partial \tau_4} = -\frac{j\omega_z^3}{\omega_p} \tag{6.109b}$$

Thus the effective zero is located at $z + \Delta z$ and is

$$z' = z + \Delta z = j\omega_z\left[1 - \left[\left(1 + \frac{d}{2}\right)\omega_p\tau_3 + \frac{\omega_z^2}{\omega_p}\tau_4\right]\right]$$

To a first-order approximation,

$$\omega_z' = \omega_z\left\{1 - \left[\left(1 + \frac{d}{2}\right)\omega_p\tau_3 + \frac{\omega_z^2}{\omega_p}\tau_4\right]\right\} \tag{6.110a}$$

and

$$\frac{1}{Q_z'} = 0 \tag{6.110b}$$

From both (6.108) and (6.110), we find that, to a first-order approximation, the values of Q_p and Q_z realized using $b = 1 + d$ are close to their ideal values. Furthermore, it should be noted that the sensitivity of the entire transfer function is zero with respect to both τ_1 and τ_2. However, there will be deviations in ω_p and ω_z, and the active sensitivities of ω_p and ω_z with respect to τ_3 and τ_4 cannot be made zero.

6.7 *A Three-Amplifier Biquad Derived from a Generalized Impedance Converter*

A three-amplifier biquad, called an *MB circuit* [13] and derived from a GIC, is shown in Fig. 6.31. The passive ω_p and Q_p sensitivities are low, as in the case of double-integrator loop filters, and both ω_p and Q_p can be controlled independently or in sequence. There is no need for a fourth amplifier to realize notch and all-pass filters as required in the case of a KHN biquad. Also, by simply setting the different conductance values of appropriate resistors to zero, one can realize all the standard second-order functions. While all the three-amplifier networks, we have considered require a resistor spread of Q_p, this network needs a resistor spread of only $4Q_p^{1/2}/H_o$. Therefore this circuit is attractive for high-Q_p realizations. When the op amps are ideal in this circuit, the transfer functions can be

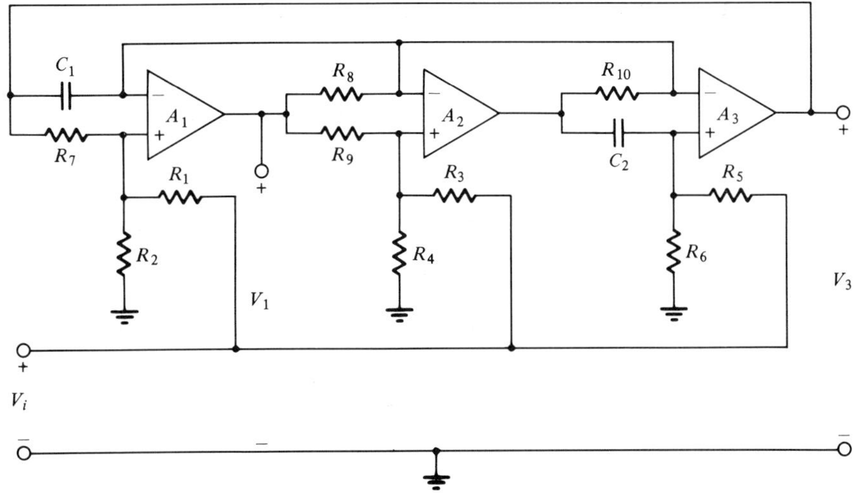

Figure 6.31 Mikhael-Bhattacharyya (MB) circuit.

derived as

$$\frac{V_1}{V_i} = \frac{b_2 s^2 + b_1 s + b_0}{a_2 s^2 + a_1 s + a_0} \tag{6.111a}$$

and

$$\frac{V_3}{V_i} = \frac{d_2 s^2 + d_1 s + d_0}{a_2 s^2 + a_1 s + a_0} \tag{6.111b}$$

where

$$a_2 = G_1 + G_2$$

$$a_1 = G_7 G_8 \frac{G_3 + G_4}{G_9 C_1}$$

$$a_0 = G_7 G_{10} \frac{G_5 + G_6}{C_1 C_2}$$

$$b_2 = \frac{G_1(G_4 + G_9) - G_2 G_3}{G_9}$$

$$b_1 = \frac{G_3 G_7 G_8}{G_9 C_1}$$

$$b_0 = G_7 G_{10} \frac{G_5(G_4 + G_9) - G_3 G_6}{G_9 C_1 C_2}$$

$$d_2 = G_1$$

$$d_1 = G_8 \frac{G_3(G_2 + G_7) - G_1 G_4}{G_9 C_1}$$

Table 6.3 Design formulas for MB circuit in Fig. 6.31

Transfer function realized	Output	Design formulas
Lowpass, $H_o \leq 2$	V_3	$R_1 = R_3 = \infty$, $R_2 = R$, $R_4 = R/\beta$, $R_5 = 2R/(\alpha H_o)$, $R_6 = R/[\alpha(1 - H_o/2)]$
Bandpass, $H_o \leq 2$	V_3	$R_1 = R_5 = \infty$, $R_2 = R$, $R_3 = 2R/(\beta H_o)$, $R_4 = R/[(1 - H_o/2)\beta]$, $R_6 = R/\alpha$
Highpass, $H_o \leq (1 + \beta)$	V_1	$R_1 = R(1 + \beta)/H_o$, $R_2 = R/[1 - H_o/(1 + \beta)]$, $R_3 = R_5 = \infty$, $R_4 = R/\beta$, $R_6 = R/\alpha$
Notch, $H_o \leq (1 + \beta)$ or $H_o \leq (1 + \beta)$ (ω_p^2/ω_z^2) if $\omega_z > \omega_p$	V_1	$R_1 = R(1 + \beta)/H_o$, $R_2 = (RR_1)/(R_1 - R)$, $R_3 = \infty$, $R_4 = R/\beta$, $R_5 = R\omega_p^2(1 + \beta)/(\alpha H_o \omega_z^2)$, $R_6 = RR_5/(\alpha R_5 - R)$
All-pass, $H_o = 1$	V_3	$R_1 = R$, $R_2 = R_3 = R_6 = \infty$, $R_4 = R/\beta$, $R_5 = R/\alpha$

Common formulas for all the designs:
$\beta = Q_p^{-1/2}$, $C_1 = C_2 = C$, $R = 1/(\omega_p C)$, $R_7 = R_9 = R$, $R_8 = R/\beta$, $R_{10} = \alpha R$

and

$$d_o = G_{10}\frac{G_5(G_2 + G_7) - G_1 G_6}{C_1 C_2}$$

The design of the various second-order transfer functions can be carried out as given in Table 6.3. Some resistor values are common to all the designs, and these values are given at the bottom of the table. Furthermore, the parameter α is arbitrary to some extent. The authors of this circuit suggest the following value of α:

$$\alpha = \frac{2}{1 + \beta} \tag{6.112}$$

where $\beta = Q_p^{-1/2}$ (given at the bottom of Table 6.3).

As we shall soon see, the choice of α as given in (6.112) almost eliminates the effect of the τ of the op amps on Q_p. This is because this choice of α makes the imaginary part of $\Delta p/p$, due to the nonzero values of τ, zero. This is similar to the case for an AM biquad circuit in Fig. 6.26*a*, and only the pole frequency ω_p is affected by the nonzero value of τ with this choice of α.

If we assume that $A_1 = A_2 = A_3 = 1/s\tau$, the denominator polynomial of any one of the designs can be obtained in terms of the parameters in

Table 6.3 as

$$D(s,\tau) = s^2 + \frac{\omega_p}{Q_p}s + \omega_p^2 + s\tau\left(e_2 s^2 + e_1 s + e_0\right) + \cdots \quad (6.113)$$

where

$$e_2 = 2(1+\beta)$$

$$e_1 s = \left[1 + \frac{1}{\beta} + \frac{2}{\beta^2\alpha}\right]\frac{\omega_p}{Q_p}$$

and

$$e_0 = \left[2 + \beta\alpha(1+\beta)\right]\omega_p^2$$

The other parameters in the polynomial are not needed now unless we are also interested in the higher-order effects of τ. Applying (3.62) to the denominator polynomial, we obtain

$$\frac{\Delta p}{p} = -\left\{\left[\beta(1-\beta-2\beta^2) + \frac{2}{\alpha}\right] + j\left[\alpha\beta(1+\beta) - 2\beta\right]\right\}\frac{\omega_p\tau}{2} \quad (6.114)$$

The imaginary part of the per-unit change can be made zero if α is selected according to (6.112). However, a better choice for α can be made to minimize the magnitude of the active pole sensitivity. This problem is left as an exercise. An empirical formula for the choice of α is given by

$$\alpha = 1.6Q_p^{1/4} \quad (6.115)$$

Example 6.4: Design a lowpass filter that meets the specifications $f_p = 20$ kHz, $Q_p = 20$, and $H_o = 1$. Choose α to minimize the sensitivity of Q_p due to the nonzero value of τ and make this choice according to (6.112). Find the per-unit change in the pole frequency, hence the ripple factor, when $\tau = 0.5/\pi$ μs.

Since $Q_p = 20$, $\beta = 0.22361$. The choice of α is to be made according to (6.112), and therefore $\alpha = 1.6345$. Selecting $C = 1$ nF, we obtain the value of R as 7.958 kΩ. Then, using Table 6.3, $R_1 = R_3 = \infty$, $R_2 = R_7 = R_9 = 7.958$ kΩ, $R_4 = R_8 = 35.588$ kΩ, $R_5 = R_6 = 9.737$ kΩ, and $R_{10} = 13.007$ kΩ. Since $\omega_p\tau = 0.02$, using (6.114), we find that

$$\frac{\Delta p}{p} = -0.01375$$

and therefore

$$\text{Ripple factor} = 2Q_p\left|\frac{\Delta p}{p}\right| = 0.55$$

■

Example 6.5: Design a bandpass filter to meet the specifications $f_p = 20$ kHz, $Q_p = 20$, and $H_o = 1$. Choose α according to (6.115), which will minimize the magnitude of the active pole sensitivity. Find the per-unit change in p and the ripple factor. Also, find the per-unit change in ω_p and Q_p.

In this example also, $\beta = 0.22361$. However, we select α according to (6.115), and it is $\alpha = 3.3836$. Choosing $C = 1$ nF and using Table 6.3, we have $R_1 = R_5 = \infty$, $R_2 = R_7 = R_9 = R = 7.958$ kΩ, $R_3 = R_4 = 71.176$ kΩ, $R_6 = 2.352$ kΩ, $R_8 = 35.588$ kΩ, and $R_{10} = 29.926$ kΩ. Next, calculating the per-unit change $\Delta p/p$, we find that

$$\frac{\Delta p}{p} = -0.00742 + j0.004786$$

and therefore

$$\text{Ripple factor} = 0.353$$

From the per-unit change, we find that

$$\frac{\Delta\omega_p}{\omega_p} = -0.00742$$

and

$$\frac{\Delta Q_p}{Q_p} = -0.1914 \qquad \blacksquare$$

Since, in this network, $\Delta p/p$ depends only on the values of ω_p and Q_p and not on the transfer function designed, we can compare the ripple factors of the two designs in Examples 6.4 and 6.5. They clearly indicate that the value of α chosen according to (6.115) gives rise to a smaller change in the magnitude of the filter response.

6.8 *Conclusions*

We have considered two basic types of multiple-amplifier biquads. The first category of filters belongs to the GIC-derived group, and the second group essentially uses the state representation of second-order transfer functions which in turn requires integrators as basic components. The circuits discussed in this chapter are particularly suitable for high-Q_p filter realization because, in these circuits, Q_p can be controlled independently using (usually) a single resistor. The GIC-derived filters were found to be optimal with certain design restrictions on the gain constants. The second group of filters had the basic problem of Q enhancement. However, by using active compensation, we found that one cannot only avoid Q enhancement in some filters but can also eliminate the first-order effects of the time constants of the op amps, as was the case in one filter. In the next chapter, we shall see how higher-order filters can be realized.

REFERENCES

1. Antoniou, A.: Realization of Gyrators Using Operational Amplifiers and Their Use in *RC*-Active Network Synthesis, *Proc. IEE*, vol. 116, pp. 1838–1850, November 1969.
2. Chiou, C. F., and Schaumann, R.: Performance of GIC-derived Active-*RC* Biquads with Variable Gains, *Proc. IEE*, Part G, No. 1, pp. 46–52, 1981.
3. Fliege, N.: A New Class of Second Order *RC*-Active Filters with Two Operational Amplifiers, *Nachrihtentech. Z.*, vol. 26, no. 6, pp. 279–282, June 1973.
4. Vogel, P. M.: Method for Phase Correction in Active-*RC* Circuits Using Two Integrators, *Electron. Lett.*, vol. 7, pp. 273–275, May 1971.
5. Natarajan, S.: Synthesis of an Actively Compensated Double Integrator Filter without Matched Operational Amplifiers, *Proc. IEEE*, vol. 68, no. 12, pp. 1547–1548, December 1980.
6. Deboo, G. J.: A Novel Integrator Results by Grounding Its Capacitor, *Electron. Des.*, vol. 15, p. 90, June 7, 1967.
7. Kerwin, W. J., Huelsman, L. P., and Newcomb, R. W.: State Variable Synthesis for Insensitive Integrated Circuit Transfer Function, *IEEE Trans. Solid-State Circuits*, vol. SC-2, pp. 87–92, September 1967.
8. Tow, J.: Active *RC* Filters—A State-Space Realization, *Proc. IEEE*, vol. 56, pp. 1137–1139, 1968.
9. Thomas, L. C.: The Biquad. I. Some Practical Design Considerations, *IEEE Trans. Circuit Theory*, vol. CT-18, pp. 350–357, May 1971.
10. Thomas, L. C.: The Biquad. II. A Multipurpose Active Filtering System, *IEEE Trans. Circuit Theory*, vol. CT-18, pp. 358–361, May 1971.
11. Akerberg, D., and Mossberg, K.: A Versatile Active *RC* Building Block with Inherent Compensation for the Finite Bandwidth of the Amplifier, *IEEE Trans. Circuits Syst.*, vol. CAS-21, pp. 75–78, January 1974.
12. Rao, K. R. et al.: An Actively Compensated Double Integrator Filter without Matched Operational Amplifiers, *Proc. IEEE*, vol. 68, pp. 534–535, April 1980.
13. Mikhael, W. B., and Bhattacharyya, B. B.: A Practical Design for Insensitive *RC* Active Filters, *IEEE Trans. Circuits Syst.*, vol. CAS-22, no. 5, pp. 407–415, May 1975.

14. Reddy, M. A.: An Insensitive Active *RC* Filter for High-*Q* and High Frequencies, *IEEE Trans. Circuits Syst.*, vol. CAS-23, no. 7, pp. 429–433, July 1976.

EXERCISES

6.1. Consider the circuit in Fig. 6.5.
(a) It is desired to simulate an inductor of 10 H. Assume that $R_1 = R_3 = R_4 = R_5 = 2$ kΩ. Find the value of C_2 for simulating such an inductor.
(b) The nonzero τ values of both op amps are $0.5/\pi$ μs. Find the per-unit changes in the values of L_{eq} and the Q value of the simulated inductor.
(c) Repeat parts (a) and (b) for $R_1 = R_3 = R_4 = 2$ kΩ and $R_5 = 4$ kΩ.

6.2. Use the type-A inductance simulation circuit in Fig. 6.5 and design the circuit to realize an inductance value of 0.1 H at $f_c = 3.8$ kHz. Assume that both op amps have $\tau = 0.5/\pi$ μs at room temperature. However, over a temperature increase of 50 °C, they track within 2%. Calculate the error in the realized inductance value in L at room temperature and find the Q factor after a temperature increase of 50°C.

6.3. The circuit in Fig. 6.11 is used to design a bandpass filter. In answering the following questions assume that $R_1 = R_5$, $R_3 = R_4$, and $C_1 = C_2$.
(a) Assume that $\tau_1 = \tau_2 = 0$, $G_c = 1/R_c = 0$. Design a filter to meet the specifications $H_o = 5$, $Q_p = 50$, and $f_p = 20$ kHz.
(b) Assume that $\tau_1 = \tau_2 = 0.5/\pi$ μs. Find the values of per-unit changes in the pole frequency and the pole Q factor due to the nonzero values of τ in the circuit designed in (a).
(c) To compensate for the effects of the nonzero values of τ_1 and τ_2 of part (b), we use the resistor R_c and assume that $G_c \neq 0$. Find the value of R_c such that $\Delta Q_p/Q_p$ due to the nonzero value of τ in part (b) is equal to zero.
(d) Determine a set of new values of R_c and R_1 using the predistortion technique such that the actual pole positions will be the desired ones when you include the nonzero values of τ. Use the initial values for R_c and R_1 as found in parts (c) and (a), respectively.

6.4. Consider the lowpass filter in Fig. 6.13*b*.
(a) Use this network to realize a lowpass filter meeting the specifications of $H_o = 5$, $Q_p = 20$ and $f_p = 20$ kHz. Assume that $R_1 = R_4$ and $R_2 = R_5$.
(b) Repeat Exercise 6.3b.

6.5. In the lowpass filter circuit in Fig. 6.13*b*, assume that the op amps are ideal and that

$$R_1 = R_2 = R_5 = \frac{1}{\omega_p C (H_o - 1)^{1/2}}$$

and

$$R_3 = \frac{Q_p}{\omega_p C} \qquad \text{and} \qquad R_4 = \frac{(H_o - 1)^{1/2}}{\omega_p C}$$

(a) Show that

$$\frac{V_o}{V_i} = \frac{H_o \omega_p^2}{s^2 + (\omega_p/Q_p)s + \omega_p^2}$$

(b) Find the element values of this circuit that meets the specifications in Exercise 6.4a.

(c) Find the active pole sensitivity if $\tau = 0.5/\pi\ \mu s$ and compare this value with the active pole sensitivity obtained in the network designed in Exercise 6.4.

6.6. Design a lowpass notch filter using the circuit in Fig. 6.13*b* to meet the specifications $f_p = 3.2$ kHz, $f_z = 4.2$ kHz, and $Q_p = 10$.

6.7. Using the KHN biquad, design a lowpass filter meeting the specifications $f_p = 4$ kHz and $Q_p = 20$.

6.8. Assume that we do not meet the specification for the gain constant requirement in the KHN filter in Fig. 6.22 and we want to minimize the magnitude of the active pole sensitivity using the free parameter R_6/R_5. Find the active pole sensitivity of this network and show that the magnitude of this active pole sensitivity can be minimized when $R_6/R_5 = 0.732$.

6.9. Consider the Tow-Thomas biquad.

(a) Realize a bandpass filter that meets the specifications $f_p = 10$ kHz, $Q_p = 2$, and $H_o = 5$.

(b) Assume that all the op amps in this circuit have the same nonzero time constants and that $\tau = 0.5/\pi\ \mu s$. Find the values of $\Delta\omega_p/\omega_p$ and $\Delta Q_p/Q_p$ in the design in part (a) due to such a nonzero value of τ.

(c) Include the effects of the nonzero value of τ and design the circuit to meet the specifications in part (a). Assume that $\tau = 0.25/\pi\ \mu s$.

(d) Another method of designing this circuit that includes the effect of the nonzero τ value is to use passive compensation as shown in Fig. 6.24. Find the value of C_c in this circuit for the design in part (a) when the value of τ is the same as in part (b). Then repeat part (b) with this capacitor included in the circuit. Compare the values of $\Delta\omega_p/\omega_p$ and $\Delta Q_p/Q_p$ obtained in parts (b) and (d).

6.10. Consider the modified TTH biquad in Fig. 6.25.

(a) Using this circuit, design a bandpass filter that meets the specifications in Exercise 6.9a.

(b) Repeat Exercise 6.9b.

6.11. Consider the AM biquad circuit in Fig. 6.26*a*.

(a) Assume that $\tau = 0$. Design a lowpass filter that satisfies the specifications $H_o = 2$, $Q_p = 20$, and $f_p = 20$ kHz.

(b) Find $\Delta p/p$ for the design when the value of $\tau = 0.5/\pi$ μs. Also, find the per-unit changes in ω_p and Q_p.

(c) Assume that predistortion is used to compensate for both the first- and second-order effects of the τ values of the op amps. Can you suggest values for Q_p and ω_p such that Q_p' and ω_p' will be 20 and 20 kHz, respectively?

6.12. The AM biquad suffers from self-oscillation when it is used for high-Q_p and/or high-ω_p realizations. A network suggested by Reddy [14] does not suffer from this problem. Otherwise this network has properties similar to those of an AM biquad. This circuit is shown in Fig. E6.12. In all the designs, assume that $R_5 = R_6 = R_7 = R_8 = R$ and $C_1 = C_2 = C$.

(a) Assume that the op amps are ideal and that $G_c = 0$. Develop network design procedures for realizing various generic second-order filter functions.

(b) Find the passive ω_p and Q_p sensitivities.

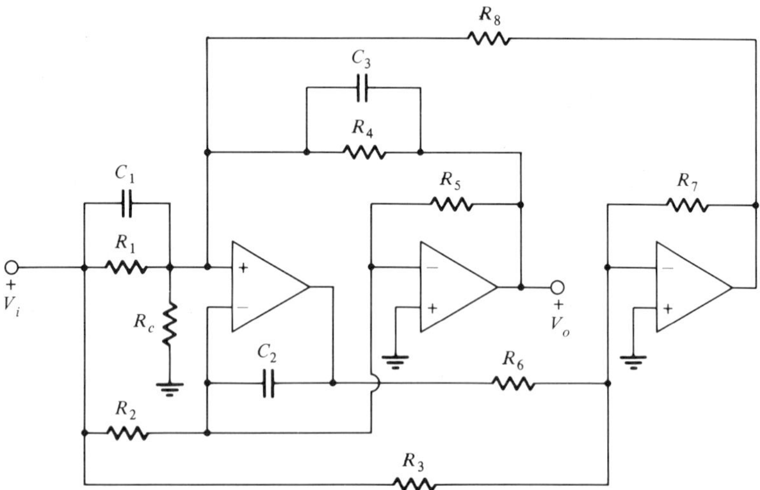

Figure E6.12 Reddy's three-amplifier biquad considered in Exercise 6.12.

(c) The basic pole-forming part of this circuit can be found by removing the components C_3, R_2, and R_3, i.e., by setting $C_3 = G_2 = G_3 = 0$. In this pole-forming circuit, assume that $G_c \neq 0$ and $\tau_1 \neq \tau_2 \neq \tau_3$. Find the active pole sensitivities with respect to τ_1, τ_2, and τ_3 as functions of R_c and the design values of ω_p and Q_p. From these sensitivities, find the total values of $\Delta Q_p/Q_p$ and $\Delta\omega_p/\omega_p$ due to the nonzero values of τ_1, τ_2, and τ_3.

(d) To eliminate the initial Q error due to the nonzero values of the time constants of the op amps, show that R_c must be approximately equal to $R_1/[(R_1/R)(2\tau_1/\tau_3 - 1) - 1]$.

6.13. Use the multiple-amplifier biquad in Fig. 6.28 to realize the transfer function

$$H(s) = \frac{s^2 + 404.26 \times 10^6}{s^2 + 2827.4s + 799.44 \times 10^6}$$

6.14. Apply the feedforward technique used in the circuit in Fig. 6.28 to the AM biquad in Fig. 6.26 and develop design procedures for various generic second-order biquads.

6.15. Repeat Exercise 6.13 using the MB circuit.

6.16. (a) Find the magnitude of $\Delta p/p$ in (6.114) as a function of α and β.

(b) For a given value of Q_p, hence β, in the MB circuit Q_p value ranging from 1 to 100, find the value of α that will minimize the magnitude of $\Delta p/p$. Use these values and show that (6.115) is approximately true.

6A

Program GICBPN

THIS APPENDIX PROVIDES the details of the program GICBPN written in GW BASIC language that can be used in IBM PC or its compatibles. This program can be used to design the GIC-derived bandpass filter in Fig. 6.11 with minimum magnitude of the active pole sensitivity and with predistortion to compensate for the effects of the nonzero time constants of the op amps used in the circuit. The procedure for using this program is simple. Just enter it and run it.

Inputs Necessary

f_p, Q_p, H_o, C ($= C_2$), R_5, and the nonzero values of the time constant of the op amps.

Outputs

Element values (see Fig. 6.11 for the notations) and the minimum magnitude of the active pole sensitivity. If desired, one can also obtain the passive sensitivities of the pole frequency and the pole Q factor for the specified design.

```
10 '
20 '
30 '                     PROGRAM LISTING OF 'GICBPN'
40 '                     --------------------------
50 '
60 '
70 'Program to design GIC-drived bandpass filter with min. active pole sens.
   along with predistortion.
80 PRINT"THIS PROGRAM CAN BE USED TO DESIGN GIC DERIVED  BANDPASS FILTER WITH MI
NIMUM   ACTIVE POLE SENSITIVITY  ALONG WITH PREDISTORTION.":PRINT
90 DIM E$(8),X(4),E(8),D(4),Z(4)
100 E$(1)="R1":E$(2)="R3":E$(3)="R4":E$(4)="R5":E$(5)="RQ":E$(6)="RC":E$(7)="C":
E$(8)="C2"
110 PRINT"INPUT THE SPECIFICATIONS FOR fp,Qp and H0 AS PROMPTED":PRINT
120 INPUT" fp in Hz";FP:INPUT" Qp";Q:INPUT" H0=";H:AL=H-1
130 PI2=8!*ATN(1):WP=FP*PI2:RE=-1/(2*Q):IM=SQR(1-RE*RE)
140 PRINT"R1,R3,RQ,RC AND R4 ARE CALCULATED BY THE PROGRAM. INPUT THE OTHER REQU
IRED PARAMETERS.":PRINT
150 INPUT"THE CAPACITANCES ARE ASSUMED TO BE EQUIVALUED.VALUE OF C";E(7):C=E(7)
160 INPUT"VALUE OF R5";E(4):R5=E(4)
170 INPUT"INPUT THE INVERSE OF THE GB PRODUCT OF THE OP.AMP.IN Hz^(-1).IN AN IDE
AL OP.AMP. IT IS ZERO.";T:T=T*FP
180 LPRINT"DESIGN OF A GIC-DERIVED BANDPASS FILTER":LPRINT STRING$(52,"_")
190 LPRINT:LPRINT"THE SPECIFICATIONS ARE":LPRINT:I=0
200 LPRINT"fp in Hz=";USING"##.#####^^^^";FP:LPRINT
210 LPRINT"Qp=";USING"##.#####^^^^";Q:LPRINT
220 LPRINT"H0=";USING"##.#####^^^^";H:LPRINT
230 LPRINT:LPRINT"RATIO fp/GB=";USING"##.#####^^^^";T:LPRINT
240 LPRINT STRING$(52,"_"):LPRINT
250 SC=WP*C:GQ=1/Q:E(5)=1/(GQ*SC)
260 J=1:J1=0:J2=1:J3=0:T2=T*T
270 'Nominal Design equations
280 AL1=SQR(H-1):G1=1/AL1:G3=G1:E(3)=(H-1)*E(4):K=G3
290 IF I=0 THEN 300 ELSE 370
300 LPRINT:E(1)=1/(SC*G1):E(2)=1/(SC*G3)
310 BE=H*(H-2)*T:GC=BE/(AL*(1+2*H*T*(1/AL1-1/(2*Q)))):E(6)=1/(GC*SC)
320 G3T=1+G3*T:AL=G3T/(H*G1*G3-G3T)
330 I=1:LPRINT"NOMINAL DESIGN PARAMETERS FOR MIN.ACTIVE POLE SENS.":LPRINT
340 FOR L=1 TO 7:LPRINT E$(L);"=";USING"##.#####^^^^";E(L):LPRINT:NEXT L
350 LPRINT STRING$(52,"_"):LPRINT
360 ' Denominator coefficients and their derivatives w.r.t. G1,GC and AL
370 IF I>0 THEN G3=K
380 G11=G1+GQ+GC:G12=G11+G3:G13=G11*G3:G3T=G3*T:AL1=AL+1
390 X(0)=AL*G1*G3:X(1)=GQ+GC+AL1*G13*T:X(2)=1+AL1*(G12+G13*T)*T
400 X(3)=AL1*T*(1+G12*T):X(4)=AL1*T2:ND=4:GOSUB 950
410 B1=-RD1:B2=-RD0
420 X(0)=AL*G3:X(1)=AL1*G3T:X(2)=AL1*T*(1+G3T):X(3)=X(4):X(4)=0:GOSUB 950
430 A11=RD1:A21=RD0
440 X(0)=0:X(1)=1+AL1*G3T:GOSUB 950
450 A12=RD1:A22=RD0
460 X(0)=G1*G3:X(1)=G11*G3T:X(2)=(G12+X(1))*T:X(3)=T*(1+G12*T):X(4)=T2:GOSUB 950

470 A13=RD1:A23=RD0:HX=AL1*(1+G3T)/(AL*G1*G3):B3=H-HX
480 A31=-HX/G1:A32=0:A33=-(1+G3T)/(AL*AL*G1*G3)
490 X=A21/A11:A22=A22-X*A12:A23=A23-X*A13:B2=B2-X*B1
500 X=A31/A11:A32=A32-X*A12:A33=A33-X*A13:B3=B3-X*B1
510 X=A32/A22:A33=A33-X*A23:B3=B3-X*B2
520 DAL=B3/A33:DGC=(B2-A23*DAL)/A22:DG1=(B1-A12*DGC-A13*DAL)/A11
530 AL=AL+DAL:GC=GC+DGC:G1=G1+DG1
540 IF(ABS(DG1/G1)<.0001 AND ABS(DGC/GC)<.0001) AND ABS(DAL/AL)<.0001 THEN 550 E
LSE 370
550 G11=G1+GQ+GC:G12=G11+G3:G13=G11*G3:G3T=G3*T:AL1=AL+1
560 X(0)=0:X(1)=AL1*G11*G3:X(2)=AL1*(G12+2*G3T*G11):X(3)=AL1*(1+2*T*G12)
570 X(4)=2*AL1*T:ND=4:GOSUB 950
580 A=RD1:B=RD0
590 X(4)=AL1*T2:X(3)=AL1*T*(1+G12*T):X(2)=1+AL1*(G12+G3T*G11)*T
600 X(1)=GQ+GC+AL1*G11*G3T:X(0)=AL*G1*G3:K=G3
610 IF J2=0 AND J3=0 THEN 620 ELSE 640
620 D(4)=X(4):D(3)=X(3):D(2)=X(2):D(1)=X(1):D(0)=X(0):J3=1
630 E(1)=1/(SC*G1):E(2)=1/(SC*G3):E(3)=AL*E(4):E(6)=1/(SC*GC)
640 X(0)=X(1):X(1)=2*X(2):X(2)=3*X(3):X(3)=4*X(4):ND=3:GOSUB 950
650 Y=RD1:R=RD0:S=SQR(((A*RE+B)^2+(A*IM)^2)/((Y*RE+R)^2+(Y*IM)^2))
660 IF J=0 THEN 680
670 J=0:K=K*1.0001:S1=S:GOTO 370
680 IF J1=2 THEN 750
690 K=K/1.0001:P1=(S-S1)*10000!/K:S7=S1:K1=K:P4=P1:P3=S7
700 IF J2=0 THEN 820 ELSE IF J1=1 THEN 800
710 IF ABS(P1)<.001 THEN 810 ELSE IF ABS(P1/K)>.1 THEN 730
```

```
720 V=1-P1/K:GOTO 740
730 IF P1>0 THEN V=.9 ELSE V=1.1
740 K=V*K:J1=2:J=1:GOTO 280
750 K=K/1.0001:P2=(S-S12)*10000!/K:S8=S1:K2=K:P4=P2:P3=S8
760 IF ABS(P2)<.001 THEN 810 ELSE IF ABS(P2)<ABS(P1) THEN 780
770 K=K1-P1*(K2-K1)/(P2-P1):K3=K1:P3=S7:GOTO 790
780 K=K2-P2*(K2-K1)/(P2-P1):K3=K2:P3=S8
790 IF ABS(K3/K-1)<.001 THEN 810 ELSE J1=1:J=1:GOTO 280
800 IF ABS(P3/S1-1)<.001 THEN 810 ELSE 740
810 K=(INT(1000*K)+1)/1000:J2=0:J=1:J1=0:GOTO 280
820 LPRINT"ELEMENT VALUES WITH MINIMUM ACTIVE POLE SENSITIVITY":LPRINT
830 LPRINT"AFTER PREDISTORTION IS APPLIED TO THE GIC-BANDPASS FILTER":LPRINT
840 FOR L=1 TO 7:LPRINT E$(L);"=";USING"##.#####^^^^";E(L):LPRINT:NEXT L
850 LPRINT"MAGNITUDE OF MIN. ACTIVE POLE SENSITIVITY=";USING"##.#####^^^^";S7*T
860 PRINT"DO YOU WANT TO KNOW THE Wp AND Qp SENSITIVITIES WITH RESPECT TO PASSIV
E         ELEMENTS? Y/N";:INPUT A$
870 IF A$="N" OR A$="n" THEN 930
880 LPRINT STRING$(52,"_"):LPRINT"Wp AND Qp SENSITIVITIES ARE":LPRINT
890 E(1)=-.5:E(2)=-.5:E(3)=.5:E(4)=-.5:E(5)=0:E(6)=0:E(7)=-.5:E(8)=-.5
900 FOR L=1 TO 8:LPRINT"Wp sens. w.r.t. ";E$(L);"=";E(L):LPRINT:NEXT L
910 E(5)=GQ/(GQ+GC):E(6)=1-E(5):E(7)=.5:E(8)=-.5
920 FOR L=1 TO 8:LPRINT"Qp sens. w.r.t. ";E$(L);"=";E(L):LPRINT:NEXT L
930 LPRINT STRING$(52,"_"):END
940 'subroutine to divide a poly. with quadratic factor of s^2+s/Q+1
950 Z(ND)=0:Z(ND-1)=0
960 FOR L=ND-2 TO 0 STEP -1
970 Z(L)=X(L+2)-Z(L+1)/Q-Z(L+2)
980 NEXT L
990 RD1=X(1)-Z(0)/Q-Z(1):RD0=X(0)-Z(0)
1000 RETURN
```

6B

Program KHNB

THIS APPENDIX PROVIDES the details of the program KHNB written in GW BASIC language that can be used in IBM PC or its compatibles. This program can be used to design the KHN biquad in Fig. 6.22 with minimum magnitude of the active pole sensitivity and with predistortion to compensate for the effects of the nonzero time constants of the op amps used in the circuit. The procedure for using this program is simple. Just enter it and run it.

Inputs Necessary

f_p, Q_p, R_4, R_6, C_1 $(= C_2)$ and the nonzero value of the time constant of the op amps.

Outputs

Element values (see Fig. 6.22 for the notations) and the minimum magnitude of the active pole sensitivity. If desired, one can also obtain the passive sensitivities of the pole frequency and the pole Q factor for the specified design.

```
10 '
20 '
30 '                               PROGRAM LISTING OF 'KHNB'
40 '                               ------------------------
50 '
60 '
70 'Program to design KHN biquad filter of Fig.6.22 with min. active pole sens.
along with predistortion.
80 PRINT"THIS PROGRAM CAN BE USED TO DESIGN KHN BIQUAD FILTER WITH MINIMUM   ACT
IVE POLE SENSITIVITY  ALONG WITH PREDISTORTION.":PRINT
90 DIM E$(8),X(5),E(8),D(5),Z(5)
100 E$(1)="R1":E$(2)="R2":E$(3)="R3":E$(4)="R4":E$(5)="R5":E$(6)="R6":E$(7)="C1"
:E$(8)="C2="
110 PRINT"INPUT THE SPECIFICATIONS FOR fp and Qp AS PROMPTED":PRINT
120 INPUT" fp in Hz";FP:INPUT" Qp";Q
130 PI2=8!*ATN(1):WP=FP*PI2:RE=-1/(2*Q):IM=SQR(1-RE*RE)
140 PRINT"R1,R2,R3 AND R5 ARE CALCULATED BY THE PROGRAM. INPUT THE OTHER REQUIRE
D PARAMETERS.":PRINT
150 INPUT"THE CAPACITANCES ARE ASSUMED TO BE EQUIVALUED.VALUE OF C1(=C2)";E(7):C
=E(7):E(8)=C
160 INPUT"VALUE OF R4";E(4):R4=E(4)
170 INPUT"VALUE OF R6";E(6):R6=E(6)
180 INPUT"INPUT THE INVERSE OF THE GB PRODUCT OF THE OP.AMP.IN Hz^(-1).IN AN IDE
AL OP.AMP. IT IS ZERO.";T:T=T*FP
190 LPRINT"DESIGN OF A KHN BIQUAD FILTER":LPRINT STRING$(52,"_")
200 LPRINT:LPRINT"THE SPECIFICATIONS ARE":LPRINT:I=0
210 LPRINT"fp in Hz=";USING"##.#####^^^^";FP:LPRINT
220 LPRINT"Qp=";USING"##.#####^^^^";Q:LPRINT
230 LPRINT:LPRINT"RATIO fp/GB=";USING"##.#####^^^^";T:LPRINT
240 LPRINT STRING$(52,"_"):LPRINT
250 SC=WP*C:J=1:J1=0:J2=1:J3=0:T2=T*T:T3=T2*T:K=SQR(3)
260 'Nominal Design equations
270 W=1/SQR(K-1):KX=1/(W*K*Q)
280 IF I=0 THEN 290 ELSE 350
290 LPRINT:E(1)=1/(SC*W):E(2)=E(1)
300 E(3)=E(4)*KX/(1-KX):E(5)=E(6)/(K-1)
310 I=1:LPRINT"NOMINAL DESIGN PARAMETERS FOR MIN.ACTIVE POLE SENS.":LPRINT
320 FOR L=1 TO 8:LPRINT E$(L);"=";USING"##.#####^^^^";E(L):LPRINT:NEXT L
330 LPRINT STRING$(52,"_"):LPRINT
340 ' Denominator coefficients and their derivatives w.r.t. k2 and w
350 WT=W*T:WT2=WT*WT
360 X(0)=(K-1)*W*W:X(1)=KX*K*W*(1+WT):X(2)=1+WT*(2+KX*K)+WT2
370 X(3)=T*(2+WT+K*(1+WT+WT2)):X(4)=T2*(1+2*K*(1+WT)):X(5)=K*T3:ND=5:GOSUB 880
380 B1=-RD1:B2=-RD0
390 X(0)=0:X(1)=W*K*(1+WT):X(2)=WT*K:X(3)=0:X(4)=0:X(5)=0:GOSUB 880
400 A11=RD1:A21=RD0
410 X(0)=2*W*(K-1):X(1)=T*(2*WT+2+KX*K):X(3)=T2*(1+K+2*K*WT):X(4)=2*T3*K:X(5)=0
420 GOSUB 880
430 A12=RD1:A22=RD0
440 AX=A11*A22-A12*A21:DKX=(B1*A22-B2*A12)/AX:DW=(A11*B2-A21*B1)/AX
450 KX=KX+DKX:W=W+DW
460 IF(ABS(DKX/KX)<.0001 AND ABS(DW/W)<.0001) THEN 470 ELSE 350
470 WT=W*T:WT2=WT*WT
480 X(0)=0:X(1)=KX*K*W*W:X(2)=W*(2+KX*K+2*WT):X(3)=2+K+2*WT*(1+K)+3*WT2*K
490 X(4)=2*T*(1+(2+3*WT)*K):X(5)=3*T2*K:ND=5:GOSUB 880
500 A=RD1:B=RD0
510 X(0)=(K-1)*W*W:X(1)=KX*K*W*(1+WT):X(2)=1+WT*(2+KX*K)+WT2
520 X(3)=T*(2+WT+K*(1+WT+WT2)):X(4)=T2*(1+2*K*(1+WT)):X(5)=K*T3
530 IF J2=0 AND J3=0 THEN 540 ELSE 560
540 D(5)=X(5):D(4)=X(4):D(3)=X(3):D(2)=X(2):D(1)=X(1):D(0)=X(0):J3=1
550 E(1)=1/(SC*W):E(2)=E(1):E(3)=KX*E(4)/(1-KX):E(5)=E(6)/(K-1)
560 X(0)=X(1):X(1)=2*X(2):X(2)=3*X(3):X(3)=4*X(4):X(4)=5*X(5):ND=4:GOSUB 880
570 Y=RD1:R=RD0:S=SQR(((A*RE+B)^2+(A*IM)^2)/((Y*RE+R)^2+(Y*IM)^2))
580 IF J=0 THEN 600
590 J=0:K=K*1.0001:S1=S:GOTO 350
600 IF J1=2 THEN 670
610 K=K/1.0001:P1=(S-S1)*10000!/K:S7=S1:K1=K:P4=P1:P3=S7
620 IF J2=0 THEN 740 ELSE IF J1=1 THEN 720
630 IF ABS(P1)<.001 THEN 730 ELSE IF ABS(P1/K)>.1 THEN 650
640 V=1-P1/K:GOTO 660
650 IF P1>0 THEN V=.9 ELSE V=1.1
660 K=V*K:J1=2:J=1:GOTO 270
670 K=K/1.0001:P2=(S-S1)*10000!/K:S8=S1:K2=K:P4=P2:P3=S8
680 IF ABS(P2)<.001 THEN 730 ELSE IF ABS(P2)<ABS(P1) THEN 700
690 K=K1-P1*(K2-K1)/(P2-P1):K3=K1:P3=S7:GOTO 710
700 K=K2-P2*(K2-K1)/(P2-P1):K3=K2:P3=S8
710 IF ABS(K3/K-1)<.001 THEN 730 ELSE J1=1:J=1:GOTO 270
720 IF ABS(P3/S1-1)<.001 THEN 730 ELSE 660
```

```
730 K=(INT(1000*K)+1)/1000:J2=0:J=1:J1=0:GOTO 270
740 LPRINT"ELEMENT VALUES WITH MINIMUM ACTIVE POLE SENSITIVITY":LPRINT
750 LPRINT"AFTER PREDISTORTION IS APPLIED TO THE KHN BIQUAD FILTER":LPRINT
760 FOR L=1 TO 8:LPRINT E$(L);"=";USING"##.#####^^^^";E(L):LPRINT:NEXT L
770 LPRINT"MAGNITUDE OF MIN. ACTIVE POLE SENSITIVITY=";USING"##.#####^^^^";S7*T
780 PRINT"DO YOU WANT TO KNOW THE Wp AND Qp SENSITIVITIES WITH RESPECT TO PASSIV
E       ELEMENTS? Y/N";:INPUT A$
790 IF A$="N" OR A$="n" THEN 860
800 LPRINT STRING$(52,"_"):LPRINT"Wp AND Qp SENSITIVITIES ARE":LPRINT
810 X=E(4)/(E(4)+E(3)):Y=1-X:Z=(E(6)-E(5))/SQR(E(5)*E(6))
820 E(1)=-.5:E(2)=E(1):E(3)=0:E(4)=0:E(5)=-.5:E(6)=.5:E(7)=-.5:E(8)=-.5
830 FOR L=1 TO 8:LPRINT"Wp sens. w.r.t. ";E$(L);"=";E(L):LPRINT:NEXT L
840 E(1)=.5:E(3)=-X:E(4)=X:E(5)=-Q*Z/(2*Y):E(6)=-E(5):E(7)=-.5:E(8)=.5
850 FOR L=1 TO 8:LPRINT"Qp sens. w.r.t. ";E$(L);"=";E(L):LPRINT:NEXT L
860 LPRINT STRING$(52,"_"):END
870 'subroutine to divide a poly. with quadratic factor of s^2+s/Q+1
880 Z(ND)=0:Z(ND-1)=0
890 FOR L=ND-2 TO 0 STEP -1
900 Z(L)=X(L+2)-Z(L+1)/Q-Z(L+2)
910 NEXT L
920 RD1=X(1)-Z(0)/Q-Z(1):RD0=X(0)-Z(0)
930 RETURN
```

6C

Program TTHN

THIS APPENDIX PROVIDES the details of the program TTHN written in GW BASIC language that can be used in IBM PC or its compatibles. This program can be used to design the modified Tow-Thomas biquad in Fig. 6.25 with minimum magnitude of the active pole sensitivity and with predistortion to compensate for the effects of the nonzero time constants of the op amps used in the circuit. The procedure for using this program is simple. Just enter it and run it.

Inputs Necessary

f_p, Q_p, H_o, R_6, C_1 ($= C_2$) and the nonzero value of the time constant of the op amps. In this case one needs to specify the type of the filter function that is designed.

Outputs

Element values (see Fig. 6.25 for the notations) and the minimum magnitude of the active pole sensitivity. If desired, one can also obtain the passive sensitivities of the pole frequency and the pole Q factor for the specified design.

```
10 '
20 '                              PROGRAM LISTING OF 'TTHN'
30 '                              ------------------------
40 '
50 'Program to design modified Tow-Thomas biquad filter in Fig.6.25 with min. ac
tive pole sens. along with predistortion.
60 PRINT"THIS PROGRAM CAN BE USED TO DESIGN MODIFIED TOW-THOMAS BIQUAD FILTER WI
TH MINIMUM   ACTIVE POLE SENSITIVITY  ALONG WITH PREDISTORTION.":PRINT
70 DIM E$(8),X(5),E(8),D(5),Z(5)
80 E$(1)="R1":E$(2)="R2":E$(3)="R3":E$(4)="R4":E$(5)="R5":E$(6)="R6":E$(7)="C1":
E$(8)="C2"
90 PRINT"DO YOU WANT TO DESIGN NONINVERTING BANDPASS(NIBP),INVERTING BANDPASS(IB
P) OR INVERTING LOWPASS(LP)?INPUT NIBP,IBP OR LP ";:INPUT A$:PRINT A$
100 IF A$="NIBP" OR A$="nibp" THEN XX=1 ELSE IF A$="ibp" OR A$="IBP" THEN XX=2 E
LSE IF A$="lp" OR A$="LP" THEN XX=3 ELSE PRINT"nibp,ibp or lp expected":GOTO 90
110 PRINT"INPUT THE SPECIFICATIONS FOR fp,Qp and H AS PROMPTED":PRINT
120 INPUT" fp in Hz";FP:INPUT" Qp";Q:INPUT" H=";H
130 PI2=8!*ATN(1):WP=FP*PI2:RE=-1/(2*Q):IM=SQR(1-RE*RE)
140 PRINT"R1,R2,R3,R4 AND R5 ARE CALCULATED BY THE PROGRAM. INPUT THE OTHER REQU
IRED PARAMETERS.":PRINT
150 INPUT"THE CAPACITANCES ARE ASSUMED TO BE EQUIVALUED.VALUE OF C1(=C2)";E(7):C
=E(7):E(8)=C
160 INPUT"VALUE OF R6";E(6):R6=E(6)
170 INPUT"INPUT THE INVERSE OF THE GB PRODUCT OF THE OP.AMP.IN Hz^(-1).IN AN IDE
AL OP.AMP. IT IS ZERO.";T:T=T*FP
180 LPRINT"DESIGN OF A MODIFED TOW-THOMAS BIQUAD FILTER":LPRINT
190 IF XX=1 THEN LPRINT" NONINVERTING BANDPASS" ELSE IF XX=2 THEN LPRINT"INVERTI
NG BANDPASS" ELSE LPRINT"INVERTING LOWPASS"
200 LPRINT:LPRINT"THE SPECIFICATIONS ARE":LPRINT:I=0
210 LPRINT"fp in Hz=";USING"##.#####^^^^";FP:LPRINT
220 LPRINT"Qp=";USING"##.#####^^^^";Q:LPRINT:LPRINT"H=";USING"###.##";H
230 LPRINT:LPRINT"RATIO fp/GB=";USING"##.#####^^^^";T:LPRINT
240 LPRINT STRING$(52,"_"):LPRINT
250 IF XX=1 THEN A4=0:A3=-1:A2=(3-H)/Q:A1=2:A0=H/Q
260 IF XX=2 THEN A4=0:A3=-H/Q:A2=-1:A1=(3+H)/Q:A0=2
270 IF XX=3 THEN A4=-1:A3=3/Q:A2=(2-H):A1=0:A0=H
280 A=1.5:
290 F=A0+A*(A1+A*(A2+A*(A3+A*A4))):F1=A1+A*(2*A2+A*(3*A3+A*4*A4))
300 AX=-F/F1:A=A+AX
310 IF ABS(AX)>.0001 THEN 290
320 K=A*A
330 SC=WP*C:J=1:J1=0:J2=1:J3=0:T2=T*T:T3=T2*T
340 'Nominal Design equations
350 W=1/SQR(K):Q1=W*(1+W*T)/(1/Q+3*T)
360 IF I=0 THEN 370 ELSE 440
370 LPRINT:E(1)=1/(SC*W):E(2)=E(1)
380 IF XX=1 THEN A=H*W/Q ELSE IF XX=2 THEN A=H*W*K/Q ELSE A=H/K
390 E(3)=E(1)/A:E(4)=Q1*E(1):E(5)=E(6)*K
400 I=1:LPRINT"NOMINAL DESIGN PARAMETERS":LPRINT
410 FOR L=1 TO 8:LPRINT E$(L);"=";USING"##.#####^^^^";E(L):LPRINT:NEXT L
420 LPRINT STRING$(52,"_"):LPRINT
430 ' Denominator coefficients and their derivatives w.r.t. Q1 and w
440 WT=W*T:WT2=WT*WT:B=1+A+1/Q1:BX=1+B:KX=K+1:KB1=(K-1)*B
450 X(0)=K*W*W:X(1)=W*(1+WT)/Q1:X(2)=1+WT*((K+2)/Q1+1-KB1+WT*(KX/Q1-KB1))
460 X(3)=T*(3+WT*(2*BX+KX/Q1)+KX*WT2*B):X(4)=T2*(KX+2+KX*BX*WT):X(5)=KX*T3
470 ND=5:GOSUB 1020
480 B1=-RD1:B2=-RD0
490 X(0)=2*K*W:X(1)=1/Q1+2*WT:X(2)=T*((K+2)/Q1+1-KB1+2*WT*(KX/Q1-KB1))
500 X(3)=T2*(2*BX+KX/Q1+WT*2*KX*B):X(4)=KX*BX*T3:X(5)=0:GOSUB 1020
510 A11=RD1:A21=RD0
520 Q2=1/Q1/Q1:X(0)=0:X(1)=-W*(1+WT)*Q2:X(2)=-(3+2*WT)*WT*Q2
530 X(3)=-T*((K+3)*WT+KX*WT2)*Q2:X(4)=KX*WT*T2*Q2:X(5)=0:GOSUB 1020
540 A12=RD1:A22=RD0
550 AX=A11*A22-A12*A21:DW=(B1*A22-B2*A12)/AX:DQ1=(A11*B2-A21*B1)/AX
560 Q1=Q1+DQ1:W=W+DW
570 IF(ABS(DQ1/Q1)<.0001 AND ABS(DW/W)<.0001) THEN 580 ELSE 440
580 IF XX=1 THEN AX=H*W/(Q*(1+T)) ELSE IF XX=2 THEN AX=H*K*W/(Q*(1+T)) ELSE AX=H
*W*W
590 IF ABS(AX/A-1)>.0001 THEN A=AX:GOTO 440
600 A=AX:WT=W*T:WT2=WT*WT:B=1+A+1/Q1:BX=1+B:KX=K+1:KB1=(K-1)*B
610 X(0)=0:X(1)=W*W/Q1:X(2)=W*((K+2)/Q1+1-KB1+2*WT*(KX/Q1-KB1))
620 X(3)=3+2*WT*(2*BX+KX/Q1+1.5*KX*B*WT):X(4)=T*(2*(K+3)+3*KX*BX*WT)
630 X(5)=3*KX*T2:ND=5:GOSUB 1020
640 A1=RD1:B=RD0
650 X(0)=K*W*W:X(1)=W*(1+WT)/Q1:X(2)=1+WT*((K+2)/Q1+1-KB1+WT*(KX/Q1-KB1))
660 X(3)=T*(3+WT*(2*B1+KX/Q1)+KX*WT2*B):X(4)=T2*(KX+2+KX*B1*WT):X(5)=KX*T3
670 IF J2=0 AND J3=0 THEN 680 ELSE 700
680 D(5)=X(5):D(4)=X(4):D(3)=X(3):D(2)=X(2):D(1)=X(1):D(0)=X(0):J3=1
```

```
690 E(1)=1/(SC*W):E(2)=E(1):E(3)=E(1)/A:E(4)=Q1*E(1):E(5)=E(6)*K
700 X(0)=X(1):X(1)=2*X(2):X(2)=3*X(3):X(3)=4*X(4):X(4)=5*X(5):ND=4:GOSUB 1020
710 Y=RD1:R=RD0:S=SQR(((A1*RE+B)^2+(A1*IM)^2)/((Y*RE+R)^2+(Y*IM)^2))
720 IF J=0 THEN 740
730 J=0:K=K*1.0001:S1=S:GOTO 440
740 IF J1=2 THEN 810
750 K=K/1.0001:P1=(S-S1)*10000!/K:S7=S1:K1=K:P4=P1:P3=S7
760 IF J2=0 THEN 880 ELSE IF J1=1 THEN 860
770 IF ABS(P1)<.001 THEN 870 ELSE IF ABS(P1/K)>.1 THEN 790
780 V=1-P1/K:GOTO 800
790 IF P1>0 THEN V=.9 ELSE V=1.1
800 K=V*K:J1=2:J=1:GOTO 350
810 K=K/1.0001:P2=(S-S1)*10000!/K:S8=S1:K2=K:P4=P2:P3=S8
820 IF ABS(P2)<.001 THEN 870 ELSE IF ABS(P2)<ABS(P1) THEN 840
830 K=K1-P1*(K2-K1)/(P2-P1):K3=K1:P3=S7:GOTO 850
840 K=K2-P2*(K2-K1)/(P2-P1):K3=K2:P3=S8
850 IF ABS(K3/K-1)<.001 THEN 870 ELSE J1=1:J=1:GOTO 350
860 IF ABS(P3/S1-1)<.001 THEN 870 ELSE 800
870 K=(INT(10*K)+1)/10:J2=0:J=1:J1=0:GOTO 350
880 LPRINT"ELEMENT VALUES WITH MINIMUM ACTIVE POLE SENSITIVITY AFTER":LPRINT
890 LPRINT"PREDISTORTION IS APPLIED TO THE MODIFIED TOW-THOMAS BIQUAD FILTER":LP
RINT
900 FOR L=1 TO 8:LPRINT E$(L);"=";USING"##.#####^^^^";E(L):LPRINT:NEXT L
910 LPRINT"MAGNITUDE OF MIN. ACTIVE POLE SENSITIVITY=";USING"##.#####^^^^";S7*T
920 PRINT"DO YOU WANT TO KNOW THE Wp AND Qp SENSITIVITIES WITH RESPECT TO PASSIV
E        ELEMENTS? Y/N";:INPUT A$
930 IF A$="N" OR A$="n" THEN 1000
940 LPRINT STRING$(52,"_"):LPRINT"Wp AND Qp SENSITIVITIES ARE":LPRINT
950 X=E(4)/(E(4)+E(3)):Y=1-X:Z=(E(6)-E(5))/SQR(E(5)*E(6))
960 E(1)=-.5:E(2)=E(1):E(3)=0:E(4)=0:E(5)=-.5:E(6)=.5:E(7)=-.5:E(8)=-.5
970 FOR L=1 TO 8:LPRINT"Wp sens. w.r.t. ";E$(L);"=";E(L):LPRINT:NEXT L
980 E(1)=.5:E(3)=-X:E(4)=X:E(5)=-Q*Z/(2*Y):E(6)=-E(5):E(7)=-.5:E(8)=.5
990 FOR L=1 TO 8:LPRINT"Qp sens. w.r.t. ";E$(L);"=";E(L):LPRINT:NEXT L
1000 LPRINT STRING$(52,"_"):END
1010 'subroutine to divide a poly. with quadratic factor of s^2+s/Q+1
1020 Z(ND)=0:Z(ND-1)=0
1030 FOR L=ND-2 TO 0 STEP -1
1040 Z(L)=X(L+2)-Z(L+1)/Q-Z(L+2)
1050 NEXT L
1060 RD1=X(1)-Z(0)/Q-Z(1):RD0=X(0)-Z(0)
1070 RETURN
```

7

Higher-Order Active Filters

THE METHODS OF DESIGNING active filters of order $N > 2$ may be classified into two major groups. In the first group, active filters are realized by simulating doubly terminated LC ladder networks. An LC ladder network, terminated with two resistances at the two ports and designed to transmit maximum power at a number of frequencies in the passband from the input port to the output port, is found to possess low-transfer function sensitivity to element variations in the passband of the filter. This fact was advanced by H. J. Orchard in 1966. There are two advantages in this approach:

1. The resulting active network retains the desirable low-sensitivity property of the LC ladder network.
2. There is a wealth of knowledge on the design of doubly terminated LC ladder networks. In fact, standard tables of such networks are available, and one can find an LC ladder network that meets almost any set of specifications.

However, the active filter simulating an LC ladder network generally requires a large number of op amps, and this increases not only the cost of the unit but also the power consumption and dissipation. Another disadvantage of this approach is that only filters based on LC ladders can be realized. These methods are useful in critical applications where filter specifications are stringent and low-sensitivity performance is of paramount importance.

The second group includes direct methods. In these procedures, the transfer function is the starting point. If one attempts to realize a given transfer function $H(s)$ using direct methods, the resulting network will be highly sensitive to component variations, and this method is not used in present-day practice. For this reason, we prefer a more practical technique involving cascade realization. In cascade realization, the network realizing a higher-order transfer function is much less sensitive than a network realized using direct methods. We briefly discussed cascade realization in Chap. 4.

In cascade realization, a given Nth-order transfer function is expressed as a product of a number of second-order (biquadratic) and, possibly one first-order, transfer functions. Each second-order transfer function may be realized using any one of the biquads discussed in the previous three chapters. If we need a first-order network, a simple RC network (with an additional op amp to provide isolation, if necessary) may be used. Alternatively, we can combine this first-order factor with a second-order factor and realize a third-order transfer function, as discussed in Sec. 4.6. If the output of each section is taken at the output of an op amp, each section will have the necessary isolation. All the sections can then be connected in cascade to realize the overall Nth-order transfer function. This is by far the most popular of all the methods of realizing higher-order active filters because (1) it has the advantage of modularity, and (2) the design methods and fabrication and tuning procedures can be standardized. Thus it becomes economical to manufacture higher-order filters using cascade realization.

The decoupling and isolation present in the cascade realization result in higher sensitivity as compared to that in the LC ladder realization. A change occurring in one section causes a shift in one pole pair, and this in turn may cause a drastic change in the magnitude response of the overall filter. The change may be severe, especially if the pole pair subject to change is a high-Q one, and the network may become useless. In LC ladder networks, there is a high level of coupling and also inherent negative feedback. Furthermore, all the pole positions of the transfer function of an LC ladder network depend on every circuit element, and the sensitivity is "spread" to every element in the network. Thus, in an LC ladder network, the sensitivity due to individual elements is less. This observation has given rise to a new set of methods of realizing higher-order filters called *coupled biquad* structures in which the modularity and building block concepts of cascade realization are retained and coupling is introduced to reduce the sensitivity of the overall network. This approach has resulted in a method called *follow-the-leader-feedback* (FLF), structure, also known as *primary resonator block* (PRB) structure. First, we shall discuss the cascade method of realizing higher-order active filters, and then we will consider other important methods.

7.1 *Cascade Realizations*

As mentioned earlier, a given transfer function $H(s)$ is realized as a voltage transfer ratio of a cascade of biquadratic filter networks. If the order N is odd, then we may need an additional first-order network. The design of a cascaded network involves two major steps:

1. *Decomposition problem*: It is necessary to decompose the given transfer function $H(s)$ into second-order (and possibly one first-order) transfer functions. When N is even, it should be of the form

$$H(s) = \prod_{i=1}^{N/2} H_i(s) = \prod_{i=1}^{N/2} H_{io} \frac{s^2 + (\omega_{zi}/Q_{zi})s + \omega_{zi}^2}{s^2 + (\omega_{pi}/Q_{pi})s + \omega_{pi}^2}$$

2. *Realization problem*: It is necessary to design each biquadratic (and, if necessary, a first-order RC network) transfer function using suitable networks. We have discussed the realization of second-order networks in the previous three chapters.

The Decomposition Problem

The decomposition problem involves three steps: (1) pole-zero pairing, (2) cascade sequencing, and (3) gain constant distribution. In the pole-zero pairing problem, each of the $N/2$ pairs of zeros (including those at the origin and at infinity) should be properly combined with the $N/2$ pairs of (usually complex) poles. We need to assign a pair of zeros to a particular pair of poles, and there are $(N/2)!$ possible combinations. The next problem involves the sequence of the cascade. The overall transfer function is not affected by the cascading sequence, and again there are $(N/2)!$ possible combinations. The final question arises as to how the gain constant of the overall transfer function should be distributed among the various second-order sections; i.e., What should the values of H_{io} be and how do we determine them? The entire decomposition problem may be tackled to minimize the overall sensitivity of the cascade sequence, however, there is little to be gained by doing this. In fact, the overall sensitivity can be minimized by designing each second-order network with minimum sensitivity. Another important criterion in filter design is the dynamic range. Thus the solution to the decomposition problem must maximize the dynamic range of the cascaded network. To do this, two objectives must be met:

1. When V_{amp} is the maximum allowed signal of the output op amp and m is the maximum required gain for the filter, the circuit should be capable of accepting a maximum input signal of

$$V_{\text{in, max}} = \frac{V_{\text{amp}}}{m}$$

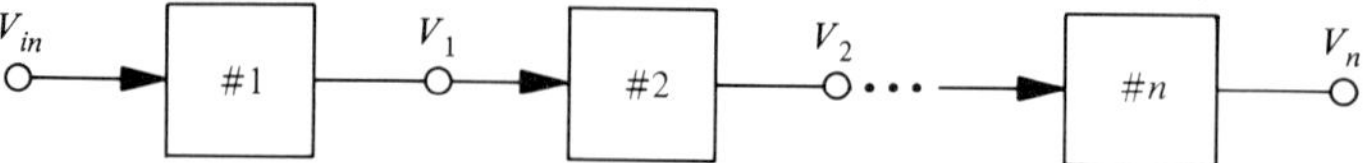

Figure 7.1 Cascade of n second-order networks with $n = N/2$, where N is assumed to be an even number.

2. The signal-to-noise ratio should be as high as possible in the entire sequence.

Consider the cascade sequence shown in Fig. 7.1. We define the transfer function

$$T_i(s) = \frac{V_i}{V_{\text{in}}} = \prod_1^i H_j(s) = \prod_1^i H_{jo} h_j(s) \tag{7.1}$$

where V_i is the ith output and $h_j(s)$ is the transfer function of the jth section with a unity gain constant. The two objectives may be met if the maximum amplitudes of all the voltages $V_1, V_2, \ldots, V_n$ are equal to V_{amp}. Thus the gain constant distribution must be made such that

$$|V_1|_{\max} = |V_2|_{\max} = \cdots = |V_n|_{\max} = V_{\text{amp}} \tag{7.2}$$

In order for the signal-to-noise ratio to be as high as possible, we must maximize the minimum of the in-band signal. What this means is that the pole-zero pairing and the cascade sequencing must be done such that the magnitude of the intermediate transfer functions $T_i(s)$ is as flat as possible over the filter passband. This is a difficult task in all but simple cases. If the order of the filter $N > 4$, it will be a tedious process to try all combinations of pole-zero pairings and all possible cascade sequences. However, we should be able to obtain simple solutions by splitting the decomposition problem into the following three steps:

1. Determine the pole-zero pairing such that each $h_i(s)$ is as flat as possible over the filter passband.
2. Find a sequence such that each $T_i(s)$ is as flat as possible.
3. Assign gain constants H_{io} such that (7.2) is satisfied.

Though systematic methods are available for solving the pole-zero pairing problem, the following rule of thumb is sufficient for all practical purposes: *High-Q poles should be combined with the closest transmission zeros*. In addition to this rule, some other considerations may also be taken into account. The availability of components and ease of construction and tuning may play an important in the pole-zero pairing.

The second step is to find the proper cascade sequence. One has to try $(N/2)!$ possible combinations before arriving at a particular sequence satisfying the second condition, which requires a lot of work. Again, a rule of thumb is sufficient: *One should place the sections with the flattest response at the input, followed by sections with less flatness, and so on. That is, the Q factors of the biquads increase progressively from the input to the output.*

Again, practical considerations may play an important role in deciding the sequence. In order to eliminate high-frequency and out-of-band signals, a bandpass or a lowpass section can be used at the input. This will avoid overloading of the subsequent sections in the cascade. Similarly, a bandpass or a highpass section can be used as the last section so that the dc offset and power supply ripples do not appear at the output of the sequence. Of course, the dc offset of the last section cannot be avoided in any case.

Invariably, the first section in the cascade is chosen to be a bandpass or a lowpass filter, or at least the flattest section possible that will have the lowest Q factor. From the remaining $N/2 - 1$ sections, the second section should be chosen to provide the flattest possible response at its output. Thus, after determining the second section, the third section may be selected from the remaining $N/2 - 2$ sections as the one that will provide the flattest possible response at its output. In this way, we will need to determine fewer combinations by trial and error, and they can usually be found almost by inspection after plotting all the individual characteristics. The question now is: What is the measure of flatness? A measure of flatness for any function $F_i(s)$ can be determined by

$$d_i = \frac{|F_i(j\omega)|_{\max}}{|F_i(j\omega)|_{\min}}$$

where

$$|F_i(j\omega)|_{\max} = \max|F_i(j\omega)| \qquad 0 \leqq \omega < \infty$$

and

$$|F_i(j\omega)|_{\min} = \min|F_i(j\omega)| \qquad \omega_L \leqq \omega \leqq \omega_H$$

where ω_L and ω_H are the lower and upper passband edge frequencies. Note that, in searching for the maximum, we use the maximum that occurs in the entire frequency range because it is this maximum that is of concern to us. However, in looking for the minimum, we use only the minimum that occurs in the passband. If the magnitude function $F_i(j\omega)$ is ideally flat, then d_i should be unity. In filtering applications, d_i cannot be expected to be unity, but the optimum choice is one where d_i is closest to unity. Also note that the gain constant of the transfer function does not play a role in determining the constant d_i because it cancels out in the determination of d_i. Thus the gain constants are not important in determining the flatness of a function, and the distribution of the gain constants can be made after finding the pole-zero pairs and the cascade sequence. Consider the following example.

Example 7.1: Elliptic approximation was used to obtain a bandpass transfer function that satisfied the following specifications: The passband extends from 1 to 1.2 kHz, and the passband ripple is 0.5 dB. The bandwidth of the stopband is 0.3 kHz, and the minimum stopband loss is at least 40 dB. The poles and zeros are

Zeros:

$$z_1, z_1^* = \pm j5571.94 \text{ (pair 1)}$$
$$z_2, z_2^* = \pm j5973.55 \text{ (pair 2)}$$
$$z_3, z_3^* = \pm j8502.26 \text{ (pair 3)}$$
$$z_4, z_4^* = \pm j7930.64 \text{ (pair 4)}$$
$$z_5 = 0, \infty \text{ (pair 5)}$$

Poles:

$$p_1, p_1^* = -190.628 \pm j7336.68, \ \omega_p = 7339.15, \ Q_p = 19.25 \text{ (pair 1)}$$
$$p_2, p_2^* = -56.0798 \pm j7548.22, \ \omega_p = 7548.43, \ Q_p = 67.3 \text{ (pair 2)}$$
$$p_3, p_3^* = -167.662 \pm j6452.80, \ \omega_p = 6454.9, \ Q_p = 19.25 \text{ (pair 3)}$$
$$p_4, p_4^* = -46.6266 \pm j6275.84, \ \omega_p = 6276.02, \ Q_p = 67.3 \text{ (pair 4)}$$
$$p_5, p_5^* = -267.646 \pm j6877.68, \ \omega_p = 6882.88, \ Q_p = 12.86 \text{ (pair 5)}.$$

It is decided to realize this transfer function as a cascade sequence using biquadratic sections. Find the pole-zero pairing and the cascade sequence that will provide the flattest possible response throughout the sequence.

Pole-zero pairing: Checking the pole Q factors, we find that the highest pole Q factor is 67.3 and that the corresponding poles should be combined with the closest zeros. Thus the pole pairs 2 and 4 are combined with the zero pairs 4 and 2, respectively. The next highest pole Q factor is 19.25, and therefore the pole pairs 1 and 3 are combined with the zero pairs 3 and 1, respectively. The remaining pole pair 5 is paired with the zero pair 5. Table 7.1 lists the different transfer functions and their pole-zero pairings.

Table 7.1 Pole-zero pairings in Example 7.1

Transfer function	Pole pair	Zero pair
H_1	1	3
H_2	2	4
H_3	3	1
H_4	4	2
H_5	5	5

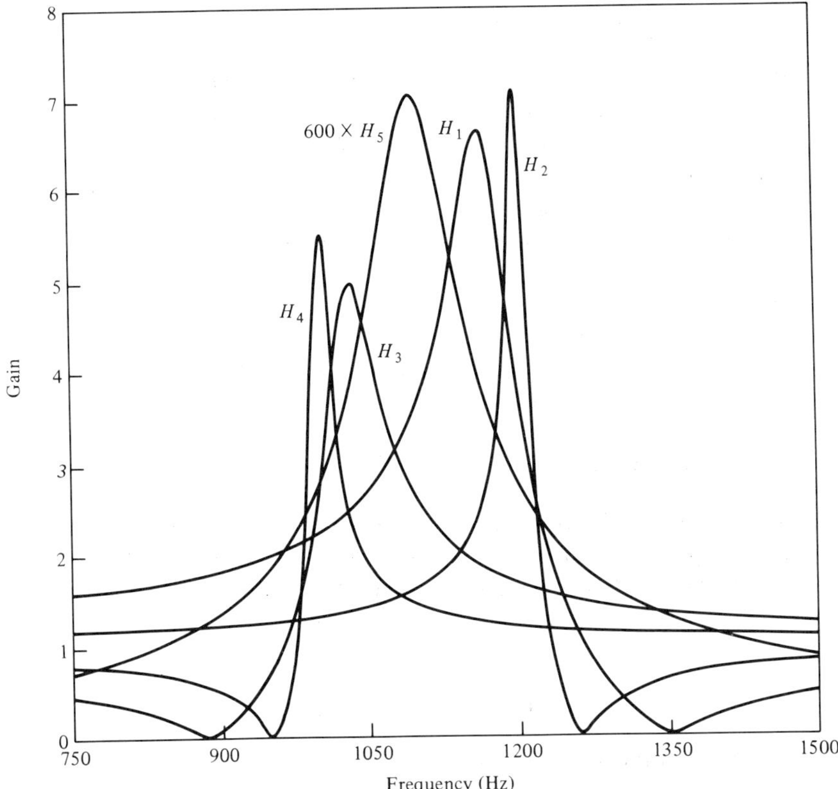

Figure 7.2 Magnitude responses of the second-order transfer functions in Example 7.1.

Cascade sequencing: The first section should be the one with the flattest possible response, and obviously it should have the lowest pole Q factor. The transfer function H_5 has the lowest pole Q factor and it should be the first section in the sequence. The second section should probably have the next lowest pole Q factor, and it may be either H_1 or H_3. To aid in making our decision we have plotted the magnitude responses of all five transfer functions with unity gain constants except for H_5. The transfer function H_5 has been multiplied by a factor of 600 arbitrarily. These characteristics are shown in Fig. 7.2. From these characteristics also, we can easily see that either H_1 or H_3 can be connected following H_5. In fact, whether we connect H_1 or H_3, the value of d at the end of the second section is the same, and it is the lowest, being equal to 4.44. Therefore we connect H_1 as the second section. Next, we have to select, from the remaining three sections, the third section such that, at the end of this section, the response will be as flat as possible. Among the remaining three transfer functions, namely, H_2, H_3, and H_4, if we connect H_4, it will make T_3 as flat as possible. Because T_2 will have its peak toward the right of the passband, we need a sharply resonating section to the left of the passband, and this is provided by H_4. If the third section is H_3, the value of d will be 3.86, and if the third section is H_4, the value of d will be only 3.20, thus indicating a flatter response. Remember that the pole Q factor of H_4 is greater than that of H_3, and thus the rule of thumb is violated. After determining the first three sections based on the principle that the response should be as flat as possible, we choose H_2 as the fourth section and, of course, the fifth section is H_3, which is a highpass notch section. We would prefer to have the last section a bandpass or a highpass section, but this is not possible because of the nonavailability of such sections. We have to be satisfied with a highpass notch section. Thus the sequence will then be H_5, H_1, H_4, H_2, and H_3. ■

Next, we shall discuss a method of distributing the gain constants. Let H_o be the gain constant of the overall transfer function. Without loss of generality, assume that $V_{in,\max} = 1$. Then we note that

$$|V_1|_{\max} = H_{1o} \max |h_1(j\omega)|$$

and

$$|V_n|_{\max} = \max |H(j\omega)|$$

Since they both must be equal to V_{amp}, we have

$$H_{1o} = \frac{\max |H(j\omega)|}{\max |h_1(j\omega)|} \tag{7.3a}$$

Next, we find that

$$|V_2|_{\max} = H_{1o} H_{2o} \max |h_1(j\omega) h_2(j\omega)|$$

and since $|V_2(j\omega)|_{\max}$ is also equal to $|V_1(j\omega)|_{\max}$, we have

$$H_{2o} = \frac{\max |h_1(j\omega)|}{\max |h_1(j\omega) h_2(j\omega)|}$$

In general,

$$H_{io} = \frac{\max |h_1(j\omega) h_2(j\omega) \cdots h_{(i-1)}(j\omega)|}{\max |h_1(j\omega) h_2(j\omega) \cdots h_i(j\omega)|} \tag{7.3b}$$

Note that $h_i(s)$ is the same as $H_i(s)$ with a unity gain constant. Of course, it is easy to compute the gain constant of the last section using

$$H_{no} = \frac{H_o}{H_{1o} H_{2o} \cdots H_{(n-1)o}} \tag{7.3c}$$

where $n = N/2$.

Table 7.2 Transfer functions as decomposed in Examples 7.1 and 7.2

Section	Pole pair	Zero pair	Gain constant
1 (BP)	$-267.646 \pm j6877.68$	$0, \infty$	535.2915
2 (LPN)	$-190.628 \pm j7336.68$	$\pm j8502.26$	0.255
3 (HPN)	$-46.6266 \pm j6275.84$	$\pm j5973.55$	0.692
4 (LPN)	$-56.0798 \pm j7548.22$	$\pm j7930.54$	0.432
5 (HPN)	$-167.662 \pm j6452.80$	$\pm j5571.94$	0.591

Evaluation of the different gain constants may be carried out using either a programmable calculator or a computer. Consider the following example.

Example 7.2: Consider the filter in Example 7.1. In order to have a maximum gain of unity in its passband, the overall gain constant must be equal to 24.124. Find the gain constants of the different sections such that a maximum dynamic range is obtained throughout the cascade.

It is given that the maximum of $|H(j\omega)|$ is equal to 1. The maximum of $|h_1(j\omega)|$ is 1/535.2915. Therefore, using (7.3*a*), we have

$$H_{5o} = 535.2915$$

Using (7.3*b*), we find that

$$H_{1o} = \frac{0.001868}{0.0073215} = 0.255$$

$$H_{4o} = \frac{0.0073215}{0.0158373} = 0.692$$

and

$$H_{2o} = \frac{0.0158373}{0.0244845} = 0.432$$

The gain constant of the last section may be calculated as

$$H_{3o} = \frac{24.124}{(535.2915)(0.255)(0.692)(0.432)} = 0.591$$

Table 7.2 lists the second-order transfer functions in the sequence in which they are connected from the input to the output, along with their gain constants. ■

The Realization Problem

The realization problem requires a design of the appropriate network, and this can be accomplished using the circuits discussed in Chaps. 4–6. However, one has to keep in mind that some circuits are useful only for low- and medium-Q_p values and others are useful even for medium- and high-Q_p values. For designing networks up to a Q_p value of 20, one may use the single-amplifier circuits discussed in Chaps. 4 and 5. The important advantage of these circuits is that they use only a single op amp, which saves on power consumption. However, when the Q_p values are in excess of 20, one has to resort to multiple-amplifier circuits discussed in Chap. 6.

Example 7.3: Obtain a cascaded network that will realize the tenth-order transfer function discussed in Examples 7.1 and 7.2.

The specifications for the different second-order sections of this tenth-order bandpass filter are given in Table 7.2. The cascade sequence has already been determined in Example 7.2. After designing suitable networks to realize each second-order transfer function in Table 7.2, we can connect these networks in cascade to realize the entire transfer function. Thus we next discuss the design of each transfer function in Table 7.2.

Section 1 (BP): The specifications for this second-order bandpass filter are given in Table 7.2, and from them, we obtain $f_p = 1095$ Hz, $Q_p = 12.86$, and $H_o = 1$. Since the Q_p value of this bandpass filter is below 20, we use the single-amplifier bandpass filter circuit in Fig. 5.10. To minimize the active pole sensitivity, we choose $\beta = 0.25$ and $n^2 = 2$. Then, using (5.39), we find the required value of m to be 0.352168. Now, choosing $C_1 = 10$ nF, we find the other parameters using (5.37), (5.38), and (5.40): $C_2 = 20$ nF, $R_1 = 124.543$ kΩ, $R_2 = 29.172$ kΩ, and $R_3 = 3.726$ kΩ. The values of R_B and R_D can be 18 and 6 kΩ, respectively.

Section 2 (LPN): The specifications for this lowpass notch filter are listed in Table 7.2, and its Q_p value is also in the medium range. Therefore, we use the single-amplifier filter in Fig. 5.13. The pertinent parameters are $f_p = 1168.1$ Hz, $Q_p = 19.25$, $f_z = 1353.1$ Hz, and $H_o = 0.255$. Choosing $C_1 = C_2 = 10$ nF, $G_A = 0.1$ mS, and $G_B = 0.1/3$ mS, we have $\lambda = \frac{1}{3}$ and $M = 1$. Then, using (5.63), we find that

$$G_1 = 174.144\ \mu\text{S}$$

Using (5.64*b*), we have

$$R_4 = 20.332\ \text{k}\Omega$$

and then, from (5.67),

$$R_5 = 8.002\ \text{k}\Omega$$

For LPN filters, $\delta = 0$, and therefore $G_6 = 0$. Using this fact and (5.65), we find that

$$R_7 = 126\ \text{k}\Omega$$

Next, from (5.66), we have

$$R_2 = 35.354 \text{ k}\Omega$$

(5.68) and (5.69) give

$$R_C = 39.216 \text{ k}\Omega$$

and

$$R_D = 13.423 \text{ k}\Omega$$

Also,

$$R_B = 30 \text{ k}\Omega$$

If the value of R_7 (126 kΩ) is considered to be a high value, then an external resistor R_X may be connected between the output and the ground (this addition will not alter the transfer function of the circuit), and a delta-to-star transformation involving R_2, R_7, and R_X may be used to reduce the impedance level (see Fig. 5.15 and explanation). However, this procedure will increase the number of resistors by 1.

Section 3 (HPN): The third section is an HPN section, and its Q_p value is 67.3, which is high. Therefore the single-amplifier filter in Fig. 5.13 is not suitable for realizing this transfer function, and it is necessary to use a multiple-amplifier circuit. The parameters for this section are $f_p = 998.9$ Hz, $Q_p = 67.3$, $f_z = 950.71$ Hz, and $H_o = 0.692$.

The two-amplifier circuit in Fig. 6.13d may be used to realize this section. But, unfortunately, the gain constant of this circuit is dependent only on the values of f_p and f_z and cannot be controlled independently. Thus the gain constant requirement of 0.692 cannot be met. Therefore, if we use this circuit, we may have to readjust the gain constants of the other sections. However, the circuit in Fig. 6.29 or 6.31 may be used to design this section, as well as the next where the gain constant requirement can also be met. The number of passive elements required to design a notch filter using the circuit in Fig. 6.31 will be higher, and therefore it is decided to use the circuit in Fig 6.29.

For the circuit in Fig. 6.29, choosing $C = 100$ nF, we require $C_3 = 69.2$ nF. Then, $R = 1/\omega_p C = 1.593$ kΩ, and $Q_p R = 107.24$ kΩ. Next, we find that $R_1 = \omega_p/\omega_z^2 C_3 = 2.542$ kΩ. Of course, $R_2 = R_3 = \infty$. The value of r can be any value, typically 10 kΩ.

A word of caution about the circuit in Fig. 6.29 is in order. Since the Q_p value is high, there may be substantial Q enhancement in the circuit even though the operating frequency is only on the order of 1 kHz. To avoid this Q enhancement and to reduce the effect of the active sensitivity on the Q factor, it is highly advisable to use the circuit in Fig. 6.30 or 6.31. It is simple, easy, and convenient to design and tune the circuit in Fig. 6.30, and it also requires fewer passive components (similar to the circuit in Fig. 6.29). Therefore we use the actively compensated circuit in Fig. 6.30, and the values of R, Q_pR, and C are the same as for the circuit in Fig. 6.29. Because of the change in notation in the circuit in Fig. 6.30, the values of C_1 and R_2 are 69.2 nF and 2.542 kΩ. In addition, the value of R/b should be 0.979 kΩ for the compensation to be successful. This completes the design of section 3 using the circuit in Fig. 6.30.

Section 4 (LPN): The various comments we made on the design of section 3 hold true in this case as well because the Q_p value of section 4 is also high. Thus, without further explanation, we design this section using the circuit in Fig. 6.30. The designed parameters are $C = 100$ nF, $C_1 = 43.2$ nF, $R = 1.325$ kΩ, $Q_pR = 89.159$ kΩ, $R_2 = 2.778$ kΩ, and $R/b = 0.897$ kΩ.

Section 5 (HPN): The last section in the sequence is an HPN filter, and its Q_p value is only 19.25. Therefore we can use the single-amplifier circuit in Fig. 5.13. Thus, choosing $C_1 = C_2 = 10$ nF, $G_A = 0.1$ mS, and $G_B = 0.1/3$ mS, we proceed to design this circuit. First, using (5.63), we obtain

$$G_1 = 153.16\ \mu\text{S}$$

Then, using (5.64b), we find

$$R_4 = 11.647\ \text{k}\Omega$$

The value of R_5 can be calculated from (5.67):

$$R_5 = 14.858\ \text{k}\Omega$$

For HPN filters, the value of δ is set to 1. Thus $G_7 = 0$ and $G_6 = G_3$, and therefore

$$R_6 = 133.08\ \text{k}\Omega$$

Then, from (5.66),

$$R_2 = 33.66\ \text{k}\Omega$$

Using (5.68) and (5.69), we obtain

$$R_C = 16.92\ \text{k}\Omega \qquad \text{and} \qquad R_D = 24.45\ \text{k}\Omega$$

Of course,

$$R_B = 30\ \text{k}\Omega$$

If the value of R_6 is considered high, we can use an additional resistor R_X as shown in Fig. 5.14 and employ delta-to-star transformation to reduce the impedance level. This procedure will not affect the realized transfer function but will increase the number of resistors by 1.

Now that we have completed the design of all five sections, we simply connect all five sections in cascade in the same sequence as they were designed. The overall circuit diagram cannot be contained in one or two pages and therefore is omitted here. ■

Example 7.4: A sixth-order bandpass transfer function was obtained using Chebyshev approximation. This transfer function meets the specifications for a passband ripple of 0.5 dB extending from 24 to 28 kHz. The pole frequency and pole Q factors are

Section 1: $f_p = 25.923$ kHz, $Q_p = 10.3451$
Section 2: $f_p = 23.959$ kHz, $Q_p = 20.7544$
Section 3: $f_p = 28.048$ kHz, $Q_p = 20.7544$.

To obtain a maximum gain of unity in the passband, the gain constant of the combined transfer function when each second-order bandpass transfer function is expressed in the form shown in Table 4.1, is 11.761676. Design a cascade network that realizes this sixth-order transfer function.

Since Chebyshev approximation gives an all-pole prototype lowpass function, the transfer function of the corresponding bandpass function will have three zeros at the origin and three zeros at infinity. The pole-zero pairing problem is very simple in such cases. We can combine the poles and zeros such that there will be a lowpass, a bandpass, and a highpass transfer function. However, for ease of realization, as well as for practical realization, we can combine the poles and zeros such that there are three bandpass transfer functions. Next, we have to decide about the cascade sequence. The first section has the lowest Q_p value, and therefore section 1 will also be the first section in the cascade. Since sections 2 and 3 have the same value Q_p value, either one of them could constitute the second section in the cascade. Therefore our cascade sequence will have the same sequence as that given above.

The third problem is to find the gain constant of each bandpass section when each one is expressed in the form shown in Table 4.1. The gain constant of the first section should be

$$H_{o1} = \frac{|H(j\omega)|_{\max}}{|h_1(j\omega)|_{\max}} = \frac{1}{1} = 1$$

In the second section the required gain constant is

$$H_{o2} = \frac{|h_1(j\omega)|_{\max}}{|h_1(j\omega)h_2(j\omega)|_{\max}} = \frac{1}{0.5365} = 1.864$$

Table 7.3 Designed circuit parameters, using the circuit in Fig. 5.10, for the circuits in Example 7.4[a]

Section	R_1 (kΩ)	R_2 (kΩ)	R_3 (kΩ)	R_D (kΩ)
1	37.589	12.527	1.419	9.06
2	44.311	13.713	1.513	9.54
3	11.224	11.714	1.406	9.54

[a] In all sections, $C_1 = 1$ nF, $C_2 = 2.2$ nF, and $R_B = 30$ kΩ.

And the gain constant of the third section is calculated to be

$$H_{o3} = \frac{11.71676}{(1.864)(1)} = 6.286$$

The maximum value of Q_p is only on the order of 20, and thus any filter circuit that uses a single op amp may be used to realize all three sections. There are two bandpass filter circuits that may be used to realize the three bandpass second-order transfer functions. One is the Sallen and Key bandpass filter circuit in Fig. 4.5, and the other is the single-amplifier bandpass circuit in Fig. 5.10. Both circuits require the same number of components. However, from the sensitivity point of view, the SAB filter in Fig. 5.10 is superior to the other circuit, and therefore we use this circuit for all three sections. The nominal design of this circuit, assuming that the op amp is ideal, can be carried out using the program SABPN, and thus the designed parameters of this circuit that meet the specifications for all the sections are given in Table 7.3.

A diagram of the entire circuit is shown in Fig. 7.3. Each section has been designed with minimum active pole sensitivity and with the assumption that the op amp time constant tends to zero. However, if the

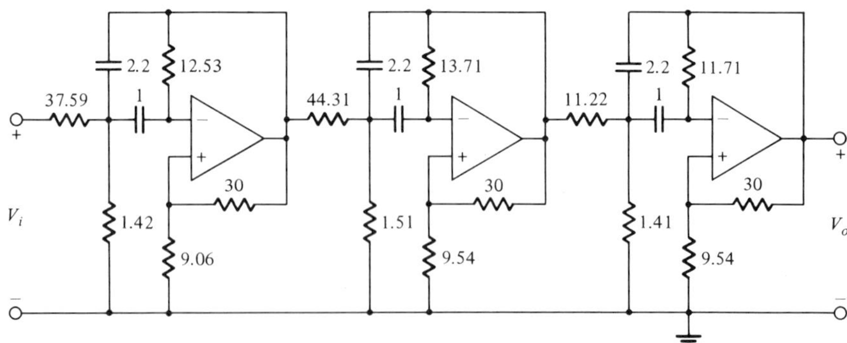

Figure 7.3 Complete realization of the transfer function in Example 7.4. All capacitors are in nF and all resistors are in kΩ.

Table 7.4 Designed circuit parameters using the circuit in Fig. 5.10 with predistortion for the circuits in Example 7.4[a]

Section	R_1 (kΩ)	R_2 (kΩ)	R_3 (kΩ)	R_D (kΩ)
1	34.363	11.965	1.361	9.06
2	40.660	13.142	1.451	9.54
3	10.149	11.149	1.344	9.54

[a] In all the circuits, $C_1 = 1$ nF, $C_2 = 2.2$ nF, and $R_B = 30$ kΩ.

op amp time constant is not zero, there will be changes in both the realized pole frequency and the pole Q factor. Therefore we may have to use predistortion in this case, because the operating frequency is in the range of 25 kHz. This is a relatively high frequency for this circuit if it uses a 741 type op amp. With the assumption that we use a 741 type op amp having a nonzero time constant of $\tau = 0.5/\pi$ μs/r, the three circuits were redesigned, and the component values are given in Table 7.4. Even though one may avoid the initial changes in the pole frequencies and pole Q factors for all three sections with the use of predistortion, the problem of active sensitivity will still be present in the circuits. However, the circuit in Fig. 5.19 may be used to realize these sections. Then, the active sensitivities of the networks can be reduced considerably. The design of the three sections can be carried out using this circuit, and this is left as an exercise for the reader. ■

7.2 *Follow-the-Leader Feedback Technique*

The next method of designing higher-order filters involves additional feedback in a cascade sequence. It is well known in control theory that negative feedback decreases the sensitivity of the output because of the changes in the circuit parameters. Thus, in this section, we provide a design technique based on a coupled biquad structure. This FLF structure is particularly suitable for the design of all-pole bandpass filters, and therefore we shall discuss only the realization of all-pole bandpass filters. However, it should be noted that, with this technique, both bandpass and bandstop filters derived from all-pole lowpass filters may be realized. This method was originally developed by Hurtig, who named it the (primary resonator block) technique [1]. At the outset, it should be pointed out that most of the procedure can be developed using the prototype lowpass function from which the bandpass or bandstop filter is derived. We will design a network that realizes the transfer function of a prototype lowpass filter using an FLF structure and then, using the appropriate network transformation, we will obtain a bandpass or a bandstop filter.

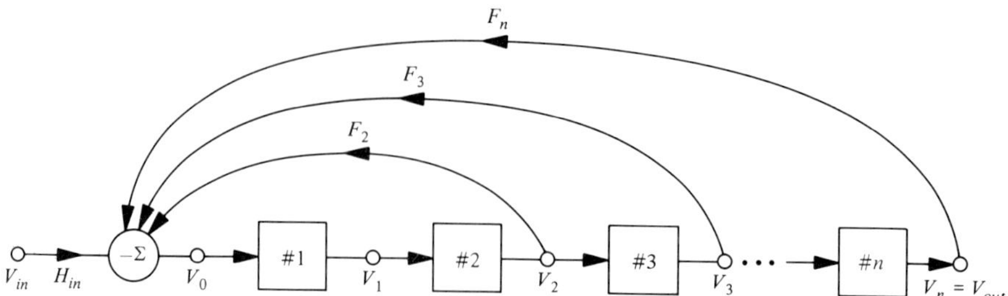

Figure 7.4 Block diagram representation of a follow-the-leader feedback structure.

Consider the block diagram representation of an FLF structure shown in Fig. 7.4. This has the companion form state variable representation. If the transfer function of each block is $H_i(s)$, simple algebra will give the overall transfer function as

$$\frac{V_{\text{out}}}{V_{\text{in}}} = \frac{H_{\text{in}} \prod_{1}^{n} H_i(s)}{1 + \sum_{2}^{n} F_i \prod_{1}^{i} H_j(s)} \tag{7.4}$$

where F_i are the feedback coefficients. For now let us assume that each transfer function is identical, having the form

$$H_i(s) = \frac{a}{s+a} \qquad i = 1, 2, \ldots, n \tag{7.5}$$

Substituting (7.5) into (7.4) and simplifying, we obtain

$$\frac{V_{\text{out}}}{V_{\text{in}}} = \frac{H_{\text{in}} a^n}{(s+a)^n + \sum_{2}^{n} F_i a^i (s+a)^{n-i}} \tag{7.6}$$

An all-pole prototype lowpass function, obtained from either maximally flat or Chebyshev approximation, has the form

$$H(s) = \frac{m a_0}{s^n + a_{n-1} s^{n-1} + \cdots + a_1 s + a_0} \tag{7.7}$$

First, comparing the coefficients of the corresponding powers of s in the denominator polynomials of (7.6) and (7.7), we obtain

$$a = \frac{a_{n-1}}{n} \tag{7.8a}$$

$$F_2 = \frac{a_{n-2}}{a^2} - \frac{n(n-1)}{2!} \tag{7.8b}$$

and F_i, in general, is

$$F_i = \frac{a_{n-i}}{a^i} - \frac{1}{(n-i)!}\left[\frac{n!}{i!} + \sum_{k=2}^{i/i-1} F_k \frac{(n-k)!}{(i-k)!}\right] \qquad i = 3, 4, 5, \ldots, n \tag{7.8c}$$

Comparing the numerator constants of (7.6) and (7.7), we have

$$H_{\text{in}} = \frac{ma_0}{a^n} \tag{7.9}$$

Obviously the value of H_{in} may be used to control the dc gain of the lowpass filter.

If one wants to design a bandpass filter, one method is to obtain a low-pass prototype function using any one of the two approximation procedures discussed in Chap. 2 and use the transformation $s = (p^2 + \omega_o^2)/Bp$ to the prototype LP function. This will result in a bandpass network function, and the next step is to realize this bandpass filter function. Another method is to realize the LP prototype network function in the form of a lowpass network first and then apply the above LP-to-BP transformation to the lowpass network. In this way, the resulting network will be a bandpass network. Thus, if we apply LP-to-BP transformation to each block in the FLF structure, the resulting network will be a bandpass network. The transfer function of each block becomes

$$H_i(s) = \frac{aBp}{p^2 + aBp + \omega_o^2} \tag{7.10}$$

If each block realizes the transfer function in (7.10) instead of the one in (7.5), then the overall network will be a bandpass filter with a center frequency of ω_o (the geometric mean of the passband edge frequencies) and a bandwidth of B. If $Q = \omega_o/B$, then the pole Q factor of each block will be nQ/a_{n-1}. Note that F_i is a constant and therefore, its value will not be affected by frequency transformation. These constants remain the same as they were earlier when we designed the lowpass function.

Assume that we want to realize a bandstop filter function with a stopband bandwidth of B and a center frequency of ω_o. Then, the design procedure will be the same as before. We start with a prototype lowpass function and after realizing this function we use the transformation $s = pB/(p^2 + \omega_o^2)$. Therefore the transfer function of each block will be

$$H_i(s) = \frac{p^2 + \omega_o^2}{p^2 + (B/a)p + \omega_o^2} \tag{7.11}$$

If the transfer function of each block is of the form of (7.11), then the resulting network will realize a bandstop transfer function derived from the lowpass prototype transfer function in (7.7).

So far, in the realization of the above filters, it has been assumed that the gain constants of all the blocks are identical and equal to unity, but this may result in overloading of the final stages in the sequence. Therefore we next consider the problem of maximizing the dynamic range of the filter sections by assigning the proper gain constant to each section.

The problem of assigning gain constants to each section may also be solved using the prototype lowpass function. Let us assign an arbitrary gain constant H_{oi} to the ith section. Then the transfer function of the ith section in the prototype lowpass filter will be

$$H_i(s) = \frac{H_{oi}a}{s + a} \tag{7.12}$$

The output of the final stage is given by

$$|V_{\text{out}}| = |H(j\Omega)|\,|V_{\text{in}}|$$

Since

$$\frac{V_{\text{out}}}{V_{n-1}} = \frac{H_{on}a}{s + a}$$

we have

$$V_{n-1} = \left(1 + \frac{s}{a}\right)\frac{H(s)V_{\text{in}}}{H_{on}}$$

and

$$|V_{n-1}| = |H(j\Omega)|\left(1 + \frac{\Omega^2}{a^2}\right)^{0.5}\frac{|V_{\text{in}}|}{H_{on}}$$

Similarly, it may be shown that

$$|V_i| = \frac{|H(j\Omega)|(1 + \Omega^2/a^2)^{(n-i)/2}|V_{\text{in}}|}{H_{on}H_{o(n-1)}\cdots H_{o(i+1)}} \qquad i = n-1, n-2, \ldots, 0$$

For maximum dynamic range, we must have

$$|V_0|_{\max} = |V_1|_{\max} = \cdots = |V_{n-1}|_{\max} = |V_n|_{\max}$$

The above requirement gives the following equation for the different gain constants:

$$H_{oi} = \frac{\left[|H(j\Omega)|(1 + \Omega^2/a^2)^{(n-i+1)/2}\right]_{\max}}{\left[|H(j\Omega)|(1 + \Omega^2/a^2)^{(n-i)/2}\right]_{\max}} \qquad i = 1, 2, \ldots, n \tag{7.13}$$

If we change the gain constants of the different sections from unity to the values given by (7.13), then the feedback coefficients must also be modified from the ones given by (7.8). Assuming that the new coefficients are denoted by F_i', then to keep the same loop gain we should have

$$F_i = F_i' \prod_{j=1}^{j=i} H_{oj}, \qquad i = 2, 3, \ldots, n$$

Therefore the new feedback coefficients are

$$F_i' = \frac{F_i}{\prod_{j=1}^{j=i} H_{oj}} \qquad i = 2, 3, \ldots, n \tag{7.14}$$

The gain constant of the input must also be modified to keep the same overall gain, and therefore it must be

$$H_{\text{in}} = \frac{ma_0}{a^n \prod_{i=1}^{n} H_{oi}} \tag{7.15}$$

Given the denominator coefficients of the prototype lowpass function, its order, and also its numerator constant for satisfying the overall required gain, the main problem is to calculate the gain constants of the various sections and the feedback coefficients. This step may be carried out using a programmable calculator or using the program PRBN (App. 7A), which also gives the polc frequency and pole Q factor of the second-order sections required for both bandpass and bandstop filters. As an application of this program in designing all-pole bandpass and bandstop filters using the PRB technique, consider the following example.

Example 7.5: Obtain a PRB design of a Chebyshev filter that satisfies the following specifications: passband ripple = 0.5 dB; passband edge frequencies = 20 kHz and 24 kHz; stopband attenuation at least 30 dB at 18 kHz and 26 kHz; maximum gain in the passband = 20 dB.

From the passband edge frequencies, we find that

$$f_o = \sqrt{(20)(24)} = 21.909 \text{ kHz}$$

and

$$B = 4 \text{ kHz}$$
$$f_{a1} = (20)(24)/26 = 18.462 \text{ kHz}$$

which is greater than 18 kHz. Therefore we choose $f_{a1} = 18.462$ kHz and $f_{a2} = 26$ kHz, and thus $f_{a2} - f_{a1} = 7.539$ kHz.

Chebyshev approximation requires a lowpass prototype filter function of order 4, and the transfer function can be obtained using the program MAXCHY from Chap. 2.

$$H_{LP}(s) = \frac{3.578469}{s^4 + a_3 s^3 + a_2 s^2 + a_1 s + a_0}$$

where $a_3 = 1.197386$, $a_2 = 1.7168667$, $a_1 = 1.0254558$, and $a_0 = 0.3790507$.

With the use of the coefficients in the above prototype transfer function and the fact that $N = 4$ and $A_p = 0.5$ dB, the required feedback coefficients and the gain constants of the sections are obtained using the program PRBN. This program also prints out the values of the pole frequency and the pole Q factor of the second-order transfer function to be realized. Note that all sections require identical pole frequency and pole Q factor values. The printout of the results will be as follows:

```
----------------------------------------------------
DESIGN OF A PRB BANDPASS FILTER
----------------------------------------------------
POLE FREQUENCY OF EACH SECTION IN Hz = 21908.9
POLE-Q FACTOR OF EACH SECTION IS 18.29728.
EACH BIQUAD IS A BANDPASS FUNCTION WITH A ZERO
AT THE ORIGIN.
* FEEDBACK COEFFICIENTS *
F(2) = 1.023961
F(3) = 0.1783369
F(4) = 0.1703671
* GAIN CONSTANT DISTRIBUTION FOR MAXIMUM
DYNAMIC RANGE *
Ho(1) = 3.634542
Ho(2) = 3.535999
Ho(3) = 3.451127
Ho(4) = 3.326623
INPUT GAIN CONSTANT, Hin = 3.020468
----------------------------------------------------
```

Since the required design parameters for all the sections are now available, it is easy to design the networks. First, we note that the Q_p value is below 20, and therefore we can use single-amplifier networks to realize these transfer functions. The bandpass filter circuit in Fig. 5.10 is well suited for the design of the four sections, which can be carried out using the program SABPN in App. 5B. The networks are designed using a 741 type op amp, and the filter design requires predistortion because the

Table 7.5 Designed circuit parameters with minimum active pole sensitivity using the circuit in Fig. 5.10 with predistortion for the circuit in Example 7.5[a]

Section	R_1 (kΩ)	R_2 (kΩ)	R_3 (kΩ)	R_D (kΩ)
1	22.225	14.619	1.802	3.16
2	22.844	14.619	1.798	3.16
3	23.406	14.619	1.795	3.16
4	24.282	14.619	1.790	3.16

[a] $C_1 = 1$ nF, $C_2 = 2$ nF, and $R_B = 10$ kΩ in all cases.

frequency range of operation is high. The resulting element values for the four sections are given in Table 7.5. Note that most of the design of follow-the-leader feedback networks may be carried out using the prototype lowpass function. The circuit designed in Example 7.5 is shown in Fig. 7.5. ■

7.3 *Component Simulation of LC Ladders*

This method uses the technique of replacing the inductors and the FDNR elements in the network with the corresponding active networks we discussed in Chap. 6. The starting point for designing these types of filters is a doubly terminated *LC* ladder prototype lowpass network. The *LC* ladders may be designed using computer methods to suit certain specifications or one may obtain these networks from standard filter tables, such as Zverev tables [2]. The low-sensitivity performance of doubly terminated *LC* ladders and the large amount of information available on the design of these filters is exploited in these methods. These same reasons also provide motivation for the techniques discussed in this and the following sections.

In the case of highpass filters, floating inductors are not necessary, and therefore each grounded inductor may be replaced with an actively simulated grounded inductor using Antoniou's GIC circuit (see Chap. 6). Lowpass filters and general bandpass filters require floating inductors. Though there are techniques for realizing floating inductors, they are not satisfactory in practice. For realizing lowpass filters, the *LC* lowpass filters are transformed into *CR* FDNR circuits using Bruton's transformation [3]. This transformation does not affect the voltage transfer functions, and the voltage transfer function of the lowpass filter remains the same. However, after this transformation is applied to an *LC* lowpass network, the network contains only grounded FDNR elements and these elements can be realized by employing active simulation of this FDNR element using the GIC circuit. For realizing the bandpass filters, however, a technique called the *Gorski-Popiel method* is available. In this technique, a group of inductors is

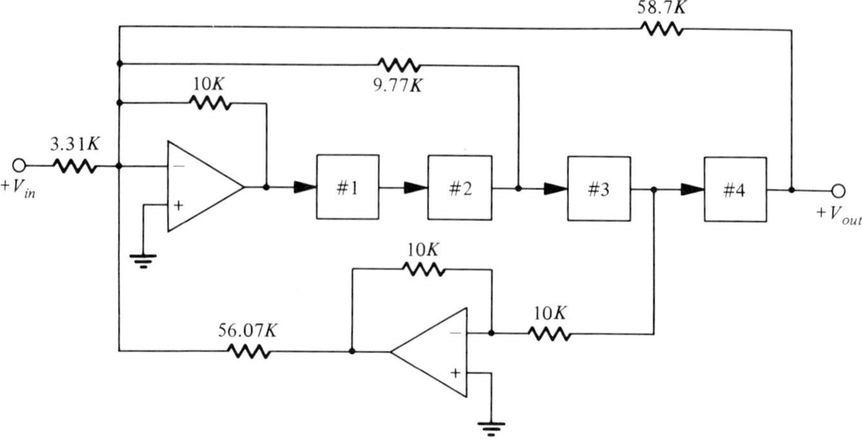

Figure 7.5 A PRB realization of the bandpass filter in Example 7.5.

replaced by another active circuit. This is again component simulation of inductors [4, 5] which uses Antoniou's GIC circuit and *RC* elements. We shall not discuss this method here.

Before we proceed with the design of active filters using component simulation technique, we shall consider the design of *LC* prototype filters. All-pole filter functions, namely, maximally flat and Chebyshev filter functions, can be realized with simple formulas. These formulas are given in Sedra and Brackett [6] and are used in conjunction with the minimum inductance networks shown in Fig. 7.6. However, the minimum capacitor ladder networks are duals of the above networks and can also be designed

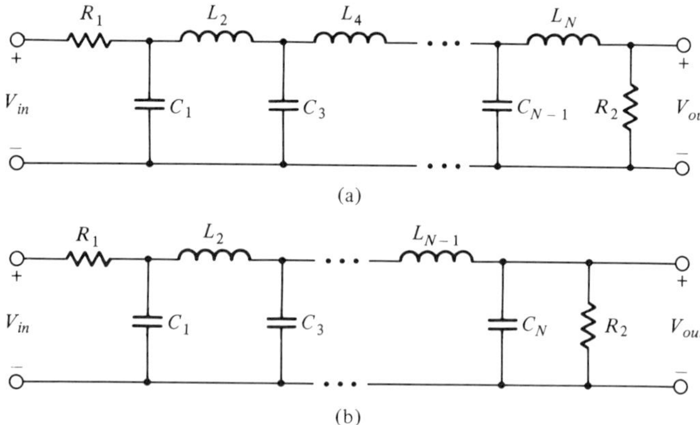

Figure 7.6 Doubly terminated *LC* ladder networks with minimum inductors for realizing the all-pole functions of order *N*. The minimum capacitor networks are the duals of the above.

using the same set of formulas. The formulas for the inductance and capacitance values of the doubly terminated network realizing an Nth-order maximally flat function are

$$C_k, L_k = 2\varepsilon^{1/N} \sin\left[\frac{(2k-1)\pi}{2N}\right] \qquad k = 1, 2, \ldots, N \qquad (7.16)$$

where ε is the ripple factor related to A_p as given in (2.11). Furthermore, both R_1 and R_2 are unity in this case.

The formulas for a Chebyshev filter are also simple. First, we evaluate h using the expression

$$h = \sqrt{\frac{1}{\varepsilon} + \left(1 + \frac{1}{\varepsilon^2}\right)^{1/2}}$$

from which we find that

$$D = h - \frac{1}{h} \qquad (7.17)$$

Then, the formulas for the inductance and capacitance values are

$$C_1 = \frac{4\sin(\pi/2N)}{DR_1} \qquad (7.18)$$

$$C_{(2k-1)}L_{2k} = \frac{16\sin[(4k-3)\pi/2N]\sin[(4k-1)\pi/2N]}{D^2 + 4\sin^2(2k\pi/N)} \qquad (7.19)$$

$$C_{(2k+1)}L_{2k} = \frac{16\sin[(4k-1)\pi/2N]\sin[(4k+1)\pi/2N]}{D^2 + 4\sin^2(2k\pi/N)}$$

$$k = 1, 2, \ldots, N/2 \qquad (7.20)$$

The end elements are

$$C_N = \frac{4\sin(\pi/2N)}{DR_2} \qquad \text{odd } N \qquad (7.21)$$

$$L_N = 4R_2\frac{\sin(\pi/2N)}{D} \qquad \text{even } N \qquad (7.22)$$

The value of R_1 is always unity, and the value of R_2 is also unity if N is odd. However, if N is even, then R_2 is not equal to unity and is found using (7.22).

The design of elliptic *LC* filters is not as straightforward, and therefore, no design formulas are provided for these filters. However, one can always obtain doubly terminated *LC* filters from filter tables such as Zverev tables. Such design tables are also available in reference [7].

Example 7.6: Design an active highpass filter derived from an *LC* ladder network that satisfies the following specifications: maximum loss in passband $A_p = 0.5$ dB at and above 4 kHz; minimum stopband loss = 40 dB at and below 2 kHz. Use elliptic approximation and terminating resistances of 10 kΩ.

The requirements of the prototype lowpass filter can be obtained from the above specifications and are $A_p = 0.5$ dB, $\Omega_p = 1$ r/s, $A_a = 40$ dB, and $\Omega_a = 2$ r/s. If we use elliptic approximation, we will need a filter on the order of 4 (this may be found using the program ELIPFT), and one can obtain the lowpass network with minimum capacitors from the filter tables in reference [7]. The reason for choosing a network with the minimum number of capacitors is that, when we obtain the highpass filter using the LP-HP transformation $p = 1/s$, all the capacitors become inductors and the inductors are transformed to capacitors. In such cases it is better to choose a network with the minimum number of capacitors in the prototype lowpass filter so that inductors are at minimum in the highpass filter. The prototype lowpass filter is shown in Fig. 7.7a. Proceeding further, we use a frequency scaling of ω_p/p to convert the prototype lowpass filter to the required highpass filter. Then, we also use a magnitude scaling factor of 10^4 to increase the impedance level such that the terminating resistors have an impedance value of 10 kΩ. Thus we have the highpass filter shown in Fig. 7.7b.

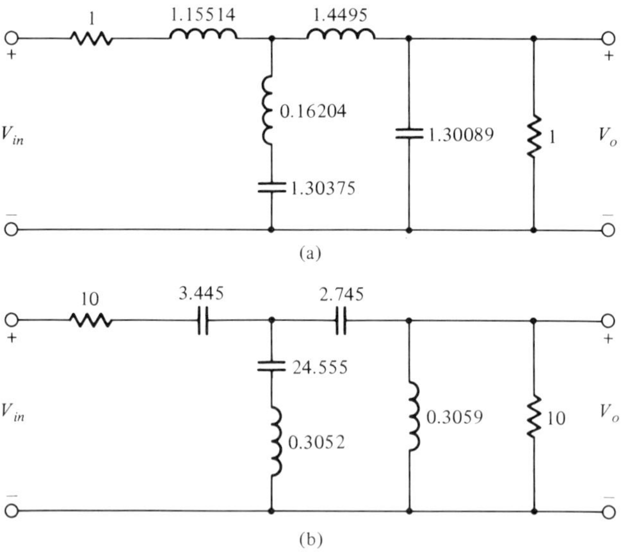

Figure 7.7 (*a*) Fourth-order elliptic prototype lowpass filter. (*b*) Passive highpass filter. *R* in kΩ, *L* in *H*, and *C* in nF.

The two grounded inductors can be replaced with simulated inductors using either the circuit in Fig. 6.5 (type A) or the circuit in Fig. 6.6 (type B). As discussed in Chap. 6, for inductor simulation, the type-B circuit in Fig. 6.6 is superior to the other one because it requires only equivalued resistors for R_2 and R_3 and no matched op amps are necessary for minimum active sensitivity. Thus the two inductors with values of 0.3052 and 0.3059 H may be simulated using the circuit in Fig. 6.6. The design requires equivalued resistors for R_2 and R_3, and we may choose 10 kΩ for each in both circuits. Furthermore, we may select any convenient value for C_2, and thus we choose a value of 10 nF in each case. The value of R_5 is chosen to be as $1/\omega_c C_2$, where ω_c is some critical frequency. The obvious choice for ω_c is ω_p. In such a case the value of R_5 in both networks is the same, and it is 3.98 kΩ. The value of R_1 alone will differ in the two networks. To simulate the inductance value of 0.3052 H, R_1 should be 7.671 kΩ, and to simulate the inductance value of 0.3059 H, R_1 should be 7.688 kΩ. Thus the final active highpass circuit designed to meet the specifications is shown in Fig. 7.8. Because the output is taken from a high-impedance node, an

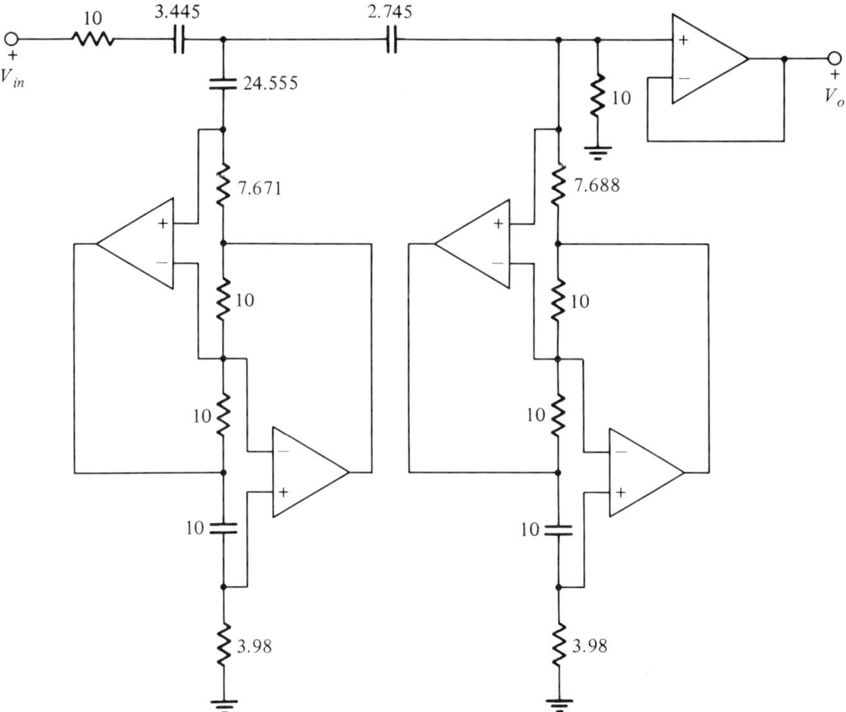

Figure 7.8 Active highpass filter realized in Example 7.6. R in kΩ and C in nF.

optional unity gain amplifier has been added to provide a low-output impedance. This op amp can be used to provide any possible gain greater than unity if necessary. ■

Next, let us consider realization of the lowpass filters using component simulation. Consider the general configuration of the fifth-order elliptic filter shown in Fig. 7.9. Since at least three of the inductors are of the floating type, we cannot use the circuit in Fig. 6.5 or 6.6. However, it is a well-known fact that magnitude scaling of a network does not affect its voltage transfer ratio. Assume that we use a magnitude scaling factor of $1/s$, where s is the complex frequency. Remember that magnitude scaling affects all the elements in the network. Also, note that this magnitude scaling is frequency-dependent. We shall see the effect of applying this scaling on the three types of elements, namely, R, L, and C.

When scaled by a factor of $1/s$, a resistive impedance of R Ω becomes an impedance of R/s. This transformed impedance represents a capacitive impedance, and the capacitance value of such a capacitor should be equal to $1/R$ farads. When we apply a magnitude scaling factor $1/s$, an inductive impedance of Ls becomes L, and this represents a resistive impedance of L Ω. The capacitive impedance of $1/Cs$ is transformed to $1/Cs^2$. This impedance is associated with an FDNR because, when $s = j\omega$, this impedance becomes a frequency-dependent negative resistance, as discussed in Chap. 6. To realize a grounded FDNR element, one may use the circuit in Fig. 6.9.

The application of this type of frequency-dependent scaling factor to a network is called *Bruton's transformation*, and it transforms an *RLC* network into a *CRD* network. In particular, applying this transformation to the network in Fig. 7.9 results in the network shown in Fig. 7.10. The voltage transfer ratio of this circuit will be the same as that in Fig. 7.9, namely, a lowpass function. Thus one may realize the circuit in Fig. 7.10 in order to realize a lowpass transfer function. In this *CRD* network, the FDNR elements are grounded elements and can be realized using the active simulation technique. Consider the following example.

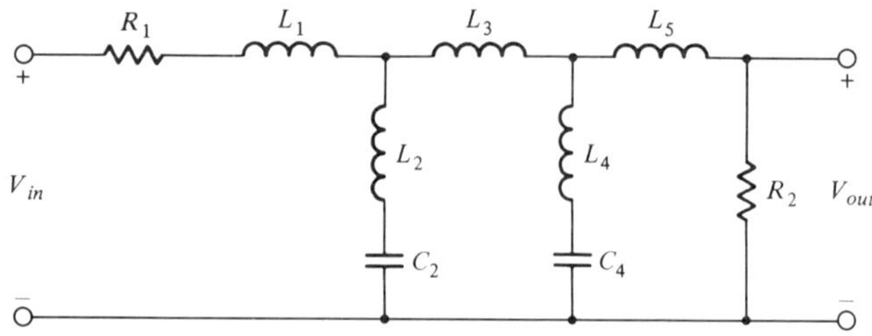

Figure 7.9 Doubly terminated fifth-order elliptic lowpass filter.

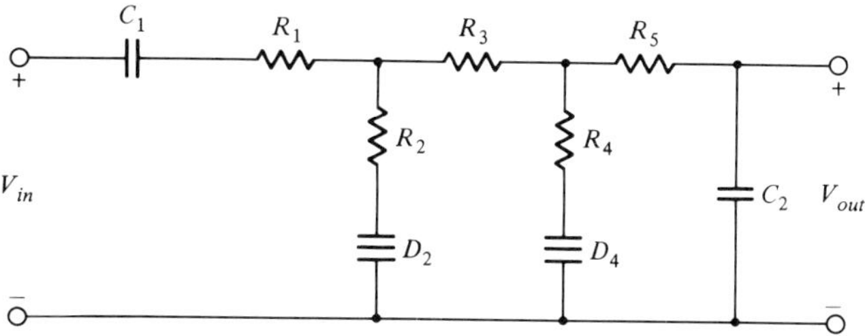

Figure 7.10 Fifth-order elliptic lowpass filter obtained from the circuit in Fig. 7.9 using Bruton's transformation.

Example 7.7: The network shown in Fig. 7.9 has the following elements: $R_1 = R_2 = 1\ \Omega$, $L_1 = 1.56789$ H, $L_2 = 0.16743$ H, $L_3 = 2.10688$ H, $L_4 = 0.46767$ H, $L_5 = 1.3372$ H, $C_2 = 1.09837$ F, and $C_4 = 0.88157$ F. This network meets the following specifications: $A_p = 0.5$ dB at and below $\Omega_p = 1$ r/s; $A_a \geqq 50$ dB at and above $\Omega_a = 1.5$ r/s. Use the above prototype network and realize an active lowpass filter with an f_p of 4 kHz.

With the frequency-dependent magnitude scaling $1/s$, the lowpass prototype network can be transformed into a *CRD* network having the same topology as that in Fig. 7.10. It will also be a prototype network whose element values are $C_1 = C_2 = 1$ F, $R_1 = 1.56789\ \Omega$, $R_2 = 0.16743\ \Omega$, $R_3 = 2.10688\ \Omega$, $R_4 = 0.46767\ \Omega$, $R_5 = 1.3372\ \Omega$, $D_2 = 1.09837\ F^2\Omega$, and $D_4 = 0.88157\ F^2\Omega$. We shall first realize the lowpass prototype *CRD* network with normalized element values and then apply frequency and magnitude scaling to obtain the appropriate f_p and impedance level. To realize the FDNR element with minimum active sensitivity, we select the element values in Fig. 6.9 as $R_2 = R_3 = R$, $R_4 = 1/\omega_c C$, and $C_1 = \omega_c D$, where ω_c is some critical frequency. Now, in our case, the critical frequency is chosen to be the cutoff frequency, which is equal to 1 r/s. In both FDNR elements, we choose $R_2 = R_3 = R$ and $C = 1$ F, where R may be any arbitrary value. Then, $R_4 = 1\ \Omega$ in both cases. The value of C_1 in the FDNR circuit in Fig. 6.9, should be 1.09837 F to realize D_2, and it should be 0.88157 F to realize D_4. Next, we scale the network using a frequency scaling of 8π kr/s so that the network will have a passband edge frequency of 4 kHz as required. To find practical values for the resistors and other elements, we also use a magnitude scaling factor of $10^5/(16\pi)$. The final network is shown in Fig. 7.11. The values of R_2 and R_3 have been arbitrarily set to 2 kΩ, and it is possible to do this, because they need only be equivalued resistors in the FDNR realizations. We have also connected two phantom resistors in addition to the required resistors. The purpose of these resistors is to provide a path for the direct currents of the noninverting input terminals of the op amps A_1 and A_3. Without these resistors, there is no path

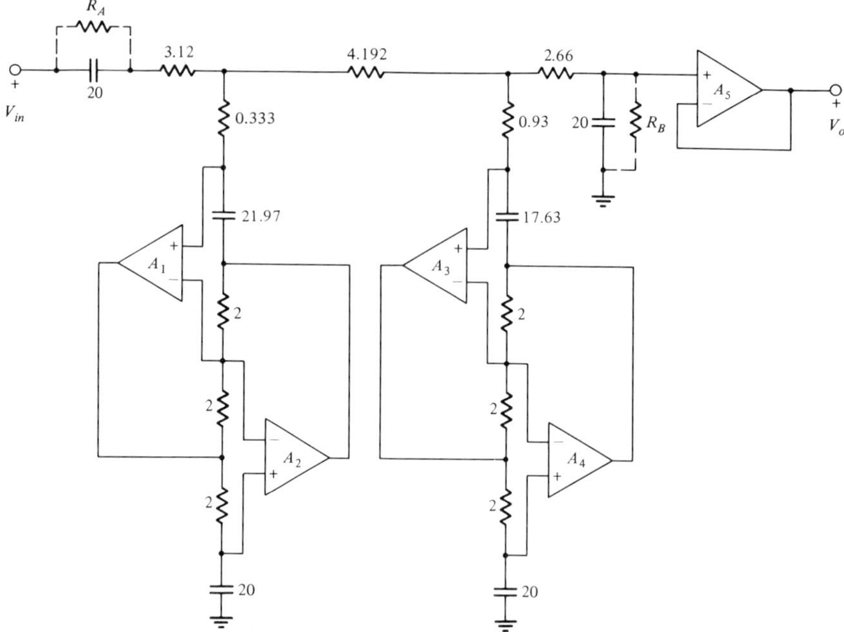

Figure 7.11 Active lowpass filter realized in Example 7.7. R in kΩ and C in nF.

for these direct currents. These two resistances should not affect the performance of the circuit otherwise. They should satisfy the inequalities

$$R_A, R_B \gg R_1 + R_3 + R_5$$

$$R_A \gg \frac{1}{\omega_p C_1} \qquad \text{and} \qquad R_B \gg \frac{1}{\omega_p C_2}$$

In addition, the dc gain of this circuit must be adjusted to 0.5, as in the original prototype circuit. Thus, choosing $R_B = 100$ kΩ (which satisfies the above inequalities), we calculate the value of R_A as 90.03 kΩ. This value also satisfies the above requirement. These two resistance values are high compared to all the other impedance values. Another practical problem must also be addressed before we leave this topic—the problem of tuning these circuits. Once we connect the simulated FDNR (or inductor) elements in the overall circuit, little tuning is possible because of the complexity of the circuits and also because the output response depends on every element in the circuit. Therefore it is advisable to tune each FDNR (inductor) element by resonating it with a known resistor (capacitor) value so that its value is known before it is connected to the overall circuit. ■

7.4 *Leapfrog Realizations*

In the previous section, we discussed *component simulation* of *LC* ladder networks. This method was suitable for the realization of highpass filters, and we may use Bruton's transformation to realize lowpass filters. The component simulation technique is not the best method for designing lowpass and bandpass filters. Therefore, in this section, we seek alternative design methods based on *operational simulation* of *LC* ladder networks. In this method, instead of replacing the inductors or FDNR elements with their active realizations, we seek an active *RC* circuit that simulates the internal working of an *LC* ladder network. Consider an inductor whose V–I relationship is given by

$$I = \frac{V}{Ls} \tag{7.23}$$

The above relationship may be simulated using two analog voltages V_2 and V_1 corresponding to the variables I and V, respectively, and using an integrator $1/Ls$ as shown in Fig. 7.12. The voltage transfer function operationally simulates the characteristic of the inductor. Thus we want to simulate the internal working of an *LC* ladder network. For this purpose we not only have to simulate the characteristics of inductors and capacitors, but we must also satisfy the complete set of circuit equations obtained by applying Kirchoff's laws at various nodes and around various loops. We seek an active network in which the node voltages are the analogs of the branch currents and voltages of the *LC* ladder network. Such an active network must consist of integrators for simulating the characteristics of the inductors and capacitors and summing amplifiers for adding and subtracting signals to satisfy the node and loop equations of the *LC* ladder networks. The resulting network corresponding to an all-pole filter is called a *leapfrog filter*, and we shall concentrate on this type of filter only. This technique may also be extended to cover other types of filters, such as elliptic filters, and it is called *signal flow graph simulation*.

The design of all-pole filters was first discussed by Girling and Good [8], and they were the first to use the term "leapfrog." We will start to develop

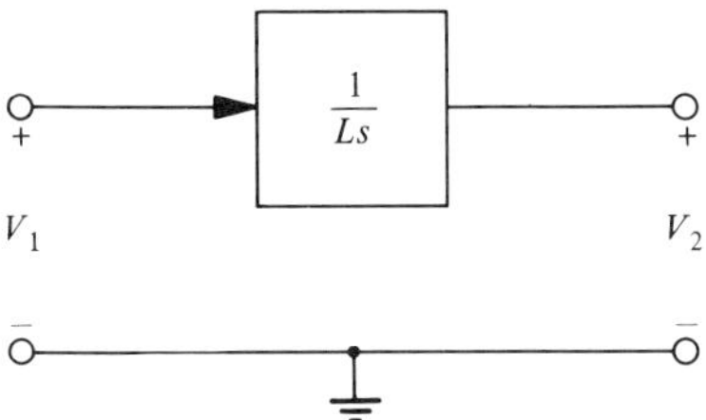

Figure 7.12 Operational simulation of (7.23).

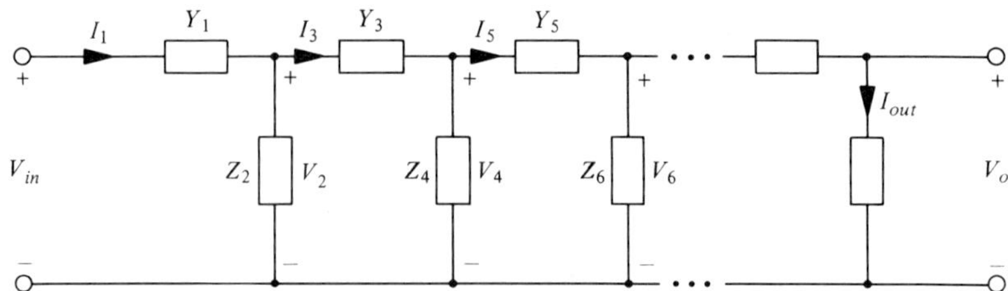

Figure 7.13 General ladder network.

this technique with the ladder network shown in Fig. 7.13. The branch relationships can be written as

$$I_1 = (V_{\text{in}} - V_2)Y_1 \tag{7.24a}$$
$$V_2 = (I_1 - I_3)Z_2 \tag{7.24b}$$
$$I_3 = (V_2 - V_4)Y_3 \tag{7.24c}$$
$$V_4 = (I_3 - I_5)Z_4 \tag{7.24d}$$
$$I_5 = (V_4 - V_6)Y_5 \tag{7.24e}$$
$$\cdots$$

Using voltage analogs for the current variables such that V_{ij} represents I_j for all j and keeping the voltage variables as they are (they do not need any modification), we rewrite (7.24) as

$$V_{i1} = (V_{\text{in}} - V_2)Y_1 \tag{7.25a}$$
$$V_2 = (V_{i1} - V_{i3})Z_2 \tag{7.25b}$$
$$V_{i3} = (V_2 - V_4)Y_3 \tag{7.25c}$$
$$V_4 = (V_{i3} - V_{i5})Z_4 \tag{7.25d}$$
$$V_{i5} = (V_4 - V_6)Y_5 \tag{7.25e}$$
$$\cdots$$

The above equations contain only analog voltages, and therefore Y_1, Z_2, Y_3, etc., can be thought of as transfer voltage ratios. These voltage ratios H_i have same functional form as Y_1, Z_2, Y_3, etc. Thus, if we realize these voltage ratios and interconnect the nodes so that (7.25) is satisfied, we will obtain operational simulation of a ladder network. Before we see how this can be done using a minimum number of components, we rewrite (7.25) in two different forms but without changing the algebraic relationships. The first set of modified equations is

$$V_{i1} = (V_{in} - V_2)Y_1 \tag{7.26a}$$
$$-V_2 = (V_{i1} - V_{i3})(-Z_2) \tag{7.26b}$$
$$-V_{i3} = (-V_2 + V_4)Y_3 \tag{7.26c}$$
$$V_4 = (-V_{i3} + V_{i5})(-Z_4) \tag{7.26d}$$
$$V_{i5} = (V_4 - V_6)(Y_5) \tag{7.26e}$$
$$\cdots$$

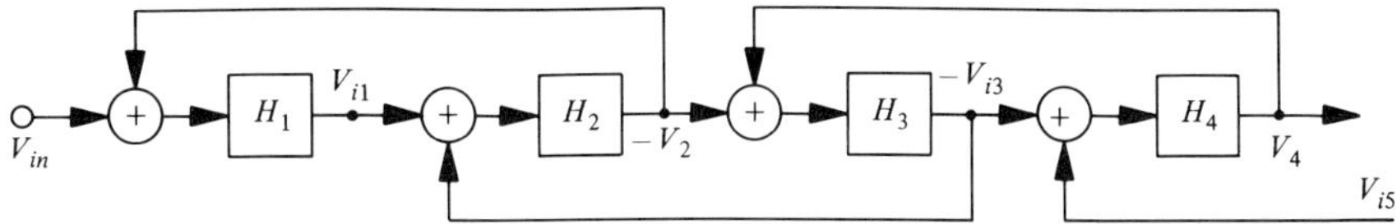

Figure 7.14 Block diagram representation of (7.26).

The above equations can be represented using the block diagram in Fig. 7.14, where

$$H_1 = Y_1 \tag{7.27a}$$

$$H_2 = -Z_2 \tag{7.27b}$$

$$H_3 = Y_3 \tag{7.27c}$$

$$\vdots$$

Alternative forms for the modification of (7.25) are

$$-V_{i1} = (V_{\text{in}} - V_2)(-Y_1) \tag{7.28a}$$

$$-V_2 = (-V_{i1} + V_{i3})(Z_2) \tag{7.28b}$$

$$V_{i3} = (-V_2 + V_4)(-Y_3) \tag{7.28c}$$

$$V_4 = (V_{i3} - V_{i5})(Z_4) \tag{7.28d}$$

$$-V_{i5} = (V_4 - V_6)(-Y_5) \tag{7.28e}$$

$$\cdots$$

This set of equations can be represented using the block diagram in Fig. 7.15, where

$$H_1 = -Y_1 \tag{7.29a}$$

$$H_2 = Z_2 \tag{7.29b}$$

$$H_3 = -Y_3 \tag{7.29c}$$

$$\vdots$$

Both forms are equivalent, and both may be used for the realization of ladders. It should be noted that each block diagram has a certain pattern, and, therefore, once we know the immittance functions of the ladder network, it is easy to obtain either type. Conventional realization of an inverting integrator requires only one op amp, whereas a noninverting integrator requires two. For this reason, we select the block diagram

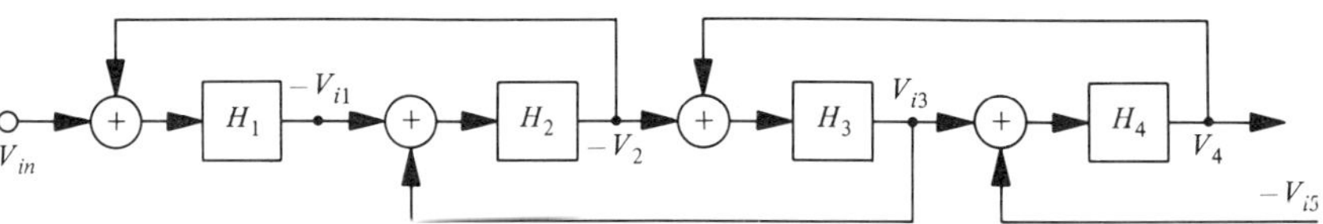

Figure 7.15 Block diagram representation of (7.28).

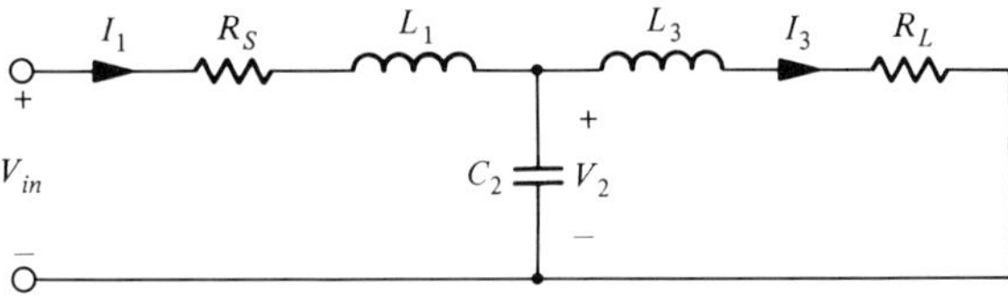

Figure 7.16 Doubly terminated third-order lowpass filter.

representation that requires a minimum number of noninverting integrators.

Consider the third-order doubly terminated lowpass filter shown in Fig. 7.16. From this network, we see that

$$H_1 = -Y_1 = \frac{-1}{L_1 s + R_S} = \frac{-1/L_1}{s + R_S/L_1} \tag{7.30a}$$

$$H_2 = Z_2 = \frac{1}{C_2 s} \tag{7.30b}$$

and

$$H_3 = -Y_3 = \frac{-1}{L_3 s + R_L} = \frac{-1/L_3}{s + R_L/L_3} \tag{7.30c}$$

Using Fig. 7.15, we can obtain the block diagram representation of the leapfrog structure for this particular ladder network, and it is shown in Fig. 7.17. Note that the output voltage V_{i3} is analogous to the output current I_3 of the ladder network, and thus V_{i3} will be the output. The block diagram representation may be physically realized using an active *RC* network. The basic active *RC* networks needed are those of inverting and noninverting integrators, as discussed in Chap. 6. By adding another resistor to the inverting integrator circuit in Fig. 6.16, we can also obtain the summing action with the use of the virtual ground property. In this way, we will not need an additional summing amplifier. We also need a noninverting integrator to realize Z_2. This may be achieved with any one of the circuits shown in Fig. 6.20. Again, using the virtual ground property of some of these circuits, we can obtain the summing action and therefore may not need a separate summing amplifier as shown in the block diagrams.

The noninverting integrator circuit in Fig. 6.20c provides an excessive phase lead due to the finite *GB* product of the op amp, and this is useful in canceling out the excessive phase lag introduced in the inverting integrator circuit in Fig. 6.16. We will use the circuits in Figs. 6.16 and 6.20c with $b = 0$ to replace the inverting and noninverting integrators, respectively. In

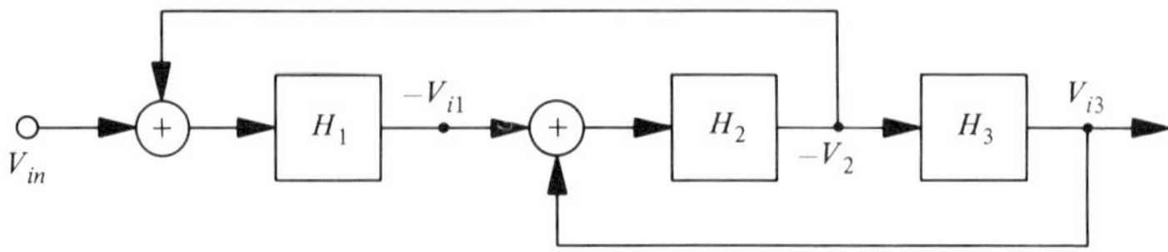

Figure 7.17 Block diagram representation of the lowpass filter in Fig. 7.16.

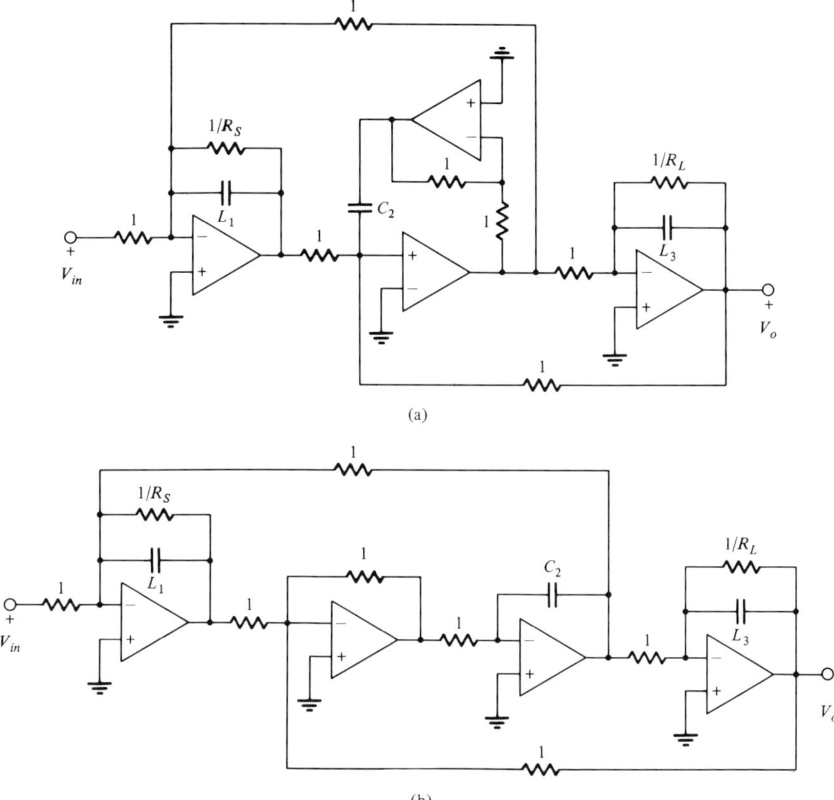

Figure 7.18 Leapfrog realizations derived from the lowpass filter in Fig. 7.16. (*a*) Using both inverting and noninverting integrators. (*b*) Using inverting integrators only.

this way the excessive phase introduced by the finite *GB* products of the op amps in every loop will be reduced to zero, and thus the overall response of the circuit will not have the *Q* enhancement as in the circuits of the double-integrator loops in Chap. 6. Thus the block diagram in Fig. 7.17 may be realized using the circuit in Fig. 7.18a. The circuit in Fig. 7.18b provides another realization of the same block diagram using the noninverting integrator in Fig. 6.20a. The circuit in Fig. 7.18b is the one that will result in the conventional method of realization. Next, we consider a numerical example to illustrate this procedure.

Example 7.8: A third-order Chebyshev filter with a 1-dB ripple may be realized with the doubly terminated network in Fig. 7.16. The prototype lowpass filter has the following element values: $R_s = R_L = 1\ \Omega$, $L_1 = L_3 = 2.0236$ H, and $C_2 = 0.9941$ F. Using this ladder network, obtain a leapfrog structure that will have a passband edge frequency of 10 kHz.

We find from our earlier discussion that the topology of the resulting active network will be the same as either Fig. 7.18a or b. The normalized element values can be found simply by substituting the element values of the prototype LC ladder network. First, we denormalize the network using a frequency scaling factor of $20\pi \times 10^3$, and thus the capacitance values will be

$$L_1 = L_3 = 32.207\,\mu\text{F} \qquad \text{and} \qquad C_2 = 15.822\,\mu\text{F}$$

Then, the impedance level of the elements is increased by a factor of 10^4. Thus we have the two practical circuits in Fig. 7.19 using the configurations of the circuits in Fig. 7.18. Note that both circuits use the same number of components, however, the effects of the finite GB products of the op amps on these two circuits will be entirely different. To prove this, we simulated the responses of these two circuits with the assumption that each op amp in both circuits has a nonzero time constant of $0.5/\pi$ μs. These responses are shown in Fig. 7.20 along with the ideally expected response for purposes of comparison. From these

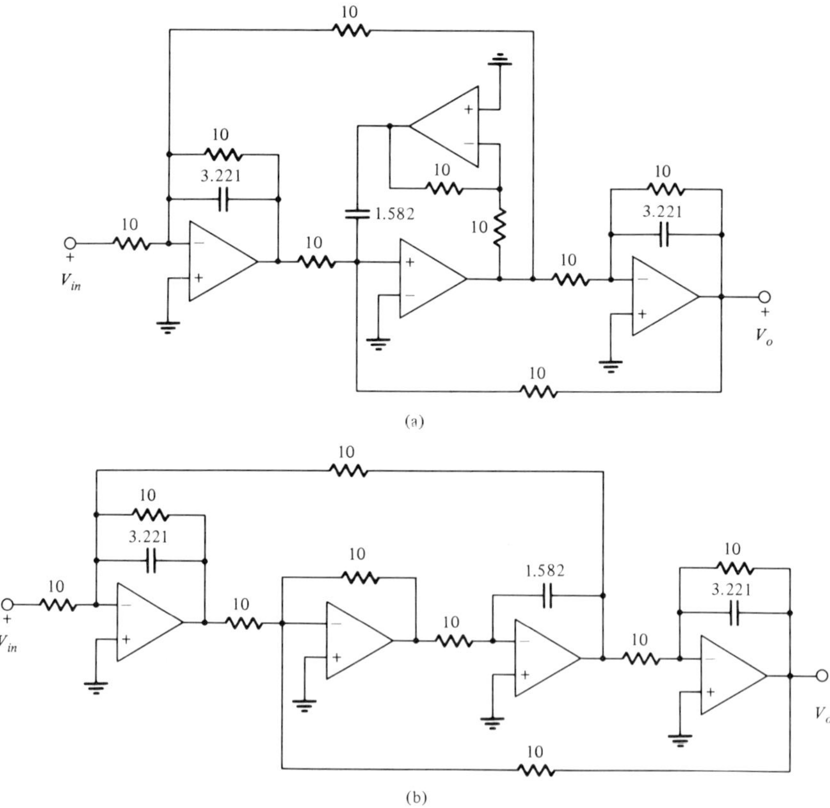

Figure 7.19 Leapfrog realizations for Example 7.8. All resistance values are in kiloohms, and all capacitance values are in nanofarads.

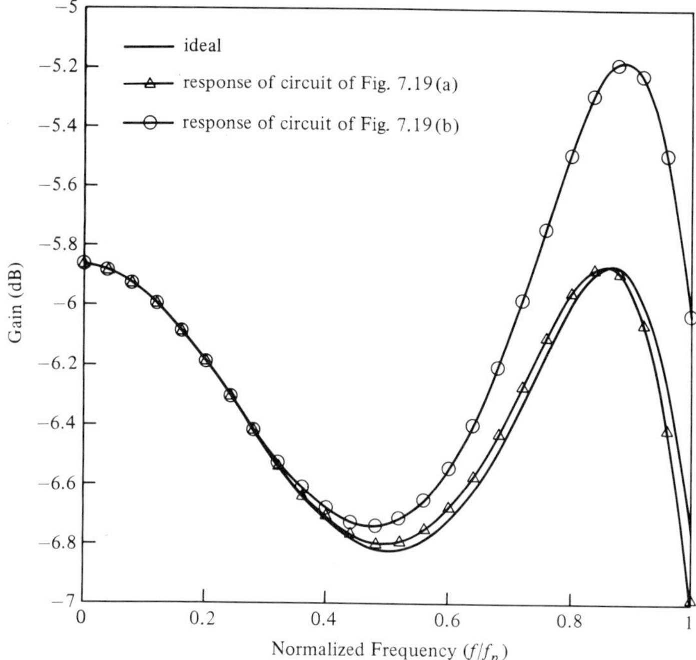

Figure 7.20 Passband frequency responses of the leapfrog realizations in Example 7.8.

expanded passband characteristics, it is clear that the response of the circuit in Fig. 7.19a follows the ideal response very closely. However, the response of the circuit in Fig. 7.19b deviates substantially from the ideal one, even though the frequency range of interest is only in the audio range. These responses also show the merits of active compensation. ■

We shall next consider the realization of bandpass filters derived from all-pole lowpass filters using the leapfrog technique. Again, the starting point is the prototype lowpass filter from which we can obtain the block diagram representation in Fig. 7.14 or 7.15. Then, we can use an LP-BP transformation to obtain the block diagram representation of the bandpass filter. The transfer function of a block in the leapfrog structure corresponding to the lowpass filter has the form $k/(s + a)$, where $a \neq 0$ for the terminal blocks and $a = 0$ for the inner blocks. In any case, when the LP-BP transformation of $s = (p^2 + \omega_o^2)/Bp$, the transfer function of the ith block is of the form

$$H_i(p) = \frac{kBp}{p^2 + aBp + \omega_o^2} \tag{7.31}$$

where ω_o is the center frequency of the bandpass filter and is the geometric mean of the passband edge frequencies. The difference between these frequencies is the bandwidth B.

Three important facts must be noted about these block transfer functions:

1. The block transfer functions are all second-order bandpass transfer functions and may be realized using the appropriate networks in Chaps. 4 through 6.
2. The pole frequencies of all the sections are same and are equal to ω_o.
3. Except for the first and last sections of the leapfrog structure, all the sections require an infinite value of Q_p, since in these sections $a = 0$. Normally a second-order bandpass section with an infinite Q_p will sustain oscillation at the pole frequency. However, in this case, the overall transfer function is stable because of the multiple negative feedbacks provided in the network.

Once we find the transfer functions of the different blocks in the structure, we must realize these transfer functions and interconnect the networks as suggested by the block diagram. Depending on the frequency range of application, we can use conventional realizations of bandpass circuits or actively compensated circuits for high-frequency filters. Some suggested circuits are those of in Figs. 5.10, 6.22, 6.23, 6.29, etc. For a high-frequency range, we can use the circuits in Figs. 5.19, 6.24, etc. The advantage of these circuits is that a summing action may also be achieved with an additional resistor. For example, consider the circuit in Fig. 5.10. The resistor R_1 can be split into two resistors R_{1a} and R_{1b} as shown in Fig. 7.21. The output V_{out} can be expressed as

$$V_{\text{out}} = H_1(s)V_1 + H_2(s)V_2 \tag{7.32}$$

where H_1 and H_2 are both bandpass transfer functions with the same f_p and Q_p values but with different gain constants H_{o1} and H_{o2}, respectively. It can be shown that H_{o1} and H_{o2} are proportional to G_{1a} and G_{1b}, respectively. Thus, if we want this network to realize an output in the form of (7.32), we must design a network as shown in Fig. 5.10, where H_o is equal to the sum of H_{o1} and H_{o2}. As shown in Fig. 7.21, the resistor R_1 is split into R_{1a} and R_{1b}, where

$$R_{1a} = \frac{H_o}{H_{o1}} R_1 \tag{7.33a}$$

and

$$R_{1b} = \frac{H_o}{H_{o2}} R_1 \tag{7.33b}$$

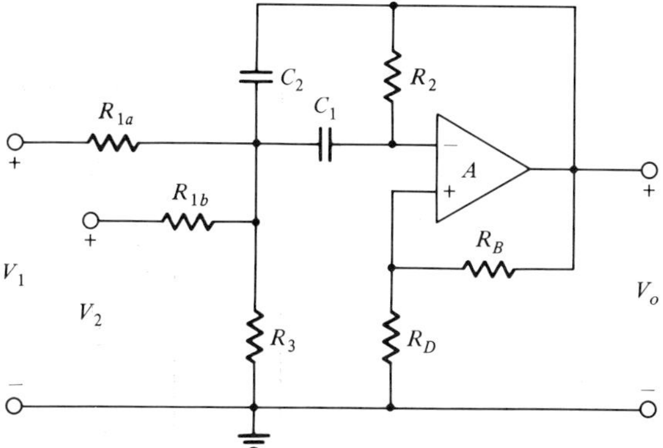

Figure 7.21 Splitting of resistor R_1 in the circuit in Fig. 5.10 for summing action.

In this way we can achieve both summing and filtering actions. This technique may be used in the network in Fig. 5.19 also. It should be noted that both these networks provide inverting outputs. We also need similar summing and filtering actions with a noninverting output in every alternate section, and we are tempted to use the Sallen and Key filter in Fig. 4.19. But, unfortunately, the summing action cannot be achieved in this network as nicely as in the circuits in Figs. 5.10 and 5.19. This is one of the reasons why we have suggested some specific networks. If we need noninverting outputs from inverting outputs, then we can use an inverting amplifier followed by an inverting bandpass section, which will provide the required noninverting output. In some networks where multiple outputs are available, it is possible to obtain both inverting and noninverting bandpass outputs. For example, this is possible in double-integrator networks. In these networks a summing action can also be achieved simultaneously with the use of an additional resistor. This is one more reason, over and above the reasons discussed in Chap. 6, for the popularity of double-integrator loops. We shall next consider a numerical example involving implementation of an all-pole bandpass filter using the leapfrog structure.

Example 7.9: A third-order Chebyshev lowpass filter with a 1-dB ripple may be realized using the doubly terminated network in Fig. 7.16. The element values of the prototype lowpass filter are same as those specified in Example 7.8. Obtain a leapfrog bandpass filter, derived from this prototype lowpass filter, with a center frequency of 26 kHz and a bandwidth of 4 kHz.

The block diagram representation of the prototype lowpass filter will be the same as the one shown in Fig. 7.16, where H_1, H_2, and H_3 are given by (7.30). The block diagram representation of the bandpass filter will also be the same, except that the transfer functions will be different

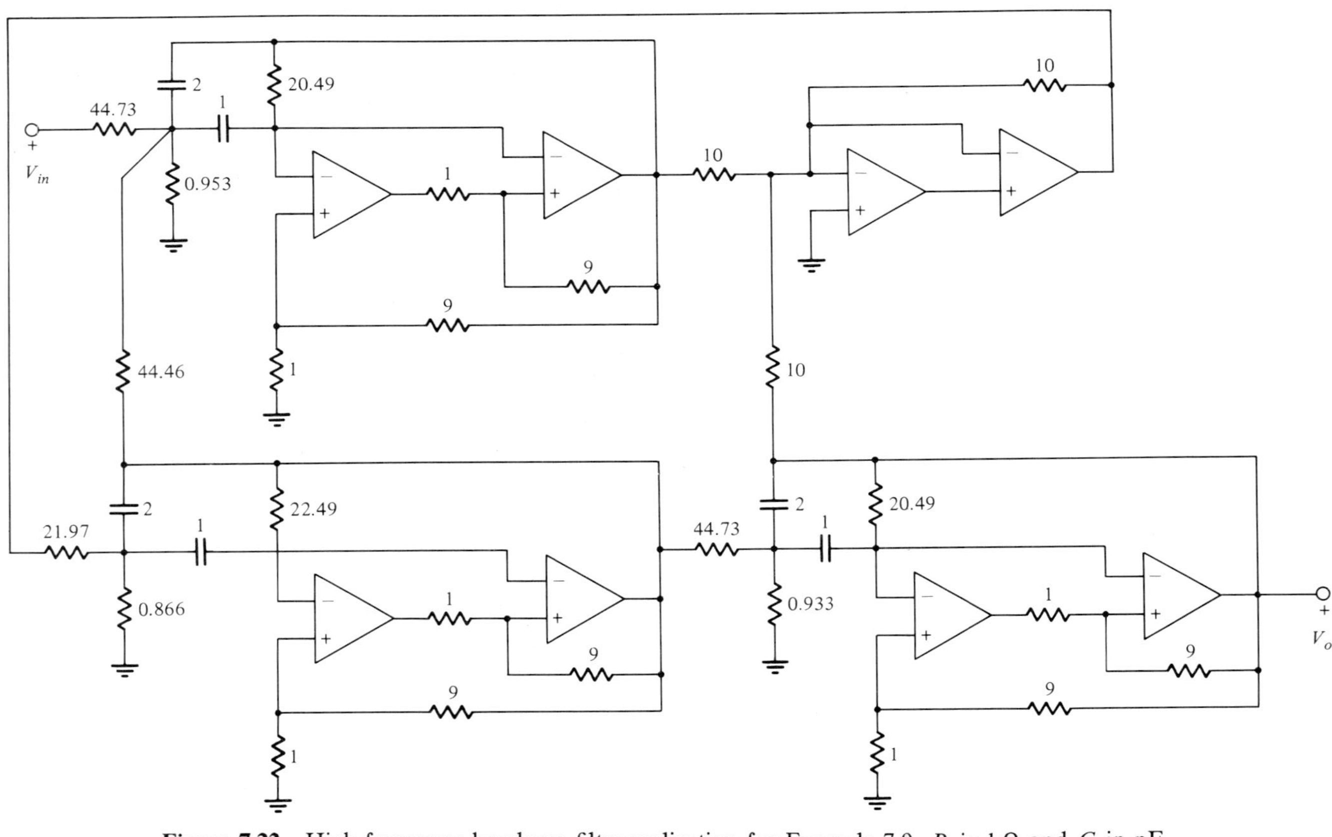

Figure 7.22 High-frequency bandpass filter realization for Example 7.9. R in kΩ and C in nF.

and are obtained with the use of a LP-BP transformation:

$$H_1(p) = \frac{-(1/L_1)Bp}{p^2 + (R_s/L_1)Bp + \omega_o^2} \tag{7.34a}$$

$$H_2(p) = \frac{(1/C_2)Bp}{p^2 + \omega_o^2} \tag{7.34b}$$

$$H_3(p) = \frac{-(1/L_3)Bp}{p^2 + (R_L/L_3)Bp + \omega_o^2} \tag{7.34c}$$

Substituting the given values, we obtain f_p, Q_p, and H_o of the second-order sections as follows:

Section 1: $f_p = 26$ kHz, $Q_p = 13.153$, $|H_o| = 1$
Section 2: $f_p = 26$ kHz, $Q_p = \infty$, $H_o^* = 25{,}252$
Section 3: $f_p = 26$ kHz, $Q_p = 13.153$, $|H_o| = 1$.

Here H_o^* cannot be expressed in the normal form because $Q_p = \infty$ in this section. It is just the numerator constant multiplying the p term.

To realize the second-order sections we can use any of the circuits suggested earlier. To limit the number of amplifiers used and the power consumption of the network, it may be preferable to use the single-amplifier network in Fig. 5.10. However, since the frequency range is high for this network, it is better to use the network in Fig. 5.19 for realizing all three sections. Since we will need a noninverting bandpass filter, we must use an additional inverting amplifier preceding this section, where summing action can also be achieved. In all three sections we will use a low value of β so that the passive sensitivities will be lower without increasing the active sensitivity too much (see Sec. 5.4). We choose $\beta = 0.1$, $C_1 = 1$ nF, and $C_2 = 2$ nF in all the sections. Then, using either the design equations (5.36) through (5.40) or the basic design equations (5.33), we can design all the sections. The complete circuit is shown in Fig. 7.22. Note that the inverting summing amplifier preceding the second section is also actively compensated so that the magnitude characteristic of the entire circuit will not be highly sensitive to the finite *GB* products of the op amps. ■

7.5 *Conclusions*

In this chapter the major techniques for the design of higher-order active filters have been presented. The cascade design is attractive in two respects. It is the most straightforward technique, and it also has the advantage of modularity. However, it suffers from high sensitivity in the passband compared to other methods. The PRB method maintains the modularity concept, while at the same time providing lower passband sensitivities than that of the cascade design. We already know that, in PRB design, the Q factors of all the sections are equal. Techniques for designing filters with unequal Q factors have been investigated to minimize the overall sensitivity of the output [9]. However, these methods are computer-aided and are also complicated. The PRB method can also be extended to design filters with finite zeros of transmission using the feedforward technique.

The second group of methods that we have considered in this chapter employs component simulation. The first of these involves element-by-element component replacement using actively simulated inductance and FDNR elements. In general, such a method requires more op amps as compared to all the other techniques discussed in this chapter. This procedure is suitable for realizing lowpass and highpass filters and is not the best method for designing bandpass filters. The leapfrog technique based on operational simulation of *RLC* networks also satisfies the concept of modularity. One of the serious difficulties of this method is that it requires second-order sections with an infinite value of Q_p when bandpass filters are to be realized. It is difficult to tune a filter with an infinite value of Q_p, and it is also difficult to tune a filter after connecting these second-order filters in the feedback configuration, because of the complexity of the filter. Though only the design of all-pole filters has been dealt with using this method, this technique can be extended to all types of filters using the signal flow graph technique. Interested readers are referred to Sedra and Brackett [6] for further details on this technique.

REFERENCES

1. Hurtig, G. III: The Primary Resonator Block Technique of Filter Synthesis, *Proc. Int. Filter Symp.*, p. 84, 1972.
2. Zverev, A. I.: *Handbook of Filter Synthesis*, John Wiley, New York, 1967.
3. Bruton, L. T.: Network Transfer Functions Using the Concept of Frequency Dependent Negative Resistance, *IEEE Trans. Circuit Theory*, vol. CT-16, pp. 406–408, August 1969.
4. Gorski-Popiel, J.: *RC*-Active Synthesis Using Positive Immittance Converters, *Electron. Lett.* vol. 3, pp. 381–382, August 1967.
5. Martin K., and Sedra, A. S.: Optimum Design of Active Filters Using the GIC, *IEEE Trans. Circuits Syst.*, vol. CAS-24, pp. 495–503, September 1977.
6. Sedra, A. S., and Brackett, P. O.: *Filter Theory and Design: Active and Passive*, Matrix Publishers, Champaign, Ill., 1978.
7. Stephenson, F. W. (ed.): *RC Active Filter Design Handbook*, The Wiley Electrical and Electronic Technology Handbook Series, John Wiley, New York, 1985.
8. Girling, F. E. J., and Good, E. F.: Active Filters. 12. The Leap-Frog or Active Ladder Synthesis, *Wireless World*, vol. 76, pp. 341–345, July 1970.

9. Laker, K. R., and Gausi, M. S.: Synthesis of a Low Sensitivity Multiloop Feedback Active *RC* Filter, *IEEE Trans. Circuits Syst.*, vol. CAS-21, pp. 252–259, March 1974.

EXERCISES

7.1. An eighth-order elliptic bandpass filter has its zeros normalized with respect to $(2\pi)(1.4)$ kr/s, as $z_1, z_1^* = \pm j0.3996$; $z_2, z_2^* = \pm j1.2501$; $z_3, z_4 = 0$; $z_5, z_6 = \infty$. The poles, normalized with the same constant, have the following pole frequencies and pole Q factors: $\omega_{p1} = 0.7106$, $Q_{p1} = 15.256$; $\omega_{p2} = 0.7911$, $Q_{p2} = 6.1920$; $\omega_{p3} = 0.9188$, $Q_{p3} = 6.7430$; $\omega_{p4} = 1.0048$, $Q_{p4} = 19.113$. It is decided to realize this filter using a cascade sequence.
(a) Find the optimum pole-zero pairing of the second-order filter functions so that the realized filter will have the maximum dynamic range possible.
(b) Determine the cascade sequence using the rule of thumb.
(c) Determine the cascade sequence using the measure of flatness.

7.2. Design the cascade bandpass filter in Exercise 7.1 using the single-amplifier filters in Chap. 5. Assume that the maximum gain required in the passband is 10.

7.3. It is decided to realize a lowpass filter using Chebyshev approximation to meet the following specifications: passband edge frequency = 3.2 kHz, passband loss ≤ 0.5 dB, stopband edge frequency = 4 kHz, and stopband loss ≥ 45 dB. To realize the lowpass filter as a cascade sequence with a maximum gain of 10, follow the steps given below.
(a) Use the program MAXCHY and obtain the transfer function required.
(b) Solve the pole-zero pairing problem. In this case it is simple.
(c) Determine the gain constant distribution assuming that the maximum gain required is 10.
(d) Use the networks in Chaps. 5 and 6 to realize the cascade sequence.

7.4. A communication system requires an equalizer for group delay compensation. This requires the transfer function

$$H(s) = \prod_{i=1}^{i=3} \frac{s^2 - (\omega_{pi}/Q_{pi})s + \omega_{pi}^2}{s^2 + (\omega_{pi}/Q_{pi})s + \omega_{pi}^2}$$

where $f_{p1} = 1.550$ kHz, $Q_{p1} = 2.55$, $f_{p2} = 2.222$ kHz, $Q_{p2} = 3.53$, and $f_{p3} = 2.894$ kHz, $Q_{p3} = 4.823$. It is decided to realize the equalizer as a cascade sequence. Use the networks in Chap. 5 to realize the filter.

7.5. Design a cascade sequence that realizes the following specifications for a bandpass filter; passband loss ≤ 0.25 dB, lower passband edge frequency = 20 kHz, upper passband edge frequency = 24 kHz, stopband loss ≥ 35 dB, lower stopband edge frequency = 18.5 kHz, and upper stopband edge frequency = 26 kHz. Use Chebyshev approximation and the network in Fig. 5.19 to realize this filter.

7.6. It is required to design a bandpass filter using a fourth-order Chebyshev prototype lowpass function and having the following specifications: equiripple passband $A_p = 0.1$ dB, center frequency $f_0 = 2$ kHz, ripple bandwidth $B = 100$ Hz, and center frequency gain = 10. Obtain a PRB design that meets the above specifications. Choose suitable networks from Chaps. 5 and 6.

7.7. Consider the specifications in Exercise 7.5.
(a) Obtain a PRB design using the network in Fig. 5.19.
(b) Assume that all the op amps in the cascade design in Exercise 7.5 and in the PRB design in this problem have finite and nonzero time constants of $0.5/\pi$ μs. Determine the maximum change in the magnitude characteristic of each design over the entire passband. You may use a circuit simulation program, such as SPICE, for this purpose.

7.8. The specifications for a highpass filter are $A_p \leq 1$ dB, $A_a \geq 45$ dB, $f_p = 4$ kHz, and $f_a = 3$ kHz. Use Chebyshev approximation to obtain a doubly terminated LC prototype. You may use formulas (7.17) through (7.22). The order N and the ripple factor ε may be obtained from the program MAXCHY or using the formulas in Chap. 2. Use the synthetic inductors in Chap. 6 to realize this filter.

7.9. Repeat Exercise 7.8 using elliptic approximation. You may use filter tables to obtain a doubly terminated LC prototype to meet the specifications.

7.10. Use filter tables and obtain a doubly terminated elliptic LC prototype lowpass filter meeting the specifications $A_p \leq 0.15$ dB, $A_a > 30$ dB, $f_p = 3.4$ kHz, and $f_a = 4.6$ kHz. Obtain an active ladder using FDNR elements.

7.11. An n-port inductive network can be simulated using a topologically identical resistive network and n GICs as shown in Fig. E7.11. This is the Gorski-Popiel ladder embedding technique. Assume that the Antoniou's GIC in Fig. 6.8 is used. Show that the impedance matrix Z, describing N, is given by $Z = ksZ'$, where Z' is the impedance matrix of the resistive network N'.

7.12. Use the technique discussed in Exercise 7.11 to design an active network simulating the L network shown in Fig. E7.12. Only two GICs are needed. Assume that the critical frequency is 3.4 kHz.

7.13. In the design of the L network in Exercise 7.12, the two GICs have slightly different k values such that $k_1 = k$ and $k_2 = k + \Delta k$. De-

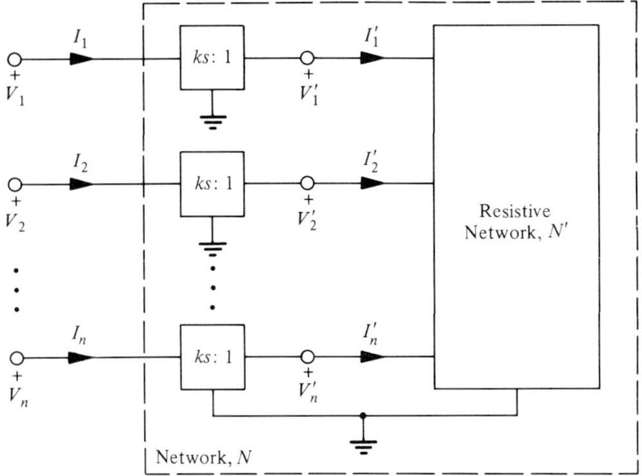

Figure E7.11 The n-port network N considered in Exercise 7.11.

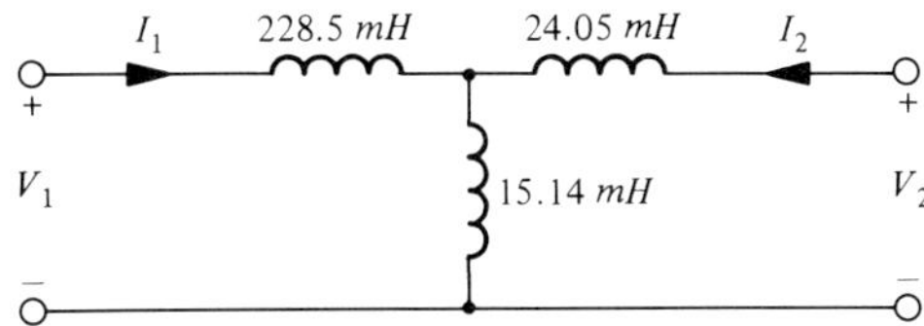

Figure E7.12 The L network considered in the Exercises 7.12 and 7.13.

termine the effects of this mismatch in the values of k on the L network.

7.14. Sketch an active realization of the LC network shown in Fig. E7.14 using the technique employed in Exercise 7.11. Use the minimum number of GICs.

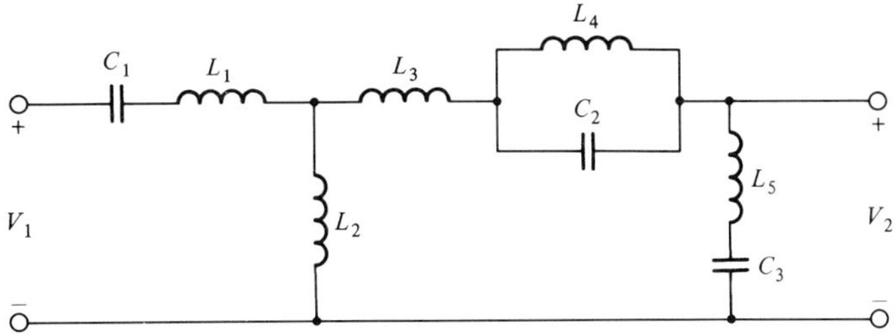

Figure E7.14 The LC network for Exercise 7.14.

7.15. Design a leapfrog circuit for realizing a third-order lowpass Chebyshev filter with $A_p \leq 0.1$ dB and $f_p = 4$ kHz. The dc gain required is 10.

7.16. Design a leapfrog bandpass filter that meets the specifications in Exercise 7.5.

7A

Program PRBN

THIS APPENDIX PROVIDES the details of the program PRBN written in GW BASIC language that can be used in IBM PC or its compatibles. This program can be used to design a higher-order bandpass or bandstop filter derived from an all-pole lowpass filter using PRB technique. This program calculates the gain constants and the feedback coefficients for maximum dynamic range. The procedure to use this program is simple. Just enter it and run it.

Inputs Necessary

(a) The type of the prototype, namely, maximally flat or Chebyshev.
(b) Order, the passband loss, the coefficients of the denominator polynomial, and the numerator constant of the prototype lowpass function.
(c) The type of filter, namely bandpass or bandstop filter, that you want to design using the PRB technique and the passband edge frequencies for bandpass filter or bandstop edge frequencies for bandstop filter.

These data can be entered through the keyboard when prompted by answering the questions for each prompt.

Outputs

(a) The pole frequency and the pole Q factors of each section of the bandpass or bandstop filter. Note that all the sections have identical values for the pole frequency and the pole Q factor.
(b) The feedback coefficients.
(c) Gain constant distribution for each section to provide maximum dynamic range in the filter.
(d) Input gain constant, H_{in}.

```
10 '
20 '
30 '                        PROGRAM LISTING OF "PRBN"
40 '
50 '
60 'PROGRAM TO DESIGN PRB FILTERS - TO CALCULATE THE GAIN CONSTANTS AND FEEDBACK
 COEFFICIENTS FOR MAXIMUM DYNAMIC RANGE.
70 LPRINT" DESIGN OF PRB FILTERS-THE GAIN CONSTANTS AND FEEDBACK COEFFICIENTS FO
R MAXIMUM DYNAMIC RANGE."
80 INPUT"THE TYPE OF FILTER,'MAX' FOR MAXIMALLY FLAT AND 'CHY' FOR CHEBYSHEV";A$

90 IF A$="MAX" OR A$="max" THEN FT1=1 ELSE IF A$="CHY" OR A$="chy" THEN FT1=2 EL
SE 80
100 INPUT"ORDER OF THE PROTOTYPE LOWPASS FILTER";N
110 INPUT"MAXIMUM PASSBAND LOSS IN DB";AP
120 E2=10^(AP/10)-1
130 DIM A(N),F(N),K(N+1)
140 PRINT"*INPUT THE DENOMINATOR COEFFICIENTS OF THE PROTOTYPE FILTER*"
150 FOR I=0 TO N-1
160 PRINT "A(";I;")=";:INPUT A(I)
170 NEXT I
180 INPUT"INPUT THE NUMERATOR CONSTANT OF THE PROTOTYPE LOWPASS FUNCTION";H
190 PRINT"DO YOU WANT TO DESIGN A BANDPASS(BP) OR BANDSTOP(BS) FILTER";:INPUT B$

200 IF B$="bp" OR B$="BP" THEN FT=3 ELSE IF B$="BS" OR B$="bs" THEN FT=4 ELSE FT
 = 0
210 IF FT = 0 THEN PRINT"'BP' OR 'BS' EXPECTED":GOTO 190
220 IF FT=3 THEN INPUT"PASSBAND EDGE FREQUENCIES IN Hz";F1,F2 ELSE INPUT"STOPBAN
D EDGE FREQUENCIES IN Hz";F1,F2
230 FP=SQR(F1*F2)
240 LPRINT STRING$(52,"_")
250 IF FT=3 THEN LPRINT"DESIGN OF A PRB BANDPASS FILTER"
260 IF FT=4 THEN LPRINT"DESIGN OF A PRB BANDSTOP FILTER"
270 LPRINT STRING$(52,"_")
280 LPRINT" POLE FREQUENCY OF EACH SECTION IS";FP
290 A1=A(N-1)/N
300 IF FT=3 THEN QP=FP/(A1*(F2-F1)) ELSE QP=FP*A1/(F2-F1)
310 LPRINT" POLE Q-FACTOR OF EACH SECTION IS";QP
320 IF FT=4 THEN LPRINT"EACH BIQUAD IS A BANDSTOP FUNCTION WITH TWO ZEROS ON THE
 IMAGINARY AXIS. THE ZEROS HAVE THE SAME MAGNITUDE OF THE POLE FREQUENCY."
330 IF FT=3 THEN LPRINT"EACH BIQUAD IS BANDPASS FUNCTION WITH A ZERO AT THE ORIG
IN."
340 'DETERMINATION OF FEEDBACK COEFFICIENTS WITH EQUAL GAIN CONSTANTS
350 F(2)=A(N-2)/(A1*A1)-N*(N-1)/2
360 FOR I=3 TO N
370 F(I)=A(N-I)/(A1^I)
380 X=N:GOSUB 890
390 Y=X
400 X=N-I:GOSUB 890
410 Z=X
420 X=I:GOSUB 890
430 W=X
440 F(I)=F(I)-Y/(Z*W)
450 X1=0
460 FOR L=2 TO I-1
470 X=N-L:GOSUB 890
480 Y=X
490 X=I-L:GOSUB 890
500 W=X
510 X1=X1+F(L)*Y/W
```

```
520 NEXT L
530 F(I)=F(I)-X1/Z
540 NEXT I
550 J=0
560 W=1:I=0:DW=.02
570 TX=(1+(W/A1)^2)^J
580 IF FT1=2 THEN 600
590 TY=1/(1+E2*W^(2*N)):GOTO 630
600 IF W<=1 THEN CW=COS(N*ATN(SQR(1-W*W)/W))
610 IF W>1 THEN NC=N*LOG(W-SQR(W*W-1)):CX=EXP(NC):CW=(CX+1/CX)/2
620 TY=1/(1+E2*CW*CW)
630 TC=SQR(TX*TY)
640 I=I+1:W=W+DW
650 IF I>2 THEN 690
660 IF I=1 THEN TC1=TC:GOTO 570
670 IF TC>=TC1 THEN TC1=TC:GOTO 570
680 IF TC<TC1 THEN W=W-3*DW:DW=-DW:GOTO 570
690 IF TC>TC1 THEN TC1=TC:GOTO 570
700 K(J)=TC1
710 IF J=0 THEN 730
720 K(J-1)=K(J)/K(J-1)
730 J=J+1
740 IF J<=N THEN 560 ELSE 750
750 KX=K(N-1)
760 LPRINT"*FEEDBACK COEFFICIENTS*"
770 FOR I=2 TO N
780 KX=KX*K(N-I)
790 F(I)=F(I)/KX
800 LPRINT"F(";I;")=";F(I)
810 NEXT I
820 LPRINT"*GAIN CONSTANT DISTRIBUTION FOR MAXIMUM DYNAMIC RANGE*"
830 FOR I=N-1 TO 0 STEP -1
840 LPRINT"Ho(";N-I;")=";K(I)
850 NEXT I
860 H=H/(A1^N*KX)
870 LPRINT"INPUT GAIN CONSTANT,Hin=";H
880 END
890 XY=1
900 IF X=0 THEN X=1:RETURN
910 IF X=1 THEN X=1:RETURN
920 FOR IX=2 TO X
930 XY=XY*IX
940 NEXT IX
950 X=XY
960 RETURN
```

8

Switched Capacitor Filters

THE QUALITY OF PERFORMANCE of active RC filters with respect to changes in environmental conditions is dependent on the control and stability of the RC products and on technology. The latter has had a tremendous influence on the design and fabrication of signal processing networks. In IC technology, the technology of the present and the future, the absolute values of the resistors and capacitors cannot be closely controlled. The IC resistors and capacitors have poor linearity and temperature characteristics. In addition, they require large amounts of "real estate" compared to the area occupied by active devices such as transistors and op amps. A category of filter realizations that has recently emerged, called *switched capacitor* (SC) *filters*, has a good potential for avoiding the disadvantages of active RC filters. Switched capacitor filters use capacitors, switches, and op amps. These elements can be realized using MOS technology, and the entire filter circuit can be realized in the monolithic form.

The basic method of realizing a switched capacitor filter is to replace a resistor in an active RC filter with a periodically switched capacitor. To understand and see the advantages of these types of filters, consider one form of the switched capacitor equivalent of a resistor [1] shown in Fig. 8.1*a*. We shall postpone a detailed consideration of this circuit until a later section. For now, to find the approximate behavior of this circuit, assume that the switch is in position 1 and the capacitor is charged instantaneously to CV_1 under ideal conditions. When the switch is thrown to position 2, the capacitor is charged to CV_2. When the switch is thrown back and forth every T seconds, there is an average charge flow, under steady-state

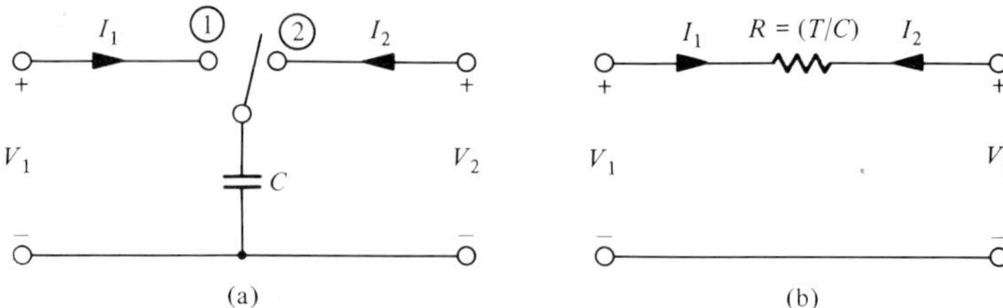

Figure 8.1 (*a*) Switched capacitor. (*b*) Its equivalent resistor.

conditions, from port 1 to port 2. This process gives rise to an average current flow from port 1 to port 2, and this average current flow is

$$I = \frac{C(V_1 - V_2)}{T} \tag{8.1}$$

If the switching frequency $f_s = 1/T$ is high compared to the signal frequency of V_1, then this circuit will behave as a floating resistor connected between V_1 and V_2. The resistance value of this equivalent resistor is

$$R = \frac{T}{C} \quad \text{ohms} \tag{8.2}$$

To see the advantages of such a switched capacitor equivalent resistor R, consider an RC product where R is obtained by switching a capacitor C_1 and another capacitor C_2. The RC product τ will then be

$$\tau = T\frac{C_2}{C_1} \tag{8.3}$$

The per-unit change in τ due to the per-unit changes in T, C_1, and C_2 will then be

$$\frac{d\tau}{\tau} = \frac{dT}{T} + \frac{dC_2}{C_2} - \frac{dC_1}{C_1} \tag{8.4}$$

Assume that T is governed externally by a clock and is controlled quite accurately. Then, the per-unit change in τ is controlled by the capacitor ratio C_2/C_1. Though it is not possible to realize C_1 and C_2 quite accurately in MOS technology, the capacitor ratios can be realized within 0.1%, and this is a very satisfactory result. The capacitors exhibit voltage coefficients in the range 20 to 100 ppm/V and temperature coefficients of 20 to 50 ppm/°C in MOS technology. However, in the capacitor ratios, these coefficients are much lower than the above values because the changes in the capacitance values cancel each other. Thus, for most practical designs using capacitor ratios, the errors due to voltage and temperature coefficients will be second-order errors, and therefore they will be insignificant. Thus switched capacitor filters are almost insensitive to the capacitor ratios. This forms the main motivation for using such switched capacitor filters.

The signals processed by the switched capacitor filters are sampled analog signals. Thus these types of filters belong in the category of sampled analog systems. As Laplace transforms play an important role in the analysis and design of linear continuous-time circuits, the Z-transform technique plays an important role in the analysis of discrete-time and sampled data circuits and systems. It is assumed that the reader is familiar with Z transforms and their applications [2], however, a review of some important concepts involving sampled data signals is provided in the next section.

8.1 *Sampled-Data Signals and Z Transforms*

An ideal sampler extracts information about an analog signal $x(t)$ at sampling instants $t = nT$. The sampling process is shown in Fig. 8.2a. The

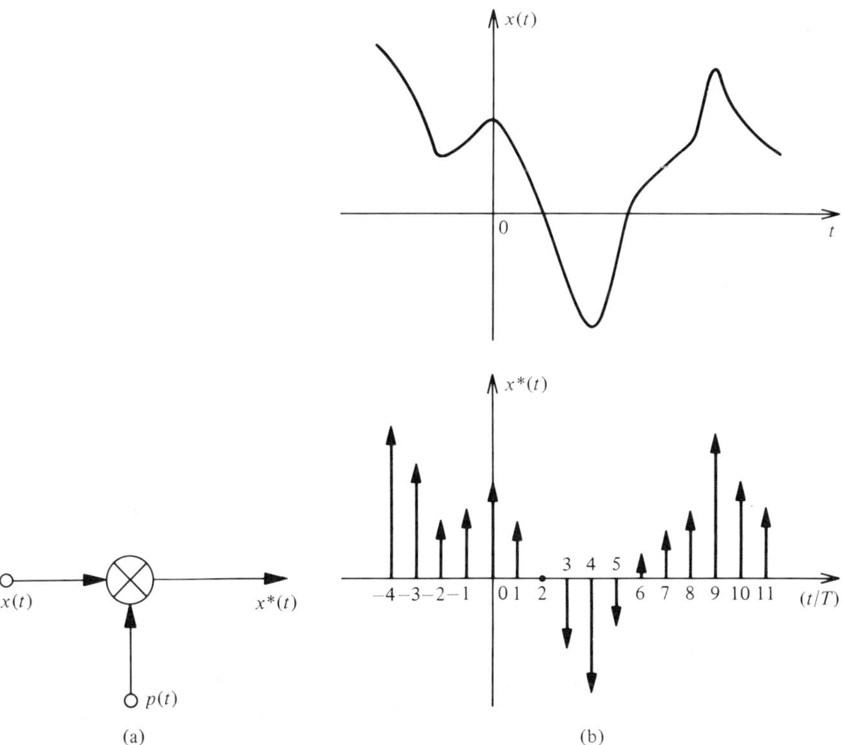

Figure 8.2 (a) Sampler. (b) Analog signal and its sampled version.

sampling signal $p(t)$ is a series of unit impulses, and thus we can write

$$p(t) = \sum_{n=-\infty}^{n=\infty} \delta(t - nT) \tag{8.5}$$

The sampling process for obtaining $x^*(t)$ is mathematically equivalent to multiplying the analog signal by $p(t)$. Thus

$$x^*(t) = x(t)p(t) = \sum_{n=-\infty}^{n=\infty} x(nT)\delta(t - nT) \tag{8.6}$$

The analog signal $x(t)$ and the sampled signal $x^*(t)$ are shown in Fig. 8.2*b*. Let us next consider how the information content of $x^*(t)$ is related to that of $x(t)$. Assume that the spectrum of $x(t)$, which is $X(j\omega)$, is as shown in Fig. 8.3*a*. Here $x(t)$ is assumed to be a band-limited signal whose maximum frequency content is ω_m such that $X(j\omega) = 0$, $|\omega| > \omega_m$. Since $p(t)$ is a periodic signal, we may use a Fourier series to represent it:

$$p(t) = \sum_{n=-\infty}^{n=\infty} C_n e^{j\omega_s t} \tag{8.7}$$

where

$$C_n = \frac{1}{T}\int_0^{T/2} p(t)e^{-j\omega_s t}\,dt = \frac{1}{T} \tag{8.8}$$

and

$$\omega_s = \frac{2\pi}{T} \tag{8.9}$$

The frequency ω_s is the sampling frequency, and the spectrum of $x^*(t)$, denoted by $X^*(j\omega)$, is then

$$\begin{aligned} X^*(j\omega) &= F[x(t)p(t)] \\ &= \tfrac{1}{2}X(j\omega) * P(j\omega) \end{aligned}$$

where $P(j\omega)$ is the Fourier transform of $p(t)$ and $*$ indicates the convolution process.

Thus we find the spectrum of $X^*(j\omega)$ to be

$$X^*(j\omega) = \frac{1}{T}\sum_{n=-\infty}^{n=\infty} X(j\omega - jn\omega_s) \tag{8.10}$$

From the above equation, we find that the baseband spectrum ($n = 0$) of an impulse-sampled signal of an analog signal is an exact replica of the spectrum of the original signal except for a multiplication factor of $(1/T)$. In addition, the sampled signal also contains translations of this spectrum centered around $n\omega_s$. Equation (8.10) is represented in Fig. 8.3*b* and *c*.

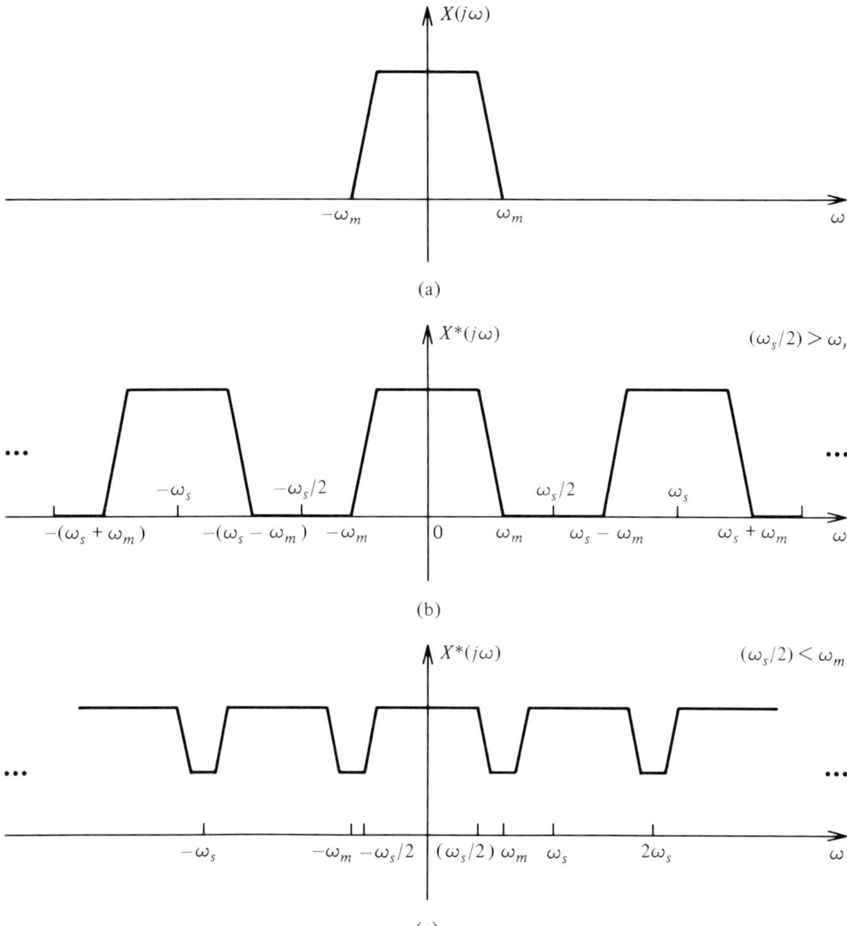

Figure 8.3 Spectra of (*a*) an analog signal. (*b*) Its sampled signal when $\omega_s/2 > \omega_m$, and (*c*) its sampled signal when $\omega_s/2 < \omega_m$.

These two cases arise when $\omega_s > 2\omega_m$ and $\omega_s < 2\omega_m$, respectively. From these figures we find that, when $\omega_s \geq 2\omega_m$, we can recover the original signal $x(t)$ without error by passing $x^*(t)$ through an ideal lowpass filter whose cutoff frequency is $\omega_{s/2}$. However, in the case of Fig. 8.3*c*, since the spectrum is distorted because of overlapping and has no relationship to the original spectrum even at the baseband, it is impossible to recover the original signal. This observation is described in *Shannon's sampling theorem:* A band-limited signal $x(t)$ with a Fourier spectrum such that $X(j\omega) = 0$, $|\omega| > \omega_m$, can be uniquely specified by the knowledge of its values at uniformly spaced time instants, $t = nT$, provided that the sampling frequency $\omega_s \geq 2\omega_m$.

When the sampling frequency is $2\omega_m$, it is called the *Nyquist rate*. Note that the signal $x(t)$ must also be band-limited. If it is not, then the spectrum of $x^*(t)$ will be distorted. When the spectrum overlaps, as in Fig. 8.3*c*, it is called *frequency aliasing*, and to avoid such aliasing the analog signals are passed through an antialiasing lowpass filter before they are sampled.

To obtain the spectrum of the sampled signal $x^*(t)$, we used one part of (8.6). To see how the Z-transform technique is useful in describing the sampled signals, we will use another part of (8.6), namely,

$$x^*(t) = \sum_{n=-\infty}^{n=\infty} x(nT)\delta(t - nT) \tag{8.11}$$

Taking the Laplace transform of the above equation, we have

$$X^*(s) = \sum_{n=-\infty}^{n=\infty} x(nT)e^{-nsT} \tag{8.12}$$

Using the substitution $z = e^{sT}$, we obtain

$$X(z) = \sum_{n=-\infty}^{n=\infty} x(nT)z^{-n} \tag{8.13}$$

In (8.13), $X(z)$ is called the Z transform of the sequence $x(nT)$ [$x(nT)$ can be thought of as a sequence of numbers as well]. In fact, (8.13) is the definition of the Z transform of the sequence $x(nT)$. If we are interested in finding the spectrum of $x^*(t)$, we just substitute $z = e^{j\omega T}$, from which the magnitude and phase spectra can be obtained. Thus, to handle discrete-time signals such as the above, the Z-transform technique can be used.

The circuits that are handled are not discrete-time circuits, nor will the signals be discrete-time signals. They are analog sampled data circuits, and the signals will also be sampled data signals. However, later in this chapter we will find that we can describe switched capacitor filters in the form of *difference equations*, as we do in the case of discrete-time systems. The input-output relationship of a switched capacitor circuit may be described by

$$y(nT) + \sum_{k=0}^{k=n} b_k y(nT - kT) = \sum_{k=0}^{k=m} a_k x(nT - kT) \tag{8.14}$$

where $x(nT)$ and $y(nT)$ are the input and output signals to the switched capacitor filter. Therefore the analysis and synthesis of switched capacitor filters can be accomplished with the same techniques as those employed for discrete-time systems and circuits using Z-transform techniques. Before we use such difference equations, we must make a change in notation. Following the conventional way of representing the difference equation, we omit T inside the parentheses in all cases. Whenever we need to, we can invoke this T. Thus, after dropping T, (8.14) may also be written as

$$y(n) + \sum_{k=1}^{k=n} b_k y(n-k) = \sum_{k=0}^{k=m} a_k x(n-k) \tag{8.15}$$

By taking the Z transform of (8.15) and using an important property of the Z transform, namely,

$$Z[y(n-k)] = z^{-k} Y(z) \tag{8.16}$$

where $Y(z)$ is the Z transform of $y(n)$, the difference equation (8.16) can be converted to an algebraic equation relating $Y(z)$ and $X(z)$:

$$Y(z)\left(1 + \sum_{k=1}^{k=n} b_k z^{-k}\right) = X(z)\left(\sum_{k=0}^{k=m} z_k z^{-k}\right)$$

The ratio of the Z transform of the output to that of the input is called the *pulse transfer function* of the circuit and is denoted by

$$H(z) = \frac{\sum_{k=0}^{k=m} a_k z^{-k}}{1 + \sum_{k=1}^{k=n} b_k z^{-k}} \tag{8.17}$$

Such a pulse transfer function can be used to describe a discrete-time system, as well as switched capacitor filters. The transfer function $H(z)$ plays the same important role as the analog transfer function $H(s)$ in the analog case. The impulse response of a discrete-time system $h(n)$ can be determined by finding the inverse Z transform of the pulse transfer function. We can also find the frequency response simply by substituting $z = e^{j\omega T}$ into (8.17) and evaluating the magnitude and phase responses of $H(e^{j\omega T})$. In this way, $H(e^{j\omega T})$ plays a role similar to that of $H(j\omega)$ in the analog domain. Thus we find many similarities between the analog and pulse transfer functions. One final note about $H(e^{j\omega T})$ is in order. The effect of sampling an analog signal is to introduce periodicity in its spectrum with a period of ω_s ($= 2\pi/T$). This is also true of $H(e^{j\omega T})$. Thus the baseband representation of $H(e^{j\omega T})$ from $-\omega_s/2$ to $\omega_s/2$ is sufficient to determine its complete characteristic. One can find the poles and zeros of the pulse transfer function $H(z)$. These poles and zeros can be marked in the z domain, and a pole-zero diagram representation of the pulse transfer function obtained.

In order to gain some insight into the stability of discrete-time circuits, let us consider the mapping of $z = e^{sT}$. Substituting $s = \sigma + j\omega$, we have

$$z = e^{\sigma T} e^{j\omega T}$$

and

$$|z| = e^{\sigma T}$$

Therefore

$$|z| \begin{Bmatrix} < 1 \\ = 1 \\ > 1 \end{Bmatrix} \qquad \begin{matrix} \sigma < 0 \\ \sigma = 0 \\ \sigma > 0 \end{matrix}$$

The poles and zeros on the $j\omega$ axis are mapped on the unit circle of the z plane. The points on the left and right halves of the s plane are mapped inside and outside the unit circle, respectively. Therefore, for the discrete-time system to be stable, the poles of $H(z)$ must lie within the unit circle of the z plane.

In analog sampled data circuits, such as switched capacitor filters, the signals are not only sampled signals but also sampled and held (S/H) signals. In fact, all the signals are sampled and held signals throughout the switched capacitor circuit whether we like it or not. The simplest form of holding the signal is called *zero-order hold*. The circuit shown in Fig. 8.4*a* is such an S/H circuit. In this circuit, $v(t)$ is the analog signal and $v_s(t)$ is the

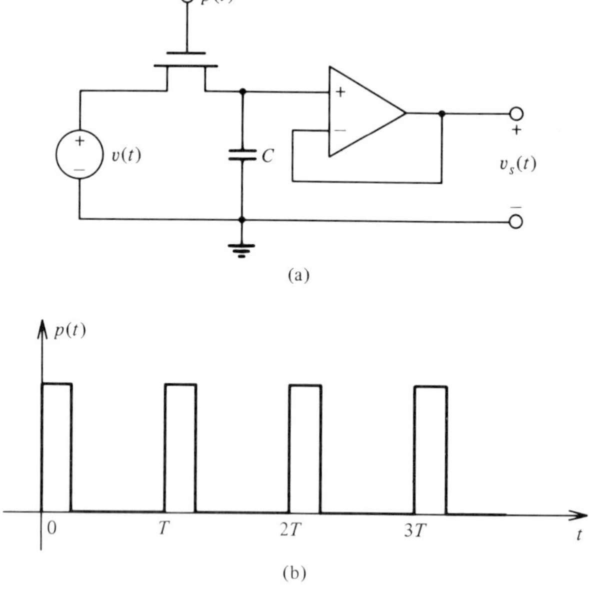

Figure 8.4 (*a*) An S/H circuit. (*b*) Waveform of the clock.

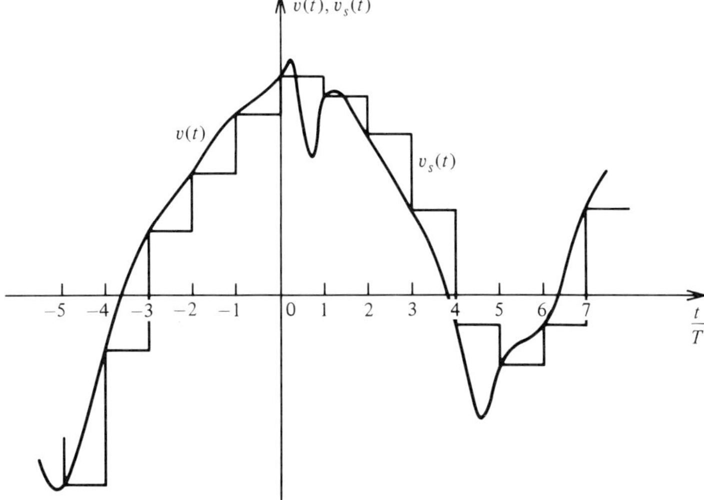

Figure 8.5 Waveforms of an analog signal $v(t)$ and the corresponding sampled and held signal $v_s(t)$.

sampled and held signal. The signal $p(t)$ is a clock with a period of T seconds as shown in Fig. 8.4*b*. When $p(t)$ becomes positive at $t = nT$, $n = 0, 1, 2, \ldots,$ the capacitor is charged almost instantaneously (assuming that the ON resistor of the MOS transistor is zero) to $v(t)$ at $t = nT$. If the time duration of this pulse is very short, the charge on the capacitor will be $Cv(nT)$. When $p(t)$ goes to zero, the MOS transistor does not conduct, and therefore the charge remains the same on the capacitor. This maintains the voltage across the capacitor as $v(nT)$ until the pulse $p(t)$ becomes positive again. A sampled and held waveform corresponding to a $v(t)$ will be as shown in Fig. 8.5. Note that $v_s(t)$ is a continuous analog signal, as opposed to the sampled signal $v^*(t)$.

The impulse response of the zero-order hold may be described by

$$h_0(t) = \begin{cases} 1/T & 0 < t \leqq T \\ 0 & \text{otherwise} \end{cases}$$

Therefore

$$H_0(s) = \frac{1 - e^{-sT}}{sT}$$

For sinusoidal excitations,

$$H_0(j\omega) = \frac{\sin(\omega T/2)}{\omega T/2} e^{-j\omega T/2}$$

Now, to see the effect of holding the sampled signal, the spectrum of $V^*(t)$

can be multiplied by the above transfer function, and thus

$$|V_s(j\omega)| = |V^*(j\omega)|\left|\operatorname{sinc}\left(\frac{\omega T}{2}\right)\right| \tag{8.18a}$$

and

$$\arg V_s(j\omega) = \arg[V^*(j\omega)] - \frac{\omega T}{2} \tag{8.18b}$$

The reconstruction process, as evidenced by the above equations, introduces frequency-dependent errors in both magnitude and phase spectra. The phase error is not a serious problem, since the additional phase is a linear function of frequency. Both these errors may be minimized with a sampling interval T such that $\omega_s \gg \omega$, where ω is the operating frequency. This is because $\sin x/x$ tends to unity as x tends to 0. However, the high-frequency spectrum of $v^*(t)$ may be heavily attenuated.

8.2 *Waveforms and the Analysis of Switched Capacitor Filters*

Before we consider various circuits of switched capacitor filters, it is important to discuss the general concepts involved in these circuits and the notation used. The block diagram representation in Fig. 8.6 shows the hardware requirements of a general system for processing an analog signal using switched capacitor circuits. We have seen the need for band-limiting the input signal before sampling and also the need for an S/H circuit. After the filtering process, an output (S/H) circuit resamples the output, and finally, a reconstruction filter smooths the output to obtain v_o. An important note about SC filters is that, in many SC networks, the S/H operation is performed internally by the circuit, and therefore the middle three blocks constitute an SC filter.

Assume that the SC filter consists only of ideal components, namely, ideal switches, ideal capacitors, and ideal independent and dependent voltage sources. The op amp can be considered a dependent voltage source, and we shall also assume that the input excitations are sampled and held voltages. The switches are MOS transistors typically controlled by a two-phase nonoverlapping clock whose duty cycle is slightly less than 50%. The switching arrangement in Fig. 8.1*a* can be realized using the circuit in Fig.

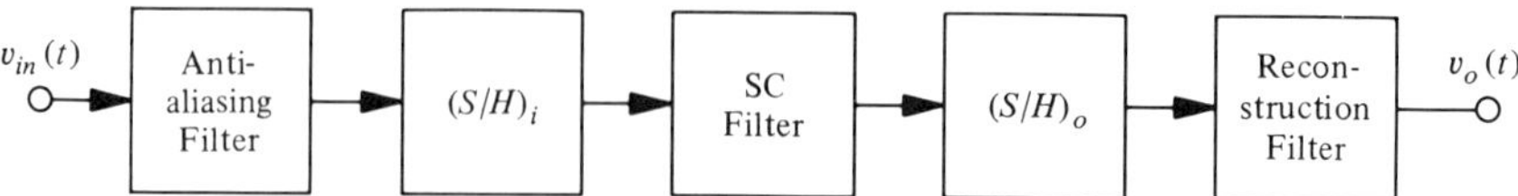

Figure 8.6 Block diagram representation of the circuits required for processing an analog signal using SC networks.

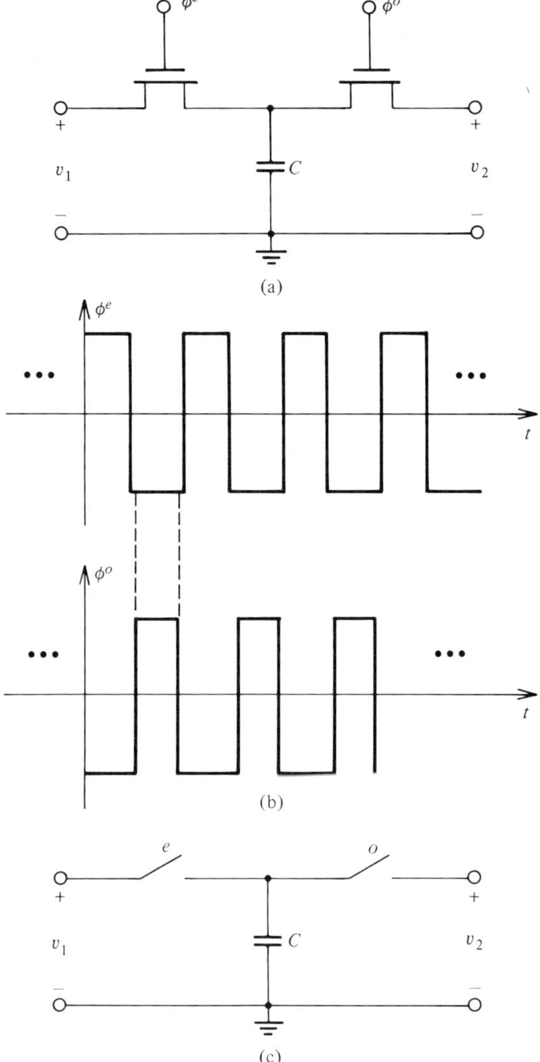

Figure 8.7 (*a*) Hardware equivalent of Fig. 8.1(*a*). (*b*) Nonoverlapping switching waveforms. (*c*) Simplified circuit representation of (*a*).

8.7*a*, which is driven by a clock as shown in Fig. 8.7*b*. Here ϕ^e and ϕ^o are the two nonoverlapping phases of the clock. A switching arrangement equivalent to that in Fig. 8.7*a* is shown in Fig. 8.7*c*, and ϕ^e or e means that the switch is closed during even phases, i.e., at $t = nT$, $n = 0, 1, 2, \ldots$. Similarly, ϕ^o or o means that the switch is closed during odd phases, i.e., at $t = (n + \frac{1}{2})T$, $n = 0, 1, 2, \ldots$. We use the switching notation in Fig. 8.7*c* throughout this book for convenience, though in the literature both types, namely ϕ^e and ϕ^o and simply e and o, are used. When both switches are open, the circuit behavior is not affected. However, if both switches are closed, improper operation of the circuits may result, which is why nonoverlapping clock phases are needed.

When the elements are assumed to be ideal, the charge transfer takes place almost instantaneously, and therefore voltage variables change in value only at discrete instants $t = nT$ and $t = (n + \frac{1}{2})T$, $n = 0, 1, 2, \ldots$. With this assumption we can use Z-transform techniques to carry out the analysis and synthesis of SC networks. The use of the Z transform implies that the signals are discrete-time signals which are impulse-sampled. Thus the Z-transform technique will predict the performance of SC filters accurately only at these discrete instants on a sample-to-sample basis. How the signals behave between the samples is not considered here. Also, if we want to examine the behavior of the signals between the samples the analysis will be more complex.

The switching action in SC networks makes them time-varying networks. A given SC network behaves as a time-invariant network during each phase interval before the switches change their positions, because the topology of the network does not change during the phase interval but only after the switching. When the switches are controlled by a biphase clock, an SC network behaves as two different time-invariant networks during the even and odd phases. These two time-invariant networks are interrelated in only one sense. The charges stored in the capacitors during one phase act as the initial conditions for the next phase period. This will be the only fact that has to be used between these two topologically independent time-invariant networks. When this approach is used, it is convenient to partition the analog sampled data signal into its even and odd components as shown in Fig. 8.8. A sampled data signal, such as the one shown in Fig. 8.8a, can be thought of as the sum of its even and odd components as shown in Fig. 8.8b and c. From this illustration it is easy to see that

$$v(t) = v^e(t) + v^o(t) \tag{8.19}$$

Assuming that we are interested in the values of $v(t)$, or for that matter $v^e(t)$ and $v^o(t)$, only at discrete times we may use the Z transforms of these signals. Thus

$$V(z) = V^e(z) + V^o(z) \tag{8.20}$$

If $v(t)$ is an S/H signal held over one full period,

$$V^o(z) = z^{-1/2}V^e(z) \tag{8.21}$$

Thus, in such cases,

$$V(z) = (1 + z^{-1/2})V^e(z) \tag{8.22}$$

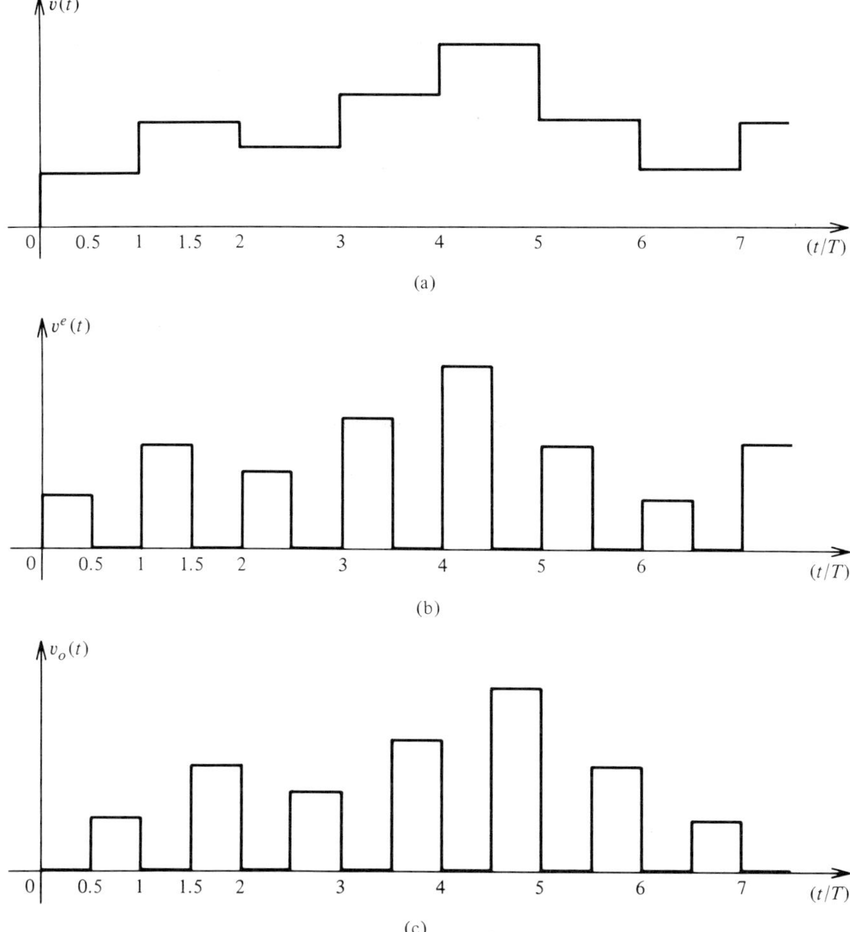

Figure 8.8 (a) Sampled and held voltage $v(t)$. (b) Its even component. (c) Its odd component.

However, if the sampling and holding is done every half-period, then (8.21) need not hold true. Therefore (8.22) cannot in general be used. But all voltage signals can usually be split into even and odd components in the case of biphase switching, and each component will have its own Z transform. This property involving separation of voltage signals into even and odd components, and the fact that for any SC network there are two time-invariant topologies corresponding to these two phases, make it possible to write two independent sets of node charge equations using the law of conservation of charge at any node. The two node charge equations, written at a given node for the two different phases are interrelated in the sense that the charges stored during the even (odd) phase act as the initial conditions for the odd (even) phase. This provides a means for combining two topologically different networks into a single equivalent circuit in the Z domain but results in increased complexity. With this approach, an n-port network will require a $2n$-port equivalent circuit [3–5].

Switched capacitor networks may be described completely in terms of charge transfer equations during the even and odd phases, using the stored charges in the capacitors which in turn can be expressed in terms of the capacitor voltages during the even and odd phases. During switching, the charges are redistributed, instantaneously satisfying the law of conservation of charge. Thus, using the law of conservation of charge, we can write node charge equations at every node of the network during the even and odd phases. This is similar to writing nodal equations for continuous-time networks using KCL. At a given node p of a network at even instants, $t = nT$, $n = 0, 1, 2, \ldots,$ we can write

$$\Delta q_p^e(nT) = \sum_{i=1}^{N_p^e} q_{pi}^e(nT) - \sum_{i=1}^{N_p^e} q_{pi}^o[(n - \tfrac{1}{2})T] \tag{8.23a}$$

and during odd phases, $t = (n + \frac{1}{2})T$, $n = 0, 1, 2, \ldots,$ we obtain

$$\Delta q_p^o[(n + \tfrac{1}{2})T] = \sum_{i=1}^{N_p^o} q_{pi}^o[(n + \tfrac{1}{2})T] - \sum_{i=1}^{N_p^o} q_{pi}^e(nT) \tag{8.23b}$$

where q_{pi}^e and q_{pi}^o are the instantaneous charges stored in the ith capacitor connected to node p during the even and odd phases, respectively, and N_p^e and N_p^o are the total number of capacitors connected to node p during the even and odd phases, respectively. The second group of terms on the right side of (8.23) is equal to the sum of the initial charges stored in these capacitors—also called the *memory charge*. The terms on the left side of (8.23), Δq_p^e and Δq_p^o, are the *incremental charges* injected into node p during the even and odd phases. If node p is an internal node, where no source of energy (either independent or dependent) is connected, these terms can be set to *zero*. However, if this node is connected to a source of energy, these terms will give the incremental charges supplied by the sources during the corresponding phase. The terms on the right side of the above equations can be found in terms of the capacitor values and capacitor voltages using the element relationship $q_i(nT) = C_i v_i(nT)$, where $v_i(nT)$ is the voltage across the capacitor C_i. These capacitor voltages can also be expressed in terms of the node voltages, and thus we can express (8.23) in terms of node voltages also. For now, in terms of the capacitor voltages, we write (8.23) as

$$\Delta q_p^e(n) = \sum_{i=1}^{N_p^e} C_i v_i(n) - \sum_{i=1}^{N_p^e} C_i v_i[(n - \tfrac{1}{2})] \tag{8.24a}$$

and

$$\Delta q_p^o(n) = \sum_{i=1}^{N_p^o} C_i v_i(n) - \sum_{i=1}^{N_p^o} C_i v_i(n - \tfrac{1}{2}) \tag{8.24b}$$

In the above equations we have omitted the T inside the parentheses for convenience. Furthermore, we have shifted (8.23*b*) by half a time unit in obtaining (8.24*b*). However, the superscripts e and o indicate the phases during which these equations are obtained.

Typically, in SC networks the sources of energy are voltage sources, and the outputs are also voltages. Therefore, if we consider an n-port network, the voltages applied at different ports may be considered independent variables and the charges supplied by the sources, namely, $\Delta q_p^e(n)$ and Δq_p^o, where p is the port, may be considered dependent variables. This is analogous to the parametric description of a continuous-time network with Y parameters, where currents are described in terms of port voltages. Fortunately, equations (8.24) are linear difference equations and Z-transform techniques can be employed to solve for the voltage variables. Thus, transforming (8.24), we obtain

$$\Delta Q_p^e(z) = \sum_{i=1}^{N_p^e} C_i V_i^e(z) - z^{-1/2} \sum_{i=1}^{N_p^e} C_i V_i^o(z) \tag{8.25a}$$

and

$$\Delta Q_p^o(z) = \sum_{i=1}^{N_p^o} C_i V_i^o(z) - z^{-1/2} \sum_{i=1}^{N_p^o} C_i V_i^e(z) \tag{8.25b}$$

where $\Delta Q_p^e(z)$ and $\Delta Q_p^o(z)$ may or may not be zero depending on the type of node, as discussed earlier, and $V_i^e(z)$ and $V_i^o(z)$ are the Z-transformed variables of the even and odd phase components of the capacitor voltage $v_i(n)$.

Example 8.1: Consider the switched capacitor network shown in Fig. 8.9*a*. Figure 8.9*b* and *c* shows the topologies of the network during the even and odd phases, respectively. Find the input-output relationship in this network in terms of the Z-transformed variables. The op amp is assumed to be ideal.

During the even and odd phases, at the input port,

$$\Delta q_1^e(n) = C_i v_i^e(n) \tag{8.26a}$$

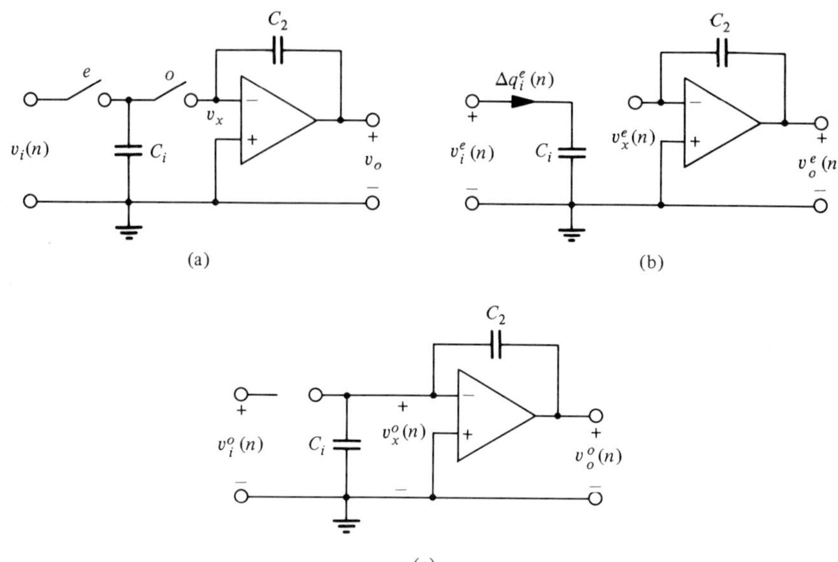

Figure 8.9 (*a*) Simple SC network considered in Example 8.1. (*b*) Its equivalent during the even phase. (*c*) Its equivalent during the odd phase.

and

$$\Delta q_1^o(n) = 0 \tag{8.26b}$$

The internal node voltage v_x is zero during both the even and odd phases if the op amp is assumed to be ideal. Furthermore, the input current into the inverting terminal is zero. At the inverting input node, no external charge is imparted during the even phase. Furthermore, there is only one capacitor C_2 connected to this node during the even phase. Now, applying (8.24*a*) we have

$$0 = C_2 v_o^e(n) - C_2 v_o^o\left(n - \tfrac{1}{2}\right) \tag{8.27a}$$

During the odd phase, two capacitors are connected to node x. We can apply the charge conservation equation of (8.24*b*), however, $v_x^o(n)$ is zero, and therefore

$$0 = -C_2 v_o^o(n) - \left[C_1 v_i^e\left(n - \tfrac{1}{2}\right) - C_2 v_o^e\left(n - \tfrac{1}{2}\right)\right] \tag{8.27b}$$

The terms inside the square bracket arise because of the initial charge stored in these two capacitors—the memory charge. Transforming (8.27) and rearranging, we obtain

$$V_o^e(z) = -z^{-1/2} V_o^o(z) \tag{8.28a}$$

and

$$C_2\left[V_o^o(z) - z^{-1/2}V_o^e(z)\right] = -C_1 z^{-1/2} V_i^e(z) \qquad (8.28b)$$

Solving for the even and odd output components in terms of the input components, we obtain

$$V_o^o(z) = \frac{-C_1}{C_2}\frac{z^{-1/2}}{1 - z^{-1}} V_i^e(z) \qquad (8.29a)$$

and

$$V_o^e(z) = \frac{-C_1}{C_2}\frac{z^{-1}}{1 - z^{-1}} V_i^e(z) \qquad (8.29b)$$

■

Note that (8.26) is not useful in finding the input-output relationship but provides information on the amount of charge supplied by the input source.

From the previous example, note that the input-output relationship depends on the phase during which the output is sampled. In this particular example, the output is influenced by only the even component of the input. This should be obvious from the circuit also, because during the odd phase the circuit is completely removed from the input. However, in general, both even and odd input components may determine the even and odd output components. Thus, if we consider a two-port network, the input-output relationship may be expressed as

$$V_o^e(z) = H^{ee}(z)V_i^e(z) + H^{eo}(z)V_i^o(z) \qquad (8.30a)$$

and

$$V_o^o(z) = H^{oe}(z)V_i^e(z) + H^{oo}(z)V_i^o(z) \qquad (8.30b)$$

In Example 8.1,

$$H^{eo}(z) = H^{oo}(z) = 0$$

$$H^{ee}(z) = \frac{-C_1}{C_2}\frac{z^{-1}}{1 - z^{-1}} V_i^e(z) \qquad (8.31a)$$

and

$$H^{oe}(z) = \frac{-C_1}{C_2}\frac{z^{-1/2}}{1 - z^{-1}} \qquad (8.31b)$$

None of these transfer functions need be equal, but they may be. Now, proceeding further, it is clear that each port of an SC network requires two ports—one for even phases and another for odd phases—if one wants to

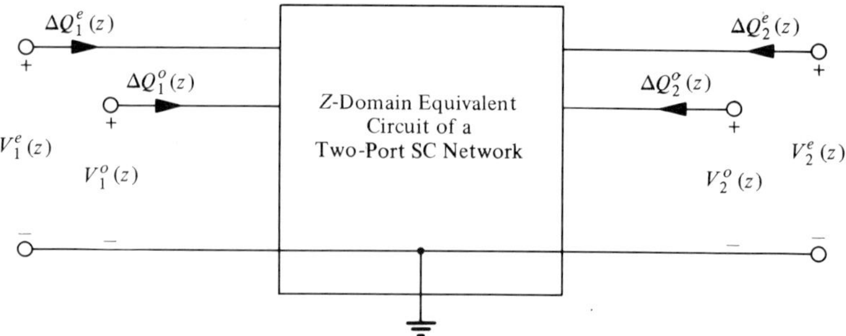

Figure 8.10 Four-port z-domain equivalent circuit of a two-port SC network.

use Z-domain equivalence. For simplicity, assume that the given SC network is a three-terminal two-port network with ground as the common terminal. Its Z-domain equivalent circuit will have four ports, as shown in Fig. 8.10.

Now, we have seen that the SC networks can be represented using Z-domain functions. As is done in the next section, we would like to represent SC networks by their corresponding Z-domain equivalent circuits (Z-domain admittance or impedance functions) so that the transfer functions of the SC networks can be obtained using conventional circuit analysis techniques. This is the technique adopted in the case of continuous-time networks, where we transform time domain networks into frequency domain equivalences and carry out the analysis and synthesis in the frequency domain.

Traditionally, for the purpose of synthesis, Z-domain transfer functions are obtained from s-domain analog transfer functions using some form of transformation. This is reasonable, because one can utilize the approximation procedures developed in Chap. 2 and no new techniques need be learned. Furthermore, the approximation procedures for analog functions are well advanced, and tables and charts are available for this purpose. In addition, we often simply simulate the performance of an analog network. A case in point involves the network in Fig. 8.9. This network is an SC integrator where the SC resistor equivalent of Fig. 8.1 has been used to replace the resistor of the analog inverting integrator in Fig. 6.15. We provide only a summary of the transformations. The reader can find more mapping details for these transformations in any book on digital signal processing, for example, Oppenheim and Schafer [2]. When these transformations are applied to an analog transfer function $H_a(s)$, three essential properties must be met:

1. The frequency response of the resulting discrete-time transfer function $H(z)$, when $z = e^{j\omega T}$, must preserve the frequency response of $H_a(s)$ when $s = j\omega$. This implies that the $j\omega$ axis of the s plane must

be mapped onto the unit circle of the z plane.

2. The transformation must preserve the stability. This implies that the left-half plane poles must be mapped onto the interior of the unit circle of the z plane.
3. The resulting transfer function $H(z)$ must be a rational function in z^{-1} (or z).

The transformations are based on the sampling process of the impulse response that the analog transfer function represents or on the numerical solutions of the differential equation from which the analog transfer function was obtained. Again, there are different ways in which a differential equation can be approximated. Therefore, depending on the type of approximation procedure, there are different names for the transformations.

Impulse-invariant transformation: In this method, the impulse response of the digital system $h(n)$ [$= h(nT)$] is equated to the impulse response $h_a(t)$ evaluated at $t = nT$. Equivalently, in terms of the transfer function, if the analog transfer function $H_a(s)$ is expressed in the form of partial fractions, such as

$$H_a(s) = \sum_{k=1}^{N} \frac{A_k}{s - p_k} \tag{8.32a}$$

then $H(z)$ is obtained as

$$H(z) = \sum_{k=1}^{N} \frac{A_k}{1 - \exp(p_k T) z^{-1}} \tag{8.32b}$$

The resulting transfer function $H(z)$ is a rational function, and if $H_a(s)$ is stable, $H(z)$ is also stable. However, $H(e^{j\omega T})$ suffers from a frequency aliasing problem, and therefore $H_a(j\omega)$ must be a band-limited function.

Backward difference (BD): The remaining methods depend on numerical approximations techniques. In the backward difference method, the differential equations are approximated using backward Euler algorithm, and the equivalent effect in the frequency domain is to replace s in $H_a(s)$ with the following function of z:

$$s = \frac{1 - z^{-1}}{T} \tag{8.33a}$$

or

$$z = \frac{1}{1 - sT} \tag{8.33b}$$

where T is the sampling period. This transformation gives rise to a rational $H(z)$, and it is also a stable transfer function. It does not suffer from the frequency aliasing problem either. However, the $j\omega$ axis of the s plane is not mapped on the unit circle of the z plane except at $z = 1$. This implies that $H(e^{j\omega T})$ approximates $H_a(j\omega)$ only when $|\omega T| \ll 1$.

Forward difference (FD): This transformation is based on the approximation of a differential equation into a difference equation using the forward Euler algorithm. The transformation used is

$$s = \frac{1 - z^{-1}}{Tz^{-1}} \tag{8.34a}$$

or, equivalently,

$$z = 1 + sT \tag{8.34b}$$

This transformation may result in an unstable transfer function $H(z)$ even though $H_a(s)$ is stable. In addition, $H(e^{j\omega T})$ approximates $H_a(j\omega)$ only when $|\omega T| \ll 1$. The condition $|\omega T| \ll 1$, in both the above cases, means that the sampling rate must be high, and it may be inefficient in many cases. Furthermore, it should be pointed out that these two transformations are highly unsatisfactory for anything but lowpass filters.

Lossless discrete integrator (LDI): This transformation is based on the equation

$$s = \frac{1 - z^{-1}}{Tz^{-1/2}} \tag{8.35a}$$

or

$$z = 0.5\left[\left\{2 + (sT)^2\right\} \pm \left\{(sT)^2\left[4 + (sT)^2\right]\right\}^{1/2}\right] \tag{8.35b}$$

This transformation satisfies conditions 1 and 2 listed earlier, but the half-power of z makes $H(z)$ a nonrational function. This problem can be eliminated by using $Z = z^{1/2}$, and $H(Z)$ will become a $2N$th-order function. $H(Z)$ will require twice as much hardware as required by other transformations.

Bilinear transformation (BL): The final transformation, very often used because of its desirable properties, is called bilinear transformation. It is based on numerical integration using Simpson's rule:

$$s = \frac{2}{T}\frac{1 - z^{-1}}{1 + z^{-1}} \tag{8.36a}$$

or

$$z = \frac{2/T + s}{2/T - s} \tag{8.36b}$$

This transformation satisfies all three properties listed earlier and therefore is very often used. It maps the entire $j\omega$ axis of the s plane onto the unit circle of the z plane, and this causes some nonlinearity in the frequency axis, known as *frequency warping*. If ω_a and ω_d are the real frequencies in the analog and digital domains, respectively, substituting $s = j\omega_a$ and

$z = e^{j\omega_d T}$, into (8.36), we obtain

$$\omega_a = \frac{2}{T}\tan\left(\frac{\omega_d T}{2}\right) \tag{8.37}$$

When $\omega_d T/2 \ll 1$, then $\omega_d = \omega_a$. However, when this condition is not met, the relationship between the two real frequencies ω_a and ω_d is nonlinear and warping occurs. This warping can be compensated by using a *prewarping* technique, as illustrated in the following example.

Example 8.2: Using bilinear transformation, obtain a pulse transfer function that meets the following specifications for a bandpass filter: $A_p \leq$ 0.5 dB, $f_{pd1} = 1$ kHz, and $f_{pd2} = 2$ kHz; $A_a \geq 20$ dB, $f_{ad1} = 0.5$ kHz and $f_{ad2} = 3$ kHz. Use Chebyshev approximation, and assume that the sampling frequency $f_s = 8$ kHz.

Using $T = 1/f_s$ and (8.37), we obtain the analog frequencies as $f_{pa1} = 1.0548$ kHz, $f_{pa2} = 2.546$ kHz, $f_{aa1} = 0.5065$ kHz, and $f_{aa2} =$ 6.1478 kHz. Calculating

$$f_{aa} = \frac{f_{pa2} f_{pa1}}{f_{aa1}} = 5.3027 \text{ kHz}$$

we find that $f_{aa} < f_{aa2}$. Therefore we choose $f_{aa1} = 0.5065$ kHz and $f_{aa2} = 5.3027$ kHz. Now, using the program MAXCHY from Chap. 2, we find the analog transfer function, which is of the following form to give a dc gain of 0 dB:

$$H(s) = \frac{86318790 s^2}{\prod_{i=1}^{2}\left(s^2 + a_i s + b_i\right)}$$

where $a_1 = 7130.849$, $b_1 = 2.39453 \times 10^8$, $a_2 = 3157.807$, and $b_2 =$ 4.695794×10^7. Next, applying the transformation of (8.36a) to the above analog transfer function, we find the pulse transfer function to be

$$H(z) = \frac{0.102558(1 - z^{-1})^2(1 + z^{-1})^2}{\prod_{i=1}^{2}\left(1 - A_i z^{-1} + B_i z^{-2}\right)}$$

where $A_1 = 0.0542925$, $B_1 = 0.625644$, $A_2 = 1.18276$, and $B_2 =$ 0.714131. One can easily verify that this transfer function satisfies the required specifications for the discrete-time filter. ■

8.3 *Synthesis Techniques*

One way of synthesizing an SC network is to start with a z-domain function and synthesize it. The z-domain function, which is usually a transfer function, can be obtained using any of the transformations discussed in the previous section. For this type of synthesis, one usually starts with a form of the network and derives its transfer function in terms of unknown network parameters, in this case, capacitance values. Then, as for active RC networks, one may use coefficient comparisons and solve for the capacitance values.

The starting point for the second method is an active RC network. Many well-known and proven active RC networks, as seen in the previous chapters, are used to realize a given analog transfer function. Replacing each resistor with a switched capacitor resistor equivalent, as shown in Fig. 8.1, we can obtain a switched capacitor network. The switched capacitor network in Fig. 8.9a was obtained in this way. Replacing the input resistor in an analog inverting integrator with the SC resistor equivalent of Fig. 8.1, we obtain the SC network in Fig. 8.9a. Thus the behavior of this network is similar to that of an integrator. This is not the only resistor equivalent, and there are many other forms of SC networks that simulate analog resistors. We shall consider them in the next section. With this method, one can synthesize an analog transfer function in the form of an active RC network and then transform it into an SC network by replacing resistors with SC resistor equivalents.

The best method of synthesizing an SC network is to adopt a combination of both the above techniques. The form of the SC network can be obtained using the second approach by replacing each resistor with its SC equivalent in terms of unknown capacitance values, and the network may be designed using z-domain transfer functions. This is possible because $H(z)$ can be obtained to meet exact specifications by using bilinear or any other transformation.

In order to learn the synthesis procedure, one has to first learn the analysis procedure. Here again, there are two approaches for SC network analysis. One can identify simple blocks of networks with necessary isolation whose transfer functions can be derived by writing only a few time domain equations and then transforming them into z-domain equations. The SC network in Example 8.1 is one such example. Then, by using the block diagram reduction technique, as in the systems approach, one can find the overall transfer function in terms of the unknown capacitance values. Here are a few problems. It is required to split the overall network into simple networks, and therefore one can only split it when there is proper isolation such as op amp output. The analysis problem is yet to be solved in the z-domain, and this will require the solution of algebraic equations. Furthermore, depending on the time at which one samples the output, the appropriate transfer functions for the individual blocks must be used.

Another method of analysis is one that is similar to the analysis of continuous-time networks, where we replace individual circuit elements by their frequency domain equivalents and analyze them using algebraic equations in the frequency domain. Then, we can solve for the network function of importance, usually a transfer function. This method may save a considerable amount of time in analyzing SC networks. The technique requires that we define impedance and admittance in the z domain, and this is what we will do now. In SC networks, the naturally arising variables are charges that are transferred during the even and odd phases, and voltages during these phase periods. Assuming that voltages are independent variables and that the transferred charges are dependent variables, we define the term "admittance" of a branch as

$$Y(z) = \frac{\Delta Q(z)}{V(z)} \tag{8.38}$$

This admittance is not the usual admittance defined for continuous-time networks, which is the ratio of the Laplace transform of the current to the Laplace transform of the voltage. For lack of another word for the ratio in (8.38), we use the term "admittance." In terms of fundamental units, it has the same dimension as capacitance. However, in the analysis of SC networks, this ratio behaves in exactly the same way as the usual admittance in continuous-time circuits. Therefore, once we find the z-domain equivalent circuits for some basic SC networks, we can simply replace the appropriate networks with z-domain equivalent circuits. Analysis of the z-domain equivalent circuit for the given SC network can be carried out in the same way as conventional circuit analysis, and no new techniques need be learned.

8.4 *Switched Capacitor Resistor Equivalents*

To form an SC network from an active *RC* network, the fundamental requirement is to replace the resistors with SC equivalents. In this section, we shall discuss some of the switched capacitor equivalent resistors and their z-domain equivalents.

The parallel SC realization of a resistor (*PSCR*): This circuit is the same as the one shown in Fig. 8.1*a*, which is reproduced here as Fig. 8.11*a* for convenience. In Sec. 8.1, we showed how this circuit behaved as a resistor for high sampling frequencies. In fact, all the circuits discussed in this section behave as resistors for high sampling frequencies, and one can find their equivalent resistance values if the need arises. However, we are not

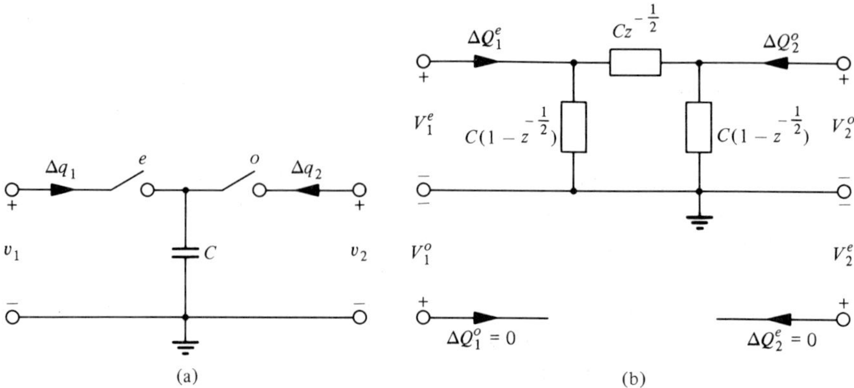

Figure 8.11 Parallel switched capacitor equivalent of a resistor (PSCR). (*b*) Its *z*-domain equivalent circuit.

interested in resistance values because our approach to the synthesis is to realize z-domain functions. Therefore it makes sense to find the four-port z-domain equivalent of this two-port SC network. First, we write the time-domain charge transfer equations:

$$\Delta q_1^e(n) = Cv_1^e(n) - Cv_2^o(n - \tfrac{1}{2})$$
$$\Delta q_1^o(n) = 0$$
$$\Delta q_2^e(n) = 0$$

and

$$\Delta q_2^o(n) = Cv_2^o(n) - Cv_i^e(n - \tfrac{1}{2})$$

Transforming the above equations, we obtain

$$\Delta Q_1^e(z) = CV_1^e(z) - Cz^{-1/2}V_2^o(z) \tag{8.39a}$$
$$\Delta Q_1^o(z) = 0 \tag{8.39b}$$
$$\Delta Q_2^e(z) = 0 \tag{8.39c}$$

and

$$\Delta Q_2^o(z) = CV_2^o(z) - Cz^{-1/2}V_1^e(z) \tag{8.39d}$$

The charge transfer equations (8.39) can be represented using a four-port network as shown in Fig. 8.11*b* [5]. An interesting point to be noted about this circuit is that two of the four ports are open and can be discarded in further calculations. This is because the input and the output are sampled only during the even and odd phases, respectively. This fact that some ports are open is not peculiar to this circuit alone but is true in many other SC networks.

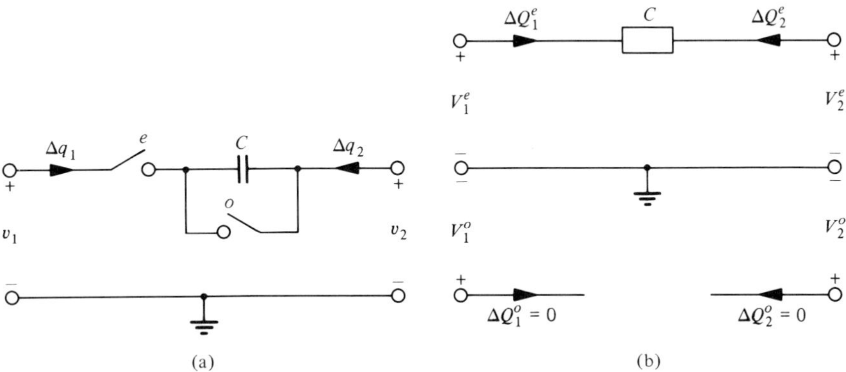

Figure 8.12 (*a*) Series SC realization of a resistor (SSCR). (*b*) Its *z*-domain equivalent circuit.

Series SC realization of a resistor (SSCR): This realization is shown in Fig. 8.12*a*. To find its *z*-domain equivalent circuit, we shall again write the time-domain equations and then transform them into *z*-domain equations. In this circuit, during the even phase, we find that

$$\Delta q_1^e(n) = Cv_1^e(n) - Cv_2^e(n)$$

and

$$\Delta q_2^e(n) = Cv_2^e(n) - Cv_1^e(n)$$

During the odd phase, there is no charge transfer from either port because the port is open and the capacitor is short-circuited. Next, transforming these charge transfer equations, we obtain

$$\Delta Q_1^e(z) = C\left[V_1^e(z) - V_2^e(z)\right] \tag{8.40a}$$

$$\Delta Q_2^e(z) = C\left[V_2^e(z) - V_1^e(z)\right] \tag{8.40b}$$

$$\Delta Q_1^o(z) = 0 \tag{8.40c}$$

and

$$\Delta Q_2^o(z) = 0 \tag{8.40d}$$

Equations (8.40) are represented by the four-port network in Fig. 8.12*b*.

Series-parallel SC realization of a resistor (SPSCR): This is a combination of parallel and series capacitances that are switched to simulate a resistor [6]. Writing the charge transfer equations directly in the *z* domain, during the even phase, we find that

$$\Delta Q_1^e(z) = C_2 V_1^e(z) - C_2 z^{-1/2} V_2^o(z) \tag{8.41a}$$

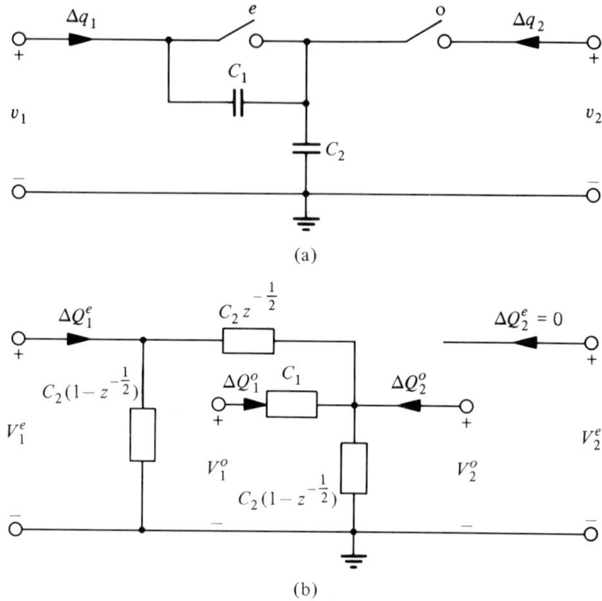

Figure 8.13 (*a*) Series-parallel SC realization of a resistor. (*b*) Its *z*-domain equivalent circuit.

and

$$\Delta Q_2^e(z) = 0 \tag{8.41b}$$

Similarly, during the odd phase, we have

$$\Delta Q_1^o(z) = C_1[V_1^o(z) - V_2^o(z)] \tag{8.41c}$$

and

$$\Delta Q_2^o(z) = C_1[V_2^o(z) - V_1^o(z)] + C_2V_2^o(z) - C_2z^{-1/2}V_1^e(z) \tag{8.41d}$$

Figure 8.13*b* shows the four-port representation of the above equations, and therefore this circuit is the *z*-domain representation of the SC circuit in Fig. 8.13*a*.

Bilinear SC realization of a resistor (*BSCR*)*:* This is the last of the basic forms of SC realization of a resistor. The SC realization of BSCR and its *z*-domain equivalent are shown in Fig. 8.14. The *z*-domain equivalent circuit can be verified from the following charge transfer equations obtained from the SC network during the even and odd phases:

$$\Delta Q_1^e(z) = C[V_1^e(z) - V_2^e(z)] - Cz^{-1/2}[V_2^o(z) - V_1^o(z)] \tag{8.42a}$$
$$\Delta Q_2^e(z) = C[V_2^e(z) - V_1^e(z)] - Cz^{-1/2}[V_1^o(z) - V_2^o(z)] \tag{8.42b}$$
$$\Delta Q_1^o(z) = C[V_1^o(z) - V_2^o(z)] - Cz^{-1/2}[V_2^e(z) - V_1^e(z)] \tag{8.42c}$$
$$\Delta Q_2^o(z) = C[V_2^o(z) - V_1^o(z)] - Cz^{-1/2}[V_1^e(z) - V_2^e(z)] \tag{8.42d}$$

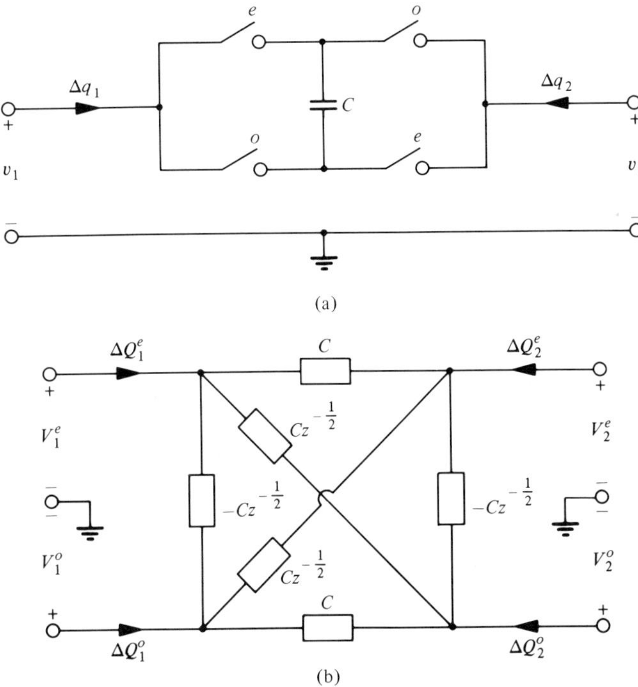

Figure 8.14 (*a*) An SC realization of a bilinear resistor. (*b*) Its *z*-domain equivalent circuit.

There are some interesting modifications of the above SC resistor equivalents. The first one is related to the circuit in Fig. 8.11*a* and is shown in Fig. 8.15*a*. It was first obtained to make the transfer admittance $Cz^{-1/2}$ in Fig. 8.11*b* negative [7]. The charge transfer equations for this SC network are

$$\Delta Q_1^e(z) = CV_1^e(z) + Cz^{-1/2}V_2^o(z) \tag{8.43a}$$

$$\Delta Q_2^e(z) = 0 \tag{8.43b}$$

$$\Delta Q_1^o(z) = 0 \tag{8.43c}$$

$$\Delta Q_2^o(z) = CV_2^o(z) + Cz^{-1/2}V_1^e(z) \tag{8.43d}$$

The above charge transfer equations (8.43) may be represented by the *z*-domain equivalent circuit in Fig. 8.15*b*. Another modified SC circuit has been obtained from the circuit in Fig. 8.12*a* and is shown in Fig. 8.16. It can easily be shown that the charge transfer equations for this circuit are the same as (8.40), and therefore the *z*-domain equivalent circuit for this SC network is the same as the one shown in Fig. 8.12*b*. The modified forms of the SC resistor equivalents of Figs. 8.15 and 8.16 have some practical advantage over those of Figs. 8.11*a* and 8.12*a*. These advantages will be discussed in a later part of this chapter. So far we have considered simple

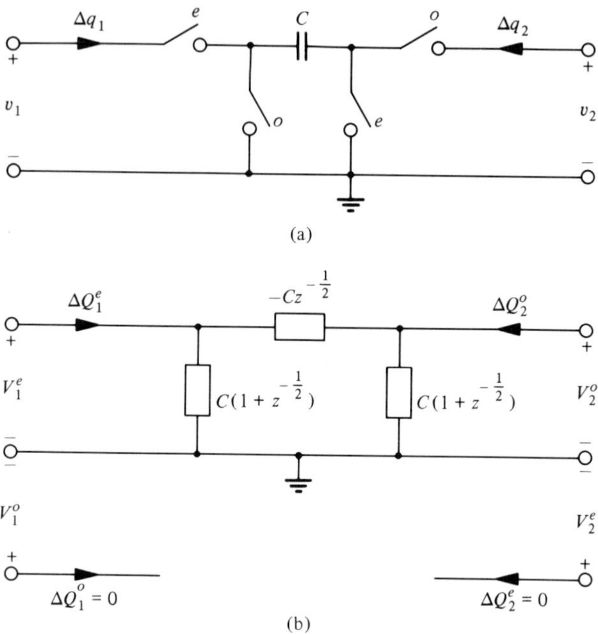

Figure 8.15 (*a*) An SC realization of an inverting PSCR. (*b*) Its *z*-domain equivalent circuit.

SC networks with two ports, which may be considered input and output ports. Another modified circuit can be obtained from the circuit in Fig. 8.15*a*. It is obtained by removing the grounded terminal of one of the *e* switches and applying another voltage to this terminal. Such an arrangement is shown in Fig. 8.17*a*, and its *z*-domain equivalent circuit is shown in Fig. 8.17*b*.

In this section, we have considered only switched capacitor elements, however, SC networks use unswitched elements as well. Some of the most important unswitched elements used in SC networks are grounded and floating capacitors, independent voltage sources, and op amps. The independent voltage sources are assumed to be sampled and held. With this

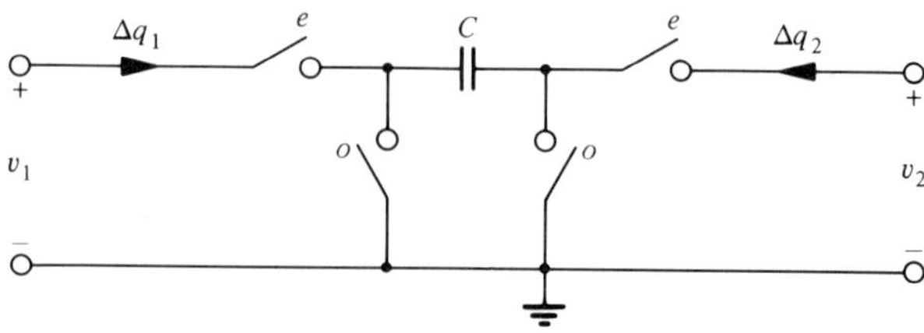

Figure 8.16 Modified form of an SSCR.

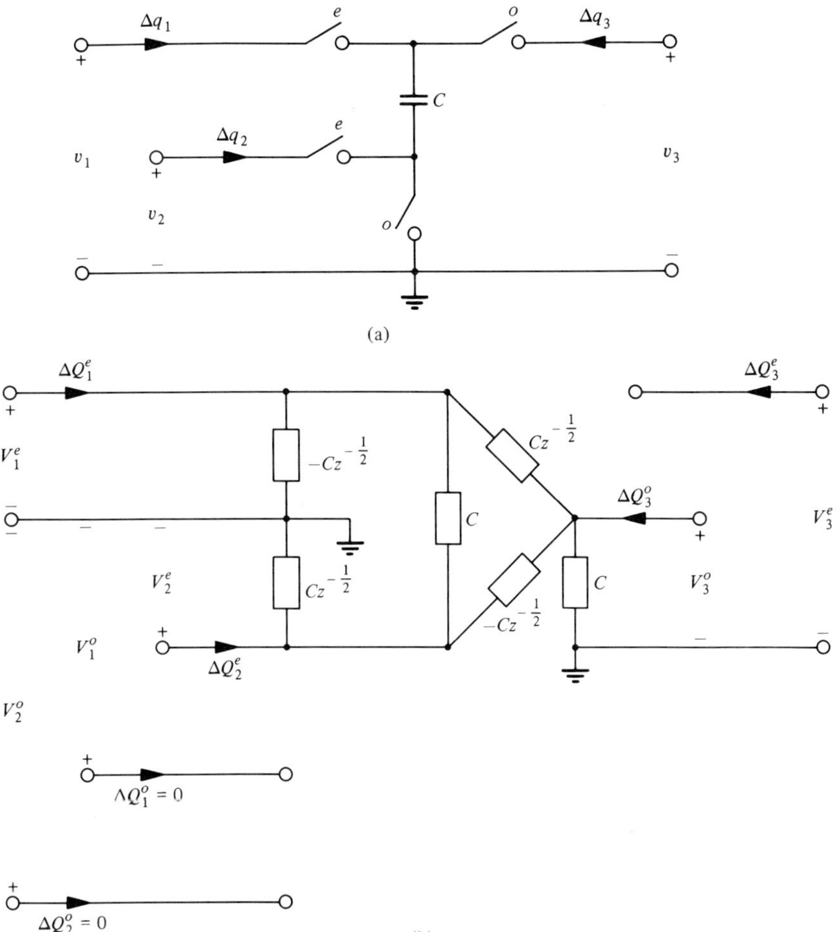

Figure 8.17 (*a*) Differential mode SC resistor. (*b*) Its *z*-domain equivalent circuit.

assumption, even the unswitched elements can be represented in the z domain. Consider the unswitched floating capacitor shown in Fig. 8.18*a*. The charge transfer equations may be written directly in the z domain:

$$\Delta Q_1^e(z) = C\left[V_1^e(z) - V_2^e(z)\right] - Cz^{-1/2}\left[V_1^o(z) - V_2^o(z)\right] \quad (8.44a)$$

$$\Delta Q_1^o(z) = C\left[V_1^o(z) - V_2^o(z)\right] - Cz^{-1/2}\left[V_1^e(z) - V_2^e(z)\right] \quad (8.44b)$$

$$\Delta Q_2^e(z) = C\left[V_2^e(z) - V_1^e(z)\right] - Cz^{-1/2}\left[V_2^o(z) - V_1^o(z)\right] \quad (8.44c)$$

$$\Delta Q_2^o(z) = C\left[V_2^o(z) - V_1^o(z)\right] - Cz^{-1/2}\left[V_2^e(z) - V_1^e(z)\right] \quad (8.44d)$$

From the above equations, one can obtain the z-domain equivalent circuit

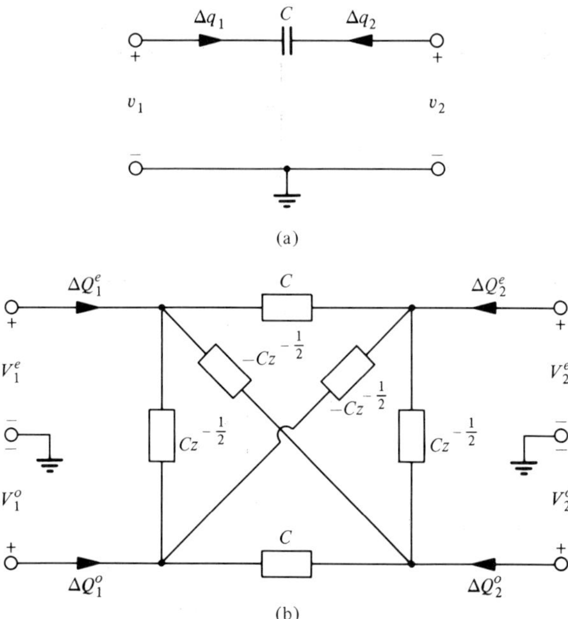

Figure 8.18 (*a*) Unswitched grounded capacitor. (*b*) Its z-domain equivalent circuit.

as shown in Fig. 8.18*b*. The grounded capacitor and its z-domain equivalent circuit are shown in Fig. 8.19. The independent and dependent voltage sources may also be split into their even and odd parts, and therefore their z-domain equivalents will be as shown in Fig. 8.20. In this figure, the z-domain equivalent of the op amp is also shown. In an ideal op amp, ΔQ_1^e, ΔQ_2^e, ΔQ_1^o, and ΔQ_2^o are all equal to zero, and ΔQ_0^e and ΔQ_0^o can be any

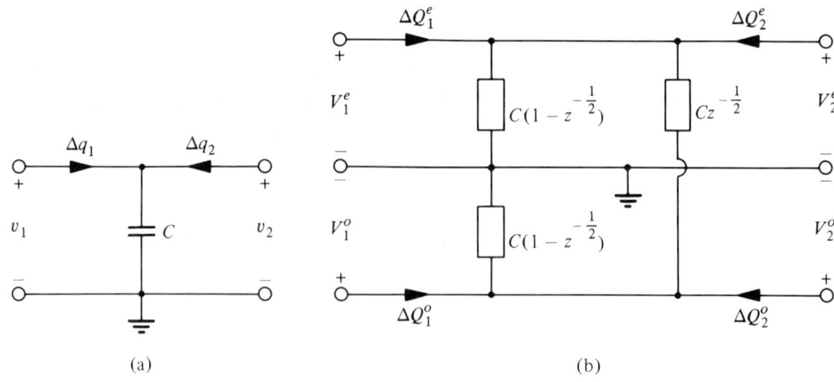

Figure 8.19 (*a*) Unswitched floating capacitor. (*b*) Its z-domain equivalent circuit.

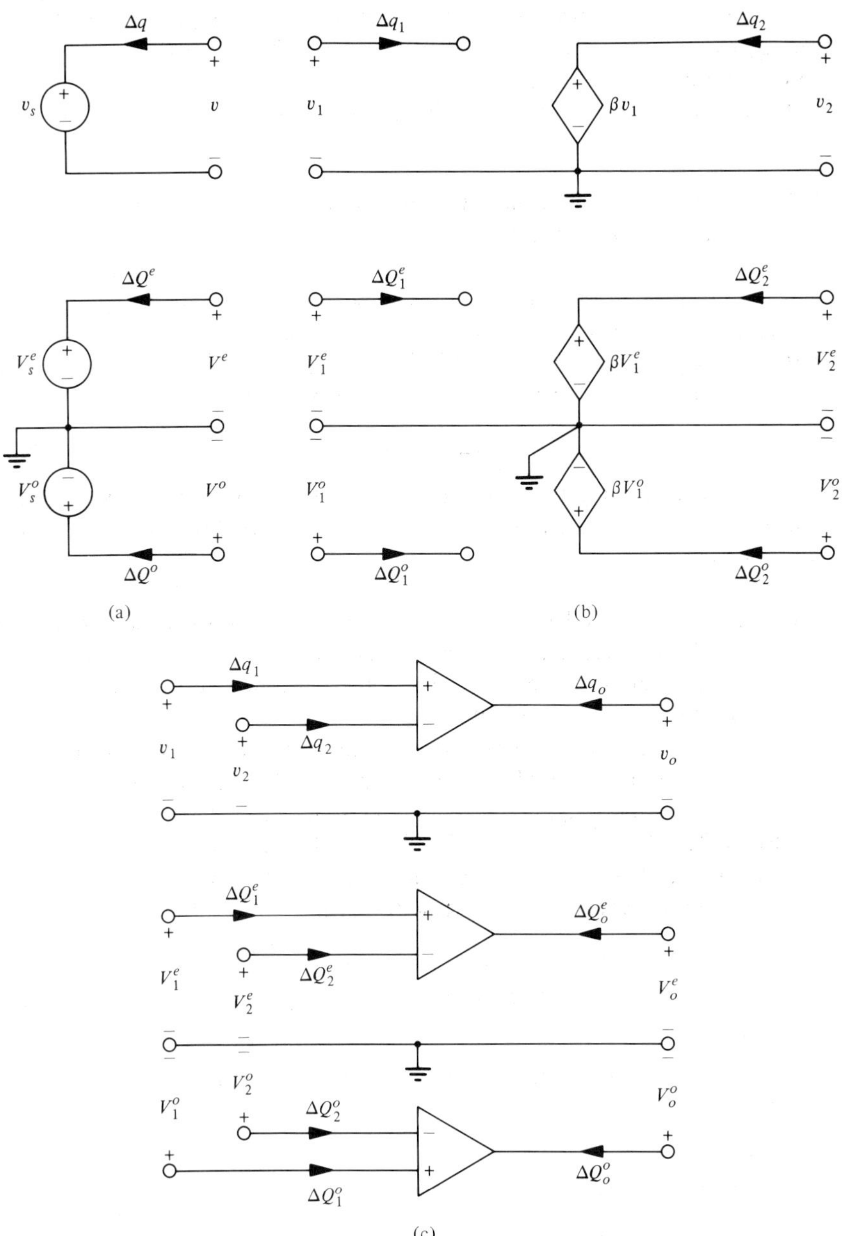

Figure 8.20 Unswitched elements and their corresponding z-domain equivalent circuits. (*a*) Independent voltage source. (*b*) Voltage-controlled voltage source. (*c*) An op amp.

arbitrary quantities because an ideal op amp can supply any amount of energy as required by the rest of the circuit. As shown earlier in this section, one can find the z-domain equivalent circuits of many such simple SC networks. In fact, a library of such equivalent circuits is provided in [5].

8.5 *Analysis of Switched Capacitor Networks Using z-Domain Equivalent Circuits*

The analysis of all SC networks may be carried out using conventional network theory once we replace these networks with their corresponding z-domain equivalent circuits. This is made possible by using the law of conservation of charge (henceforth denoted by QCL),

$$\sum \Delta Q_i(z) = 0 \tag{8.45}$$

at any node and using the definition of admittance given by (8.38). We shall consider some simple examples of SC networks to show how this can be accomplished.

Example 8.3: Consider the simple lowpass RC network shown in Fig. 8.21a. A switched capacitor circuit has been obtained as shown in Fig. 8.21b, replacing the resistor with a PSCR. Find the transfer functions V_o/V_i during all possible phases.

The switched and unswitched capacitors can be replaced by the z-domain equivalents of Figs. 8.11b and 8.19b, respectively. This results in the z-domain equivalent circuit in Fig. 8.22 for the SC network in Fig. 8.21b. Note, from Fig. 8.22, that V_i^o does not have any effect on the outputs, and therefore we need to evaluate only two transfer functions in this case. Applying QCL at nodes A and B, we obtain

$$\begin{aligned} C_1 z^{-1/2}(V_o^e - V_i^e) + Cz^{-1/2}(V_o^e - V_o^o) \\ + (C_1 + C)(1 - z^{-1/2})V_o^e = 0 \end{aligned} \tag{8.46a}$$

and

$$Cz^{-1/2}(V_o^o - V_o^e) + C(1 - z^{-1/2})V_o^o = 0 \tag{8.46b}$$

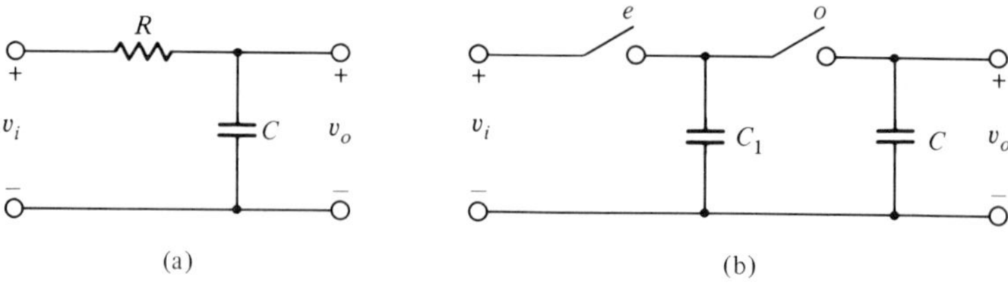

Figure 8.21 Lowpass networks for Example 8.3. (a) An RC network. (b) An SC network.

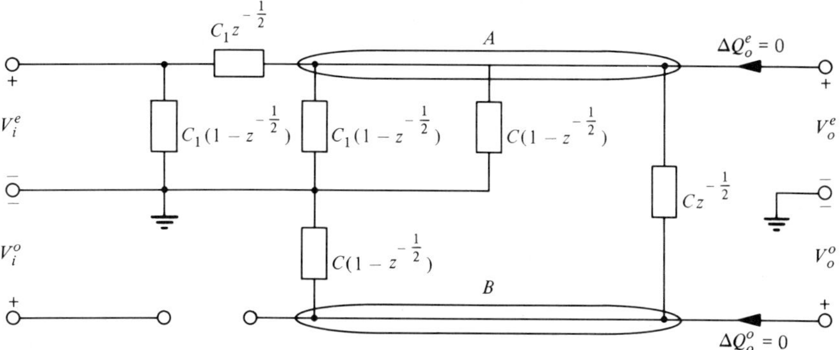

Figure 8.22 The z-domain equivalent circuit of the SC network in Fig. 8.21.

From (8.46b), we find that

$$V_o^o = z^{-1/2} V_o^e \tag{8.47}$$

The above equation is obvious from the SC network also. Substituting (8.47) into (8.46a), we obtain

$$H^{ee}(z) = \frac{V_o^e}{V_i^e} = \frac{az^{-1/2}}{1 - bz^{-1}} \tag{8.48}$$

where $a = C_1/(C_1 + C)(< 1)$ and $b = 1 - a$. The other transfer function of interest is

$$H^{oe}(z) = \frac{V_o^o}{V_i^e} = \frac{az^{-1}}{1 - bz^{-1}} \tag{8.49}$$

Of the two transfer functions for this network, if we sample the output during the even phase, we must use (8.48) to find the output. However, if we sample the output during the odd phase, we must use (8.49). Next, we shall show that this network is indeed a lowpass network. The RC network has the transfer function

$$H(s) = \frac{1}{1 + RCs} \tag{8.50}$$

With the use of any of the transformations discussed earlier, except one, $H^{ee}(z)$ cannot be obtained from (8.50). However, using FD transformation in (8.50), we can obtain the form in (8.49). Thus we apply FD transformation to (8.50) and have

$$H(z) = \frac{T}{RC} \frac{z^{-1}}{1 - (1 - T/RC)z^{-1}} \tag{8.51}$$

Comparing (8.51) with (8.49), we find that $a = T/RC$. In terms of the sampling frequency, $f_s = 1/T$, and the 3-dB frequency of the RC network, $\omega_p = 1/RC$, we have

$$a = \frac{\omega_p}{f_s} \tag{8.52}$$

The above equation is the design equation for the SC lowpass filter circuit in Fig. 8.21. For example, if $f_s = 8$ kHz and $f_p = 1$ kHz, using (8.52), we find that the capacitance ratio C_1/C is equal to 3.66.

Inverting and noninverting integrators play important roles in the design of second- and higher-order active RC filters. We have already considered a SC integrator, as shown in Fig. 8.9. To see that this is an inverting integrator, assume that we apply FD transformation to an analog transfer function $-k/s$ to obtain the form of the transfer function (8.31*b*). This means that, if we sample the output of this circuit during the even phase, it will become a FD integrator. Similarly, if we apply LDI transformation to a similar analog transfer function $-k/s$, we will obtain (8.31*c*). Thus, in the circuit in Fig. 8.9, if we sample the output during the odd phase, the same circuit can be considered a LDI integrator. As discussed earlier, bilinear transformation is better than any of the other transformations. Therefore we consider another SC network, shown in Fig. 8.23*a*, that uses the bilinear SC resistor in Fig. 8.14*a*. The z-domain equivalent of this circuit is also shown in Fig. 8.23*b*. Applying QCL at nodes X and Y, which are at virtual ground potentials, and solving for $V_2^e(z)$ and $V_2^o(z)$, we obtain

$$\begin{bmatrix} V_2^e(z) \\ V_2^o(z) \end{bmatrix} = -\frac{C_1/C}{1-z^{-1}} \begin{bmatrix} 1+z^{-1} & 2z^{-1/2} \\ 2z^{-1/2} & 1+z^{-1} \end{bmatrix} \begin{bmatrix} V_1^e(z) \\ V_1^o(z) \end{bmatrix} \tag{8.53}$$

Two of the above transfer functions are

$$H^{ee}(z) = \frac{-C_1}{C} \frac{1+z^{-1}}{1-z^{-1}} \tag{8.54a}$$

and

$$H^{oo}(z) = \frac{-C_1}{C} \frac{1+z^{-1}}{1-z^{-1}} \tag{8.54b}$$

If the output is sampled at both the even and odd phases, then the transfer function will be

$$H(z) = \frac{V_2^e + V_2^o}{V_1^e + V_1^o} = \frac{-C_1}{C} \frac{1+z^{-1/2}}{1-z^{-1/2}} \tag{8.55}$$

From (8.55), we find that the effective sampling rate has been doubled in

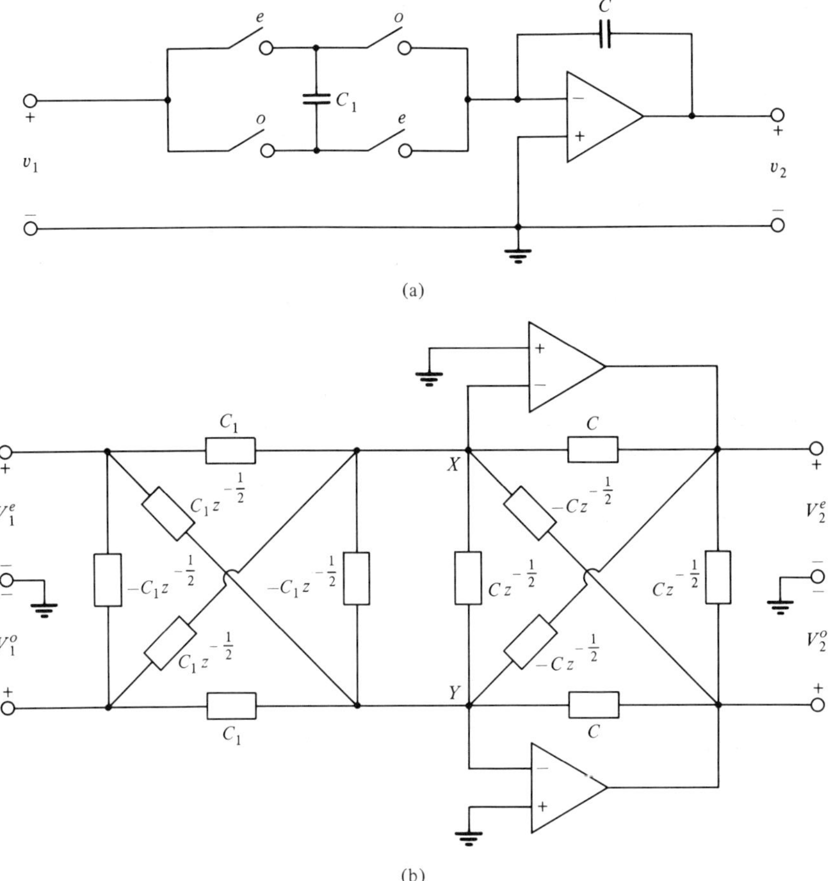

Figure 8.23 (*a*) Bilinear SC integrator. (*b*) Its *z*-domain equivalent circuit.

this circuit. We also find that even-odd symmetry has been achieved. We can define $Z = z^{1/2}$, and then the above equation becomes

$$H(z) = -\frac{C_1}{C}\frac{1 + Z^{-1}}{1 - Z^{-1}} \tag{8.56}$$

It is obvious from (8.56) and (8.54) that this circuit is a bilinear integrator. ■

By replacing the resistors in active *RC* networks with SC resistor equivalents, it is possible to generate a host of SC networks of all possible types. Depending on the type of SC resistor used, a lowpass active *RC* network can be converted to a lowpass SC network using any appropriate transformation. Similarly, all types of filters can be converted to their corresponding SC networks. However, there are some practical considerations to be taken into account that are relevant to SC networks. First, there must be at least one unswitched capacitor in the feedback path of the op amp directly connecting the output to the inverting input of the op amp so that there is continuous negative feedback. Otherwise, the op amp may be driven to saturation or locked into oscillation. Second, every node should be connected through either switched or unswitched capacitors to voltage sources of either type or to ground terminals. The most important consideration is the effect of parasitic capacitances. In MOS capacitors, there is always a nonlinear parasitic capacitance, between the bottom plate and the substrate, on the order of 15 to 20% of the nominal value of the capacitance. If these capacitors are not properly connected, the parasitic capacitance may not only cause errors in the transfer function coefficients but also introduce nonlinear distortion in the signal. To prevent this, the bottom plate of every capacitor should be connected, either directly or through a switch, to the ground terminal, to a virtual ground terminal, or to a voltage source. We shall soon consider the effects of other (but small) parasitic capacitances. Next, to avoid spurious noise voltages and sensitivity due to parasitic capacitances, it is preferable to have the noninverting input terminal of the op amp connected to a constant potential. If all the above practical considerations are applied to SC networks, all (including the bilinear integrator) but a few SC network topologies will be found unsuitable for high-quality stringent applications.

In addition to bottom plate parasitic capacitance, there is a parasitic capacitance, on the order of 0.1 to 1% of the nominal value, from the top plate to the ground. The switches are also realized using MOS transistors, and there are interelectrode capacitances in these transistors also. The parasitic capacitances associated with analog switches must also be taken into account. Consider the SC integrator circuit in Fig. 8.9. In this circuit the effects of the bottom plate parasitic capacitances have been completely eliminated. However, the effects of the top plate parasitic capacitances are still present. Consider a circuit, reproduced in Fig. 8.24, where the switches

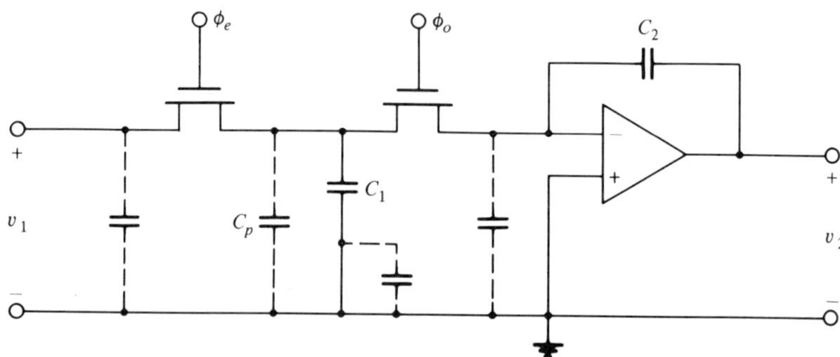

Figure 8.24 Inverting SC integrator with parasitic capacitors.

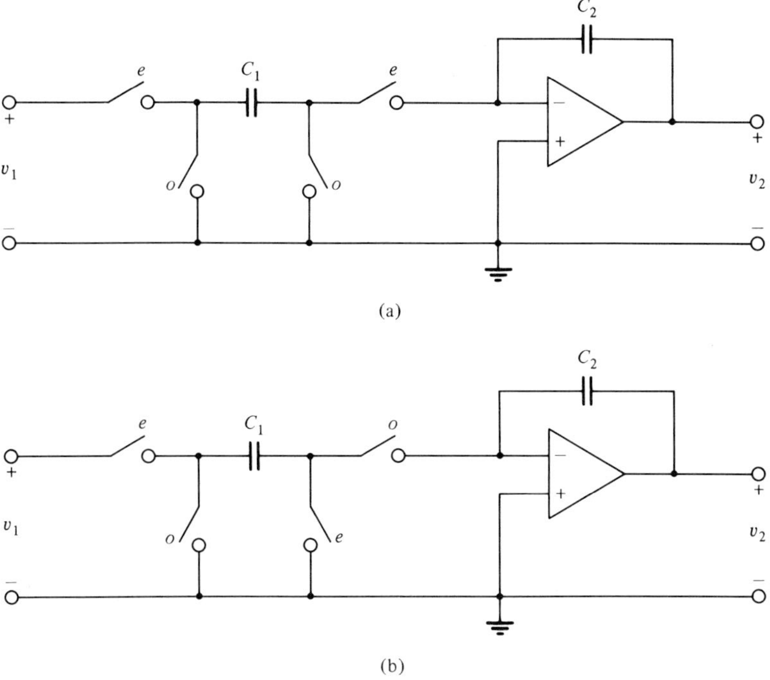

Figure 8.25 Pair of stray-insensitive integrators. (*a*) Inverting. (*b*) Noninverting.

have been replaced with MOS transistors and all possible parasitic capacitances, except gate-to-drain and gate-to-source capacitances have been included. None of the parasitic capacitances, except C_p, affects the charge transfer and therefore the overall transfer function. Here C_p includes the top plate parasitic capacitances and two of the analog switch capacitances. The transfer functions of this circuit, of course, can be derived as

$$H^{ee}(z) = \frac{-(C_1 + C_p)}{C_2} \frac{z^{-1/2}}{z^{1/2} - z^{-1/2}} \tag{8.57a}$$

and

$$H^{eo}(z) = \frac{-(C_1 + C_p)}{C_2} \frac{1}{z^{1/2} - z^{-1/2}} \tag{8.57b}$$

In precision applications, the errors in the magnitude and phase of the transfer functions are objectionable, and therefore such circuits are not suitable. Researchers have developed SC integrator circuits in which charge transfer equations are not affected by parasitic capacitances, and their operations are stray-insensitive. Two such circuits are shown in Fig. 8.25. The first (Fig. 8.25*a*) is an inverting integrator, and the second (Fig. 8.25*b*) is a noninverting integrator. Note that these two integrator circuits use the

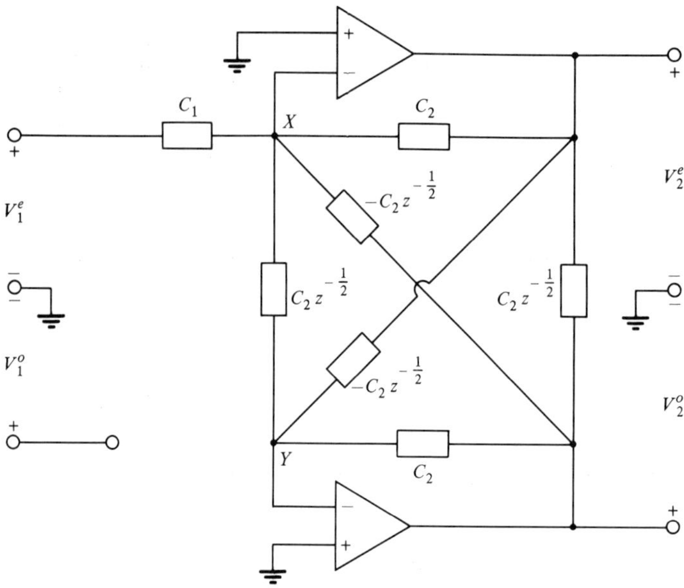

Figure 8.26 The z-domain equivalent circuit for the inverting integrator in Fig. 8.25a.

SC resistor equivalents of Figs. 8.16a and 8.15a, respectively. It may be shown that, in these circuits, the transfer functions are totally independent of stray capacitances between any node and ground, since the charge transfer to C_2 is not affected by these capacitances [8]. Because of these desirable properties most practical SC realizations of filter functions use these two integrator circuits, or circuits derived from them, as basic building blocks. Although SC networks have been developed using other forms of SC resistors, they have been found to be unsuitable for precision applications. The z-domain equivalent of the inverting integrator circuit in Fig. 8.25a is shown in Fig. 8.26. The transfer functions of this circuit can be derived, by using QCL at the nodes X and Y, as

$$H^{oe}(z) = \frac{V_2^o(z)}{V_1^e(z)} = \frac{-C_1}{C_2} \frac{z^{-1/2}}{1 - z^{-1}} \tag{8.58a}$$

and

$$H^{ee}(z) = \frac{V_2^e(z)}{V_1^e(z)} = \frac{-C_1}{C_2} \frac{1}{1 - z^{-1}} \tag{8.58b}$$

The z-domain equivalent of the noninverting integrator circuit in Fig. 8.25b is shown in Fig. 8.27, from which the transfer functions of this circuit may

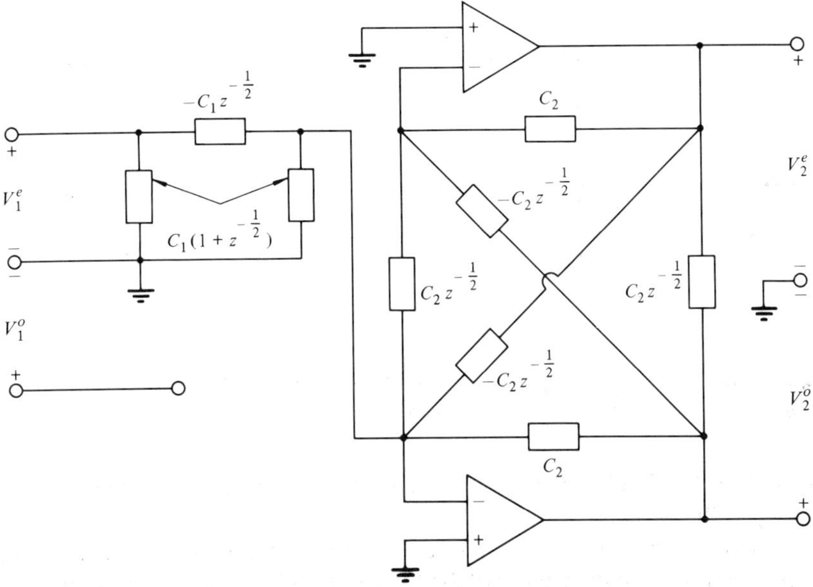

Figure 8.27 The z-domain equivalent circuit for the noninverting integrator in Fig. 8.25b.

be derived as

$$H^{oe}(z) = \frac{V_2^o(z)}{V_1^e(z)} = \frac{C_1}{C_2}\frac{z^{-1/2}}{1 - z^{-1}} \tag{8.59a}$$

and

$$H^{ee}(z) = \frac{V_2^e(z)}{V_1^e(z)} = \frac{C_1}{C_2}\frac{1}{1 - z^{-1}} \tag{8.59b}$$

Let us next consider the situation where these integrator circuits are used as basic building blocks. These building blocks are driven by voltage sources and op amps, and in addition, we are interested in determining the transfer function of the overall circuit. Even if we are interested in finding some internal variables, they will be outputs of the op amps. In such cases the block diagram or signal flow graph representation of these circuits will be most useful. This representation of the overall circuit can be obtained in a matter of minutes by using signal flow graph representations of the basic building blocks, since the loading problem does not arise in these cases. Thus the transfer function of the overall network can be obtained quickly by using signal flow graph techniques or by writing some simple algebraic equations. The signal flow graph representations of the two integrators are shown in Fig. 8.28.

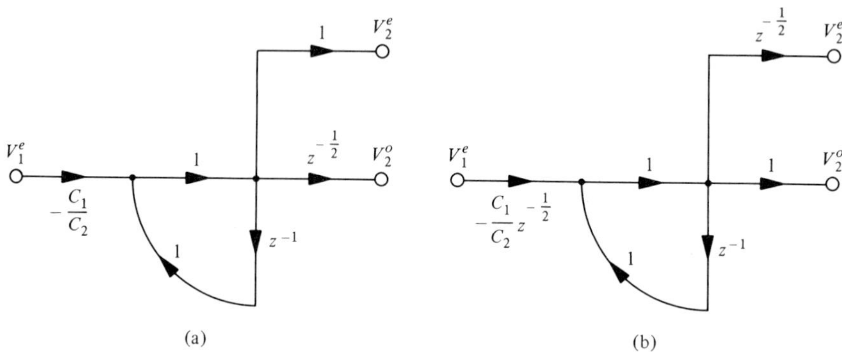

Figure 8.28 Signal flow graphs of the integrators in Fig. 8.25. (*a*) For the inverting integrators. (*b*) For the noninverting integrators.

Example 8.4: Find the transfer functions of the two lossy integrators in Fig. 8.29.

Considering the inverting lossy integrator in Fig. 8.29*a*, we find that the circuit, excluding the switched capacitor C_3, is the inverting integrator in Fig. 8.25*a*. Now, when we include this capacitor C_3, we find that the even phase component of the output is fed back into the inverting input terminal of the op amp. Using this fact and the signal flow graph in Fig. 8.28*a*, we obtain the complete signal flow graph representation of the lossy integrator in Fig. 8.29*a* as shown in Fig. 8.30*a*. Similar arguments lead to the signal flow graph representation in Fig. 8.30*b* for the lossy integrator in Fig. 8.29*b*. The transfer functions of the two lossy integrators can be obtained from these signal flow graphs and are

Inverting integrator in Fig. 8.29*a*:

$$H^{ee}(z) = \frac{V_2^e}{V_1^e} = \frac{-C_1/(C_2 + C_3)}{1 - C_2 z^{-1}/(C_2 + C_3)} \tag{8.60a}$$

and

$$H^{oe}(z) = \frac{V_2^o}{V_1^e} = \frac{-C_1/(C_2 + C_3)z^{-1/2}}{1 - C_2 z^{-1}/(C_2 + C_3)} \tag{8.60b}$$

Noninverting integrator in Fig. 8.29*b*:

$$H^{ee}(z) = \frac{V_2^e}{V_1^e} = \frac{C_1 z^{-1}/(C_2 + C_3)}{1 - C_2 z^{-1}/(C_2 + C_3)} \tag{8.61a}$$

and

$$H^{oe}(z) = \frac{V_2^o}{V_1^e} = \frac{C_1/(C_2 + C_3)z^{-1/2}}{1 - C_2 z^{-1}/(C_2 + C_3)} \tag{8.61b}$$

■

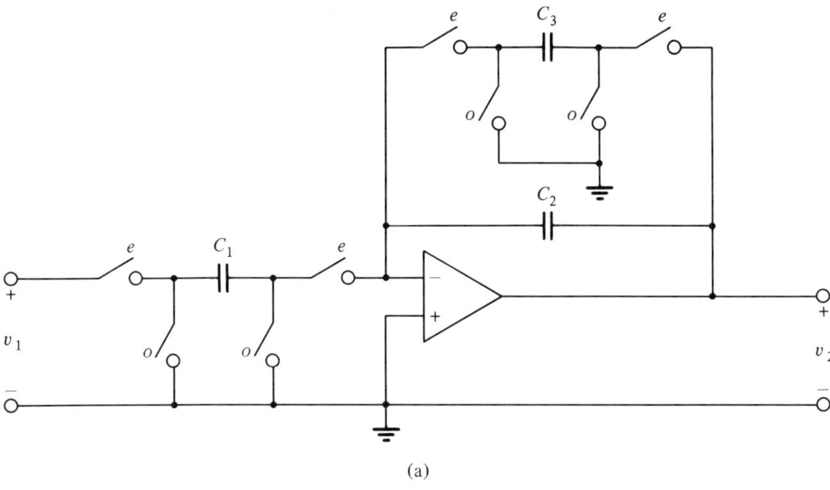

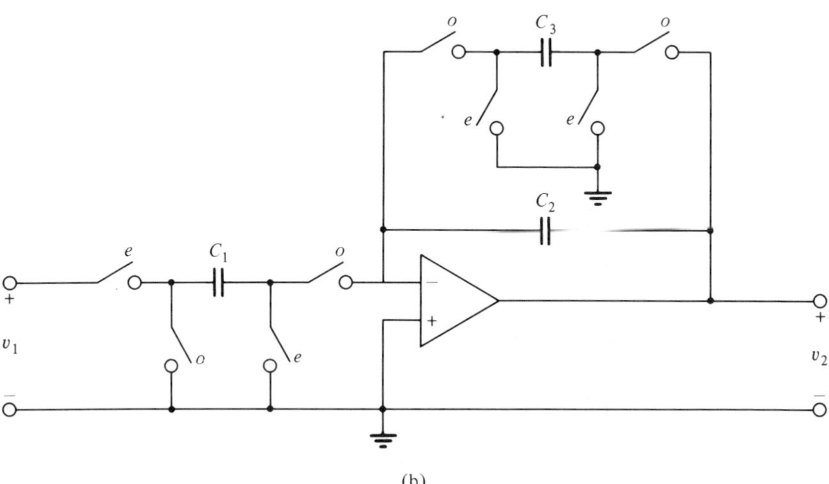

Figure 8.29 Stray-insensitive lossy integrators. (*a*) Inverting. (*b*) Noninverting.

The two circuits discussed in Example 8.4 act as first-order lowpass filters when the sampling frequency is much higher than the maximum frequency of interest. Any one of the two networks can be used as a lowpass filter. Except for the additional linear phase and, perhaps, another 180° phase shift in the case of inverting transfer functions, every form of the transfer function provides the same magnitude response. Thus consider

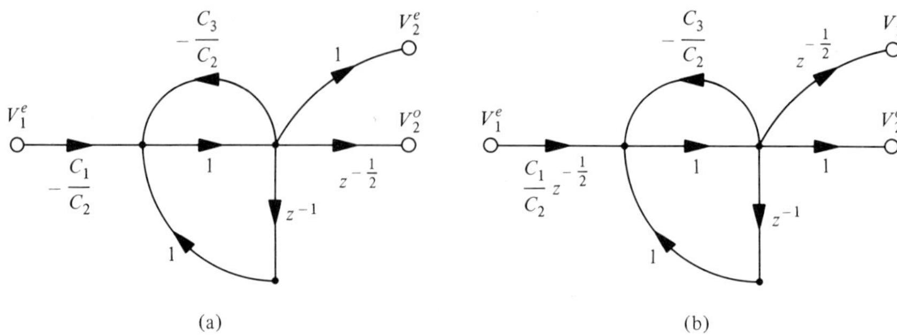

Figure 8.30 Signal flow graphs for the lossy integrators in Fig. 8.29. (*a*) For the inverting integrators. (*b*) For the noninverting integrators.

the transfer function (8.61*a*):

$$H^{ee}(z) = \frac{a}{z - b} \tag{8.62}$$

where $a = C_1/(C_2 + C_3)$ and $b = C_2/(C_2 + C_3)$.

When $\omega T \ll 1$, using $\cos(\omega T) \simeq 1$ and $\sin(\omega T) \simeq \omega T$, we have

$$H^{ee}(e^{j\omega T}) \simeq \frac{a/(1 - b)}{1 + j\omega T/(1 - b)} \tag{8.63}$$

It is clear from (8.63) that these circuits behave as lowpass filters whose dc gain is $a/(1 - b)$, and the 3-dB frequency is approximately equal to $(1 - b)/T$. The design of such lowpass filters is illustrated in the following example.

Example 8.5: Design a first-order lowpass filter with a cutoff frequency of 4 kHz and with a dc gain of 10. The sampling frequency is 128 kHz.

Let us first obtain the values of a and b using the approximate expression for the 3-dB frequency. Thus we have

$$\frac{(1 - b)}{T} = \omega_p = 8000\pi$$

Solving the above equation, we find that $b = 0.8037$. Equating the dc gain to $a/(1 - b)$, we obtain $a = 1.9635$. Using these values of a and b, we find that $C_3/C_2 = 0.2443$ and $C_1/C_2 = 2.443$.

A more accurate design is possible. Substituting $z = e^{j\omega T}$ into (8.62), we find that

$$H^{ee}(e^{j\omega T}) = \frac{a}{[\cos(\omega T) - b] + j\sin(\omega T)}$$

and

$$|H^{ee}(e^{j\omega T})|^2 = \frac{a^2}{1 + b^2 - 2b\cos(\omega T)}$$

The dc gain is still same as before, namely, $a/(1 - b)$, and therefore $a^2 = 100(1 - b)^2$. At the 3-dB frequency, $\omega = 8000\pi$, the magnitude squared function relative to its dc value must be equal to 0.5, and we also obtain another equation:

$$\frac{(1 - b)^2}{1 + b^2 - 2b\cos(\pi/16)} = 0.5$$

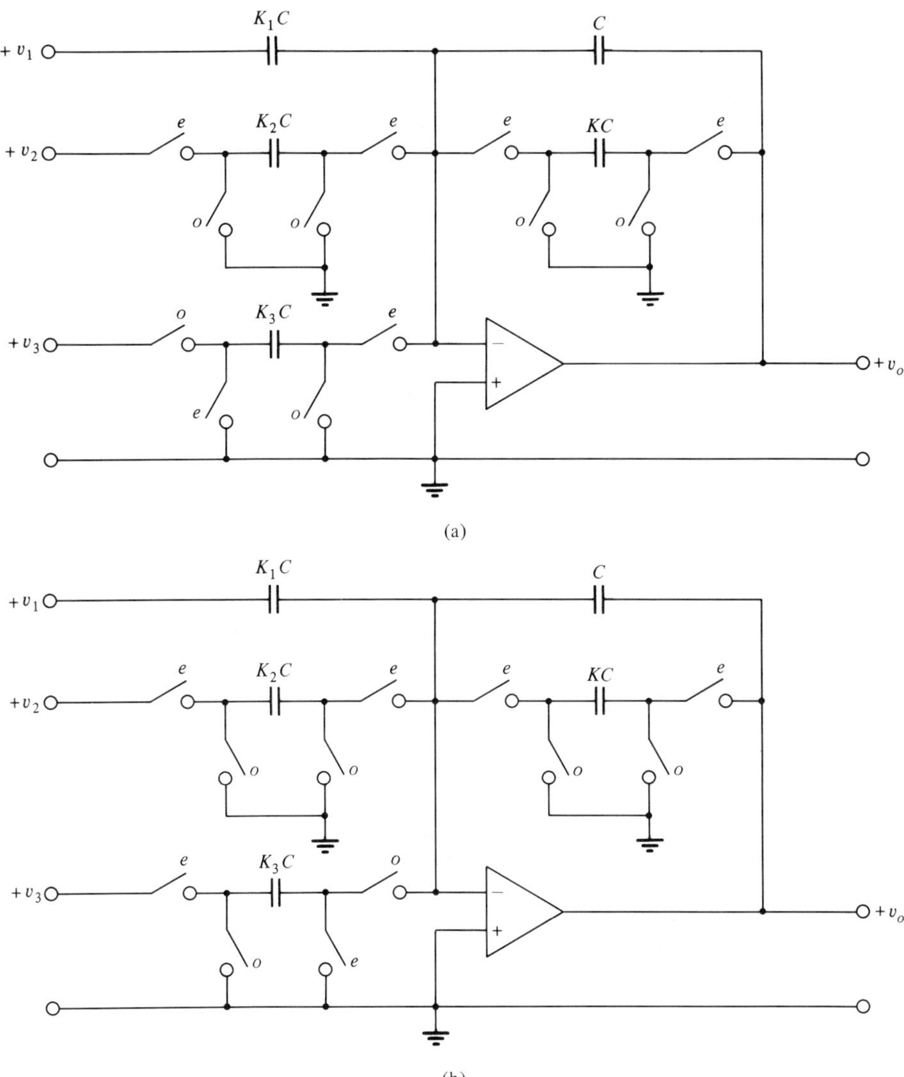

Figure 8.31 Two stray-insensitive building blocks.

The solution of the above equation gives two values for b, and we must choose the one that is less than unity for stability reasons. Thus we find that b is equal to 0.82224. Using this value of b in the equation for a, we find that $a = 1.7776$. From these values of a and b, we find the capacitance ratios to be

$$\frac{C_3}{C_2} = 0.2162 \quad \text{and} \quad \frac{C_1}{C_2} = 2.162$$ ■

The final SC networks we shall consider in this section are two building blocks used in developing second- and higher-order filters. These networks are shown in Fig. 8.31. In these circuits, v_1, v_2, and v_3 are the input signals and v_o is the output signal. After replacing each subnetwork with its z-domain equivalents, we can find the describing equations with the use of QCL. For the circuit in Fig. 8.31a, the describing equations are

$$V_o^e = -KV_o^e + z^{-1}V_o^e - K_1(1 - z^{-1})V_1^e - K_2V_2^e + K_3z^{-0.5}V_3^o \quad (8.64a)$$

and

$$V_o^o = z^{-0.5}V_o^e - K_1V_1^o + K_1z^{-0.5}V_1^e \quad (8.64b)$$

For the circuit in Fig. 8.31b, the describing equations can be obtained as

$$V_o^e = -KV_o^e + z^{-1}V_o^e - K_1(1 - z^{-1})V_1^e - K_2V_2^e + K_3z^{-1}V_3^e \quad (8.65a)$$

and

$$V_o^o = z^{-0.5}V_o^e - K_1V_1^o + K_1z^{-0.5}V_1^e + K_3z^{-0.5}V_3^e \quad (8.65b)$$

The above equations describing the building blocks in Fig. 8.31a and b can be represented in the form of signal flow diagrams, and these are provided in Figs. 8.32a and b, respectively.

8.6 *Second-Order Switched Capacitor Filters*

In the previous section we considered the analysis of SC networks using z-domain equivalent circuits. In this process we considered some stray-insensitive SC networks that are important in practice. In order to simplify the analysis of the networks we consider in this and the following sections, we also obtained simplified flow graph representations of these building blocks. In this section we shall see how these building blocks can be used in developing second-order filters.

Theoretically it is possible to generate many second-order filters by replacing the resistors in active RC filters with their SC resistor equivalents. However, in order that they be useful in practice, they must also be

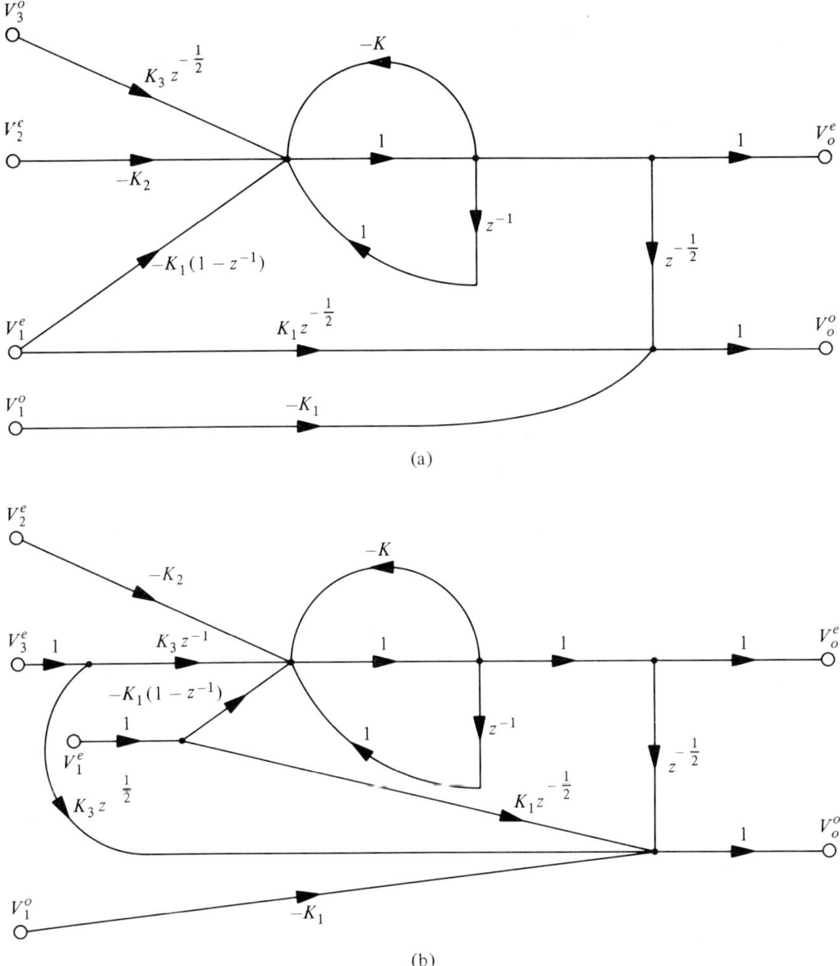

Figure 8.32 Signal flow graphs for the two building blocks in Fig. 8.31. (*a*) For the SC circuit in Fig. 8.31*a*. (*b*) For the SC circuit in Fig. 8.31*b*.

stray-insensitive. Furthermore, as may be observed from the previous section, the ratios of the capacitance values determine the transfer function coefficients. Therefore the absolute values of the capacitances may be very small, on the order of a few picofarads. It is well known that, in IC technology, the smaller the capacitance value, the smaller the area required for fabricating it. Thus, with the use of small capacitance values, the area of the chip can be reduced considerably. Thus there is no necessity for limiting the number of capacitors used in SC networks, as in active RC networks, to two capacitors per second-order section (canonical realizations). Also, MOS op amps consume very little power as compared to op amps fabricated using bipolar technology, and therefore, limiting the number of op amps used to a minimum (namely, one op amp) does not mean much in SC networks. From this discussion, we conclude that one of the best ways to realize second-order (as well as higher-order) filters is to use the stray-insensitive building blocks of the previous sections which are just integrators in which a summing action can also be achieved simultaneously. Especially, when the capacitance values are very small, insensitivity to stray capacitances is very important in order for the coefficients of the transfer functions to be closely controlled and the sensitivity of the overall filter to be kept low. The obvious method is to use these stray-insensitive integrators in a double-integrator loop similar to the ones considered in Chap. 6. At this juncture, note that both noninverting and inverting integrators can be realized using a single op amp where the noninverting input terminals can be grounded. Thus a SC double-integrator loop can be realized using two op amps. The arbitrary zeros of transmission can be realized using a feedforward technique.

There are many possible structures that realize general second-order z-domain transfer functions [9–13]. The references cited here are some of the earlier contributions. Assume that we want to realize a prewarped second-order analog transfer function. Thus consider the general second-order analog transfer function

$$H_A(s) = \frac{a_2 s^2 + a_1 s + a_0}{s^2 + b_1 s + b_0} \tag{8.66}$$

Applying bilinear transformation to the above transfer function, we obtain the pulse transfer function

$$H(z) = \frac{n_0 + n_1 z^{-1} + n_2 z^{-2}}{d_0 + d_1 z^{-1} + d_2 z^{-2}} \tag{8.67}$$

where

$$n_0 = a_2 + \frac{a_1 T}{2} + \frac{a_0 T^2}{4}$$

$$n_1 = -2a_2 + \frac{2a_0 T^2}{4}$$

$$n_2 = a_2 - \frac{a_1 T}{2} + \frac{a_0 T^2}{4}$$

$$d_0 = 1 + \frac{b_1 T}{2} + \frac{b_0 T^2}{4}$$

$$d_1 = -2\left(1 - \frac{b_0 T^2}{4}\right)$$

and

$$d_2 = 1 - \frac{b_1 T}{2} + \frac{b_0 T^2}{4}$$

A general stray-insensitive SC biquad [12] for realizing second-order transfer functions of the form of (8.67) is shown in Fig. 8.33*a*. The zeros of transmission are realized by the capacitors K_1C_1, K_2C_1, M_2C_2, and M_3C_2. An efficient realization of the same network is shown in Fig. 8.33*b*. In this network similarly switched capacitors are connected with a single switch, and thus many switches can be omitted. In the circuit in Fig. 8.33*a*, 28 switches are required. When sharing of switches is allowed, as in the network in Fig. 8.33*b*, we need only 12 switches, and thus there is a considerable saving. This also means that we can save the same number of switching transistors and that the total area of the chip will also be reduced. The stray-insensitive nature of the original network is not destroyed by such switch sharing. The flow graph representation of the SC network in Fig. 8.33 can be obtained from the flow graphs of the basic building blocks discussed in the previous section and is shown in Fig. 8.34. Two op amp outputs are available—V_1 and V_2. In both cases we find that

$$V_1^o = z^{-0.5} V_1^e \qquad \text{and} \qquad V_2^o = z^{-0.5} V_2^e \tag{8.68}$$

Thus the even components of both these outputs are sufficient to describe the complete outputs of both V_1 and V_2. Sampling the even components and holding the output over one full clock cycle, we can obtain the complete output. For now, let us also assume that the input is sampled and held over one full clock period. This means that

$$V_{\text{in}}^o = z^{-0.5} V_{\text{in}}^e \tag{8.69}$$

If (8.68) and (8.69) hold true, then one transfer function relating the even input and output components is sufficient for each output. From the flow graph in Fig. 8.34, we can find the transfer functions

$$H_1^{ee} = \frac{V_1^e}{V_{\text{in}}^e} = \frac{\begin{array}{r}[-K_1(1+M) + M_2(K_3+K_4)] + z^{-1}[K_1 + K_2(1+M) \\ -M_3(K_3+K_4) - M_2K_4] + z^{-2}(K_4M_3 - K_2)\end{array}}{(1+M) - z^{-1}[2 + M - M_1(K_4+K_3)] + z^{-2}(1 - M_1K_4)} \tag{8.70}$$

$$H_2^{ee} = \frac{V_2^e}{V_{\text{in}}^e} = \frac{-[M_2 - z^{-1}(-K_1M_1 + M_2 + M_3) + z^{-2}(-M_1K_2 + M_3)]}{(1+M) - z^{-1}[2 + M - M_1(K_4+K_3)] + z^{-2}(1 - M_1K_4)} \tag{8.71}$$

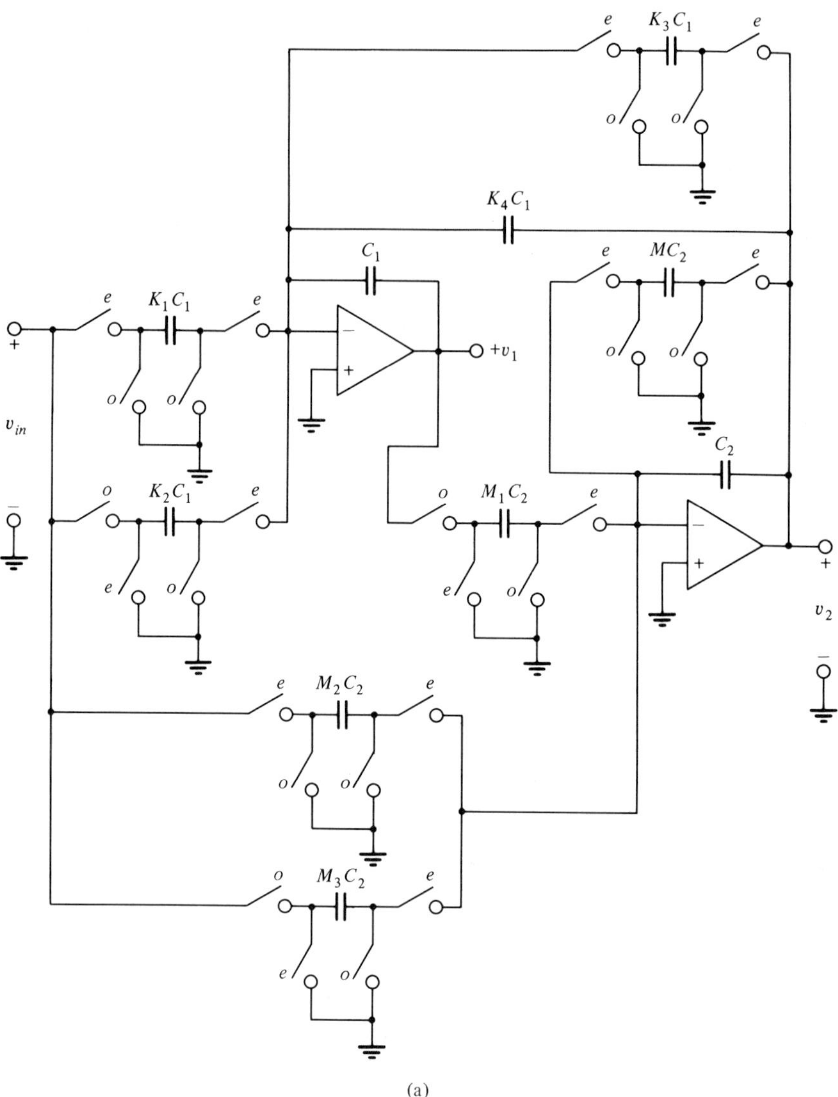

(a)

Figure 8.33 (*a*) General stray-insensitive SC biquad (type 1). (*b*) Its minimum switch configuration.

Note that (8.70) gives a noninverting transfer function, while (8.71) gives an inverting one. The poles are controlled by the capacitance ratios M, K_3, and K_4, which are capacitor ratios in the feedback paths. The zeros of transmission are mainly controlled by the capacitance ratios in the feedforward paths in both transfer functions. There are six coefficients to be determined, and there are eight capacitance ratios to choose from. Thus two

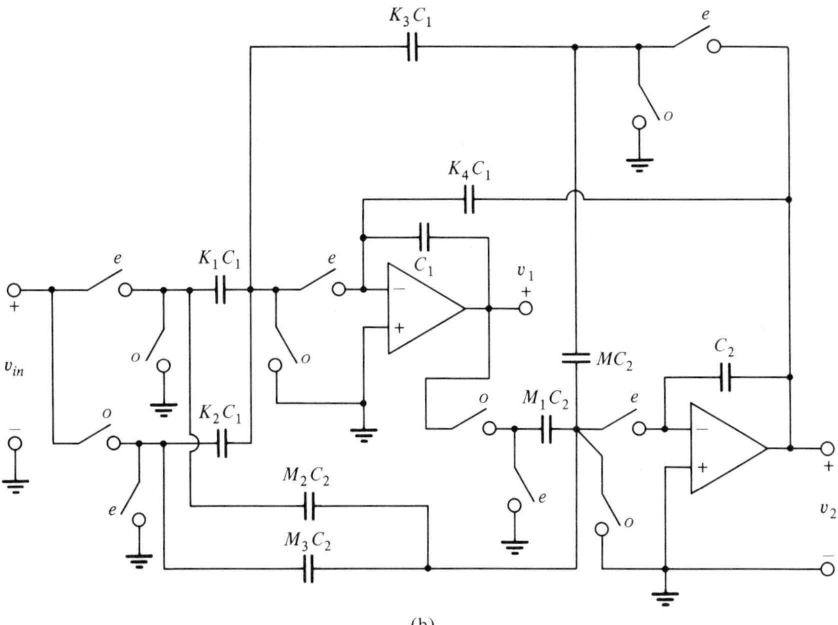

(b)

Figure 8.33 Continued.

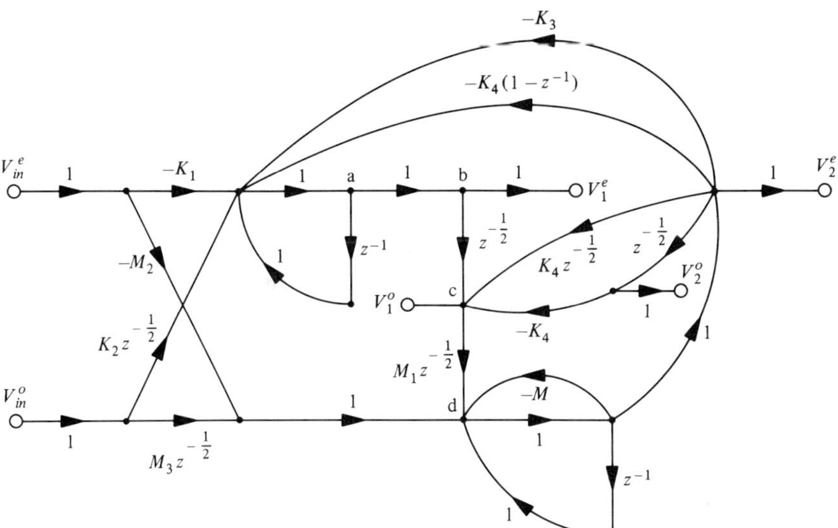

Figure 8.34 Signal flow graph for the SC biquad in Fig. 8.33.

capacitance ratios can be chosen arbitrarily. One of the convenient arbitrary choices is $M_1 = 1$. Both K_4 and M provide the damping in this circuit, and therefore another convenient choice is either $M = 0$ or $K_4 = 0$. Fleisher and Laker, the authors of this circuit, call it an *E circuit* if $M = 0$ and a *F circuit* if $K_4 = 0$. Therefore four possible designs may be carried out using this circuit. There is no basic, definite procedure for designing an arbitrary transfer function of the form of (8.67). After selecting an appropriate transfer function and the desired output, one equates the coefficients of the transfer function in either (8.70) or (8.71) with known coefficients in (8.67). Then, one can solve for the remaining six capacitor ratios. To see that these circuits are practically useful, the final step is to adjust the capacitor ratios to maximize the dynamic range for the best performance but without affecting the realized transfer function. Before we consider a numerical design example, we shall illustrate the method of adjusting the capacitor ratios for best dynamic range.

The best dynamic range can be achieved when the maximum outputs of both V_1^e and V_2^e are equal to the maximum voltage the op amp can handle, say V_{amp}. To adjust the voltage level of V_1 without affecting the transfer function H_2^{ee}, we multiply H_1^{ee} by an appropriate constant so that H_1^{ee} becomes, say KH_1^{ee}. Without loss of generality, assume that $K > 1$. This means that we want to increase the signal level of V_1^e. To achieve this, we divide the capacitance value of C_1 (the feedback capacitor in op amp A_1) by a factor of K, and to keep the signal at the same level as it was before, we also decrease the capacitance value of M_1C_2 to M_1C_2/K. This is equivalent to multiplying the unit transmittance branch between a and b by K and dividing the branch transmittance of $M_1z^{-0.5}$ by the same constant K in the signal flow graph in Fig. 8.34. Note that the transfer function H_2^{ee} is not affected by this procedure. Thus, if we want

$$H_1^{ee} \rightarrow KH_1^{ee}$$

we simply scale the capacitance values of the capacitors connected to the output of op amp A_1. In this case, we scale the capacitances as follows:

$$(C_1, M_1C_2) \rightarrow \left(\frac{C_1}{K}, \frac{M_1C_2}{K}\right)$$

Similarly, if we want to adjust V_2^e without affecting V_1^e, such that

$$H_2^{ee} \rightarrow TH_2^{ee}$$

we scale the capacitances connected to the output of the op amp A_2, which are C_2, MC_2, K_3C_1, and K_4C_1. Thus we have

$$(C_2, MC_2, K_3C_1, K_4C_1) \rightarrow \left(\frac{C_2}{T}, \frac{MC_2}{T}, \frac{K_3C_1}{T}, \frac{K_4C_1}{T}\right)$$

At this point, another practical point must also be considered. To minimize mismatch errors and improve capacitor ratio accuracy, capacitors are often made up of arrays of identical *unit geometry capacitors* [14]. Each unit geometry capacitor has a value of anywhere between 0.5 and 2 pF. Then, in order to obtain the appropriate capacitance value (> 1), the unit geometry capacitors are interconnected in parallel. In our circuit, the capacitance with the least value can be a single unit geometry capacitance, and the values of all other capacitors can be adjusted accordingly. In fact, in our circuit, there are two groups of capacitors whose values can independently be adjusted: $(C_1, K_1C_1, K_2C_1, K_3C_1, K_4C_1)$ and $(C_2, MC_2, M_1C_2, M_2C_2, M_3C_2)$. Note that each group of capacitors is connected to the inputs of the op amps A_1 and A_2, respectively. By scaling both groups with a specific constant, both transfer functions will be unaffected because the transfer functions depend on the ratios of the capacitors. In fact, we have already used this scaling for the circuits in Fig. 8.33. Next we shall consider the design of a second-order lowpass notch function.

Example 8.6: Consider the prewarped analog transfer function

$$H_A(s) = \frac{H_B(s^2 + a_0)}{s^2 + b_1 s + b_0}$$

where $a_0 = 158.1677 \times 10^6$, $b_0 = 128.0767 \times 10^6$, and $b_1 = 566.6$. Assume that the constant H_B is chosen to give a unity gain at zero frequency. That is, $H_B = b_0/a_0 = 0.8097526$. Also, assume that the sampling frequency is 128 kHz. Find the pulse transfer function using bilinear transformation. Realize such a transfer function using the circuit in Fig. 8.33 with (a) $M = 0$ (an E circuit) and (b) $K_4 = 0$ (a F circuit). Both circuits are to be realized using the output of V_2.

Bilinear transformation will result in a pulse transfer function of the form of (8.67), where the coefficients are $n_2 = n_0 = 0.8083381$, $n_1 = -1.608891$, $d_0 = 1$, $d_1 = -1.987807$, and $d_2 = 0.9955919$.

(a) Design of the E Circuit ($M = 0$): Comparing the coefficients in (8.71) with those of the transfer function to be realized, and with $M = 0$ and $M_1 = 1$, we have

$$2 - (K_3 + K_4) = 1.987807$$

$$1 - K_4 = 0.9955919$$

$$M_2 = 0.8083381$$

$$M_2 + M_3 - K_1 = 1.608891$$

and

$$M_3 - K_2 = 0.8083381$$

Note that we will be realizing an inverting transfer function. There are six parameters to choose from only five equations. In fact, one of the three parameters K_1, M_3, or K_2 may be selected arbitrarily. If the arbitrary parameter is chosen as zero, the need for one capacitor will be eliminated. The only choice that can be made in this way, and that makes all the capacitance values positive, is $K_2 = 0$. Then, we obtain $K_1 = 0.007784738$, $K_2 = 0$, $K_3 = K_1$, $K_4 = 0.004408129$, $M = 0$, $M_1 = 1$, and $M_2 = M_3 = 0.8083381$. The values of C_1 and C_2 can be chosen as unity. This will be the initial design of the E circuit. With the above element values, we find that $|H_1^{ee}|_{\max} = -9.3$ dB and $|H_2^{ee}|_{\max} = 11.595$ dB. Therefore, in order to increase the signal level of $|V_1^e|$ without affecting $|V_2^e|$, we divide the values of C_1 and M_1C_2 by a factor of $10^{(11.595+9.3)/20} = 11.08537$. Dividing the capacitance value C_1 by 11.08537 is equivalent to multiplying all the K values by the same factor. Thus the dynamically adjusted values are $K_1 = 0.0862966$, $K_2 = 0$, $K_3 = K_1$, $K_4 = 0.04886572$, $M = 0$, $M_1 = 0.0920903$, and $M_2 = M_3 = 0.8083381$. Again, the values of C_1 and C_2 are unity. Rescaling the set of capacitances connected to the inverting input terminals of each op

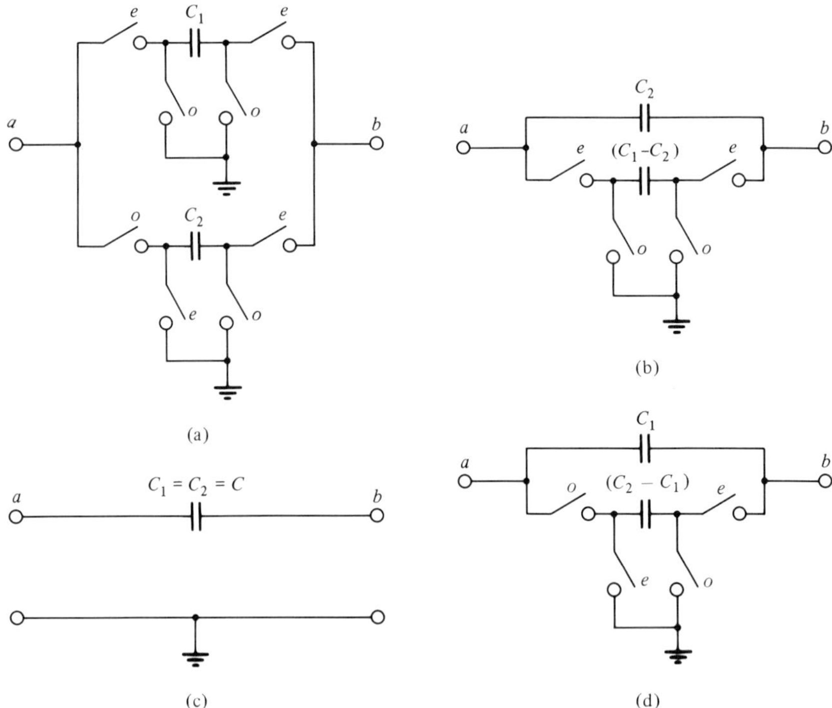

Figure 8.35 (*a*) An SC network configuration and its equivalent circuits. (*b*) $C_1 > C_2$. (*c*) $C_1 = C_2$. (*d*) $C_2 > C_1$.

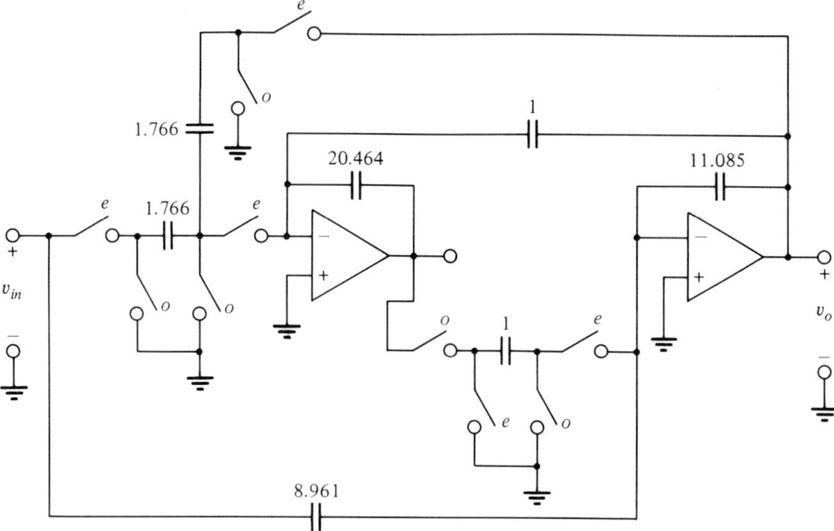

Figure 8.36 Final realization of the transfer function in Example 8.6—E circuit.

amp such that the minimum capacitance value is unity in each group, we have the final design values $K_1C_1 = 1.766 = K_3C_1$, $K_2C_1 = 0$, $K_4C_1 = 1$, and $C_1 = 20.464$; $MC_2 = 0$, $M_1C_2 = 1$, $M_2C_2 = M_3C_2 = 8.961$, and $C_2 = 11.085$. In addition to the switch simplification discussed with reference to the SC circuits in Fig. 8.33*a* and *b*, depending on the capacitance values of the circuit, additional simplifications can be effected as illustrated in Fig. 8.35. Since $M_3C_2 = M_2C_2$, we may use the switching simplification in Fig. 8.35*c*, and the E circuit designed will be as shown in Fig. 8.36.

(b) Design of the F circuit ($K_4 = 0$): Following the same procedure as in the previous design, which is straightforward, we obtain the parameters of the initial design of the F circuit as $K_1 = K_3 = 0.007819206$, $K_2 = K_4 = 0$, $M = 0.004427646$, $M_1 = 1$, and $M_2 = M_3 = 0.8119171$. Again, the values of C_1 and C_2 are arbitrary and may be unity for convenience. The dynamically adjusted final parameter values are $C_1 = 11.707$, $K_1C_1 = K_3C_1 = 1$, and $K_2C_1 = K_4C_1 = 0$; $C_2 = 225.85$, $MC_2 = 1$, $M_1C_2 = 20.674$, and $M_2C_2 = M_3C_2 = 183.37$.

There are two important criteria on the basis of which the SC circuits are compared: (1) the sensitivity of the output with respect to the capacitor ratios, and (2) the sum of all the capacitance values—the *total capacitance* (TC). When SC networks are implemented in IC form, the larger the value of TC, the larger the amount of real estate required on the IC chip, hence it will be costlier. It is easy to verify that the magnitudes of the coefficient sensitivities, and therefore the f_p and Q_p, sensitivities, in these networks are less than or equal to unity. In the above example, the TC value required in the E circuit is approximately $\frac{1}{10}$ the value required in the F circuit. But this is not a general conclusion about E and F circuits, since the F circuit may be preferable in other realizations. The design of generic forms of second-order transfer functions is described in the form of a table by Fleisher and Laker in [12]. ■

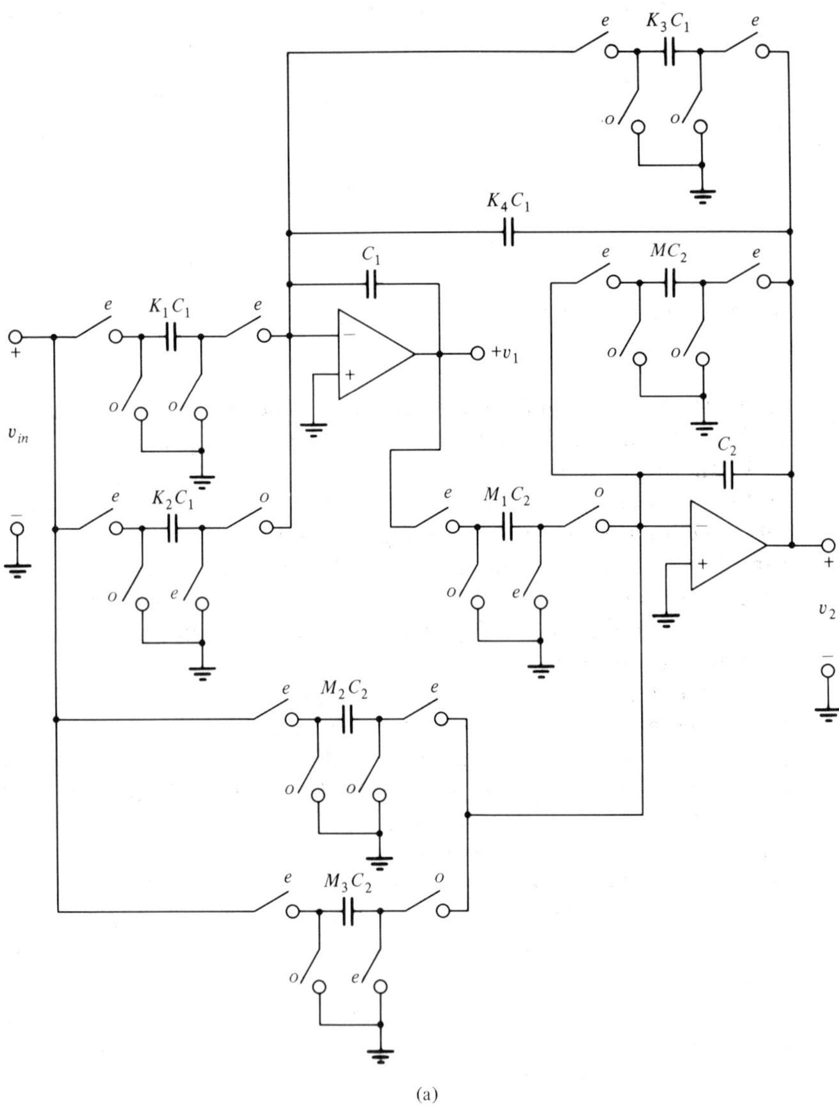

(a)

Figure 8.37 (*a*) General SC biquad. (*b*) Minimum switch configuration for the same SC biquad.

In the circuit in Fig. 8.33, it was assumed that the input signal was sampled and held over a full clock cycle. This restriction can be removed if the output signal is made independent of the input signal during the odd phase. This may be done by reversing the switching phases of some of the switches, as shown for another general SC circuit (called type 2) in Fig. 8.37. Note that the odd phase component of the input signal has no influence on the output at all because the output is sampled at the even phase only. This circuit uses the basic building blocks in Fig. 8.31*b*, and therefore the signal flow graph of the network in Fig. 8.37 may be obtained quickly, as shown in Fig. 8.38. The two transfer functions, V_1^e/V_{in}^e and

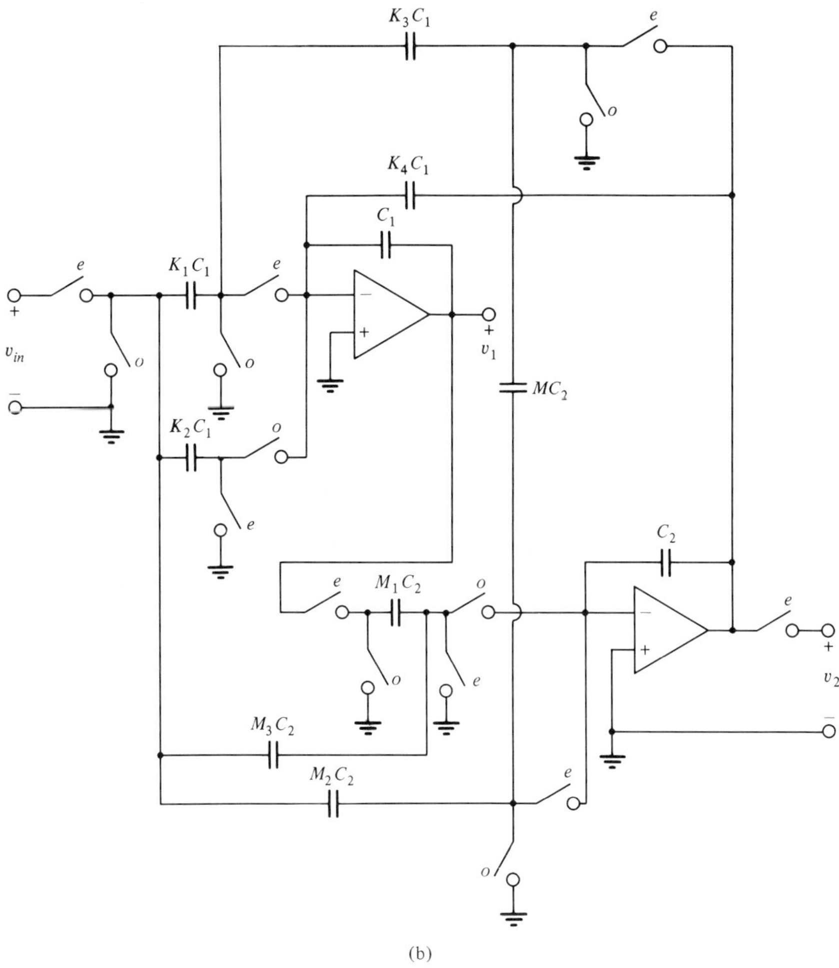

(b)

Figure 8.37 Continued.

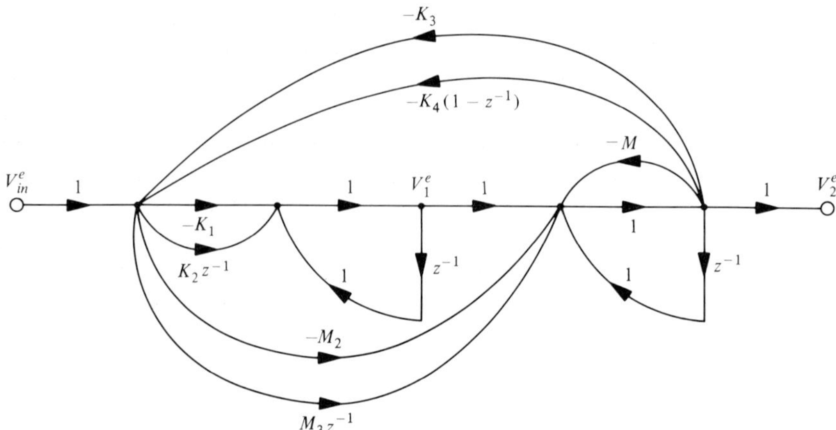

Figure 8.38 Signal flow graph for the SC network in Fig. 8.37. (Odd phase components are not shown.)

V_2^e/V_{in}^e, for this network can be obtained from this flow graph and can be shown to be exactly the same as those given by (8.70) and (8.71), respectively. Therefore the design steps for this network will be the same as before. Consider a design example for the network in Fig. 8.37.

Example 8.7: A switched capacitor bandpass filter is to be designed using the network in Fig. 8.37. The bandpass filter should have a center frequency of 1 kHz and $Q_p = 20$. The desired center frequency gain is 20 dB, and the sampling frequency is 8 kHz. Use bilinear transformation to obtain the discrete-time transfer function.

In the discrete-time domain, the 3-dB frequencies must be $f_{d1} = 0.9753$ kHz and $f_{d2} = 1.0253$ kHz. Using prewarping on these two frequencies, we obtain $f_{a1} = 1.0250$ kHz and $f_{a2} = 1.0846$ kHz. The center frequency of the prewarped analog filter function is then

$$f_{pa} = \sqrt{f_{a1}f_{a2}} = 1.0549 \text{ kHz}$$

Then, the second-order analog transfer function is

$$H_a(s) = \frac{10B_a s}{s^2 + B_a s + \omega_{pa}^2}$$

where B_a is the bandwidth and is equal to $\omega_{a2} - \omega_{a1}$. Substituting, we obtain the analog transfer function as

$$H_a(s) = \frac{3681.2s}{s^2 + 368.12s + 43.9292 \times 10^6}$$

Next, using bilinear transformation, we find the discrete-time transfer function to be

$$H(z) = \frac{0.1926(1 - z^{-2})}{1 - 1.3869z^{-1} + 0.96148z^{-2}}$$

If the output is taken at V_2, we will obtain an inverting output. Further assuming that $M = 0$ (E circuit) and $M_1 = 1$ and equating the coefficients of the numerator and denominator of the above transfer function to those in (8.71), we have

$$M_2 = 0.1926$$

$$-K_1 + M_2 + M_3 = 0$$

$$M_3 - K_2 = -0.1926$$

$$2 - (K_3 + K_4) = 1.3869$$

and

$$1 - K_4 = 0.96148$$

Again, there are more parameters to choose than there are equations and there is a choice of one free parameter. The simplest choice for this parameter is $M_3 = 0$. Thus the initial design values for the capacitance

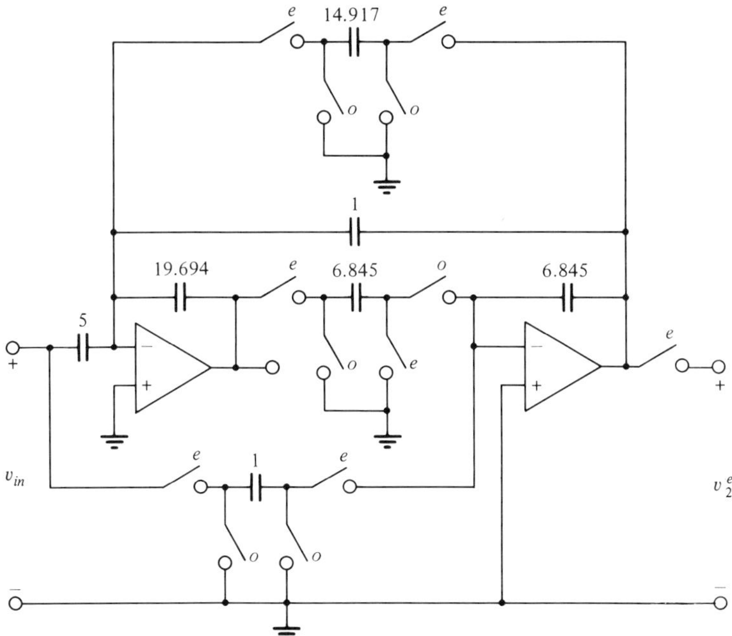

Figure 8.39 An SC network realization of the bandpass filter in Example 8.7.

ratios are $K_1 = K_2 = 0.1926$, $K_3 = 0.57458$, $K_4 = 0.038519$, $M = 0$, $M_1 = 1$, $M_2 = 0.1926$, and $M_3 = 0$. Since $K_1 = K_2$, the combinations K_1C_1 and K_2C_1 may be replaced with an unswitched capacitor (Fig. 8.35*c*). In this way, $K_1 = K_2$ can also be achieved. After adjusting the capacitance ratios for maximum dynamic range as in Example 8.6, the capacitance values are $K_1 = K_2 = 0.2539$, $K_3 = 0.7574$, $K_4 = 0.05078$, $M = 0$, $M_1 = 1$, $M_2 = 0.1461$, and $M_3 = 0$. Finally, making the minimum value of the capacitance unity in each group of capacitances, we find a new set of values: $K_1C_1 = K_2C_2 = 5$, $K_3C_1 = 14.917$, $K_4C_1 = 1$, and $C_1 = 19.694$; $MC_2 = 0$, $M_1C_2 = 6.845 = C_2$, $M_2C_2 = 1$, and $M_3C_2 = 0$.

The final circuit diagram of this bandpass realization should be as shown in Fig. 8.39. The total capacitance value in this circuit is $55.3C_u$, where C_u is the unit capacitance value. ■

Another stray-insensitive biquad circuit [13] capable of realizing any second-order transfer function is shown in Fig. 8.40*a* and its flow graph in

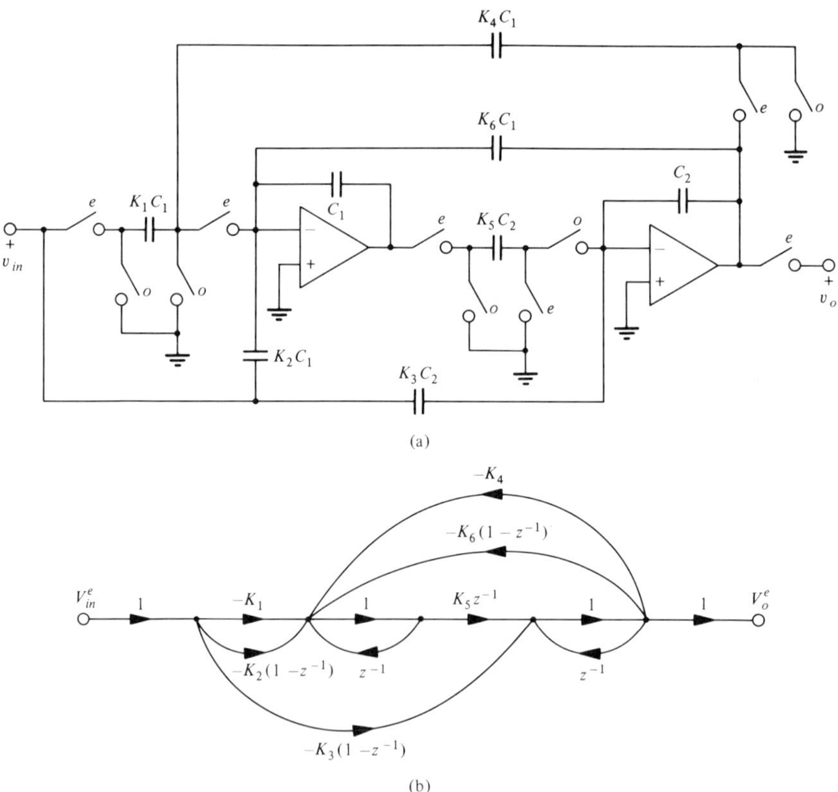

Figure 8.40 (*a*) Stray-insensitive SC biquad capable of realizaing any second-order function. (*b*) Its signal flow graph (odd phase components are not shown).

Fig. 8.40*b*. Assume that the output is sampled at the even phase. Furthermore, the odd phase input signal has no influence on the even phase output signal in this circuit. One of the advantages of this circuit is that it uses only 10 switches as opposed to the more than 10 switches required by the previous circuits. The design procedure for this circuit, with a maximum dynamic range, is very straightforward, as we shall soon see. This circuit provides an inverting output, and its transfer function can be derived as

$$H^{ee}(z) = \frac{V_o^e}{V_{\text{in}}^e} = \frac{-\left[K_3 + z^{-1}(-2K_3 + K_1K_5 + K_2K_5) + z^{-2}(K_3 - K_2K_5)\right]}{1 + z^{-1}(-2 + K_4K_5 + K_5K_6) + z^{-2}(1 - K_5K_6)} \tag{8.72}$$

We shall next consider the design of various standard biquads using the circuit in Fig. 8.40*a*. The standard analog second-order transfer function is given in (8.66), and the corresponding bilinearly transformed discrete-time transfer function is given by (8.67). Comparing the coefficients in (8.67) with those in (8.72), we obtain the equations for the capacitance ratios K_1 through K_6. Thus we obtain only five equations, while we have to select six parameters. Instead of choosing the free parameter arbitrarily, we can use a condition for maximizing the dynamic range, $K_4 = K_5$. Then, solving these equations, we obtain

$$K_4 = K_5 = T\sqrt{mb_0} \tag{8.73a}$$

$$K_6 = b_1\sqrt{\frac{m}{b_0}} \tag{8.73b}$$

$$K_3 = m\left(a_2 + \frac{a_1T}{2} + \frac{a_0T^2}{4}\right) \tag{8.73c}$$

$$K_2 = a_1\sqrt{\frac{m}{b_0}} \tag{8.73d}$$

and

$$K_1 = a_0T\sqrt{\frac{m}{b_0}} \tag{8.73e}$$

where

$$m = \frac{1}{1 + b_1T/2 + b_0T^2/4} \tag{8.73f}$$

The advantage of this circuit can be seen from the simplicity of the design procedure for maximum dynamic range and with a minimum number of switches as compared to the previous circuits. Further, note that the capacitor ratios can be obtained directly from the prewarped analog trans-

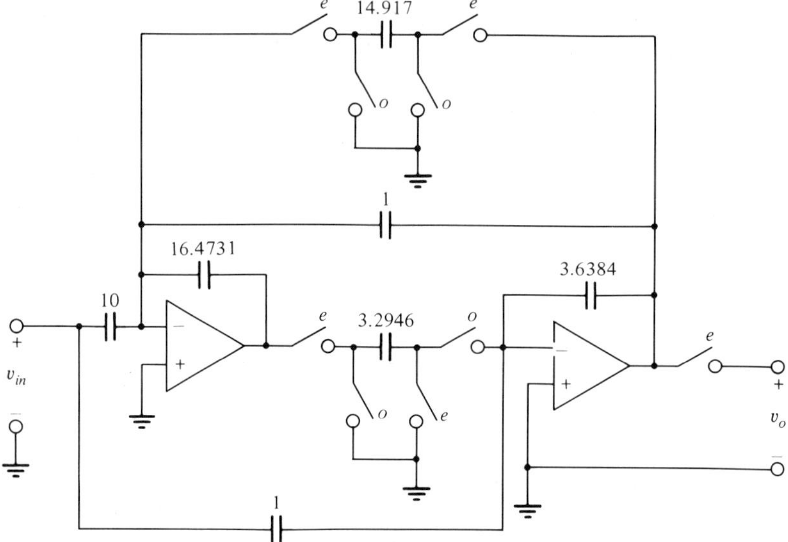

Figure 8.41 Circuit realization for Example 8.8.

fer function without bilinear transformation. The formulas for various second-order biquads are only particular cases of (8.73).

Example 8.8: Consider the design of the analog transfer function in Example 8.7. It is a bandpass filter transfer function, already prewarped, for realizing a center frequency of 1 kHz and a Q_p of 20. The sampling frequency is 8 kHz. Obtain a design for realizing the same transfer function using the circuit in Fig. 8.40*a*.

From the analog transfer function, we find that, $a_2 = a_0 = 0$, $a_1 = 3681.2$, $b_1 = 368.12$, and $b_0 = 43.9292 \times 10^6$. The value of T is equal to 0.125 ms. Using the above values in (8.73), we find the capacitance ratios directly to be $K_4 = K_5 = 0.905522$, $K_6 = 0.0607051$, $K_3 = 0.274849$, $K_2 = 0.607051$, and $K_1 = 0$. There are two groups of capacitors in the above design: $(C_1, K_2C_1, K_4C_1, K_6C_1)$ and (C_2, K_3C_2, K_5C_2). Choosing the minimum capacitance value in each group as the unit capacitance and rescaling the others, we find the two groups of capacitance values to be $(C_1, K_2C_1, K_4C_1, K_6C_1) = C_u(16.4731, 10, 14.917, 1)$ and $(C_2, K_3C_2, K_5C_2) = C_u(3.6384, 1, 3.2946)$. ■

The circuit diagram for this realization is shown in Fig. 8.41. The total capacitance required is only 50.32, which is less than that needed for the circuit in Fig. 8.39. However, we cannot make such a comparison based on only one example. Also note that we need only 8 switches (excluding the switch at the output, which may or may not be necessary), as opposed to a

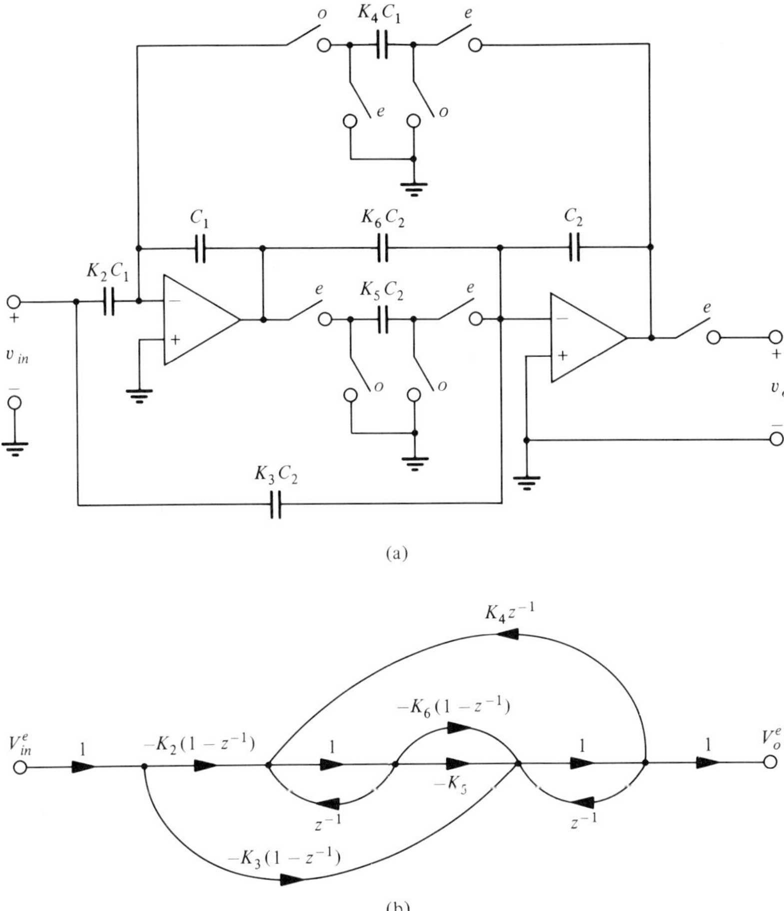

Figure 8.42 (*a*) Stray-insensitive noninverting bandpass SC filter. (*b*) Its signal flow graph.

minimum of 10 switches needed in the circuit in Fig. 8.39. Another factor that is also important in SC networks is the *element spread*, which is the maximum-to-minimum capacitance ratio. Again, comparing the element spread in the SC circuits in Figs. 8.39 and 8.41, we find the circuit in Fig. 8.41 to be superior.

When realizing higher-order bandpass filters in the next section, we will need not only inverting bandpass filters (a particular case of the above network as designed in the previous example) but also noninverting bandpass filters. Such a noninverting bandpass filter circuit is shown in Fig. 8.42*a*. This circuit is also stray-insensitive. The even phase output is not affected by the odd phase input component, and its simplified flow graph representation is shown in Fig. 8.42b. The transfer function of this circuit can be found to be

$$H^{ee}(z) = \frac{V_o^e}{V_{\text{in}}^e} = \frac{(1 - z^{-1})[(K_2K_5 + K_2K_6 - K_3) + z^{-1}(K_3 - K_2K_6)]}{1 + z^{-1}(-2 + K_4K_5 + K_4K_6) + z^{-2}(1 - K_4K_6)} \tag{8.74}$$

The output of this circuit is noninverting. The design equations can be obtained in the same way as for the previous circuit. Again, for maximum dynamic range, it is required that $K_4 = K_5$. Thus the capacitance ratios are

$$K_4 = K_5 = T\sqrt{mb_0} \tag{8.75a}$$

$$K_6 = b_1\sqrt{\frac{m}{b_0}} \tag{8.75b}$$

$$K_3 = a_1T\frac{1 + 2b_1/(b_0T)}{2} \tag{8.75c}$$

and

$$K_2 = \frac{a_1}{(m/b_0)^{1/2}} \tag{8.75d}$$

The value of m is still the same as that given by (8.73f).

We shall next consider the realization of a third-order transfer function that will be stray-insensitive. Such a network will be useful in realizing higher-order transfer functions of odd order in cascade form. Such a network is shown in Fig. 8.43 [15]. Assuming that we sample the even phase component of the output, we find that the odd phase input signal has no influence on the output. Then we have only one transfer function:

$$H^{ee}(z) = \frac{V_o^e}{V_{\text{in}}^e} = -\frac{N_0 + N_1z^{-1} + N_2z^{-2} + N_3z^{-3}}{D_0 + D_1z^{-1} + D_2z^{-2} + D_3z^{-3}} \tag{8.76}$$

where

$$N_0 = A_5$$
$$N_1 = -(3A_5 - A_2B_6)$$
$$N_2 = 3A_5 - B_6(A_3 + A_2 - A_0B_5)$$
$$N_3 = -(A_5 - A_3B_6)$$
$$D_0 = 1$$
$$D_1 = -(3 - B_2B_6)$$
$$D_2 = 3 + B_6(B_0B_5 - B_2 - B_3)$$

and

$$D_3 = -(1 - B_3B_6)$$

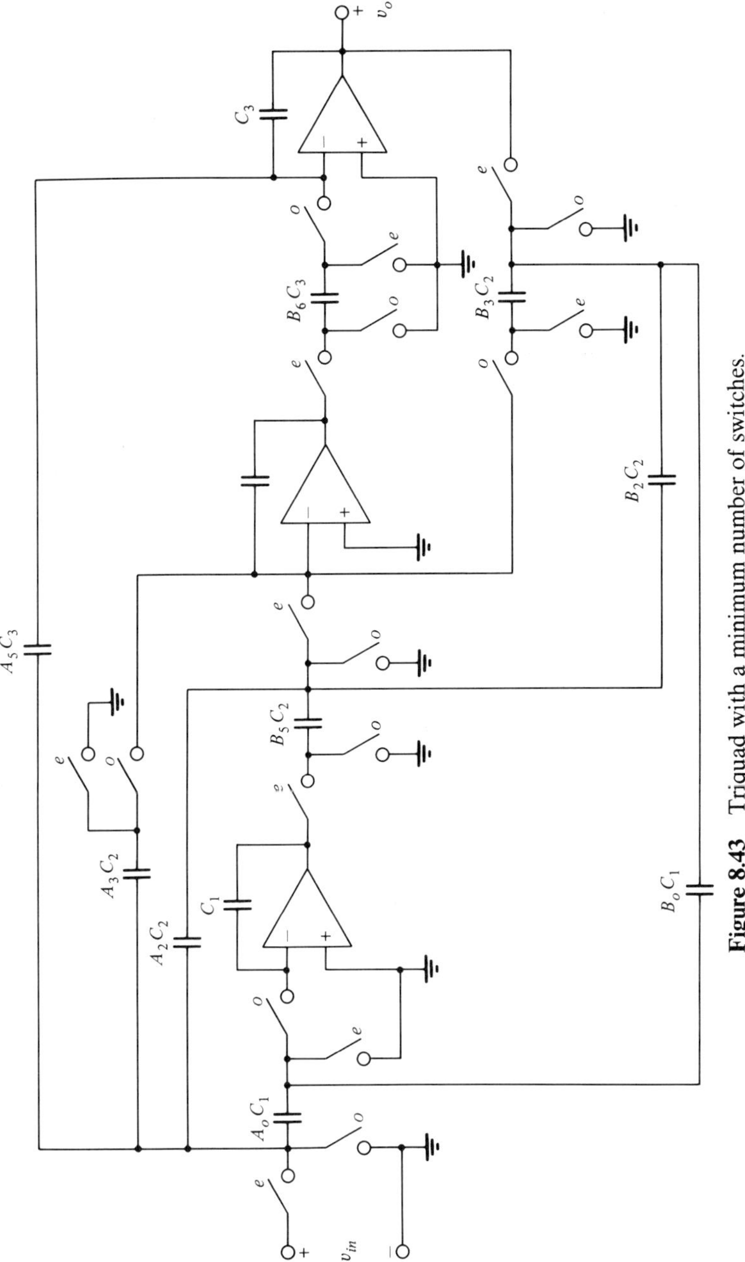

Figure 8.43 Triquad with a minimum number of switches.

The design of this circuit is similar to those discussed earlier and involves comparing the coefficients of the transfer functions to (8.76) and solving the equations for the capacitor ratios. However, the usefulness of this circuit is very much limited because the element spread may become large.

8.7 *Higher-Order Switched Capacitor Filters*

We discussed the realization of higher-order ($N > 2$) active RC filters in Chap. 7. All the techniques described for active RC networks may be used for the realization of SC filters also. The only precaution one may have to take is to follow the prewarping procedure necessary in the case of SC filters. We have used this prewarping technique in almost every example presented in this chapter. There are many exact procedures for designing higher-order SC filters [16–18]. The references cited are only a few recent ones, and a considerable amount of research has preceded them. The reader is referred to the literature for descriptions of all the different techniques, since we shall not discuss them all here. We shall consider only two examples of designing higher-order SC filters. The first illustrates the simplest and most straightforward method, cascade realization. Because of the low sensitivity of LC simulation techniques, we provide a second example based on the leapfrog technique. This method employs operational simulation of an LC network as discussed in Chap. 7.

Example 8.9: It is desired to design a lowpass filter satisfying the following specifications: $A_p = 0.5$ dB, $f_p = 1$ kHz; $A_a \geq 30$ dB, $f_a = 1.5$ kHz. Use Chebyshev approximation to obtain an SC filter using the cascade approach. Assume that the sampling frequency is 8 kHz.

We will use bilinear transformation to obtain the pulse transfer function from the analog transfer function. Hence, prewarping will be necessary. Thus the passband and stopband edge frequencies are prewarped and are $f_{pa} = 1.055$ kHz and $f_{aa} = 1.702$ kHz. The following transfer function satisfies the analog domain specifications:

$$H_A(s) = \frac{2.28997 \times 10^{18}}{(s + 2401.73)(s^2 + 1484.35s + 45.51284 \times 10^6)(s^2 + 3886.08s + 20.94937 \times 10^6)}$$

Next, using bilinear transformation on the above analog transfer function, we obtain the discrete-time transfer function:

$$H(z) = H_1(z)H_2(z)H_3(z) \tag{8.77a}$$

where

$$H_1(z) = \frac{0.1305165(1 + z^{-1})}{1 - 0.7389669z^{-1}} \tag{8.77b}$$

$$H_2(z) = \frac{0.061775(1 + z^{-1})^2}{1 - 1.386211z^{-1} + 0.6333093z^{-2}} \tag{8.77c}$$

and

$$H_3(z) = \frac{0.1399265(1 + z^{-1})^2}{1 - 1.29426z^{-1} + 0.8539664z^{-2}} \tag{8.77d}$$

One way of realizing the transfer function $H(z)$ is to combine $H_1(z)$ and $H_2(z)$ and realize the third-order function using the triquad. However, it may be shown that, in such a triquad, the element spread will be high. For this reason we attempt to realize $H(z)$ using two biquads and a first-order network. The second-order transfer functions may be realized using the network in Fig. 8.40, and we shall discuss this later. For now, consider the problem of realizing (8.77*b*). We have not discussed a stray-insensitive network that realizes a transfer function in the form of (8.77*b*). The network in Fig. 8.44 realizes a transfer function if the input is sampled and held over one period. If the input signal is sampled and held in Fig. 8.44, the transfer function can be shown to be

$$H^{ee}(z) = \frac{V_o^e(z)}{V_{\text{in}}^e(z)} = \frac{-K}{1 + M}\,\frac{1 + z^{-1}}{1 - z^{-1}/(1 + M)} \tag{8.78}$$

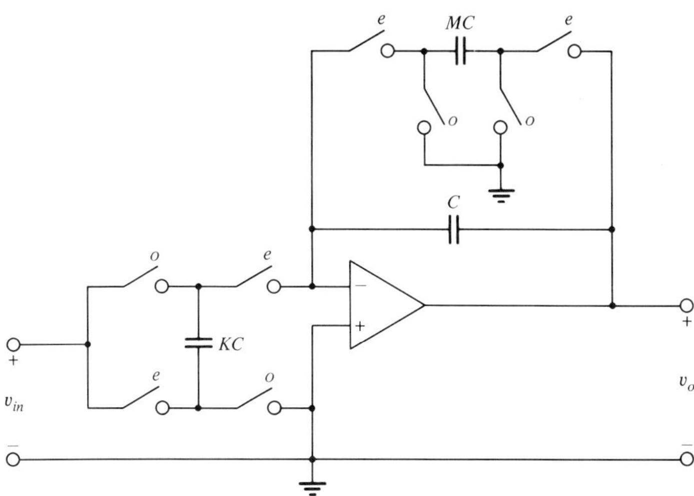

Figure 8.44 An SC network realization of the transfer function in (8.78) in Example 8.9.

Now, comparing (8.77b) and (8.78), we obtain the equations for the two parameters K and M. Solving these equations and rescaling the capacitances, we have $(KC, MC, C) = (1, 2, 5.662)$. This completes the design of the first-order network.

We shall next consider the realization of second-order sections, and the network in Fig. 8.40a may be used for this purpose. The element values for the two sections are

For realizing $H_2(z)$: $K_4 = K_5 = K_1 = 0.49709$, $K_6 = 0.73768$, $K_3 = 0.061774$, and $K_2 = 0$

For realizing $H_3(z)$: $K_4 = K_5 = K_1 = 0.74814$, $K_6 = 0.1952$, $K_3 = 0.13993$, and $K_2 = 0$.

Next, we adjust the capacitance values such that in each group of capacitances the minimum value is unity. Thus the required networks are as shown in Fig. 8.45. By connecting the networks in Figs. 8.44, 8.45a, and 8.45b, we can realize the given specifications. ■

In the active RC literature, it is well known that low-sensitivity bandpass filters can be realized using a wide variety of methods, and we described these methods in the previous chapter. We shall next discuss the realization of higher-order SC filters of all-pole types using the leapfrog technique in which the working of LC ladders is simulated. Consider the LC ladder network in Fig. 7.16 and the operational simulation of this circuit shown in Fig. 7.17, where the block transfer functions H_1, H_2, and H_3 are given by (7.30). Switched capacitor networks may be used to realize these block transfer functions. However, there is one difference if SC networks are used. The LC prototype networks must be found to meet the prewarped specifications and not the given specifications. This requirement must be met because SC networks realize z-domain functions. Since the given specifications are to be met by the SC networks, we have to prewarp them to obtain the specifications in the analog domain. We can always find a prototype LC ladder network that meets the analog domain specifications. Once this difference is taken into account, the procedure is similar to those given in Chap. 7. For example, if we want to realize a bandpass filter, we may apply LP-BP transformation to the block transfer functions and realize them using the bandpass SC networks in Fig. 8.40 and 8.42. The realization of a bandpass SC filter using the leapfrog technique is illustrated in the next example.

Example 8.10: The specifications of a bandpass filter are as follows: passband, 0.9 to 1.2 kHz, equiripple with a maximum deviation of 0.5 dB; upper stopband, $A \geq 30$ dB, $f \geq 1.5$ kHz; lower stopband, $A \geq 30$ dB, $f \leq 0.8$ kHz; clock frequency $f_s = 8$ kHz. Obtain a SC network that meets the above specifications.

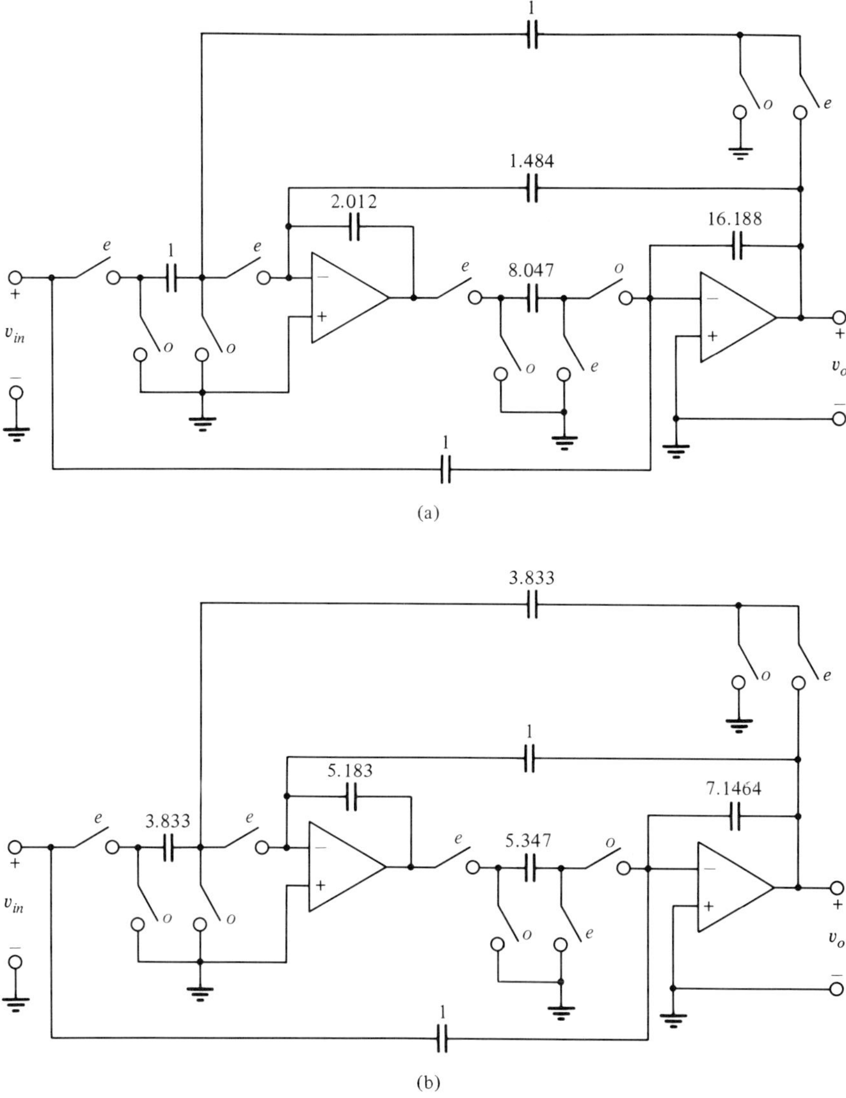

Figure 8.45 The SC network realizations of the second-order transfer functions in Example 8.9. (a) Realization of $H_2(z)$. (b) Realization of $H_3(z)$.

The first step is to prewarp the above specifications, and therefore we obtain the following specifications in the analog domain: passband, 0.9394 to 1.297 kHz, equiripple with a maximum deviation of 0.5 dB; upper stopband, $A \geq 30$ dB, $f \geq 1.702$ kHz; lower stopband, $A \geq 30$ dB, $f \leq 0.8274$ kHz. From the above set of specifications, we need to determine the specifications for a prototype lowpass filter. To do this,

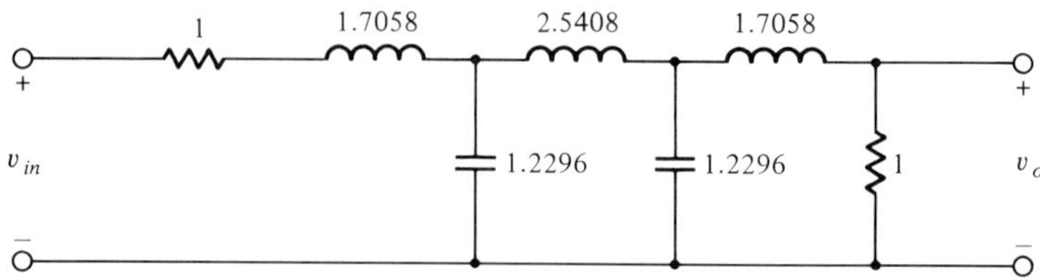

Figure 8.46 Prototype lowpass *LC* ladder network meeting the specifications in Example 8.10. *R* in Ω, *L* in *H*, *C* in *F*.

we follow the procedure in Chap. 2 and find that $\Omega_a = 1.804$. Therefore we need to find a doubly terminated LP prototype *LC* ladder filter network to meet the following specifications: passband, 0.5 dB, equiripple; stopband, 30 dB, $\Omega_a = 1.804$. We need a fifth-order network to meet the above specifications, and it is shown in Fig. 8.46.

The block diagram representation of this *LC* network shown in Fig. 8.47 was obtained using the procedure in Chap. 7. The transfer functions of the different blocks needed to realize the *LC* prototype are

$$H_5(s) = H_1(s) = \frac{-0.58624}{s + 0.58624}$$

$$H_4(s) = H_2(s) = \frac{0.81327}{s}$$

and

$$H_3(s) = -\frac{0.39358}{s}$$

Using the LP-BP transformation

$$s \to \frac{p^2 + 48.121 \times 10^6}{2249.7p}$$

we obtain the required transfer functions for the bandpass filter:

$$H_1(p) = H_5(p) = \frac{-1318.9p}{p^2 + 1318.9p + 48.121 \times 10^6}$$

$$H_2(p) = H_4(p) = \frac{1829.6p}{p^2 + 48.121 \times 10^6}$$

and

$$H_3(p) = \frac{-885.44p}{p^2 + 48.121 \times 10^6}$$

We use the stray-insensitive networks in Fig. 8.40 to realize the transfer

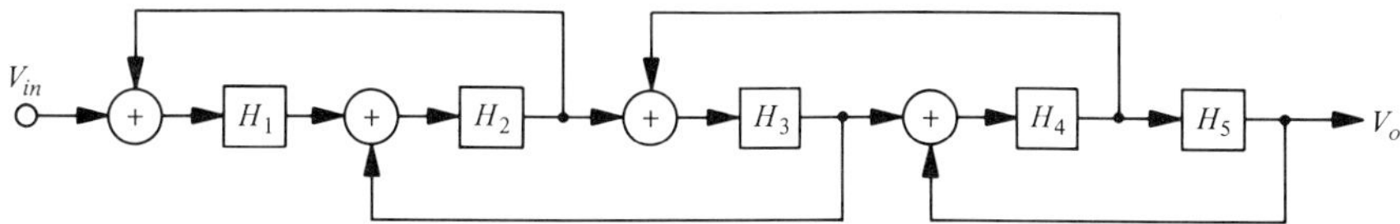

Figure 8.47 Block diagram representation of the leapfrog realization for Example 8.10.

functions H_1, H_3, and H_5, and in Fig. 8.42 to realize the transfer functions H_2 and H_4.

Design of $H_1(p)$ [and $H_5(p)$]: Here $a_2 = a_0 = 0$, $a_1 = 1318.9$, $b_1 = 1318.9$, and $b_0 = 48.121 \times 10^6$. Substituting into (8.73), we obtain $K_4 = K_5 = 0.7693$, $K_2 = K_6 = 0.1687$, $K_3 = 0.06489$, and $K_1 = 0$. Rescaling the capacitances, we have $(C_1, K_2C_1, K_4C_1, K_6C_1) = C_u(5.928, 1, 4.56, 1)$ and $(C_2, K_3C_2, K_5C_2) = C_u(15.41, 1, 11.86)$.

Design of $H_3(p)$: Here $a_2 = a_0 = 0$, $a_1 = 885.44$, $b_1 = 0$, and $b_0 = 48.121 \times 10^6$. Substituting into (8.73), we obtain $K_4 = K_5 = 0.7956$, $K_6 = 0$, $K_3 = 0.046584$, $K_2 = 0.11711$, and $K_1 = 0$. Rescaling the capacitances, we have $(C_1, K_2C_1, K_4C_1) = C_u(8.539, 1, 6.794)$ and $(C_2, K_3C_2, K_5C_2) = C_u(21.467, 1, 17.079)$.

Design of $H_2(p)$ [and $H_4(p)$]: These are noninverting transfer functions, and therefore we use the network in Fig. 8.42. The designed element values are $(C_1, K_2C_1, K_4C_1) = C_u(3.479, 1, 2.768)$ and $(C_2, K_3C_2, K_5C_2) = C_u(8.741, 1, 6.955)$.

Now that we have designed the modular blocks, we are ready to interconnect these modular networks. The overall network will be a large circuit because we will have to interconnect five different second-order sections. This interconnection is left as an exercise for the reader to complete this circuit. ■

8.8 *Conclusions*

In this chapter we have discussed the theory and design of switched capacitor filters. We have covered as much material as necessary so that one can design stray-insensitive second- and higher-order SC networks. Notably missing in this chapter is the sensitivity analysis of these networks due to various capacitor ratios and due to the finite gain and gain-bandwidth products of the op amps used in the networks. The passive sensitivity analysis due to capacitor ratios is similar to the case of active *RC* networks. The effects of the finite *GB* products of the op amps in SC networks can also be found though it involves a little more work [19, 20].

REFERENCES

1. Hosticka, B., Broderson, R., and Gray, P. R.: MOS Sampled Data Recursive Filters Using a Switched Capacitor, *IEEE Journal of Solid-State Circuits*, vol. SC-12, no. 6, pp. 600–608, December 1977.
2. Oppenheim, A. V., and Schafer, R. W.: *Digital Signal Processing*, Prentice-Hall, Englewood Cliffs, N.J., 1975.
3. Kurth, C. F., and Moschytz, G. S.: Nodal Analysis of Switched Capacitor Networks, *IEEE Trans. Circuits Syst.*, vol. CAS-26, no. 2, pp. 93–104, February 1979.
4. Kurth, C. F., and Moschytz, G. S.: Two-Port Analysis of Switched Capacitor Networks Using Four-Port Equivalent Circuits, *IEEE Trans. Circuits Syst.*, vol. CAS-26, no. 3, pp. 166–180, March 1979.
5. Laker, K. R.: Equivalent Circuits for Analysis and Synthesis of Switched Capacitor Networks, *Bell Syst. Tech. J.*, vol. 58, pp. 727–767, March 1979.
6. Rahim, C. F. et al.: A Functional MOS Circuit for Achieving the Bilinear Transformation in SC Networks, *IEEE J. Solid-State Circuits*, vol. SC-13, no. 6, pp. 906–909, December 1978.
7. Broderson, R. W., Gray, P. R., and Hodges, D. A.: MOS Switched Capacitor Filters, *Proc. IEEE*, vol. 67, pp. 61–75, January 1979.
8. Martin, K.: Improved Circuits for the Realization of Switched Capacitor Filters, *Proc. ISCAS, Tokyo*, pp. 756–759, 1979.
9. Jenkins, W. K., Trick, T. N., and El-Masry, E.: New Realizations for Switched Capacitor Filters, Twelfth Annual Asilomar Conference on Circuits, Systems and Computers, pp. 694–698, November 1978.
10. Szentirmai, G., and Temes, G. C.: Switched Capacitor Building Blocks, Thirteenth Annual Asilomar Conference on Circuits, Systems and Computers, pp. 542–549, November 1979.
11. El-Masry, E. I.: Stray-Insensitive Active Switched Capacitor Biquad, *Electron. Lett.*, vol. 16, pp. 480–481, June 1980.
12. Fleisher, P. E., and Laker, K. R.: A Family of Active Switched Capacitor Biquad Building Blocks, *Bell Syst. Tech. J.*, vol. 58, pp. 2235–2269, December 1979.
13. Martin, K., and Sedra, A. S.: Exact Design of Switched Capacitor Bandpass Filters Using Coupled Biquad Structures, *IEEE Trans. Circuits Syst.*, vol. 27, pp. 469–475, June 1980.
14. Grebene, A. B.: *Bipolar and MOS Analog Integrated Circuit Design*, John Wiley, New York, 1984.
15. Allen, P. E., and Sinencio, E. S.: *Switched Capacitor Circuits*, Van Nostrand Reinhold, New York, 1984.
16. Attaie, N., and El-Masry, E. I.: Multiple-Loop Feedback Switched Capacitor Structures, *IEEE Trans. Circuits Syst.*, vol. 30, pp. 865–852, December 1983.

17. Hokenek, E., and Moschytz, G. S.: Design of Parasitic-Insensitive Bilinear-Transformed Admittance-Scaled SC Ladder Filters, *IEEE Trans. Circuits Syst.*, vol. 30, pp. 873–888, December 1983.

18. Datar, R. B., and Sedra, A. S.: Exact Design of Stray-Insensitive SC Ladder Filters, *IEEE Trans. Circuits Syst.*, CAS vol. 30, pp. 888–898, December 1983.

19. Sedra, A. S., and Martin, K.: Effects of Op Amp Finite Gain and Bandwidth on the Performance of Switched Capacitor Filters, *Proc. IEEE ISCAS*, vol. 1, pp. 321–325, April 1980.

20. Bhattacharyya, B. B., and Raut, R.: On the Analysis of Active SC Networks Considering the Finite Values of DC Gain and *GB* Product of the Op Amps, *Proc. 1983 IEEE ISCAS*, vol. 1, pp. 64–67, May 1983.

EXERCISES

8.1. Consider the SC network in Fig. 8.21. Using the charge conservation equations in the time domain, prove (8.48) and (8.49).

8.2. Repeat Exercise 8.1 for the network in Fig. E8.2 and shown that

$$H^{oe}(z) = \frac{V_2^o(z)}{V_1^e(z)} = \frac{(1-\alpha)z^{-1/2}}{1-\alpha z^{-1}}$$

and

$$H^{ee}(z) = \frac{V_2^e(z)}{V_1^e(z)} = \frac{1-\alpha}{1-\alpha z^{-1}}$$

where $\alpha = C_2/(C_1 + C_2)$.

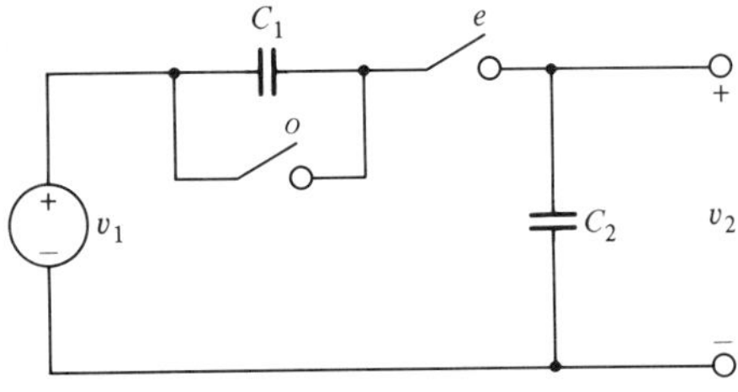

Figure E8.2 The SC network for Exercise 8.2.

8.3. Using the impulse-invariant response method, obtain a pulse transfer function from the following continuous-time transfer functions. Assume that the sampling frequency is 10 r/s.

$$\text{(a) } H(s) = \frac{1}{(s+1)(s^2+s+1)}$$

$$\text{(b) } H(s) = \frac{s^2 - 3s + 3}{s^2 + 3s + 3}$$

Can the impulse-invariant method handle 8.3*b*?

8.4. Using bilinear transformation, obtain an elliptic pulse transfer function that meets the following lowpass filter specifications: passband, $A_p \leq 0.5$ dB, $0 \leq \omega \leq 1000$ r/s; stopband, $A_a \geq 30$ dB, $\omega \geq 2000$ r/s. Assume that the sampling frequency is $\omega_s = 4000$ r/s.

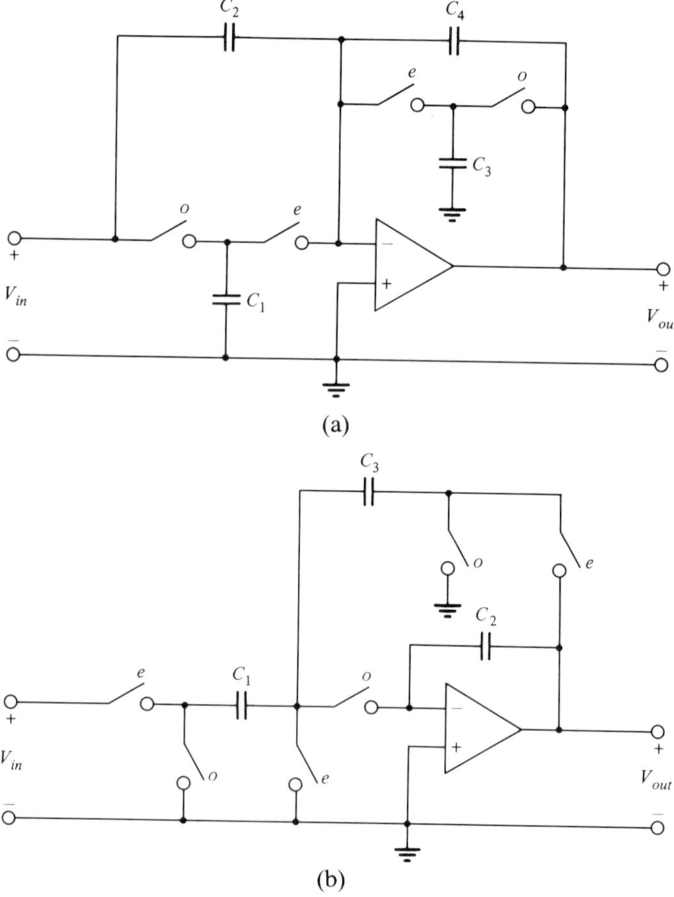

Figure E8.6 SC networks for Exercise 8.6.

8.5. Using bilinear transformation, obtain a Chebyshev pulse filter transfer function that meets the following highpass filter specifications: stopband, $A_a = 40$ dB, $\omega_a \leq 400$ r/s; passband, $A_p = 0.5$ dB, $\omega_p = 800$ r/s. Assume that the sampling frequency is $\omega_s = 4000$ r/s.

8.6. Use z-domain equivalent circuits and obtain the appropriate transfer functions for the SC circuits in Fig. E8.6.

8.7. Consider the active RC bandpass filter circuit in Fig. 5.4. Obtain the equivalent SC network using the SC resistor equivalent of Fig. 8.1a. Find the transfer function $H^{ee}(z)$ for the network in terms of the unknown capacitance values.

8.8. Use the SC network in Exercise 8.7 to design a SC bandpass filter that meets the specifications $f_p = 3.75$ kHz and $Q_p = 20$. Assume that the sampling frequency is 108 kHz.

8.9. With the use of the SC second-order filter network in Fig. 8.33, a lowpass filter that meets the specifications $f_p = 1.5$ kHz and $Q_p = 10$ is to be realized. Assume that $f_s = 8$ kHz. The required dc gain is 10.
(a) Find the dynamically adjusted E and F circuits.
(b) Find the value of the total capacitance in each case.

8.10. Realize an SC notch filter with a prewarped set of specifications $\omega_p = 12576.5$ r/s, $Q_p = 22.2$, $\omega_z = 11317.1$ r/s, and $H_0 = 1$. Assume that the sampling frequency is 128 kHz. Use the circuit in Fig. 8.37 for realization. Obtain both E and F circuits. Find the total capacitance values in each circuit.

8.11. Repeat Exercise 8.10 using the circuit in Fig. 8.40a and compare the total capacitance value with that obtained in Exercise 8.10.

8.12. Realize the transfer function in Exercise 8.9 using the network in Fig. 8.40a. Find the total capacitance value and compare this value with that obtained in Exercise 8.9.

8.13. Using the cascade design approach, obtain an SC bandpass filter that meets the following specifications: passband, 1.5 to 2 kHz, equiripple with a maximum deviation of 0.5 dB; upper stopband, $A_a \geq 25$ dB, $f \geq 4$ kHz; lower stopband, $A_a \geq 30$ dB, $f \leq 0.8$ kHz. Assume that the sampling frequency is 16 kHz.

8.14. Using the cascade design approach, obtain an SC bandstop filter that meets the following specifications: passbands, equiripple with a maximum deviation of 0.25 dB; lower passband, $f_{p1} = 0.8$ kHz; upper passband, $f_{p2} = 4$ kHz; stopband, $f_{a1} = 1.5$ kHz, $f_{a2} = 2$ kHz. Assume that the sampling frequency is 32 kHz.

Bibliography

In addition to the references given in the text, the author has used ideas from other research material of his own and many others. The references are numerous, and no attempt has been made to list all of them. Following is a list of books that discuss various topics covered in this book.

Allen, P. E., and Sinencio, E. S.: *Switched Capacitor Circuits,* Van Nostrand Reinhold, New York, 1984.

Antoniou, A.: *Digital Filters: Analysis and Design,* McGraw-Hill, New York, 1979.

Chirilian, P. M.: *Analysis and Design of Integrated Electronics Circuits,* Harper & Row, Cambridge, MA, 1981.

Ghausi, M. S., and Laker, K. R.: *Modern Filter Design—Active-*RC *and Switched Capacitors,* Prentice-Hall, Englewood Cliffs, NJ, 1981.

Gray, P. R., and Meyer, R. G.: *Analysis and Design of Analog Integrated Circuits,* 2nd ed., John Wiley, New York, 1984.

Grebene, A. B.: *Bipolar and MOS Analog Integrated Circuit Design,* John Wiley, New York, 1984.

Huelsman, L. P., and Allen, P. E.: *Theory and Design of Active Filters,* McGraw-Hill, New York, 1981.

Johnson, D. E.: *Introduction to Filter Theory,* Prentice-Hall, Englewood Cliffs, NJ, 1976.

Luenberger, D. G.: *Linear and Nonlinear Programming,* 2nd ed., Addison Wesley, Reading, MA, 1984.

Mitra, S. K.: *Analysis and Synthesis of Linear Active Networks,* John Wiley, New York, 1969.

Moschytz, G. S., *Linear Integrated Networks: Design,* Van Nostrand Reinhold, New York, 1975.

Natarajan, S.: *Some Design and Optimization Techniques for Extending the Operating Frequency Range of Active-*RC *Filters,* Ph.D. Thesis, Concordia University, Montreal, 1978.

Roberge, J. K.: *Operational Amplifiers,* John Wiley, New York, 1975.

Schaumann, R., Soderstand, M. A., and Laker, K. R. (Eds.): *Modern Active Filter Design,* IEEE Press, New York, 1981.

Sedra, A. S., and Brackett, P. O.: *Filter Theory and Design: Active and Passive,* Matrix, Portland, OR, 1978.

Sedra, A. S., and Smith, K. C.: *Micro-Electronics Circuits,* Holt, New York, 1982.

Stephenson, F. W. (Ed.): RC-*Active Filter Design Handbook,* John Wiley, New York, 1985.

Temes, G. C., and Mitra, S. K., *Modern Filter Theory and Design,* John Wiley, New York, 1973.

Vlach, J., and Singhal, K.: *Computer Methods for Circuit Analysis and Design,* Van Nostrand Reinhold, New York, 1983.

Zevrev, A. I.: *Handbook of Filter Synthesis,* John Wiley, New York, 1967.

Index